AF880461

The History of Fossils Over Centuries

Maurizio Forli • Andrea Guerrini

The History of Fossils Over Centuries

From Folklore to Science

Springer

Maurizio Forli
Independent Researcher, Gruppo
Archeologico e Paleontologico Livornese
Museo di Storia Naturale del Mediterraneo
Leghorn, Italy

Andrea Guerrini
Independent Researcher, Gruppo
Archeologico e Paleontologico Livornese
Museo di Storia Naturale del Mediterraneo
Leghorn, Italy

ISBN 978-3-031-04686-5 ISBN 978-3-031-04687-2 (eBook)
https://doi.org/10.1007/978-3-031-04687-2

© The Editor(s) (if applicable) and The Author(s), under exclusive license to Springer Nature Switzerland AG 2022
This work is subject to copyright. All rights are solely and exclusively licensed by the Publisher, whether the whole or part of the material is concerned, specifically the rights of reprinting, reuse of illustrations, recitation, broadcasting, reproduction on microfilms or in any other physical way, and transmission or information storage and retrieval, electronic adaptation, computer software, or by similar or dissimilar methodology now known or hereafter developed.
The use of general descriptive names, registered names, trademarks, service marks, etc. in this publication does not imply, even in the absence of a specific statement, that such names are exempt from the relevant protective laws and regulations and therefore free for general use.
The publisher, the authors and the editors are safe to assume that the advice and information in this book are believed to be true and accurate at the date of publication. Neither the publisher nor the authors or the editors give a warranty, expressed or implied, with respect to the material contained herein or for any errors or omissions that may have been made. The publisher remains neutral with regard to jurisdictional claims in published maps and institutional affiliations.

This Springer imprint is published by the registered company Springer Nature Switzerland AG
The registered company address is: Gewerbestrasse 11, 6330 Cham, Switzerland

Once again to Maria Teresa e Rita

Preface

When in 2013 I was given for Christmas the book of Professor Giancarlo Baronti *Between children and dirty waters: Immersions in the collection of amulets of Giuseppe Bellucci*, I would not have imagined that, 4 years later, I would have been inspired to write a book on fossils, their meaning, their understanding over time, and their multiple nature according to particular keys of interpretation, which go beyond the common way of thinking about them.

The content of that book, in addition to fascinating me, was therefore the primary cause of the drafting of this text. To all this has joined my interest in fossils and paleontology in general, some multiform naturalistic knowledge that I have acquired over the years and a friendship grown and consolidated in the last decade, with Andrea Guerrini, the other author who has given me the pleasure of writing these lines. Our collaboration is therefore the other cornerstone of the formation of the work. But which "work" are we going to present to you?

After this preamble, either you are curious or you have already closed the book. So, hoping for the first hypothesis, I tell you at once that we have chosen to begin our studies by doing a bibliographical and above all iconographical research aimed at finding the first printed images of the objects that today we know as *fossils*: who was the first person to illustrate an ammonite, a shell, a shark tooth, and so on? Since the field of study is so vast, we have therefore limited ourselves, without missing any exceptions, to carrying out research on invertebrates in general, because it is they who will accompany us both in the history of understanding the true nature of fossils, and in the complex relationship existing between "academic" and "popular" science up to the present day, in which the latter has almost always been transformed into an unconscious habit.

We finally decided to conclude the treatment of each group when the ancient, petrified organisms were recognized as such and thus assigned and included in the present classification.

We could have found a thousand other ways to explain our thought, but the path we have chosen, reporting extensive original quotations more or less ancient and plausible, is the one that seemed the best we could go. So, to sum up, we propose a "collection of ancient prints," starting from the late fifteenth century, dedicated to the most common groups of invertebrates, fossils systematically described according to current scientific knowledge, and "fossils" instead considered medicinal substances or special stones with apotropaic properties. Lateral points of view that make you curious but also better understand how all topics can and should be addressed and understood totally. There is no contradiction in dealing in the history of paleontology also with paintings, archaeological finds, legends, invertebrate anatomy, fossils and folklore, and magic. They are all aspects of our knowledge, even inner, determined by science, knowledge as such, collective and individual. Starting from the atoms that make up the matter of books, one can arrive at the psychology of the author, the reader, and their place in the universe. An absurd path? probable, but, if probable, it can be done.

Not being historians of paleontology, for the treatment of the argument, we have naturally availed ourselves of the opinion of the scholars who have made it their profession. In conclusion: we have tried to use "the art of not inventing anything" to invent this whole of ours, consisting of the pages that will follow.

Prato, Italy Maurizio Forli

Presentation

Maurizio Forli and Andrea Guerrini proposed me to present this manuscript with the typical modesty (but also the enthusiasm ... and the infinite patience!) of those who have not followed the *Cursus Honorum* of academics but the Sacred Fire of their own passion, born almost always during childhood or the years of youth, and of those who have accumulated a profound knowledge of fossils through endless explorations of gullies, quarries, and rock faces, adventurous collections of specimens dug out of the sediment that has preserved them for millions of years, long and tiring days of work to wash and sift even the smallest specimens, removing the cemented matrixes to isolate even apparently insignificant fragments and recompose entire specimens, and many hours of their free time dedicated to the microscope to measure, determine, and catalogue the fruit of their labors and to study the scientific literature.

Precious artisans of paleontology and irreplaceable connoisseurs of their own territory (in both cases especially the rich Tertiary lands of Tuscany), the so-called "paleontophiles" like Forli and Guerrini are fundamental collaborators of researchers for the richness of their collections, which they willingly make available (and much more often than is believed they give to museums), and for the glance of the country naturalist, increasingly rare in the career researcher, oriented towards the "maximum systems" and affected by bulimia from publication, under the banner of the Anglo-Saxon publish or perish.

The nearly 400 pages of this journey through the history of fossil interpretation are written with the same spirit and punctiliousness that the authors themselves devote to determining each and every specimen in their collections.

What follows is not an essay on the history of science, although it is precisely the history of palaeontology that we are talking about; rather, the historical-scientific literature is a tool for contextualizing the immense repertoire of textual, iconographic, and anecdotal sources and creating a solid basis for the narrative.

Because this is narration: the authors participate in first person in this journey through the centuries, when they stimulate the reader to imagine the character of the "sanpaolaro" (a small pearl among the many that are discovered among these pages) and immediately afterwards entering directly into the scene with the description and the photo of a precious souvenir made by the Procidan grandmother of one of the authors.

A narrative device that through the tasty "inserts" recurs up to the afterword—in my opinion with great effectiveness—through which the reader is put at ease until he realizes that he no longer feels awe (respect yes!) for Pliny, Leonardo and Cuvier that mark the path, but can abandon himself to the imaginative world made of magical symbols, religious and archaic rites, the atmosphere almost Proustian and welcoming family memories.

The originality of this work lies above all in this skillful interweaving of the natural sciences and the humanities, from archaeology to literature and popular traditions, which makes it interesting both for the encyclopedic specialist, who can find the answer to many questions here and perhaps even stimulate new ones, and for those who are approaching these subjects for the first time as simply curious.

In this effort to cover all the points of view through which fossils have been interpreted over time, Maurizio and Andrea are ideally close to the Renaissance scholars widely quoted throughout the text, Agricola, Gessner, Calzolari, Mattioli, and, above all, Ulisse Aldrovandi, not so much the partly reworked *Museum Metallicum* or the printed *Monstrorum Historia to which he* owes his alternate posthumous fortune but the one who personally edited the autograph *Ornithologiae* and above all the *Theatrum Naturae*, the impressive project of representing nature in all its forms and artifices that he tried to recreate in his museum.

Happy reading!

Alessandro Ceregato

Acknowledgments

We would like to thank all those who have supported us with pictures, advice, as well as direct and indirect contributions, and apologize to those we may have forgotten to mention.

We are grateful to Professor Gian Battista Vai (Bologna) for his advice and critical reading of the text, his encouragement to complete the final draft, and for providing us the contacts of the editor. Alessandro Ceregato (Bologna) for the critical revision, the precious comments and suggestions, and the writing of the benevolent preface; Mario Giovanni Santini (Prato), author of the first insert, who introduced us to the psycho-anthropological point of view of thinking about magic and science; and Giacomo Santini for the re-reading of the text in English and some suggestions; moreover, in alphabetical order, we list all the other important contributors: Lucia Angeli (Pisa), Lucia Angiolini (Milano), Stefano Bartolini (Firenze), Cesare Bogi (Livorno), Enrico Bonino (Liegi, Belgium), Arnaud Brignon (Bourg La Reine, France), Mauro Brunetti (Navas del Selpillar, Spain), Carlo Chirli (Tavarnelle, Firenze), Nicola Cosanni (Geocollection; Silvi, Teramo), Massimo Cresti (San Casciano, Firenze), Bruno Dell'Angelo (Genova), Matteo Riccardo Di Nicola (Novara e Vercelli), Stefano Dominici (Firenze), Simone Farina (Pisa), Misa Forli (Prato), Marina Forlizzi (Milano), Piero Frediani (Castelfiorentino, Firenze), Federico Gerace (Firenze), Mario Girgenti (Palermo), Romano Guerra (Bologna), Walter Landini (Pisa), Giuseppe Manganelli (Siena), Damiano Marchi (Pisa), Giuseppe Marramà, Andrea Micheli (Perugia), Savino Montanari (Spilamberto, Modena), Angela Oleandri (Bologna), Daniele Pagli (Capraia e Limite, Firenze), Romeo Perrotta (Prato), Diego Pieruccioni (Pisa), Gian Luigi Pillola (Cagliari), Salvatore Platania (Montelupo, Firenze), Francesco Pusateri (Palermo), Stefano Puzzuoli (Pisa), Ermanno Quaggiotto (Vicenza), Paolo Russo (Venezia), Francesco Ruggiero (Agliana, Pistoia), Carlo Smriglio (Roma), Valeriano Spadini (Lucignano, Siena), Chiara Sorbini (Pisa), Giovanna Auletta (Procida, Napoli), Giovanni Todesca (Signa, Firenze), Maria Teresa Tommasi (Prato), Franco Valente (Venafro, Isernia), and Angelo Vazzana (Reggio Calabria).

For all the material of the Bellucci Collection, very important for the writing of our text, our gratitude goes to Mariangela Turchetti and Tiziana Caponi, of the

National Archaeological Museum of Umbria and of the Hypogeum of the Volumni, Necropolis of Palazzone (Perugia).

We would also like to thank a number of Institutions that have given us their precious and indispensable contribution:

Ministry for Cultural Heritage and Activities and Tourism; Regional Directorate Museums of Umbria; the National Archaeological Museum of Umbria in Perugia; the Museum of Palazzo Poggi at the University of Bologna; the Museum of Natural History at the University of Pisa; Schatzkammer und Museum des Deutschen Ordens (Vienna); Società Italiana di Malacologia (Napoli); Società Italiana di Paleontologia (Milano); and Rivista Italiana di Paleontologia e Stratigrafia (Milano).

All the digital libraries that have made their ancient texts available to the public, without which we could not have completed our study. We thank and refer to the attributions given in the captions or throughout the work.

Finally, we are grateful to Aaron Schiller, Associate Editor of Earth Sciences, Geography, and Environment, Springer New York, and his colleagues, Zoe Kennedy, Herbert Moses, Tessy Priya, and Henry Rodgers, for their help in the translation from Italian to English, the final preparation of the text, and the subsequent editorial process. All very friendly and competent; to them we dedicate this special thanks.

Andrea and Maurizio
(Livorno & Prato, 17 February 2022)

Contents

Short Biographical Notes

Maurizio Forli (Rosignano M.mo, Leghorn, 1954). Independent researcher, always been interested in natural sciences, with preferences for paleontology and malacology. Various graphic and artistic experiences. Curator of the organization and staging of numerous expositions, conferences and cultural initiatives, he has so far published, over forty scientific articles of fossil malacology and numerous others in various magazines and journals. Co-author of books on the Italian Plio-Pleistocene malacofauna, particularly from Tuscany, has recently published with Andrea Guerrini "Storia Naturale della Toscana" and with another co-author "The family Rissoidae Gray, 1847 from Miocene to Present-day in the Mediterranean basin" (Danaus Editions). Active member of the Gruppo Archeologico e Paleontologico Livornese, he is member of the Italian Paleontological Society and of the Società Italiana di Malacologia.

Andrea Guerrini (Leghorn 18/2/1968). Radiology Technician and graduated in Natural Sciences in 2007 (University of Pisa), scholar of Paleontology, he attended the online courses BipedalismX Course: The Science of UprightWalking – Dartmouth University (Prof. Jeremy DeSilva), ASM246x Human Origins Course – Arizona State University (Prof. Don Johanson) and Forensic Archaeology and Anthropology – Durham and Teesside University. He is a member of the Italian Paleontological Society, Società Toscana di Scienze Naturali and an active member of the Gruppo Archeologico e Paleontologico Livornese. He is the author of papers on palaeontology and on paleoradiology. He is co-author of the book of palaeontology "Storia Naturale del Pianeta Terra", Vol. I and II (Pacini Publisher) and "Storia Naturale della Toscana" published with Maurizio Forli (Danaus Editions).

Chapter 1
Toad's Eyes, Devil's Nails, Snake Tongues...

Abstract The discovery in the past of strange and unknown petrified objects has given rise to many legends and superstitions since the remotest antiquity. Before recognizing fossils as the remains of ancient beings who lived in the more or less remote past, scholars of natural things had to struggle at least until the beginning of the nineteenth century. At the same time, popular culture has continued to maintain the ancient knowledge to this day, remembering part of the ancient rituals. We report some testimonies.

Keywords Fossilia · Fossil · Toad's eyes · Devil's nails · Snake tongues

Toad's eyes, devil's nails, snake's tongues, etc. this list of ingredients, apparently fanciful and repugnant, could be the beginning of a magic recipe like the ones we sometimes hear about in films or read in fairy tales or in novels. From the point of view of this text, in reality we are dealing with nothing more than fossils, called by the name by which they were known hundreds of years ago, feared, or appreciated, according to their use for different purposes.

The discovery in the past of strange and unknown petrified objects has given rise to many legends and superstitions since the most remote antiquity, as people could not explain their origin. In historical age, from the Egyptians to the Romans, the studies and the magic religious experiences, or linked to a sort of primitive medicine, have made that a lot of these news, often borrowed through fanciful tales more and more improbable, have given origin to myths, today often brought back to their original meaning, once understood and correctly interpreted the fossils connected to them.

An almost inexhaustible source of the most varied information, which has strongly influenced the research carried out since the medieval alchemists, is the *Naturalis Historia of* Pliny the Elder, a Roman author of the first century AD (23–79 AD), who in his work in more than 30 volumes, bringing together from various sources the knowledge of his time and his own observations, laid the foundations for the continuation of this knowledge.

Beliefs and superstitions related to particular stones endowed with strange shapes, among which our fossils, strengthened and developed in the European area,

© The Author(s), under exclusive license to Springer Nature Switzerland AG 2022
M. Forli, A. Guerrini, *The History of Fossils Over Centuries*,
https://doi.org/10.1007/978-3-031-04687-2_1

especially during the Middle Ages, often with conflicting interpretations, linked to aspects of religiosity. In this context, contrary to the positive virtues usually attributed to gems and similar stones, a whole series of petrified remains can be found which, precisely because of their appearance and mysterious genesis, were often interpreted as manifestations of the "Evil One" and, as such, played a role both in defeating and searching for him, even though they were often used as medical matter, antidotes or amulets.

These particular stones characterized by special shapes, crystals, simple rocks, or fossils, due to their external morphology, according to a common use at the time that saw in things a mysterious ambivalence, *an* opposite duplicity (*natura ancipite*), were also considered bearers of signs that revealed properties hidden in them by nature or by the intervention of the saints and therefore endowed with symbolic meanings and magical and therapeutic virtues. In the field of medicinal magic, a method of treatment based on the use of plants in particular, but also of minerals and fossils, according to their similarity of shape and color with parts of the human body, has been very popular since ancient times. There would be a link between all natural things, including the stars, which would bind those with similar functions or with the same appearance, recognizable by particular common signs, which would ensure that, for example, a plant with parts resembling parts of the human body, can be effective and therefore used to cure or protect the organs with which there is similarity or, on the contrary, spoil them permanently leading to death. From these concepts was perfected in the course of time the so-called theory of signatures on which physicians and official apothecaries were based. In the poorer and less educated social classes, culturally backward and superstitious, folk medicine was handed down mainly by some women who prepared remedies with herbs or other substances and often acted as midwives. It was they who, in case of particular conditions of hardship in the communities where they practiced, such as famine, plague, or episodes considered criminal, risked being identified with those mythological figures much feared in the collective imagination, endowed with occult magical powers derived from the pact with the devil, called witches. In recent times, still in the 1960s and 1970s, for direct testimony (FM), my paternal aunt Settimina was called to cure small diseases or remove the evil eye, similarly to others who continue to practice their medicine even in our days. We report a rare case in which is testified the use of fossil shells for these popular practices:

> *Da noi sono sempre delle donne, delle donne di una certa età; come anche per la medicina dell'occhio, dove si usa un fossile, si fa una preghiera sopra un bicchiere con questo fossile, generalmente sulla testa o a fianco di chi deve essere pulito, della persona a cui deve essere tolto l'occhio, diciamo, e sono sempre donne che lo fanno e sempre di una certa età. Alcune situazioni non vengono mai chiarite; ci sono sempre quei, come ti posso dire, quei piccoli misteri.*
>
> [...] *il malocchio è un qualcosa che energeticamente tu vedi da determinati segni. Io lo sai come lo vedo? Quando butto la pietra, questo fossile nell'acqua, si formano un tot di numeri di bolle e a seconda della grandezza e dell'ubicazione della bolla viene fuori il tipo anche di malocchio.* (De Bernardi 2015: 322, 333)

With us it is always women, women of a certain age; as also with evil eye medicine, where a fossil is used, a prayer is made over a glass with this fossil, generally on the head or beside the person who is to be cleaned, of the person whose evil eye is to be removed, let us say, and it is always women who do it and always of a certain age. Some situations are never cleared up; there are always those, as I can tell you, those little mysteries.

[...] *the evil eye is something that energetically you see by certain signs. Do you know how I see it? When I throw the stone, this fossil in the water, a number of bubbles form and depending on the size and location of the bubbles, the type of evil eye also emerges.*

Some fossils were believed to be components of magic potions or evil in themselves, but, as well as being taken in various ways to cure and heal illnesses and diseases, they were used as amulets with multiple properties, for example, to defeat the devil or remove the evil eye. This way of understanding and using fossils as magical or medicinal ingredients has gradually diminished and been lost over the last three centuries. On the contrary, popular knowledge and superstitious practices oriented to the use of stones and minerals in general to prevent and cure various ailments or for their beneficial astral influence are increasingly spreading, especially through the web. A sort of cultural revival that finds millions of admirers and users of these natural substances, believed to be bearers of who knows what positive properties, perhaps linked to the signs of the zodiac, in such a way that follows more or less slavishly the medieval and even more ancient knowledge.

This can only be surprising because today it is widely demonstrated that certain magical properties, often even therapeutic properties of said stones or fossils, were and are absolutely nonexistent... yet the general culture is considerably higher than that of medieval and modern populations. With the economic, scientific, and cultural progress, we would expect a less diversification of the religious cults and a decrease of the superstitious beliefs, but probably, the decrease of the dominant religious practice is counterbalanced by an increase of the superstition, meant as regression to previous stages, to which the religion has replaced and superimposed itself, instead of a parallel reorganization of both.

The study of fossils implies more or less close links with different sciences such as palaeontology, geology, physics, anthropology, ethnology, philosophy, evolutionism, which, developing more or less at the same time, have allowed the elaboration of current knowledges.

In the notes that follow, compiled also and above all by consulting ancient books, scientific articles, and news on the web, we gather information and notes on many fossils unknown to most people, known only to lovers and scholars of paleontology, geology, or natural sciences, in order to convey what little we know, in a unicum reasoned built by assembling critically, and we hope in a relevant way, dozens and dozens of sources, according to today's scientific canons even if, sometimes, we have lingered on those *magic* and *curious* things that still attract many of us, aware that however they are also an important cultural heritage that is part of our history, trying to look at fossils as a subject, visible not only from the naturalistic side but also from other, apparently distant, that lead the thought to investigate in unexpected fields to a superficial and unidirectional vision and understanding (Fig. 1.1).

Fig. 1.1 Internal molds of fossil bivalve shells, carried in pockets as lucky charm in some Italian regions (e.g., Sardinia and Tuscany). Average shells width 5 cm. (Photo Maurizio Forli)

References

De Bernardi M (2015) Marking healing: ethnosemiotics of a magical act: semiotic reading of the practice of "signatures" in the transcultural perspective relative to the Apennine area of the Italian region of Emilia Romagna. Social anthropology and ethnology. Université Toulouse le Mirail – Toulouse II, 2015. English

Chapter 2
Fossilia and Fossils: Considerations on Their Understanding Over the Centuries

Abstract At present it is quite obvious that fossils are the remains of ancient organisms, but it was not always so. The conception of the origin of fossils has been the subject of antithetical interpretations, as early as the classical age, which can be summarized to simplify into two main categories sometimes not clearly separated, called inorganic genesis and organic genesis. These interpretations have been intertwined with each other, prevailing now one or the other, until the confirmation of organic genesis during the eighteenth century at least for the most recognizable and widespread organic remains. It was mainly in Italy that these new currents of thought took shape in the course of the sixteenth century and then spread later throughout Europe, which will prove fundamental for the development of the natural history studies. In this respect, in the chapter we will try to make a brief overview by mentioning the various authors and their works, who contributed most to the *quaestio de fossilibus* and thus also to the development of the embryonic stages of earth sciences. It is in their publications that we also find the first printed illustrations concerning fossils, the recurring theme of our text. These authors, in particular of the sixteenth and seventeenth centuries, were still far from the concept and definition of the term paleontology, used for the first time in 1822 by de Blainville, but they are to be considered, among the founders of this discipline. It was only early of the nineteenth century that the term fossil took on its current meaning and toward the end of the nineteenth century, with the rise of evolutionary ideas and the awareness of the expansion of the past, that fossils were definitively considered and studied as remains of ancient populations subject to the same laws of biology that regulate today's living beings.

Keywords *Fossilia* · Fossils · History · Understanding over the centuries

For many of us, it is quite obvious that fossils are the remains of ancient organisms, but it has not always been so. This concept has been formed and consolidated with difficulty over time, starting from the beginning of the modern age, as a consequence of different paths and sometimes opposing ideas on the same subject, although based on the same knowledge.

© The Author(s), under exclusive license to Springer Nature Switzerland AG 2022

M. Forli, A. Guerrini, *The History of Fossils Over Centuries*,
https://doi.org/10.1007/978-3-031-04687-2_2

There are countless ancient and current writings on the subject by more or less important philologists, philosophers, and historians of the natural sciences, investing the religious and scientific aspects of human knowledge, with the many implications that follow.

On the other hand, a hint of how our predecessors considered fossils, we believe it is essential to clarify the concepts according to which, even in the first printed texts, these are, by naturalists who began to study them, putting them in some order, identified as single lithified objects and as such included in different groups depending on whether they were whole organisms or their partial remains. It is not infrequent, in fact, that invertebrates such as sea urchins, corals, or shells are listed and described in different categories depending on the appearance they took in the case they were whole or fragmented.

Human beings have the ability to lead back to known images, and therefore to objects or people "present" in our mind, those structures, shapes, and random appearances that every now and then are proposed to us. We believe that this is the mental mechanism that has transformed the casts of bivalves into ox hearts or vaginal openings, the discoidal shells of macroforaminifera into coins or lentils, some corals into petrified mushrooms, the teeth of sharks and the rostrums of belemnites into points of lightning, the spirals of ammonites into snakes, and so on.

Following the trend of influences that naturalistic thought underwent over the centuries, the concept of the origin of fossils has been the subject of antithetical interpretations, which can be summarized into two main categories called inorganic genesis and organic genesis. These interpretations have been intertwined with each other, prevailing now one or the other, until the final confirmation of organic genesis during the second half of the seventeenth century.

The current vision, accredited in the scientific community, of a planet Earth as an open and dynamic system, in continuous biological and geological evolution, represents the culmination of a series of intellectual processes and objective findings that have accelerated since the mid-nineteenth century, a period in which naturalistic and philosophical concepts, basic for today's understanding and comprehension of natural history and the world in general, became explicit and consolidated. Although they were used in prehistoric times as pendants, for the production of jewelry or as tools, it is believed that a conscious interest in fossils in Western Europe cannot be prior to the pre-Socratic philosophers, although the contents derived from written evidence, present as quotations and fragments, are of uncertain interpretation:

> *These authors were the first to begin to examine them concretely, and particularly because of their common occurrence, they referred mainly to petrified shells. Free from all prejudices, they drew from them conclusions of importance to natural history rarely departing from the truth even as to their petrification and actual nature.*
>
> *It is certain that Xenophane* […] *remarked as a particular phenomenon, that petrified shells are found in the bosom of the Earth and in the mountains, and that from the quarries of Syracuse were taken remains of all species of marine animals. Thus he concluded that all these places had been covered by marine waters. Herodotus, one of the oldest historians,* […] *knew the shells in the mineral kingdom, so well, that of those which were found on the Egyptian mountains, he believed he could affirm that the sea had withdrawn there little by little. Others, mainly Eratosthenes, Xanthus of Lydia, Straton and Strabo drew the same*

> *conclusions about the petrified shells, concerning all the land, now inhabited,* [...] *They speak of* Pectinites *and other remains of marine animals, naming the places where they were found, thus proving the great knowledge they had* [...] *Pausanias put the petrified shells among the most remarkable things he saw in his travels, as it seems very clear from a passage of his work. It seems evident that in Greece, the lovers of Nature knew the petrified shells, that they considered them as natural bodies, which afterwards were changed into stone, and although they were not able to determine the causes of this change, they did not leave, even with little knowledge, to make very important cosmological observations and conclusions.* [Walch 1768–1773: pp. 17–18 (translation from the original in French)]

In addition to this view of the nature of petrified bodies, it is possible to find, in the lapidaries and treatises on medicine and natural history of the classical age, another particular view of these objects. In this regard, we remember Aristotle's thought [IV century BC (384–322)], with the theory of celestial influences, which is opposed to that of the abovementioned authors according to which petrified bodies would be spontaneous products of the earth, through a more or less mysterious *vis formativa* activated by "dry exhalations" of the earth itself, produced by the action of solar rays. On this theory, expressed in the *Meteorologica, is* based a writing, the *Perí líthon*, by his pupil and successor Theophrastus (371–287 BC), which had a lot of credit in medieval times:

> *the Perí líthon (On the stones) provides a detailed description of sixty different types of solid materials, meaning minerals, gems, clays, rocks, earths and even pearls, all solid materials that can be extracted in mines, quarries or however from the subsoil, excluding metals. It is the first mineralogical treatise ever written (ca. 315 B.C.) and it is also the first to be printed (February 1497, by Aldo Manuzio).* (Mottana and Napolitano 1997)

Also related to this interpretation of their nature, because of the contradiction inherent in having organic appearance, but stony substance, some of them assume properties and a magical dimension-therapeutic, becoming special stones also as components of traditional pharmacopoeia. In this regard, we remember the *Naturalis Historia* of Pliny where the emphasis is also placed on the symbolic meaning of their morphology and the ways in which they would have acquired these virtues.

This Plinian literature, together with other works such as Lucretius' *De rerum natura* (first century BC) and Seneca's *Naturales quaestiones* (first century AD), defined as "meteorological" because it recalls the homonymous work of Aristotle, arrives as compendiums or extracts until the early Middle Ages, influencing also the opinions of the masters of the cathedral schools such as the Spanish theologian and archbishop Isidore of Seville (560–636) and the Anglo-Saxon Christian monk and historian Bede the Venerable (673–735). This type of writings, although sometimes enriched by the experiences of the authors, in general remains works lacking in organicity and limited in the area of Plinian derivation (Fig. 2.1).

As early as the sixth century, monastic communities had introduced the copying of ancient manuscripts into their *work*, but a great process of mediation between classical culture and the West was carried out from the eighth century onward by the Arabs. The Arabs, by promoting the process of translation and commentary on the philosophical and scientific works of the Greeks, also in the conquered lands, particularly in Spain, promoted knowledge throughout the Latin West (Fig. 2.1).

Fig. 2.1 On the left, Aristotle, the philosopher also known as the Stagirite, from his hometown of Stagira, in the Peloponnese, and on the right, Gaius Pliny the Second, better known as Pliny the Elder. Medallions with portraits, modified after *Icones veterum aliquot, ac recentium medicorum, philosophorumque elegiolis suis editae by Jànos Zsàmboki*, Latinized as Janus Sambucus (1574) [https://babel.hathitrust.org/cgi/pt?id=gri.ark:/13960/t0dv92j6f&view=1up&seq=8 (Public Domain)]

Regarding the origin of petrified bodies, in the course of time, also other theories found credence beyond that of Aristotle as, for example, the *vis plastica* of Avicenna (980–1037) according to which the petrified organic remains were nothing but abortions of nature that would have been able to shape them but not to infuse them with life:

> *Avicenna, in the tenth century, contributed not a little to enriching and consolidating this absurdity, [...] in what he called* Vis Plastica. *As Aristotle supposed an equivocal or spontaneous generation, and it was natural that the advocates of this idea should suppose a certain virtue. A virtue which was commonly called* Vis Plastica *and in regard to the Mineral Kingdom* Vis Lapidifica. (Walch 1768–1773: p. 19)

However with the Middle Ages, a long phase opened in which in the first part of this period, conditioned by the progressive disintegration of the Roman Empire, by the invasions and migrations of different peoples, there were no favorable conditions for the development of studies or researches and the interest toward natural phenomena was very scarce. The dominant philosophy, based on biblical arguments, in particular derived from the first chapters of the *Book of Genesis*, was that the history of the earth was short-lived and the expression of a "recent" creation.

Around the twelfth century, again in the West, there was a general awakening from the cultural point of view, the first universities were founded, and through *scholasticism*, which was medieval philosophy and theology, there was a mixture of ancient Greek, Latin, and Arabic science and the universal hegemony of Christian culture. There is always a sharp separation in approaching the study due to the dualism between popular religious culture and that of the learned in the universities, which leads them to profoundly different conclusions. In the first case, it is believed that through the intervention of the saints, animals, plants, and stones are endowed with magical and symbolic virtues, while the wise interpret and explain these properties through the dictates of the philosophy of Platonic, neo-Platonic, hermetic, and Aristotelian concepts, the doctrine of signatures of Plinian derivation, the arrangement of the stars, and their influence between the macrocosm and microcosm of man. In a divinely created universe in which all parts are connected through a dense network of hidden affinities, with the whole fused into a coherent whole, correspondences can be manifested by similarities and analogies not only between microcosm and macrocosm but also between heaven and earth, between animals and plants, and between living and nonliving entities. The nascent science is presented as natural magic.

On the basis of this, therefore, easily identifiable marine bodies, especially shells, are considered traces of the universal flood, while others, although still marine but difficult to conceive, are believed to be particular figurative stones, produced in the place where they were found by celestial phenomena or by the ground itself, in peculiar astral conditions, on the basis of what has been said above. In this context the Dominican bishop Albertus Magnus (~1206–1280) had an important role in separating the sources of discussion of theology from those of philosophy, differentiating their respective fields and attributing to the latter the study of nature in its various articulations. To this interest he applied what was then a new methodology of study, experimenting and investigating directly into the causes operating in nature, no longer relying solely on the authority of ancient writers and their presumed veracity. Despite this, his opinion on the formation of fossils, starting from the Aristotelian legacy, based on a *virtus mineralis lapidifica formative* infused by the stars, was harshly criticized by Agricola in his *De ortu et causis subterraneorum* (Ottaviani 2000):

> *His* formative Virtus *is the* Vis Plastica *of others. He himself admits that this formative virtue has no particular name, and that each of its supporters believes himself entitled to give a name to his fantasy* [...] *Nevertheless Albert the Great, does not entirely deny the possibility of the petrification of animals, which would be done by means of a petrifying virtue, which he called* Mineral Virtue. (Walch 1768–1773: p. 19)

The agreement between nature and the Bible, but with the action of celestial forces, is also evident in a writing of the contemporary Ristoro d'Arezzo, the *Composition of the World*, manuscript in several copies of 1282 remained unpublished until 1849, according to which all inorganic bodies derive from celestial influences that guide and determine the mixture of elements that form the material diversity. In dealing with the origin of mountains, he believes that the shells, which he calls *fish bones*, contained in them, derive from the biblical Flood, which occurred by the conjunction of water stars, but also from the previous presence of seawater or their similarity:

> *And now we see the generation of mountains* [...] *And also the mountain can be because of the water of the deluge, which being the water of the deluge, and covering the earth and remaining on the earth, because of the wind or other reason, can take the earth from one place and put it in another; therefore it is the nature of the water, if it is kept on the earth, to make the mountain and the valley, and it is its nature to leave the earth mountainous and*

> *valley-like* [...]. *And we have already found and excavated, almost to the sum of a very large mountain, many fish bones, which we call snails, and such they call niches: which were similar to those of the painters, in which they hold the colors* [...] *And that area where there are these mountains, where there is the sand and the fish bones, is a sign that in that area was already the sea, or water like the sea* [...]. (Restoro 2: 6, 8 in Narducci 1859: pp. 85–86)

Another testimony on the same subject concerning ancient Tuscany, after Ristoro, is in Giovanni Boccaccio (1313–1375). In *The Filocolo* (1336), Boccaccio, in an allegorical way, highlights the presence of a great quantity of fossil shells in the hills around Certaldo, his native town, attributing them to the Flood:

> *Nella fruttifera Italia siede una picciola parte di quella la quale gli antichi, e non immerito, chiamarono Tuscia, nel mezzo della quale, quasi fra bellissimi piani, si leva un picciolo colle, il quale l'acque, vendicatrici della giusta ira di Giove, quando i peccati di Licaon meritarono di fare allagare il mondo, vi lasciò secondo l'opinione di molti, la quale reputo vera, però che ad evidenzia di tale verità si mostra il picciolo poggio pieno di marine cochiglie* [...] *biancheggianti tutte*. (Squarzafico 1497, 5: p. 8)
>
> *In the fruitful Italy sits a small part of that which the ancients, and not undeservedly, called Tuscia, in the middle of which, almost between beautiful plains, stands a small hill, which the waters, avengers of the just wrath of Jupiter, when the sins of Lycaon deserved to flood the world, left there according to the opinion of many, which I believe to be true, however, as evidence of this truth shows the small hill full of sea shells* [...]. *whitening all.* (Squarzafico 1497, 5: p. 8)

With the progressive estrangement of Islamic civilization from the West, which began in the late Middle Ages, a new cultural phase developed. There was an obsessive search to recover the originals of ancient Greek texts, even those that were obscure or thought to be lost, in order to be able to make new translations to replace the medieval ones, in the certainty that a translation into correct Latin, starting directly from the original language, was more valuable than one from the twelfth or thirteenth century, considered to be corrupted by the mediation of the Arabic language in these versions [...] (Boas 1973: p. 12). By the middle of the fifteenth century, this great research had practically come to an end, and the fall of Constantinople in 1453 at the hands of the Turks closed the door on the surest and richest source of such works (Ibid. 17). All this allowed the scholars of the fifteenth and sixteenth centuries the correct interpretation of the texts, thus also developing the scientific culture, especially in the pharmaceutical medical field, but also naturalistic.

At the beginning of the sixteenth century, Leonardo da Vinci had conceived, with intuition and logic, original ideas both on the nature of fossils and on geological phenomena. To him we owe their rational interpretation in a paleoenvironmental and geological context that was at least two centuries ahead of the history of geological sciences (Raffi and Serpagli 1993). These concepts are expressed by notes, annotations, and drawings scattered in numerous manuscript sheets and in particular in the collection called *Codex Leicester*, also known as the Hammer Codex, where memories of experiences and observations made while travelling through the hills and valleys of Tuscany, Romagna, or the Po Valley often appear. Leonardo rejected the ideas of those who believed that niches, i.e., shells, were formed by a particular plasma power of the soil or the stars. He observed that it would not be possible to find together in the same place, as in fact happens, whole or shattered specimens of young and old *niches*, distinguishable by age on the basis of the number of growth lines on the surface of the shell, as Leonardo himself discovered, sometimes mixed with fish bones and teeth, recognized as such, thus affirming the animal origin of the remains called *lightning bolts* or *snake tongues*, decisively rejecting also the diluvial theory with numerous arguments (Fig. 2.2):

> *se non fossero stati accumulati lungo un antico lido dalle onde del mare* […] *d'altri nichi dentro alli nichi interi* […] *non si troverà infra loro ossa e denti di pescie, li quali alcuni dimandano saette, e altri lingue di serpenti* […] *se il diluvio fu per pioggia che ingrossò i fiumi o per ringorgamento del mare,* […] *li nicchi, come cosa grave non son sospinti dal mare alli monti, né tirati a sé dalli fiumi contro al corso delle loro acque.* (Codice Leicester, probabilmente scritto tra il 1504 e il 1510, in particolare foglio 9 r)

> *if they had not been accumulated along an ancient shore by the waves of the sea* […] of other niches inside whole niches […] one *will not find among them bones and teeth of fish, which some call lightning, and others snakes' tongues* […] *If the flood was by rain that swelled the rivers or by the rising of the sea,* […] *the niches, as a heavy matter, are not pushed from the sea to the mountains, nor pulled to themselves by the rivers against the course of their waters.* (Leicester Codex, probably written between 1504 and 1510, in particular folio 9 r)

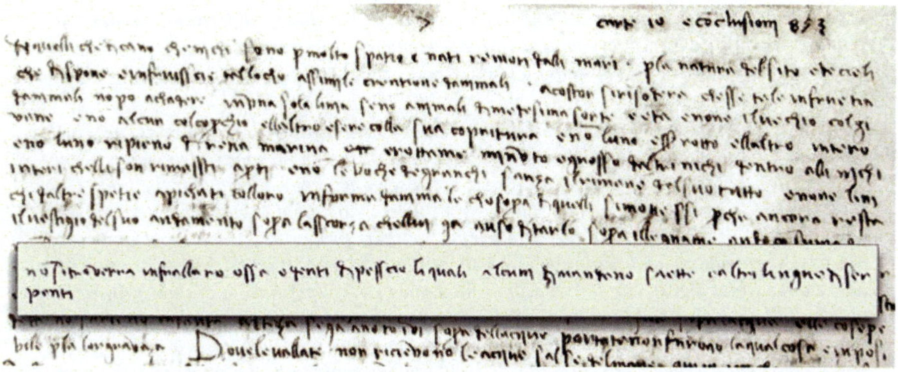

Fig. 2.2 Fragment from the Leicester Codex, folio 9 r, line nine and the beginning of the tenth. The sentence has been inverted horizontally to be legible according to common usage (modified) (Public Domain)

These Leonardian considerations, like many other of his observations and studies, did not spread much because they were unknown or little known among his contemporaries and had no influence on the establishment of paleontology, from his time to the second half of the nineteenth century when the codices were found. Nevertheless, it seems that similar ideas, as well as Galilean teachings, somehow circulated in Roman cultural circles in the first half of the seventeenth century at the time frequented by the Messina painter Agostino Scilla, who will be influenced, at least for the method of analysis adopted, in the drafting of his treatise, *La Vana speculazione disingannata dal senso* (Giallombardo 2016: pp. 37–38).

Because of the following that it had on his contemporaries and later scholars, the most accredited explanation of the term *fossilia*, with the indication among them of the remains of ancient lithified organisms, even if not recognized as such by the author and therefore believed to be nonorganic still following the Aristotelian legacy, is generally attributed to Georg Bauer (or Pawer) (1495–1555), also known by the Latinized name of Georgius Agricola, a German scientist considered the father of mineralogy:

> *Quelle cose fossili poi, che ò sono un rimedio, et una medicina al corpo; ò che l'affliggono e tormentano; prendendosi ò col mangiare, ò col bere; ben che anche esse patiscano, hanno nondimeno una segnalata virtu e potentia di fare. Ma quelle, che sono un rimedio al corpo, il curano parte per le loro proprietà, che hanno in tutto l'essere lor: parte per qualche qualità, che cagiona la sanita. Di quelle che per la proprieta vagliono; alcune repugnano, et ostano à i veleni; alcune altre ne guariscono i morbi* […] *Altre che vanno a finire à guisa di turbine, sono torte, come lumache; la qual figura in alcune pietre si vede* […] *Alcune hanno d'ogni parte un certo che sporto in fuori, come un porro; come è la Mirmecia. Altre rappresentano figure di altre cose; come il corno di Ammone rappresenta il corno; il Tephrite mostra la effige d'una nuova luna cornuta* […] *Pare un core di bue, la bucardia* […] *rappresentano figura di lentechia, la arena de l'Aphrica; et alcune pietruzze presso le piramidi de l'Egitto; et in un colle della Cappadocia. Di più; lo stelechite è simile à un tronco d'albero: il belemnite, à una saetta; la gemma che chiamano chalazia; à un grandine: la pietra giudaica, à la ghianda.* (Tramezzino 1550)

> *Those fossil things, then, which are either a remedy or a medicine for the body, or which afflict and torment it, whether by eating or drinking, although they may also suffer, nevertheless have the virtue and power to do so. But those which are a remedy for the body cure it partly by their properties, which they have in their whole being: partly by some quality which causes health. Of those that are useful for their properties, some repel and resist poisons; some others heal diseases* […] *Others that end up like a whirlwind, are twisted, like snails; which figure can be seen in some stones* […] *Some have a certain protruding outward, like a leek, like Myrmecia. Others represent figures of other things; as the horn of Ammon represents the horn; the Tephrite shows the effigy of a new horned moon* […] *It seems an ox hearth, the bucardia* […] *represent the figure of lentil, the sands of Africa; and some pebbles near the pyramids of Egypt; and in a hill of Cappadocia. Moreover; the stelechite is similar to a tree trunk: the belemnite, to a thunderbolt; the gem they call chalazia, to a hailstone: the Judaic stone, to the acorn.* (Tramezzino 1550)

Moreover, Agricola expresses his own idea of the meaning of *fossilia*, which is of excavated things (*fossor* is the Latin etymon corresponding to our *digger*) concluding that under this denomination are included those bodies whose constituent substances can be traced back to earth, stone, and metal, summarizing everything in a synoptic table (Fig. 2.3):

Fig. 2.3 Tramezzino Michele (De la natura de le cose fossili, 1550: 185v). The subdivision of *fossilia* proposed by Agricola [https://archive.org/details/hin-wel-all-00002012-001 (not copyright found)]

> *Ma perché si faccia questa dottrina più piana e più chiara, mostrerò quali siano quelle cose, che sono prime e più generali in questa materia, e che complettono tutte le cose fossili. Ma quante e quali queste si siano, non vi si trovano gli scrittori concordi: perché (come Aristotele ci insegna) sono due le spetie de' corpi, che dentro la terra si generano; cio è i fossili, che egli chiama ne la lingua sua Oricti: e quelli, onde i metalli si generano, che egli metalleuti chiama.* […] *Ma ò che Aristotele seguisse questa comune usanza di parlare, ò pure che per altra cagione questa distensione si facesse; egli non avertì che ogni materia, onde i metalli si fanno; non è meno fossile, che le terre di pregio, e le pietre si siano: il che essendo cosi come io dico; assai chiaro si vede, che il corpo fossile è più generale, e comprende in se la terra, la pietra, il metallo.* (Tramezzino 1550)

> *But in order that this doctrine may be made easier and clearer, I will show which are the first and most general things in this matter, and which include all things fossil. But there is no consensus among writers as to how many of these things there are, because (as Aristotle teaches us) there are two types of bodies that are generated in the earth, namely, fossils, which he calls in his language Oricti, and those from which metals are generated, which he calls metalleuti.* […] *But whether Aristotle followed this common custom of speaking, or whether for some other reason this distintion was made, he did not say that every matter, from which metals are made, is no less fossil than valuable earth and stones are: which, being so as I say; is very clear that the fossil body is more general, and includes in itself earth, stone and metal.* (Tramezzino 1550)

As it is clear from Agricola's explanatory text and from the proposed scheme, the broader category of *fossilia*, i.e., inanimate underground bodies, therefore all-inclusive, is divided into subcategories, among which, at the last level of the scheme, there is a grouping including stone, gem, marble, and rock, but without any reference to petrous bodies in the shape of animate things, which would have allowed us to recognize fossils as we understand them today, even though they are indicated in some way in the text, as in the passage above. For greater clarity, therefore, we point out that conceptually Agricola's fossils are not the *fossils* as we understand them today, even if they are unconsciously contained in them. Archaeological finds were also included in these categories of unearthed materials and are sometimes depicted in the same publication together with natural things, as did, for example, Gessner and Aldrovandi. For a more detailed discussion of Agricola's contribution, see Nicoletta Morello (Vai and Caldwell 2006).

Many scholars believe that the Protestant Swiss physician and naturalist Conrad Gessner (1516–1565) of Zurich, Switzerland, has the merit of explicating *fossilia* in his writings and numerous illustrations, including those of animal and plant remains. Considered by many to be the greatest naturalist of his century and a forerunner of paleontology, he would acquire this merit in his last literary work *De omni rerum fossilium genere* (1565). This work dedicated entirely to this set of petrous objects, although a prodrome to a larger, though never finished and published, due to the death of the author, plays an important role in the development of the study of *fossilia* to fossils; Gessner's book introduced several innovations that will be taken up by his contemporaries and subsequent writers, although they were included in the *Index Librorum Prohibitorum*, a list of works prohibited for the content deemed wholly or partly prohibited, drawn up by the Roman Church in the fight against heresies.

If it is true that other texts already existed that dealt with these stone bodies, and some of the names used can be found in the works of classical authors up to medieval lapidaries, Gessner's book is the first in which the illustrations are used systematically. In addition, the names of *fossils* are written not only in Latin but also in Greek and German, in order to expand the knowledge and facilitate the recognition. Also the already mentioned, Agricola deals with these topics, but the absence of illustrations made then difficult the identification of the objects he was describing. Another new element found in Gessner's treatise is the encyclopedic approach with which, while acknowledging the value of the ancients, he emphasizes personal and direct observation of nature. These two aspects form the basis for another innovation contained in this book, the direct reference to a personal collection of artifacts, considered the first collection of these natural objects according to today's concept, that of his friend Johann Kentmann (1518–1574) (Fig. 2.4).

Fig. 2.4 On the left, a portrait of Georg Bauer, the Georgius Agricola of the Latins and, on the right, a portrait of Konrad Gesner, or Conrad Gessner, his Latinized name being *Conradus Gesnerus*. [Modified after https://babel.hathitrust.org/cgi/pt?id=gri.ark:/13960/t0dv92j6f&view=1up&seq=8 (Public Domain)]

Obviously the question of fossils could not be separated from the hypotheses on the layers in which they were found and the origin of the earth's crust, so the dominant intent, in addition to classify them, was to explain a fundamental characteristic of these objects, the petrification, the being of stone, and their mode of formation. Other sixteenth-century authors, important for having given their contribution to the dispute about the nature of fossils and other natural phenomena such as the origin of earthquakes, mountains, thermal waters, etc., were Girolamo Fracastoro (~1476–1553), Gerolamo Cardano (1501–1576), Gabriele Falloppio (1523–1562), and Andrea Cesalpino (1519–1603), in whose books on these subjects, however, illustrations of fossils are absent.

In Renaissance treatises, in the Latin West, the Greek term *Orycta* was replaced by the Latin term *fossilia*, a substitution then due not to a different meaning of the definitions but to the different language used.

It is above all in Italy that during the course of the sixteenth century, these new currents of thought took shape, later spreading throughout Europe, which would later prove to be fundamental for the development of the science called natural history. Also during the Renaissance, the interest in spices, essences and plants in general, led to the founding of the first botanical gardens and an increase in studies on medicine, famous among the authors who dealt with it, the Sienese Pietro Andrea Mattioli who translated into Italian Dioscorides' *De Materia Medica*, published for the first time in 1544 in Venice. Some fossils were also occasionally used as ingredients in the pharmacopoeia of the time (Mattioli, 1548). This climate of new awareness and attention obtained through new knowledge acquired also personally through direct experience and observation, tending to fill the gaps inherited from the authors of the past, continued in the first half of the seventeenth century where numerous naturalistic works were produced, such as monographs, encyclopedias, and descriptions of collections and museums, even if a general and unified nomenclature had not yet been achieved:

> *In this century the systematic division of petrified shells into genera and species was not yet known. From the time of Mercati the names conchites, ostracite, ceramite, ctenite, muricite, buccinite, dentalite, strombite, cochlite and ammonite, of which Pliny had made use, were known and in use, but without any proper classification, so that petrified shells were well placed among the fossils, but were as much among the class of idiomorphic stones as in that of figured stones and sometimes in another. The particular species of petrified shells were not then examined more closely, as was done in the following century.* (Walch 1768–1773: p. 21)

To the process of circulation of ideas in general, and on fossils in particular, contributed in an obvious way the introduction of printing with movable type that allowed a more rapid spread of writings through printed volumes, in greater quantities and at lower costs, first without illustrations and then more and more richly illustrated, to consider the reproduction of the image almost more important than the description. In the first prints, the fossils are reproduced in a rather inaccurate, sometimes banal way, despite the fact that in the Renaissance, the figurative and pictorial arts were at a very high level of quality and expression. In the following seventeenth-century naturalistic studies benefited from more accurate illustrations which, together with the first true ideas about the formation of fossils, allowed research to continue for further investigation which was fully developed in the following century.

In the 1600s, iconography became an indispensable part of the text, a new practice of communication between writers and readers, an indispensable visual support for correctly identifying what had previously been handed down in the form of names, which often became increasingly obscure and indecipherable as the translations and copies of the oldest texts changed hands, or the countless natural finds hitherto unknown, from the "new worlds" discovered with the expansionist, commercial, and even naturalistic explorations that multiplied from the discovery of the Americas. This figurative mode with illustrations of remarkable quality and precision, even colored by hand, was widely used in herbaria, for example, making it easy to recognize the plants from which the natural principles were extracted, essential ingredients in medical preparations and sources of considerable earnings. Perhaps the most striking example of this new way of representation, and at the same time of visual cataloguing, is the so-called *Museo Cartaceo* edited by Cassiano dal Pozzo (1588–1657), a vast collection of drawings and prints, put together with the intention of creating a sort of encyclopedia of ancient and natural things, such as landscapes of ancient Rome, archaeological remains, animals, plants, minerals, and fossils.

The years between the sixteenth and seventeenth centuries are therefore crucial for the development of earth sciences. Those who contributed most to the scientific debate through their works, simple treatises, or real encyclopedic collections, on the *quaestio de fossilibus*, were mainly the already mentioned Agricola and Gessner, the Bolognese Ulisse Aldrovandi (1522–1605), the Neapolitans Ferrante Imperato (1550–1631) and Fabio Colonna (1567–1640), the Dane Niels Stensen, Italianized as Niccolò or Nicola Stenone (1638–1686), the Messina-born Agostino Scilla (1629–1700), and the Palermo-born Paolo Silvio Boccone (1633–1704). Very often the first illustrations referring to the most common fossils are found in their works; in particular, through their studies on glossopetres, stone cusps, until then believed to be human or snakes' tongues, and now correctly identified with sharks' teeth, Colonna and Scilla will be decisive to clarify the organic origin of these and other fossils, while Stenone, with the ideas expressed in *De solido intra solidum naturaliter contento dissertationis prodromus* (1669), will make it possible to understand the principles of formation of geological strata and of "solids," whether mineral or fossil, within them, deducing that at the time of deposition the first body was already solid while the rock was still incoherent sediment.

In the process of distinguishing fossils between the 1500s and 1600s, we should not underestimate the contribution given by the spread of the so-called *Wunderkammer* (chamber of wonders or cabinet of curiosities), where the most heterogeneous objects were grouped together, as long as they were fantastic or exotic, both artistic and natural, including many fossils by virtue of their magical-naturalistic ambivalence. In general, these particular rooms were the prerogative of kings, nobles, or exponents of the wealthier classes. Nevertheless, the materials collected, and sometimes even the buildings, were fundamental for the creation of what in many cases would become museums, as we understand them today, with objects conserved, arranged, and exhibited using scientific methods. Often in the volumes concerning *naturalia*, these environments are depicted, museums *in nuce*, owned by or accessible to the authors. Famous among the others is the one of the naturalist apothecary Ferrante Imperato, shown in the frontispiece of his *Historia Naturale*, and above all that of Ulisse Aldrovandi, who, starting from the middle of the sixteenth century, gathered a multitude of animals, plants, minerals, and natural objects that constituted a real large museum that is still partially preserved in Bologna, unlike the collections of his colleagues and friends (*amicorum*) which were soon lost, including that of Gessner, of which only a few remains are currently available:

> *We begin by presenting the history of the petrified shells of the seventeenth century* [...] *If in this century numerous lovers of nature had not begun to make collections of the bodies of the mineral kingdom,* [...] *much time would have elapsed,* [...] *before this considerable gap in natural history could be filled and this part of mineralogy made complete.* [...] *But it is thus that the collections of fossils and minerals, which were made in this century, showed first of all how imperfect the knowledge of bodies foreign to the mineral kingdom was. This began chiefly when catalogues and descriptions of this kind of collection of natural bodies began to be published. In the preceding century we had but two, but from this appeared several considerable catalogues and numerous detailed descriptions. The beginning was equally in Italy, and again in Verona, which then possessed a special interest in the Gabinetto Calceolario, of which a very extensive description was published in the year 1622. Twenty years later Besler's description of the museum was published (1642). In the year 1652 Wormius's museum; in the year 1633 Spener's; in the year 1666 Septala's; in the year 1669 The King of Denmark's museum; in the year 1674 Gottorp's; in the year 1678 Kirker's; in the year 1687 Grew's; in the year 1695 Pettiver's.* (Walch 1768–1773: p. 21)

Are the Anglo-Saxon authors in particular who had to solve the problems of identifying those stone remains left by extinct organisms such as belemnites, crinoids, and trilobites. Because of their geological and temporal distribution, these fossils are present only in the most ancient rocks, lithified sediments over 100 million years old, which are most widespread in the countries of the central belt from the East to the West of Europe up to present-day Great Britain:

> *The fossils that were preserved in these collections,* [...] *convinced more than one to give more attention to the productions of the surroundings of their cities and countries, and since these researches were not useless, numerous works of oryctography appeared in this century, which contributed infinitely to knowledge* [...]. *The writings of Charleton* [...] of *Lackmund* [...] *of Lister, of Plot, of Merret* [...], *and mainly those of Lhwyd, sufficiently prove the attention, with which it has been collected,* [...] *all that could be curious in relation to the Mineral Kingdom, and above all, what has been judged worthy of this attention,* [...] *in spite of the prejudices that still reigned about their origin. Nevertheless, it began in this century to publish particular treatises on petrification and petrified bodies, which, although they were founded almost all on false principles, and which gave for the most part false reasons* [...] *served at least to teach the lovers of nature to know the different petrified bodies, even if they did not take them for what they were.* (Walch 1768–1773: p. 22)

Between the end of the 1600s and the beginning of the 1700s, in addition to the scholars mentioned above and others that we will meet in the continuation of our text, we should mention the Englishman Robert Hooke (1635–1703) who also considers the petrifications of unknown beings as the remains of extinct organisms and not as freaks of nature.

Even in the second half of the eighteenth century, during the period of greatest development of the ideas brought by the Enlightenment, there remains a minimum degree of indetermination in distinguishing "our" fossils from *fossilia*, even if their origin is clear, at least for the most recognizable and widespread organic remains:

> *On appelle fossiles en général toutes les substances qui se tirent du sein de la terre. Souvent on se sert indistinctement du nom de fossiles & de celui de mineraux, pour désigner les mêmes substances. C'est ainsi que l'usage veut que l'on dise le règne minéral, & non pas le règne fossil.* [...] *On distingue deux especes de fossiles, 1°. ceux qui ont été formés dans la terre, & qui lui sont propres; on les appelle fossiles natifs. Tels sont les terres, les pierres, les pierres preciéuses, les cristaux, les métaux, &c. 2°. ceux qui ne sont point propres à la terre, que l'on appellee fossils étrangers à la terre. Ce sont des corps appartenans, soit au regne mineral, soit au regne vegetal: tells que le coquilles, les ossemens de poisons & de quadrupeds, les bois, les plantes, &c. que l'on trouve ensevelis dans les entrailles de la terre où ils ont été portés accidentellement.* (Diderot and D'Alembert 1765, 7: pp. 211–212)

> *Fossils are generally called all substances which are extracted from the sinus of the earth. Often the name of fossils and minerals is used indiscriminately to designate the same substances. It is for this reason that custom wants to say the mineral kingdom and not the fossil kingdom.* [...] *there are two kinds of fossils: 1. those which have been formed in the earth and which are proper to it; they are called native fossils. Such are the earth, stones, precious stones, crystals, metals,* etc. *2°. Those which are not at all proper to the earth; these are called fossils foreign to the earth. These are bodies belonging both to the mineral and vegetable kingdoms: such as shells, the bones of fishes and quadrupeds, woods, plants,* etc., *which are found buried in the bowels of the earth, where they have been brought accidentally.*

To define the practice of the study of fossils, we continued to use, derived from the Greek, the word *Orittologia* or *Orittografia*, in use since the end of the seventeenth century. In the definition given by Diderot and D'Alembert's (*Encyclopedie*, 1765, 11: p. 677), we read that:

> *To define the practice of the study of fossils we continued to use the word Oryctology derived from the Greek orycta; under this generic name is included the doctrine of salts, sulphurs, marbles, common stones, precious stones, and metals.*

In this sense oryctology still differs from the modern meaning; the concept of fossil continues to be more extensive as it still includes all mineralized bodies that are extracted from the earth. From the eighteenth century, this term and its broad concept were used at least until the first half of the nineteenth century, as shown, for example, by James Parkinson (1822) *Elements of Oryctology: Introduction to the Study of Fossil Remains*; Stefano Borson (1821–23) *Saggio di Orittografia Piemontese*; Nicola Da Rio (1836) *Orittologia Euganea….* From the thirties of the same century, oryctology was finally replaced by paleontology, which more strictly means the study of ancient beings, and by the new ideas that represent it.

Palaeontology: the term that gave rise to the homonymous science was coined in the second decade of the nineteenth century: this definition is used for the first time by Henri-Marie Ducrotay de Blainville (1777–1850) holder of the chair of natural history at the *Muséum National d'Histoire Naturelle* in Paris, after Jean-Baptiste Lamarck and then of that of comparative anatomy after Georges Cuvier, simply as the title of a paragraph intended to summarize some scientific works that had as their exclusive object the description of fossilized organized bodies for geological use (*Journal de Physique*, 1822: liv) (Fig. 2.5).

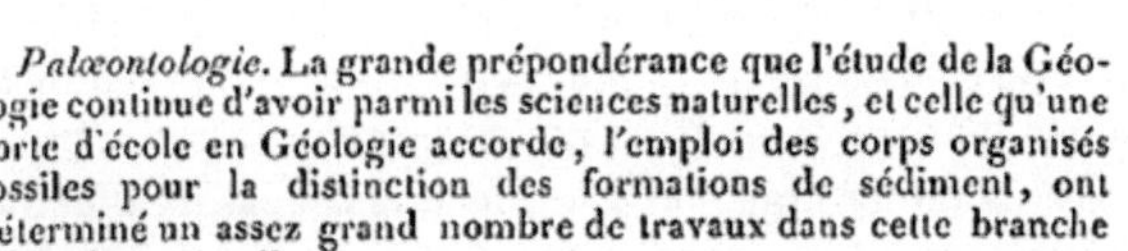
Palæontologie. La grande prépondérance que l'étude de la Géologie continue d'avoir parmi les sciences naturelles, et celle qu'une sorte d'école en Géologie accorde, l'emploi des corps organisés fossiles pour la distinction des formations de sédiment, ont déterminé un assez grand nombre de travaux dans cette branche d'histoire naturelle.

Fig. 2.5 Blainville H. -M. D. de, *Journal de Physique*. 1822: liv. Detail of the paragraph in which the French scholar uses the word paleontology. [https://www.biodiversitylibrary.org/bibliography/6637 (Public Domain Mark)]

In reality, the Renaissance and later attempts, up to the 1700s and even beyond for some conservative scientists, to recognize these natural objects as deriving from the remains of former living beings, only came to fruition in the course of the nineteenth century, finally finding a way to express what is currently the concept of fossil, with an unambiguous meaning, shared and accepted by science, albeit one that is constantly being updated. Even if during all the 1700 the petrified remains had been almost all recognized as what remained of ancient organisms, attributing their death and transformation mostly to the universal flood or to other similar catastrophes variously repeated, admitting in this also the process of extinction (Georges

Cuvier, 1769–1832), it was only toward the end of the 1800, with the affirmation of the evolutionistic ideas and the awareness of the dilatation of the past, that the fossils were definitively considered and studied as remains of ancient populations subject to the same laws of the biology that regulate the actual living beings, becoming in this way, the testimony of their continuous diversification and transformation in the course of the time.

And now? In general, fossils, in today's meaning, are the remains of organisms that lived during geological eras, variously preserved and transformed, mostly petrified, lithified, in sedimentary rocks. Commonly they appear in the form of casts, reproducing totally or not, the internal surface and volume of the organism, their external—evidence of biological activity, such as footprints or traces related to movement, trails, footprints, eggs, residues produced by the digestive system, or various types of intestinal contents such as gastrolites, or coprolites, and fecal pellets, excrement of vertebrates and invertebrates.

The term fossil is also used in a wider sense to indicate fossilized vegetable substances containing or not organisms, for instance tree resin transformed into amber, and traces of various phenomena which have left evidence of their action. It is in this way that the expressions fossil wave or fossil rain are used, to indicate, respectively, the traces of the wave motion and the imprints of the raindrops impressed on a surface, which later became rocky.

After death, starting from a hypothetical condition of integrity of the organism, its preservation and transformation in fossil will be more or less probable according to various factors. The most important of these, but not the only ones, are the biological, chemical, physical, and sedimentological ones. The chain of events culminating in fossilization is divided into four phases that follow one another in time: the death of the organisms, the activities that affect the remains until their natural burial, the final burial, and the fossilization itself. The processes of fossilization, involving both organic and mineralized matters, are carried out with chemical reactions and physical phenomena that are very different and more or less complex. Generally the formation of a fossil is rather rare, and to occur it is necessary that the remains of an organism are removed as quickly as possible to the set of those biological, mechanical, and chemical processes that tend to destroy them. With the exception of exceptional cases such as wind deposits, volcanic layers, landslides, cave deposits, encapsulation in bitumen, or tree resins, the burial of remains and their fossilization takes place in a watery environment where the sedimentation processes frequently prevail over the destructive ones.

The chemical-mineralogical composition and body structure of the organism are also important. The presence in animals or plants of more resistant and hard parts, both mineralized and not, increases their possibility of transformation into fossils. A skeleton, consisting of parts joined by organic tissue, only in case of rapid burial and special chemical and physical conditions can be preserved completely, in connection, while generally are found in sediments the individual parts as in the case, among invertebrates, of echinoderms (platelets, ossicles) and bivalve mollusks (separate shells). Also in the constitution of the same organism, the degree of mineralization of the various parts is discriminating; just think of the otoliths, small bones of the auditory apparatus, and the teeth, ichthyodontoliths, which are frequently found in the sediment as the only remains of the fish that possessed them.

The conditions affecting fossilization come into play as soon as the organism ceases to live. The conservative action in an asphyctic underwater environment and with sedimentation of fine materials is carried out above all by subtracting the body from putrefaction, from the action of bacteria and oxygen, guaranteeing at the same time a quiet and regular covering by the sedimentary material. It is easy to understand how a body not transported by currents, deposited in a quiet place and with little or no oxygen, with regular sedimentation, can fossilize integrally and better than another one that instead finds itself in the opposite conditions. This explains, among other things, why are more abundant the fossil remains of marine organisms or however of aquatic environment, than those of terrestrial or flying animals.

A good sedimentary covering is the main cause of the conservative action, even if it is not enough to bury an organism to reach a fossil. Other factors, especially chemical ones, which put it in balance with and in the sediment in which it is immersed, have a significant part in the transformation process. The more the encompassing sediment is formed by fine particles, the more the fossil will be preserved in detail.

Synthesizing, fossilization is a long process that through numerous modifications leads to the conservation, usually in rocks or in more or less incoherent sediments, of an organism or of parts of it, as a consequence of determined circumstances and conditions that take place in the environment in which the organism itself finds itself when it is still alive, or when it has just died. At this point it is well to clarify that the fossils found in rocks are not formed in them, but it is the rocks that become such by undergoing transformations together with the remains they contain.

But what are the fossils useful nowadays? Paleontology, which studies them, makes use of them to know, first of all, the existence of the forms of life of the past, from the most remote times, going back to some billion years ago, up to the most recent ones. In short, without them we would not know anything about the vital manifestations that have preceded us on our planet; associated with sediments they are used as indicators of time and space, the so-called guide fossils; their associations serve for the reconstruction of ancient environments; they have practical applications in industry, both direct and indirect, such as coal or microfossils used for oil research. Furthermore, the variations of ancient organisms, confirmed by their fossilized remains, have contributed and continue to contribute substantially to the elaboration of theories on the evolution of life. The most important finds are on display in the numerous museums all over the world dedicated to them, and the cultural and commercial activities related to them are an important part of the economy and of human knowledge..., and this is only a part of what concerns them. Some fossils are still used in traditional medicine in some regions of the earth, but they have lost their importance as therapeutic remedies and especially as objects endowed with magical powers. Reminiscences more or less conscious remain in their use, still today very diffused, under form of pendants or parts of jewels of various shapes, worn like lucky charms or more often and simply, for their aesthetic beauty (Figs. 2.6, 2.7, 2.8, and 2.9).

Fig. 2.6 Portrait of the Museum by Ferrante Imperato. Antiporta of the *Historia naturale* in the version of 1672. [https://doi.org/10.3931/e-rara-11175/ **(Public Domain Mark)]**

Fig. 2.7 View of the museum by the Veronese Francesco Calzolari. From the relative catalogue compiled by Cerutus and Chioccus in 1622 (modified). [https://doi.org/10.3931/e-rara-50568 (Public Domain Mark)]

Fig. 2.8 ***Museum Wormianum seu Historia Rerum Rariorum*** (1655). **[Courtesy Biblioteca Digital del Real Jardín Botánico, Madrid (Spain)** https://bibdigital.rjb.csic.es/idurl/1/14142)]

Fig. 2.9 **The Cospiano Museum. The remaining finds are currently on display, together with those of the Aldrovandi collection, in Palazzo Poggi, in Bologna (Legati** 1677) [https://babel.hathitrust.org/cgi/pt?id=gri.ark:/13960/t7dr4608w&view=1up&seq=30&skin=2021 **(Public Domain Mark)]**

References

Boas M (1973) The scientific renaissance 1450–1630. Feltrinelli, Milan
Borson S (1821–23) Saggio di Orittologia Piemontese. Mem R Acad Sci 25:180–229; 26:298–364
Cerutus B, Chioccus A (1622) Musaeum Franc. Calceolari iun Veronensis. Apud Angelum Tamum, Veronae. https://doi.org/10.3931/e-rara-50568/. Public Domain Mark
Da Rio N (1836) Orittologia Euganea. Tipografia Cartallier, Padova
de Blainville HMD (1822) Journal de Physique, de Chimie, d'Histoire Naturelle et des Arts, avec des planches en taille-douce. Tome XCIV. Bachelier, Paris
Diderot D, D'Alembert JLR (1765) Encyclopédie ou Dictionnaire raisonné des sciences, des arts et des métiers... Tome Huitieme. Samuel Faulche & Compagnie, Libraires & Imprimeurs, Neufchastel
Gessner C (1565) De omni rerum fossilium genere, gemmis, lapidibus, metallis, et huiusmodi, libri aliquot, plerique nunc primum editi. Tiguri. https://doi.org/10.3931/e-rara-4176/. Public Domain Mark
Giallombardo F (2016) Agostino Scilla (1629–1700) and the visual culture of historia, between antiquaria and natural history. PhD thesis, University of Palermo
Imperato F (1672) Historia naturale di Ferrante Imperato napolitano. Nella quale ordinatamente si tratta della diversa condition di minere, pietre pretiose, & altre curiosità [...]. Presso Combi & La Noù, Venetia. ETH-Bibliothek Zürich, Rar 607. https://doi.org/10.3931/e-rara-11175/. Public Domain Mark
Legati L (1677) Museo Cospiano annesso a quello del famoso Ulisse Aldrovandi e donato alla sua Patria dall'Illustrissimo Signor Ferdinando Cospi. Giacomo Monti, Bologna.
Mattioli MP (1548) Il Dioscoride dell'eccellente dottor medico m. P. Andrea Matthioli da Siena; co i suoi discorsi, da esso la seconda volta illustrati, et diligentemente ampliati: con l'aggiunta del sesto libro de i rimedi di tutti i veleni da lui nuovamente tradotto, & con dottissimi discorsi per tutto commentato. Vincenzo Valgrisi, at the workshop of Erasmus, Vinegia
Mottana A, Napolitano M (1997) Il libro "Sulle pietre" di Teofrasto. Prima traduzione italiana con un vocabolario dei termini mineralogici. Rend Lincei Sci Fis Nat s. 9 8:151–234
Narducci E (1859) La composizione del mondo di Ristoro d'Arezzo, testo italiano del 1282, pubblicato da E. Narducci. Tipografia delle Scienze Matematiche e Fisiche, Roma
Ottaviani A (2000) Il fascino indiscreto delle nature ancipiti: un saggio della "istoria naturale" nei secoli XVII e XVIII. Giornale Critico della Filosofia Italiana, Anno LXXIX (LXXXI) (2–3):316–369
Parkinson J (1822) Outlines of oryctology. An introduction to the study of fossil organic remains; especially those found in the British strata: intended to aid the student in his inquiries respecting the nature of fossils, and their connection with the formation of the earth. Printed for the Author, London
Raffi S, Serpagli E (1993) Introduzione alla paleontologia. UTET, Torino
Sambucus J (1574) Icones veterum aliquot, ac recentium medicorum, philosophorumque, elogiolis suis editae, opera I. Ex Officina Christophori Plantini, Antverpiae. https://babel.hathitrust.org/cgi/pt?id=gri.ark:/13960/t0dv92j6f&view=1up&seq=8 (Public Domain)
Squarzafico G (1497) Giovanni Boccaccio. The Philocolus. Incomincia il libro primo di Florio & di Bianzafiore chiamato Philocolo... Impresso in Venetia: per maestro Antonio da Gusago Bresano, nel MCCCCLXXXXVII ad XXII novembrio
Stenone N (1669) De Solido Intra Solidum Naturaliter Contento Dissertationis Prodromus. Typographia sub signo Stellae, Florentia
Tramezzino M (1550) Di Giorgio Agricola De la generatione de le cose che sotto la terra sono e de le cause de' loro effetti e nature, lib. V; De la natura di quelle cose che dalla terra scorrono, lib. IIII; De la natura de le cose fossili e che sotto la terra si cauano, lib. X; De le minere antiche e moderne, lib. II; Il Bermanno, o De le cose metallice, dialogo/recato tutto hora dal latino in buona lingua volgare. In Vinegia: per Michele Tramezzino. https://archive.org/details/hin-wel-all-00002012-001 (not copyright found)

Vai GB, Caldwell GWE (eds) (2006) The origins of geology in Italy. Geological Society of America, Special Paper 411

Walch JEI (1768–1773) Walch Johann Ernst Immanuel: Die Naturgeschichte der Versteinerungen: zur Erläuterung der Knorrischen Sammlung von Merkwürdigkeiten der Natur. gedruckt mit Felsseckerischen Schrifften, Nürnberg. ETH-Bibliothek Zürich, Rar 399. https://doi.org/10.3931/e-rara-10246/. Public Domain Mark

Worm O (1655) Museum Wormianum, seu Historia rerum rariorum, tum Naturalium, quam Artificialium, tam Domesticarum, quam Exoticarum; quae Hasniae, Danorum in aedibus Authoris Servantur. Iohannem Elsevirium, Lugduni Batavorum. https://doi.org/10.5962/bhl.title.63844 (not in copyright)

Chapter 3
Fossils, Stones, and Magic (Mario Santini)

Abstract Reflections on the meaning and use of "critical thinking" and the intuitive-magic one, the "thinking of origins," referring to fossils, as symbols of the origins of life.

We can realistically think that magical thought represents the thought "of the origins," the direct and simplest answer to the needs and fears of man in the "state of nature," and that it is partially replaced by a different type of thought, more adherent to empirical reality with the evolution of knowledge, but also of existential security that is not always linked to "certainty." Certainty, in fact, is a particularity of thought that is not of science or philosophy but of religion through faith, or of magical thought that precisely has the effect of recovering subjective certainties, where science seeks verifiable or falsifiable "hypotheses."

Keywords Considerations · Magical thinking · Critical thinking · Fossils

We can probably think that magical thought represents the thought "of the origins," the direct and simplest answer to the needs and fears of man in the "state of nature," and that it is partially replaced by a different type of thought, more adherent to empirical reality with the evolution, not only of knowledge but also of existential security that is not always linked to "certainty." Certainty, in fact, is a coloring of thought that is not of science or philosophy but of religion through faith, or of magical thought that precisely has the effect of recovering subjective certainties, where science seeks verifiable or falsifiable "hypotheses."

The mental structure of *Homo sapiens* is essentially based on basic functions:

- The emotional activity
- Cognitive activity

Functions that are always and constantly integrated in the reality of the mind.

Cognitive activity finds its basis in the high tension of exploratory "motivation" that is expressed through an innate curiosity, which naturally demands to be satisfied and is supported by the fantastic and symbolizing capacity to mentally and momentarily exit from the "empirical" present. The process is supported by

© The Author(s), under exclusive license to Springer Nature Switzerland AG 2022
M. Forli, A. Guerrini, *The History of Fossils Over Centuries*,
https://doi.org/10.1007/978-3-031-04687-2_3

emotional activation that gives value to what is sought and found. All this is common and often sufficient to ensure gratification and satisfaction of the exploratory need at a first level of integration between knowledge and explanation of events.

But the world of nature is made up of objects and evident facts immersed in and emerging from an ocean of mystery and often of dissatisfaction, which feed a continuous exploratory tension, because our mind bears mystery badly and only if it is possible to dissolve it, using two traditional and usual ways:

- A first process harnesses intuitive, fantastic abilities, anticipation and symbolization. If this is not enough, the explanations and imaginative, metaphysical, and esoteric solutions, identified as real, are increased. This is the "magic solution".
- A second process uses the analytical and critical capacity that is satisfied with a reactivation of exploration, verification, and search for answers. This is the "scientific" solution.

Thus, two forms of thinking are activated that maintain a complementary relationship:

- The "slow thinking" because it is reflective, analytical, theorizing, "critical" that is expressed in science
- The "fast" thought, more archaic, intuitive, fantastic, which can be expressed in creativity

Both, however, are not autonomous; they maintain a bond of complementarity that is broken when one prevails over the other in an absolute way.

The innate need to explain and interpret is applied and exercised on all things and events. On the spontaneous level, it is activated in particular by facts and objects which by their form, exceptionality, place themselves outside common and everyday use, into which use brings the explanation.

The search for a satisfactory explanation has two origins; the first is the innate exploratory curiosity, and the other is the perception of an existential insecurity that always requires to be organized. The organization of answers to insecurity can follow either the rational or the fantastic path, as long as a satisfactory answer is identified on a personal level.

Certain objects or facts of nature, in addition to being mysterious in themselves, take on value and attract attention as much as they do because of their appearance, position, or other factors, influencing a normal perception, as is also the case with "fossils" or "crystals." It is in this way that they take on a "metaphysical," "symbolic" value, a thought meaning that emotionally can be, assuming an auspicious or inauspicious meaning, hardly indifferent, which goes beyond their real meaning. Even the scholar, the collector, for example, lives, in a healthy and virtuous way, his activity as a gratifying and therefore reassuring confirmation of his personal identity. This is also the process of building talismans, beliefs, and the effect of magical thinking that creates objects, facts, and theories that we are programmed to believe, building "beliefs" that tend to have a viral spread. This process then, once constructed, becomes a competitor to critical thinking.

Magical thought actually enters as a component of everyday reality, and the explanations that support it resemble oneiric thought, partly released from the linearity of time and space. After having given these brief definitions, we can extrapolate the characteristics of magical thought, which in the end is a syncretic way of thinking, not linear, but recursive and parallel, which exploits symbolic relationships of similarity rather than differences.

Magic belongs to man, perhaps since the beginning. It has always belonged to man's culture, even in its most concrete aspects: the most rationalistic thinking has always been imbued with magical thinking. As critical thinking always uses intuition, it is therefore opportune to ask ourselves, as far as possible, what significance this has on an anthropological level, and how magical thought functions and how it is integrated both in everyday life and in the culture of man, of each individual man and of the community, particularly in our hyper-technological era. Magic is in the history of man the demonstration that man, however he organizes himself, structurally, mentally, and in behavior, is always projected out of a linear logic, out of the total comprehensiveness of the world and of himself, and therefore always having to deal with the dark side of the world, which is represented both by what happens in the outside world, among things, and by the unpredictable and uncontrollable tangle of his own emotions.

Man finds himself immersed in a universe that is only partly reachable by objective knowledge, but which is often representable only through "symbols." The symbol belongs to a form of thought that is not strictly linear; it represents a kind of "short circuit" in the rational linearity. From a psychological point of view, magic has its origin in the feeling of the precariousness of existence. In the absence of means of control of external reality, and of the internal one, brought by the emotions, the first form of control of reality is represented by "magic" that opens up a mythical horizon, in that dimension that De Martino defines as "metahistorical" in which everything is already foreseen, described, and anticipated in its good end. This means that when real life subjects us to negative, painful, threatening, and uncontrollable situations, magic makes us rest in a thought for which, with rituals and formulas, which can also be made only of thought, negativity can be reabsorbed and controlled, through an access to a dimension where all good, as well as all evil, is possible. The magic acts in a dimension of thought that is linear and fantastic; the linearity is in thinking that if you do "so" will happen "that," the fantasy is to enter fully into a dimension that is that of "as if." Through the magic act, and therefore in the magic thought, there are two effects, which make us overcome the "real world": the first effect is that of "destoricization"; the second effect is that of placing existence in a horizon of representations where everything is stable and controllable.

Jean Piaget has normally highlighted a form of "magical thinking" in children, characterized by the fact that, like primitive man, the child is led to consider all things in the world, even inanimate things, endowed with life and intentionality. For its part, psychoanalysis, starting with Freud, identifies the origin of magical thinking in infantile omnipotence,.characterized by the fact that the child, like the adult, tends to explain and manage his relationship with the world by projecting both fears and containment and control identified in omnipotent forces and figures. This control, however, is possible through an "operative" type of thinking that gives concreteness to unconscious fantasies, through complexes of thought in which the mystery is united to the explanation through links that do not take into account the categories of space and time, which bind us to reality, and that find synthesis in a "symbol" or in a "ritual," which serve to represent and manipulate "the dark side of the world."

For Jung, magic is born from the identification with unconscious forces that belong to the archetypal unconscious, which are organized in complexes that are autonomous from the ego complex, which, if it is not overwhelmed, tries to control even by identifying. For metaphysics, as in magic, for example, the hypotheses awaiting verification are part of that thought that suspends a "critical" judgment of reality, which is assumed in its manifestations. In both these examples, there is a leap, which eludes and interrupts the linearity of an empirical and critical logic. It is significant to think that if magical thought is identified with the "philosophy" of the populations most adherent to the state of nature, as in the shamanic culture, it is reduced almost exclusively, in Western culture, to a reductive rituality and consumption. It seems almost not to belong, or contained in the mythical and religious sphere, to the Greek culture, so much so that it had to import its name, the Greek word *magheia* refers to the science of the *magoi*, the magicians who were of the culture of the Medes and Persians. The irrational for the Greeks and then for the Romans is all organized in myth and religion. Magical thought, however, seems to reappear, to take shape and consistency, in moments of crisis, or of rapid and substantial cultural transformation, just think, in this regard, of the Hellenistic culture and the late Roman Empire on the one hand, and the Renaissance on the other.

Let's try now to outline what are the fundamental characteristics of this way of thinking from the psychoanthropological point of view, in order to see how much it is represented today in our culture, in our vision of the world, and in the management of everyday life. A first characteristic is represented by the fact that magical thinking is constantly linked to needs, fears, and uncertainties, and this is true precisely because it is often accompanied by an experience of security and certainty, and this is why it represents an original mental form.

A second aspect is that the categories of space and time are not used in a linear and progressive way. The dark side of the world is always located outside the time we live in and outside the space in which we move. For example, a fossil or a natural crystal, a stone of unusual color and shape, belongs to another world.

The third aspect corresponds to the fall of the limits of identity, of distinction, of duality. Magical events and magical actions always take place in an existential field in which the I and the Thou, between the I and an object, are indistinct, or even united. The shaman in his rites identifies himself and becomes that certain spirit, that certain animal, that certain power. The mediumistic trance is an obvious manifestation of this process. Magical thought, so similar to dream thought and delusional productions, is possible, and maintains its meaning if it is inserted in a cultural and collective context that gives it meaning and contains it, keeping it in a field of recognition, and it works precisely because it is expressed within "containers," such as ritual, individual, and collective.

A fourth essential element is represented by the contextualization, if the context is lacking, or is different, magical thinking can become an expression of pathology, to give an extreme example, a shaman in our culture risks delirium or at least a personality disorder. It is no coincidence that in the "New Age movement," which in my opinion well embodies the crisis of certainties of our historical period, in the climate of the philosophy of globalization, everything "magical" that man has produced is recovered, rationalized, intellectualized, consumed, and marketed.

Today in our "culture," everyone is able to recognize and distinguish a fossil from another stone and no longer falls into archaic and naive explanations; it is however possible that he does not escape the magic of the "talisman" for which a small trilobite or a crystal kept in his pocket can be a source of reassurance. This process is not to be considered regressive and immature, because "critical thinking" is nourished by the intuitive-magic thinking that is the "thinking of the origins" as the fossils are the symbol of the origins. The important thing is that their "magical" meaning be filtered by a mature "critical thought" that can make them, with legitimate dignity, both an object of science and an expression of art (Figs. 3.1 and 3.2).

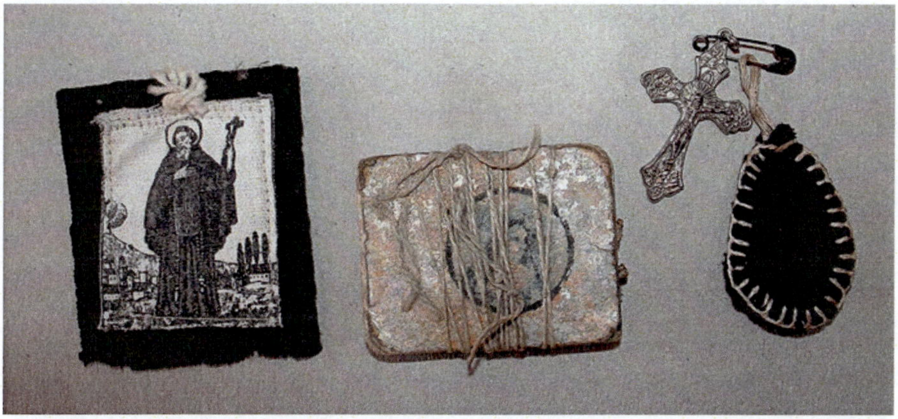

Fig. 3.1 Between magic and Christian devotion: two scapulars and a brief, small bags worn or hung on clothes with pins, containing sacred images and various materials considered protective against adversity and the evil eye in general (from Procida, Napoli, 1930s and 1950s). (Photo Maurizio Forli, courtesy private collection)

Fig. 3.2 A compartmentalized tray with modern lucky charms, a more or less unconscious atavistic heritage, on sale during a current mineral, fossil, and shell exhibition. Mother-of-pearl crosses and horns, *Turbo* and *Bolma rugosa* opercula (Santa Lucia eyes), pieces of amber, shark teeth, and arrowheads (*Glossopetrae* and *Ceraunie*). (Photo Maurizio Forli)

Chapter 4
The Classification of Living Beings: Carl Nilsson Linnæus and Systematics

Abstract Simple lines of explanation on that division of science that classifies living beings or fossils, called systematics. In particular on the method of classification introduced since 1735 by Carl von Linné based on some guiding characters with which organisms can be grouped by similarity, creating hierarchies of gradually larger sets: from the single species to the phylum.

Keywords Linnæus · Taxonomy · Binomial nomenclature

As in the course of the text we shall find ourselves quite often speaking of fossils, using their scientific name, we think it necessary to give these brief lines of explanation to those who may be uninformed about that division of science which classifies the organisms, called systematics (Fig. 4.1).

© The Author(s), under exclusive license to Springer Nature Switzerland AG 2022

M. Forli, A. Guerrini, *The History of Fossils Over Centuries*,
https://doi.org/10.1007/978-3-031-04687-2_4

Fig. 4.1 Carl von Linné in a painting by Alexander Roslin from 1775. Nationalmuseum Stockholm. [https://it.wikipedia.org/wiki/File:Carl_von_Linn%C3%A9.jpg (Public Domain Mark)]

It is up to it to study and define their hierarchies and to make them explicit through taxonomy, defining the criteria and the most suitable methodology to reflect as much as possible their natural order, through univocal classification, carried out according to the rules of nomenclature.

By his mental conformation, man is inclined to arrange the objects that surround him with a certain criterion and order. For example, if he possesses or must manage a discreet collection of books, he will tend to place them on the shelves of his library, grouping them by subject, placing historical books on one level, literary books on another, scientific books on another, and so on. In this way he will be able, if necessary, to quickly and safely find what he needs to read, consult, or study. Attempts at classification have existed since ancient times, and the lists of stone materials, gems, rocks, or fossils that were used and reported, for example, in some classical Greek texts, follow an order dictated by the similarities or differences of the various elements.

In the view of nature, the same criterion has been adopted, namely, the "degree of similarity." The concretization of a practical method applicable to the various categories of organisms and natural objects, such as fossils derived from them, is due chiefly to *Carl Nilsson Linnæus* (1707–1778), known to most simply as *Linnaeus*, from the Latinized form of the name *Carolus Linnæus*, a Swedish botanist, who wished to bring order to the great confusion which still existed in his time in regard to the classification and nomenclature of living things.

Before Linnaeus, a species was identified by a series of words, made up of a name to which adjectives and attributes were added until it seemed to the author to be exhaustively and effectively characterized; since there was no established and recognized criterion, no name was fixed and universally accepted.

It is with the tenth edition of his *Systema Naturae*, later taken as a reference for the beginning of zoological nomenclature, published in 1758, that systematics acquired the scientific form still used today. In this text Linnaeus described and listed the organisms then known, assigning to each of them a double name. This method, called binomial nomenclature, giving the name from the combination of two terms, the *Genus* and the *species*, is still the standard convention used in systematics for the naming and classification of organisms, both actual and fossil. For systematics, the species is the hierarchically lowest obligatory taxonomic level. But what is the concept of species?

Today there are many definitions of species and different methods of classification, some of them very complex, in continuous change as the organisms are studied not only on a morphological basis. For example, the paleontological species may not coincide, in whole or in part, with the corresponding actual species, and then it becomes difficult to define it with a univocal criterion. The classical definition of biological species is that applicable to a group of similar individuals, male and female populations, which can reproduce among themselves and whose offspring are equally fertile.

For the Kingdom Animalia, with the method of binomial nomenclature, in the Genus are included the similar species, more genera can be grouped in *Family*, more families in *Order*, *more* orders in *Class*, and more classes in *Type* or *Phylum*. If, therefore, we wanted to classify and describe the animal commonly called mussel or mussels, to which, as we shall see later, speaking of bivalves, some curious interpretations are connected, we should write that, scientifically, today this mollusk is classified as follows:

Phylum Mollusca
Class Bivalvia
Order Mytilida
Family Mytilidae
Genus *Mytilus*
Species *Mytilus edulis*

More correctly, it is written in italics *Mytilus edulis* L., 1758, followed by the name of the person who has established such species, in this case L. stands for Linnaeus; 1758 is, on the contrary, the year of publication of the study where the corresponding species has been determined and described for the first time, in a correct way, following certain rules accepted by the scientific community. The name and the date are at times enclosed between brackets, just for the application of one of these criteria, which foresees their use, if the genus to which the species belongs is changed, by a scholar subsequent to the first one, keeping this one scientifically valid. In case the new species is to be attributed to two authors, the citation will be double, e.g., *Murexul zezae* Brunetti and Forli (2010); in case of more authors, as it will happen sometimes to find in the following text, the Latin expression *et alii* (and others) will be used, abbreviated with et al., e.g., *Haliotis bertinii* Forli et al. (2003).

In his work, the great Swedish botanist has brought back the expressions and the references to the naturalists who have preceded him, and who have so much contributed to the formation of the knowledge of the time, constituting, in spite of the appearance, a *continuum* with the past even if his *Systema Naturae* can be considered the starting point of a new way of understanding and interpreting the zoological and botanical world, applicable also to the world of the fossils as ex-living beings.

Among the many books written by Linnaeus, there are some dealing with the vegetables used as ingredients in medicinal compounds, in particular in the volume published the first time in 1749 with the title *Materia medica, liber 1. de plantis*. In a later reprint, published after the author's death (fourth edition in 1782), including also minerals, soils, and fossils, some of these are still reported as effective medicinal components (Fig. 4.2).

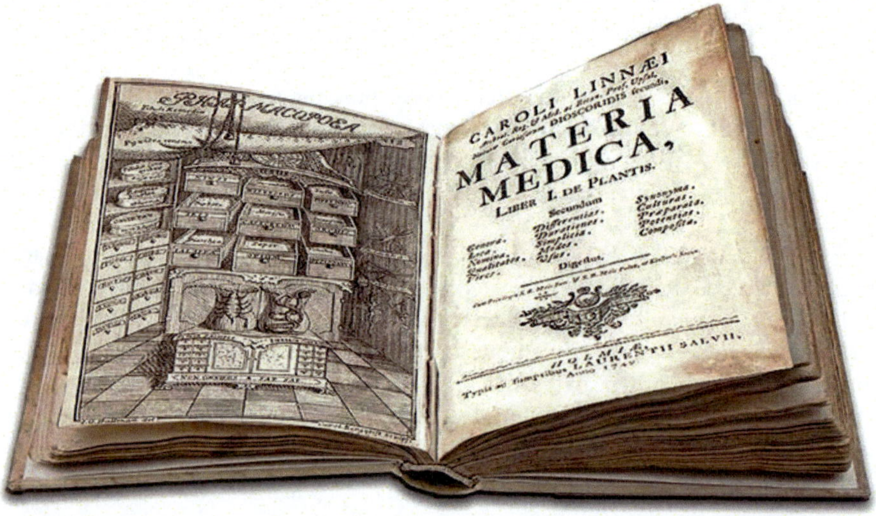

Fig. 4.2 Linnaeus C., 1749. ***Materia Medica. Liber I. De Plantis***. [Graphic compilation Maurizio Forli. Modified after https://commons.wikimedia.org/wiki/File:Titlepage_to_%27Materia_Medica%27_Wellcome_L0051527.jpg (licensed under the Creative Commons Attribution 4.0 International)]

References

Brunetti MM, Forli M (2010) Murexul zezae n. sp. from the Italian Pliocene (Gastropoda: Muricidae). Boll Malacol 46:4–8

Forli M, Dell'Angelo B, Ciappelli F, Taviani M (2003) A new species of Haliotidae (Mollusca, Vetigastropoda) in the Italian Pliocene. Boll Malacol 38(9–12):149–154

Linnæus C (1758) Systema Naturae per Regna Tria Naturae, secumdum Classes, Ordines, Genera, species, cum characteribus differentiis, synonimis, locis. Tomus I. Editio decima, Reformata. Laurentii Salvii, Holmiae. https://www.biodiversitylibrary.org/bibliography/542. Public domain

Linnæus C (1782) Materia Medica. Editio Quarta Auctior, curante Io. Christ. Dan. Schrebero. Wolfgangum Waltherum, Leipzig and Erlangae

Chapter 5
Giuseppe Bellucci and his Collection of Amulets, in Perugia (Maurizio Forli)

Abstract Considerations on the collection of amulets Giuseppe Bellucci preserved in the National Archaeological Museum of Umbria, in Perugia. This collection of great interest is constituted from hundreds and hundreds of objects endowed in the past of "magic" value; value becomes them inherent through the possessor, for which such object assumes such properties to make it become worthy of cult and respect. In short, for an amulet to be such it, it must have been believed and used as a "magical object," thaumaturgic; it is not enough that it has the appearance.

Keywords Bellucci Giuseppe · Collection · Amulets · Museum Perugia

Over 20 years ago, Giuseppe Bellucci's collection of amulets was permanently exhibited in the rooms specially set up in the National Archaeological Museum of Umbria, in Perugia…, and we knew absolutely nothing about it! We owe our discovery of the existence of the most important national collection of amulets, assembled by its former owner, from the second half of the nineteenth century until the second 20 years of the following century (1870–1920) to the reading of an essay by Prof. Giancarlo Baronti (*Tra bambini e acque sporche* … 2008). Professor Giuseppe Bellucci (Perugia, 1844–1921), according to what is reported in the texts that deal with his biography, was a professor and rector of the University of Perugia, devoting himself to studies of chemistry, palethnology, ethnography, and folklore. Scholar and collector of prehistoric antiquities, according to what he himself said, he discovered his interest in amulets trying to acquire stone artifacts, especially flint tips and polished axes, for him nothing more than ancient tools, archaeological remains, considered instead by their discoverers and owners, remains of lightning and, as such, protective objects against the same, hence the idea of forming a collection that would document the comparison between the habits and traditions of primitive peoples and those of present-day populations. As a collector, at the beginning he probably thought more of accumulating material than of giving it a "profound" meaning, an importance that he fully understood later. Being economically well off, compared to the then subordinate classes, he didn't have to spend a lot to buy what

© The Author(s), under exclusive license to Springer Nature Switzerland AG 2022
M. Forli, A. Guerrini, *The History of Fossils Over Centuries*,
https://doi.org/10.1007/978-3-031-04687-2_5

interested him, even if, as reported by Baronti, and testified in his own writings, sometimes he complains about having "spent too much" and made big sacrifices to get the things he wanted. This is a way of emphasizing and giving more importance to one's own collection, typical of a collector who is proud of what he owns: it happens even nowadays, in the common talk among collectors of any kind of objects.

But what interested us so much among the amulets in the Bellucci collection and why? Naturally, in the context of this book, it is the fossils and their apotropaic use, a use in which they maintain their identity as natural objects and at the same time acquire a supernatural dimension, reflecting their past intentions, borrowed in the form of superstitions and popular beliefs, partially preserved up to our days, in which today, without knowing their use and meaning, they are reduced to more or less insignificant pendants and jewels depending on their shape. As excellently expressed by Prof. Baronti, a concept valid for all amulets in general, they are instead *cultural "fossils" coming from a distant past* (op.cit. p. 17), memories now almost vanished of ancient magical and religious manifestations, mute testimonies, but that know and can "speak" to those who approach them aware of their essence.

It is not the objects themselves that are important, but the value we attribute to them. In the case of amulets, it is the "magical" value that becomes inherent in them through the possessor, for whom such an object takes on properties that make it worthy of worship and respect. In a few words, for an amulet to be such, it must have been so, it must have been believed and used as a magical object," thaumaturgic; it is not enough that it has the appearance.

Let's make an explicative example: in the figure we propose below, some shark teeth are represented, mounted in various ways, obviously belonging to the collection; all amulets are authentic because they are certified by Bellucci himself, who had finally well understood the difference between an ornament and a corresponding amulet. Among these, in the center, there is a simple pendant, a jewel that is aesthetically the same but conceptually different, in order to explain the difference between ornaments and amulets. Your mind will easily understand the substantial difference: as a simple pendant, it will be out of place compared to the rest of the context and therefore simply considered "different," not relevant. Our artifact has never had the substance of an amulet and therefore cannot be considered as such. This is, in our opinion, the fundamental point to understand for those who will visit the Bellucci collection. You will not find yourself in front of an exhibition of more or less whimsical and extravagant bijoux, but you will have the opportunity to admire hundreds of unique pieces cloaked in a magical and religious symbolism, now almost completely lost and, for this reason, unrepeatable and unrecognizable without the necessary awareness.

The use of fossils from this point of view therefore reflects the ancient beliefs and properties attributed to them when their true nature was unknown and they were cloaked by a certain something indefinite. These particular amulets therefore become indispensable tools of knowledge in the intellectual journey we have undertaken in the extended understanding of fossils and for this reason, when present, will be treated in the respective chapters (Fig. 5.1).

Fig. 5.1 Top from left, fossil shark tooth (*Carcharodon carcharias*), labial (anterior) and lingual (posterior) views, blunt and with root system shaped to fit a metal ligature. From Castelleone, Deruta, province of Perugia. Late nineteenth century (coll. G. Bellucci, M.A.N.U. (PG). Inv. 728 B; 22 × 25 × 7 mm). Below, lingual and labial view of a living shark's tooth, ivory, bound in silver. Chieti. Late nineteenth and early twentieth century (coll. G. Bellucci, M.A.N.U. (PG). Inv. 5022 B). These amulets were called "tongues of St. Paul" because their power was assimilated to the thaumaturgical power of the Saint who had freed the island of Malta from snakes, after being bitten by a viper without being harmed, and for this reason they were considered powerful antivenoms. (Reproduced by kind permission of the Ministry of Cultural Heritage and Activities and Tourism, Regional Directorate Museums of Umbria – National Archeological Museum of Umbria). In the center, a contemporary pendant of an Oligocene shark tooth, fossil from Morocco, shaped and bound in silver (mm 30 × 20 × 8) (central image photo and graphic composition Maurizio Forli)

Reference

Baronti G (2008) Tra bambini e acque sporche. Immersione nella collezione di amuleti Giuseppe bellucci. Morlacchi editore

Chapter 6
Quaestio de Fossilibus: Glossopetres, Snake Tongues and Ceraunids

Abstract The beginning of the chapter is dedicated to the basics of the classification of fishes with cartilaginous skeleton (Chondrichthyes), with particular reference to the fossil species and their most abundant remains: the teeth. Especially the largest ones, belonging to the genera *Carcharocles* and *Carcharodon*, have been believed since ancient times to be petrified human tongues and for this reason called *glossopetrae*, as testified by Pliny the Elder. In the second half of the 1500, with the greater circulation of the ideas about natural things and the diffusion of the printing with mobile characters, some authors, first of all Conrad Gessner, compared the fossil remains to the corresponding present sharks without however expressing clearly on their derivation from living beings. They were the scientists Fabio Colonna, Niccolò Stenone, and Agostino Scilla, who pronounced themselves decidedly in such sense, respectively for experimental way, for reasoning, and for morphological comparison, opening the way to the successive dissertations on the organic genesis of some *fossilia*. Nevertheless, for a whole series of economic interests directly related to their flourishing trade, partly linked to the cult of St. Paul, both as real drugs and amulets, the glossopetrae, or the more modest "snake tongues" (the fossil teeth of sharks of smaller size) remained sought after and appreciated as counter poisons, considered by popular wisdom, special objects, however, having special therapeutic and magical properties, without a clear distinction of the two functions.

Keywords Fossils · History · Glossopetrae · Shark teeth · Disputes · Amulets

© The Author(s), under exclusive license to Springer Nature Switzerland AG 2022

M. Forli, A. Guerrini, *The History of Fossils Over Centuries*,
https://doi.org/10.1007/978-3-031-04687-2_6

Fig. 6.1 In the popular imagination, until after the Middle Ages, among other things, the common fossilized teeth of sharks, because of their similarity to the forked tongue of snakes, were generally believed to be their petrified tongues and were believed to have fallen from the sky during lunar eclipses or thunderstorms. The find was also seen as evidence of the presence of dragons or similar ferocious beasts. [melanic *Vipera aspis aspis* (Linnaeus, 1758), Piemonte (province VCO), Italy. Photo credit Matteo Riccardo Di Nicola] (graphic composition Maurizio Forli)

With the generic term of fishes, we mean a wide and varied group of vertebrates characterized, almost always, by having the body covered with scales, by being equipped with fins and by breathing through the gills. Exclusively aquatic, even if some of them are able to live externally for not too long periods, they have a wide distribution both in marine waters and in fluvial and lacustrine waters. They can be subdivided into two great groupings, utilized for easily distinguishing them, those with cartilaginous skeleton (Chondrichthyes) and those with bony skeleton (Osteichthyes), while their real and proper classification is rather complex and extensive. The fossil remains of these vertebrates most commonly found are their teeth. The fact that they are almost always found isolated has meant that in the past they were considered objects in themselves and not parts of more complex organisms. It is from them that we shall begin the discussion of the question of fossils. It is in fact around the process of formation and the meaning of the *Glossopetrae* that during the 1600s, the scientific controversy on their origin, organic or inorganic, is centered and developed, with all that follows about their attribution to the animal kingdom or not (Fig. 6.1). The Chondrichthyes are characterized by a cartilaginous skeleton or, more precisely, by a skeleton with cartilaginous matrix, mineralized by prismatic crystals of hydroxyapatite. Their first fossil evidences, represented by isolated remains of microscopic dermal scales and teeth, are found in soils dating back to the Upper Ordovician-Lower Silurian (Sansom et al. 2001). The most archaic remains of Chondrichthyes, belonging to the Cladoselachida, are known from the Upper Devonian, while from the subsequent Carboniferous, the radiation of the group begins. In the late Paleozoic appear the Neoselachii, whose ancestors of the modern forms are known from the Jurassic. The cartilaginous fishes living nowadays are well distinguishable externally, at first sight, from the other pisciform vertebrates, for being equipped with uncovered and well-evident gill slits, usually five, placed on the sides of the head or under it, for having the nostrils and the mouth opening ventrally, for the rough skin and for the particular shape and disposition of the fins. Generally, sharks have a tapered body with a high hydrodynamics, while that of the rays is roundish and flattened, with big pectoral fins starting from the head, with long whiplike tails, and live, except for some species, close to the marine substratum. The typical cartilaginous skeletal structure, less heavy and rigid than the bony one, helps their movements while swimming and floating. As they do not have a swim bladder, they regulate their buoyancy through their liver. This last, besides forming a great percentage of the body mass, contains an oil, the squalene, with low specific weight. Its efficacy is however limited, and the sharks must swim and move the fins to avoid sinking. Behind the eyes, there is a hole, the spiracle, connected to the respiration, and utilized as a pump. It is particularly developed in the benthic species and in particular in the rays where it is utilized in lieu of the mouth for sucking the water and oxygenate the gills, in such a way not to let enter debris, while in the pelagic species, it is reduced or even absent. They all have a very developed sense of smell and taste, live mainly in the marine environment, but some of them, among the sharks as well as among the rays, may penetrate also for long time in the inland freshwaters. The outer covering of the body of these fishes is called tegument and is formed by a deeper part, the dermis, and by a superficial one, the epidermis, from which come out very numerous small hooked scales, which render the skin extremely rough, especially if touched opposite to their orientation, which is toward the tail. On the oral opening, the cutaneous denticles transform in real and true teeth, placed on several rows and with increasing dimensions from the inside to the outside, in such a way that the last row is the most developed and prominent. It is only this outermost row that is functional; the innermost ones are different stages of development, useful for the gradual replacement of worn-out teeth. The shapes of the teeth, like in the other vertebrates, reflect the alimentary habits of these fishes, essentially predators, pointed, triangular with sharp margins, serrated or not in the sharks, flattened, and placed in battery in the rays. Due to their extremely resistant structure and their abundance, also the great quantity of fossil remains of selachians in general, is formed by teeth. As will be customary in our text, we provide the scientific classification of a fossil protagonist of the class of living things we are dealing with contextually, in this case what is believed to have been the largest and deadliest shark ever to appear in the seas, the megalodon, which lived from about 16 million years ago (Ma) during the Middle Miocene and presumably became extinct at the beginning of the Pleistocene, about 2.6 Ma (Pimiento and Clements 2014). This classification is still controversial as there is no agreement among scientists, as to whether or not the megalodon should be placed in the same line as the present-day white shark [*Carcharodon carcharias* (L., 1758)]. As now most of the scholars lean toward a different origin, we stick to this opinion:

Phylum Chordata
Class Chondrichthyes
Order Lamniformes
Family Otodontidae
Genus *Carcharocles*
Species *Carcharocles megalodon* (Agassiz, 1843) †
(Figs. 6.2, 6.3, 6.4, 6.5, 6.6, 6.7 and 6.8)

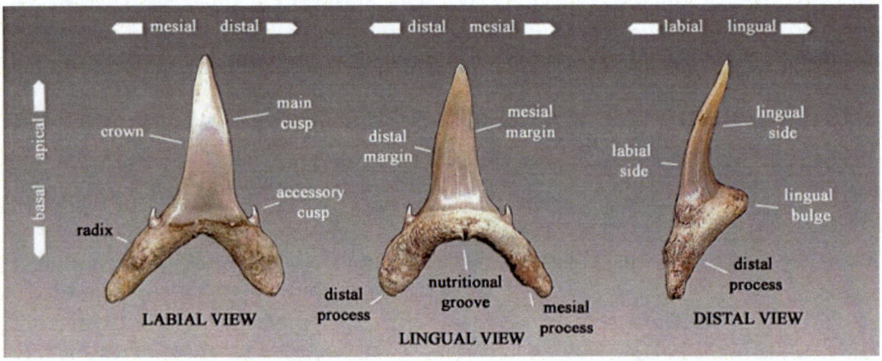

Fig. 6.2 Orientation axes and terminology of a tooth of *Carcharias acutissimus* (Agassiz, 1843). [modified after Manganelli and Spadini (2003)]

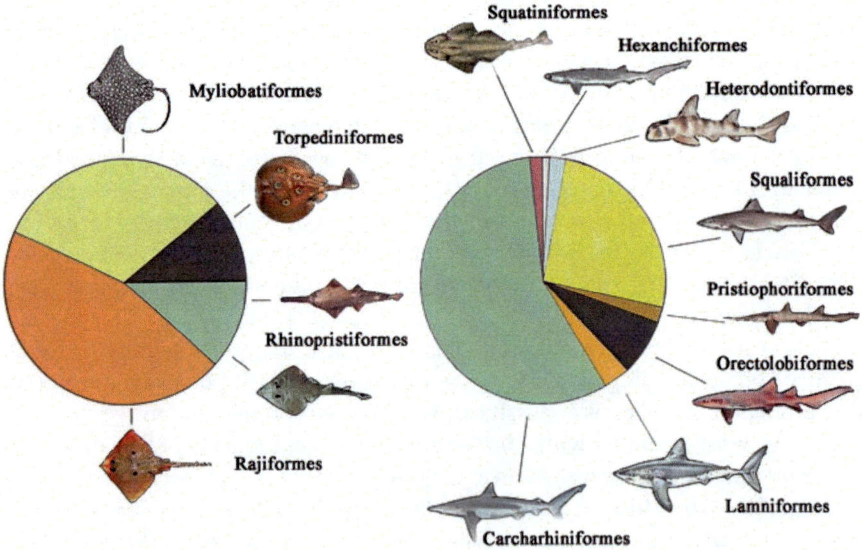

Fig. 6.3 Elasmobranchs, subdivision into orders for raimorphs (five groups) and squalimorphs (eight groups). [modified after Manganelli and Spadini (2003)]

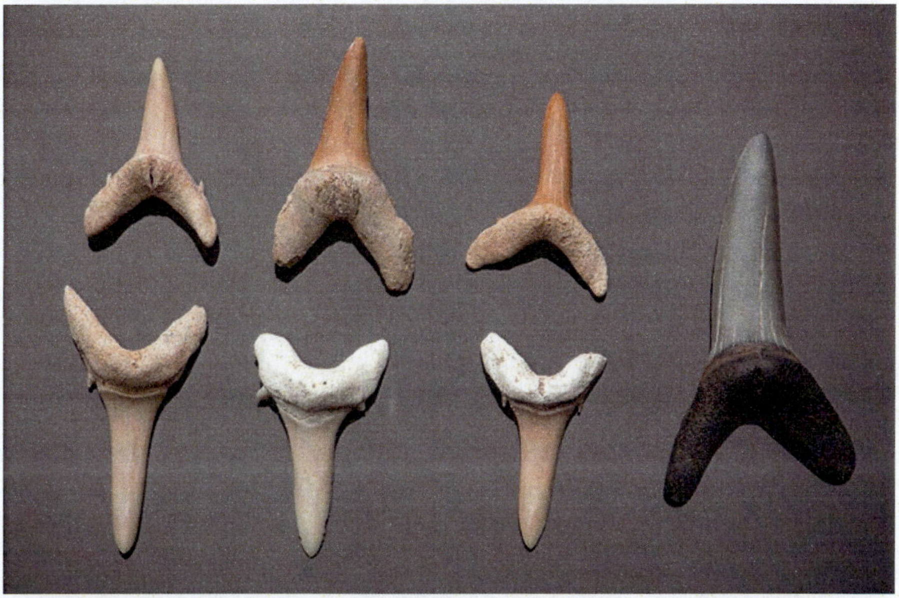

Fig. 6.4 Fossil teeth of *Odontaspis* spp. (Khouribga, Morocco), medium length ~3 cm, età ~50 Ma. The morphology of these particular teeth, sharp forked, made them look like petrified snake tongues or, if the root was missing, bird tongues. (Photo credit Giovanni Todesca)

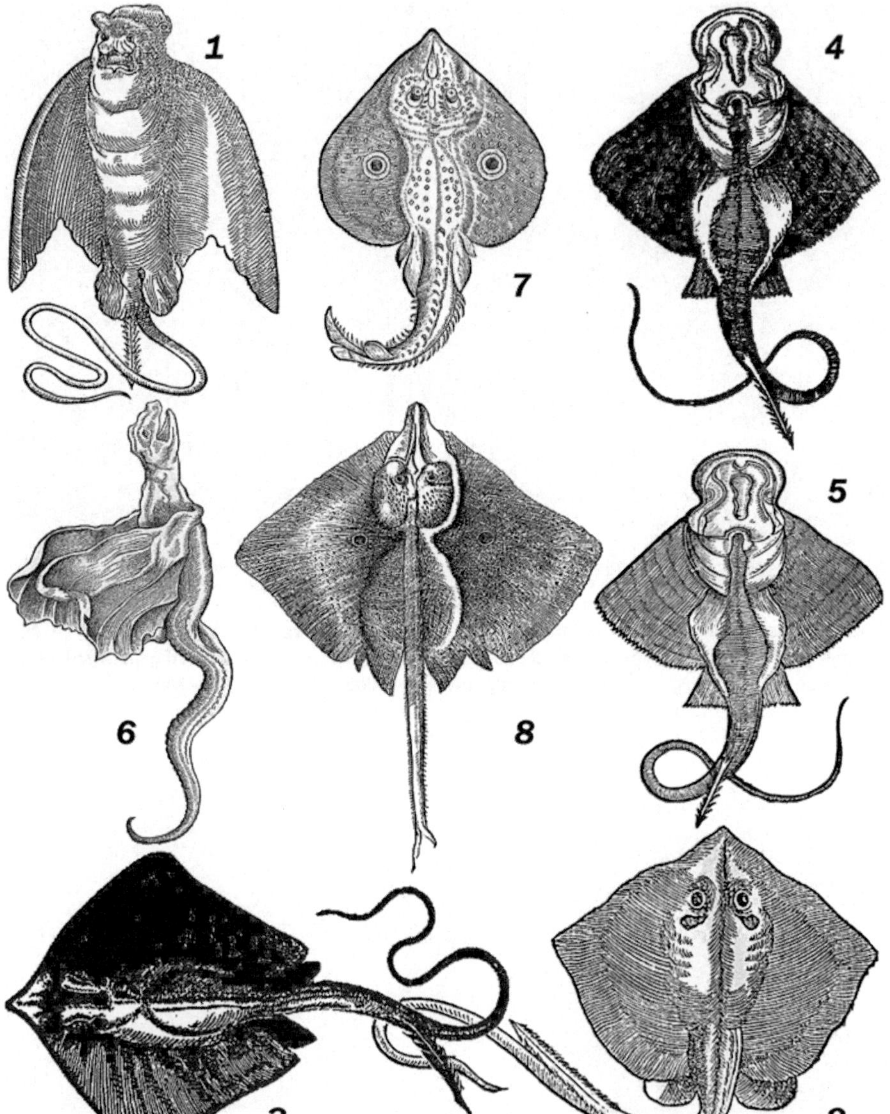

Fig. 6.5 Raiform fishes from early printed texts: Bellon (*De aquatilibus*, 1553) (1) Sea eagle (p. 97), (2) Parsnip (p. 95). Rondelet (*Libri de piscibus marini*, 1554). (3) Parsnip (p. 331). (4) Second species of Parsnip (p. 338). Gessner (1558 *De Aquatilium*). (5) Sea eagle modified after Rondelet (p. 88). (6) Dried breed in the shape of a sea snake (p. 150). Gessner (1560 *Nomenclator Aquatilium Animantium*). (7) Torpedine (p. 132). (8) Oculata race, modified after Cornelius Sittardo (p. 133). [graphic composition Maurizio Forli. (Credit from: (1–2) https://www.biodiversitylibrary.org/bibliography/5765 (Public domain). (3–4) https://www.biodiversitylibrary.org/bibliography/64229 (not in copyright). (5–6) https://doi.org/10.3931/e-rara-16853 / Public Domain Mark (7–8) https://doi.org/10.3931/e-rara-48090 / Public Domain Mark)]

Fig. 6.6 Squaliforms and their teeth: Rondelet (*Libri de piscibus marini*, 1554). (1) Lamia (p. 393). Gessner (*Nomenclator Aquatilium Animantium*, 1560). (2) Canis carcharias (p. 151). (3) Dried specimen of Canis carcharias (p. 152). Scilla (*De corporibus marinis lapidescentibus*, 1759). (4) Tooth of *Isurus* sp. (pl. I). (5–6) Teeth and figure of gray shark (*Notidanus griseus*) (pl. I and XXVIII). (7–9) Hammerhead sharks (*Sphyrna tudes*) and its teeth (pl. XXVIII) [graphic composition Maurizio Forli. (Credit from: (1) https://www.biodiversitylibrary.org/bibliography/64229 (not in copyright). (2) (3). https://doi.org/10.3931/e-rara-48090 / Public Domain Mark (4–9) personal copy Maurizio Forli]

Fig. 6.7 *Carcharocles megalodon* (Agassiz, 1843), Miocene, height ~10 cm. In particular, these large teeth, due to their shape reminding the human tongue, were those called *Glossopetrae*. (Photo credit Giovanni Todesca)

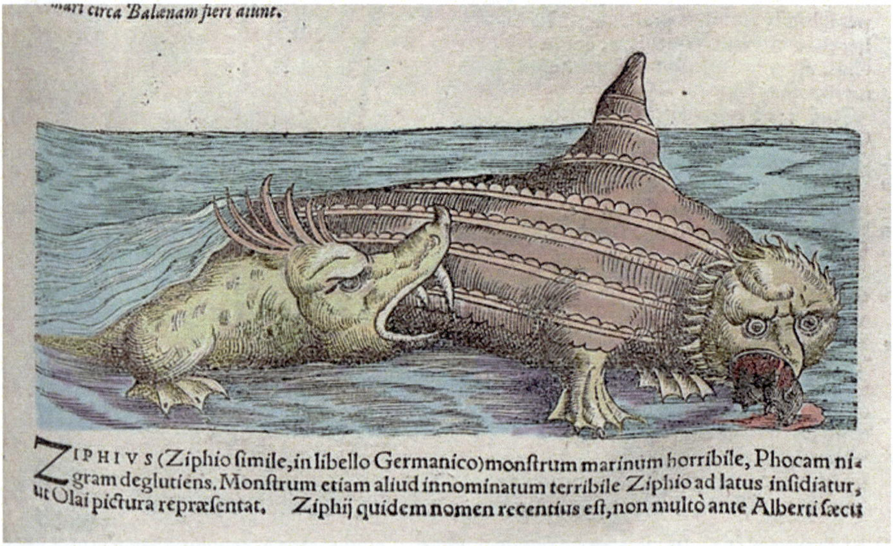

Fig. 6.8 Two sea monsters published by Gessner (*Historiae Animalium Liber IIII, De Aquatilibus*, 1558: p. 118) reproduced from Olaf Manson (Olao Magno, *Carta marina et Descriptio septemtrionalium terrarum*, 1539). The one on the right, in the act of eating a pinniped, is the so-called Ziphio, a being derived from an orca, a shark, and a swordfish. The one on the left who is attacking him, a sort of crested boar with webbed legs, more than likely inspired by a walrus, is unnamed. [https://doi.org/10.3931/e-rara-16853 (Public Domain Mark)]

The word shark is enough to make us imagine a sea monster a few meters longer than necessary, with large teeth sharp as razors, endowed with a monstrous strength, voracious, and ruthless. Today we can attribute the formation of this frightening collective memory for the most part to the cinema, or to other technological means of cultural diffusion, but a few centuries ago, the popular and even erudite imagination was even more impressed and terrified by these beings. Many of these big fishes were represented as monstrosities; often hybrids between several marine creatures, part whale, part fish, and part dragon. Their fossils, especially the teeth, were thought to be completely different things, like human tongues, snakes forked tongues, and remains of petrified thunderbolts, know as Glossopetres and Ceraunids, respectively:

> *Glossopetra, linguae similis humanae, in terra non nasci dicitur, sed deficiente luna caelo decidere, selenomantiae necessaria. Quod ne credamus, promissi quoque vanitas facit: ventos enim ea comprimi narrant.* (Pliny, *Naturalis Historia*, XXXVII: 59)
>
> *The glossopetra, which resembles the tongue of man, is said not to be born in the earth, but to fall from the sky in eclipses of the moon, and to be indispensable in selenomancy. But I do not believe the vain promise that it can calm the winds.*

The classical definition of glossopetra was ascribed to the large flattened fossil teeth of the greater shark, the megalodon or the white shark. On the contrary, those of sharks of others species, being mostly narrow and pointed, were commonly called snake tongues.

The first printed representation of glossopetrae is found in André Thevet, *Cosmographie de Levant* (1554: p. 204), a tooth of *Carcharocles*, cut at the base, from the Maltese rocks:

> *Au surplus, se treuve en cette Isle une certaine espece de langues, & à mon avis doivent etre de serpens. Les uns disent que cèst une dét de Lámier, ce que ie ne puis croire. Ce langues ici se trouvent dens les rochers & grans quartiers de pierres, aglaties & congelees, lesquelles sont si bien polies, & dételees tout á l'enviró, qu'un bon ouvrier feroit bien empeché d'en faire de semblables. Elles sont souverainement bonnes cótre le venin, qui me fait coniecturer que ce sont langues de serpens.*
>
> *Moreover, in this island there is a certain species of tongue, and in my opinion it must be that of a snake. Some say this is a Lámier (shark) tooth, which I cannot believe. These tongues here are found in rocks and large blocks of stone, agglomerated and frozen, which are so well smoothed, and serrated all round, that a good workman would easily be prevented from making the like. They are extremely good against poison, which leads me to suppose that they are tongues of snakes*

This is the iconography proposed by Gessner, in the *Historiae Animalium Liber IIII* (1558: p. 210) used also in the little following work *Nomenclator Aquatilium Animantium* (1560: p. 153). In the latter, however, the fossil tooth is not an isolated figure but is in correlation with a corresponding living organism, a shark with similar teeth, in what can be considered a sort of first paleontological reconstruction based on morphological comparison. The image of the shark juxtaposed to the tooth is taken from *Libri de piscibus marinis, in quibus verae piscium effigies expressae sunt* (1554) by the great French contemporary zoologist, Guillaume Rondelet (1507–1566) who in his own text already mentions the Maltese ichthyodontolites referring to the figure in question without providing, however, an image of them (Fig. 6.9).

Fig. 6.9 The first printed representations of the juxtaposition between glossopetres and sharks (1) Rondelet (*Libri de piscibus marinis*, 1554: p. 393). (2) Thevet (*Cosmographie de Levant*, 1555: p. 208). (3) Gessner [*Nomenclator Aquatilium Animantium*, 1560: p. 153] [graphic composition Maurizio Forli. (Credit from: (1) https://www.biodiversitylibrary.org/bibliography/64229 (not in copyright), (2) https://static.lib.virginia.edu/rmds/gordon/gordonimages/Gordon1554_T54/ (copyright not evaluated), and (3) https://doi.org/10.3931/e-rara-48090 / Public Domain Mark)]

In *De omni rerum fossilium*… (1565), the fossil is the same as that reproduced later under A, as evidenced by the self-citation in the text, when Gessner depicts three ichthyodontolites of galeoids, numbering them respectively with A, B, and C subdivided into 1, 2, and 3 (1565: p. 162) (Fig. 6.10):

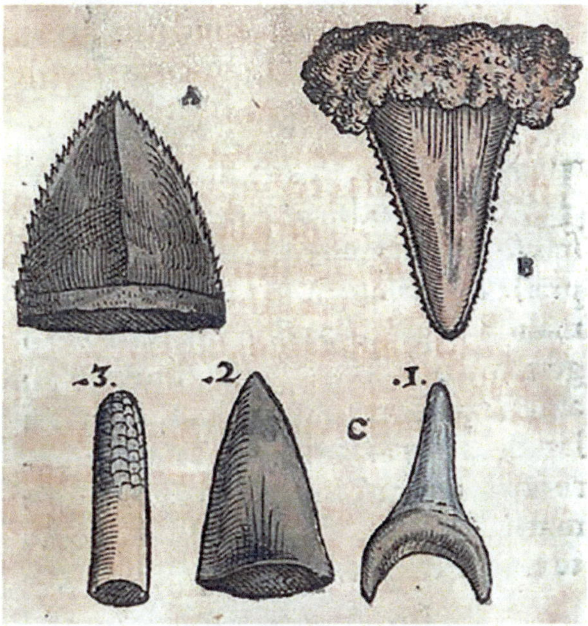

Fig. 6.10 The glossopetres illustrated by the Swiss author Gessner (*De omni rerum fossilium genere*, 1565: p. 162). [https://doi.org/10.3931/e-rara-4176 (Public Domain Mark)]

> *De eo qui in A. formula exprimitur, plura leges in Volumine aquatilium nostro, ubi de Cane carcharia & Lamia agimus, pag. 210. In formula C. ad numerum primum expressus lapis, similis est caeteris, substantia, duritie et spledore*: [...] *& vero Lamiae aut carchariae denti* [...] *In eadem formula media figura Glossopetram latiusculam & fine radice sua repraesentat. Tertia verò lapidem Cornu serpetis à quibusdam dictum* [...].

The first is always the same tooth of *Carcharocles*, triangular in shape with serrated edges, cut at the root, which Gessner himself, on the next page 163 attributes to *Cane carcharia*; at letter B is represented a tooth of *Carcharodon*, at Figure C 1 a tooth of *Isurus*, at C 2 that of a *Lamna*, while the one at Figure C 3, usually called *snake horn* or *Cerastes*, in reference to the horny protuberance of the homonymous horned viper [*Vipera ammodytes* (L., 1758)], has been identified as a fragment of a sea urchin spine (Etter and Schmidt 2019, 2020). The teeth of Lamniformes, such as that of *Isurus* (C1), probably mostly lacking their roots, were also believed to be petrified woodpecker tongues; this belief, for example, is reported by Johannes Kentmann (1518–1574), a German naturalist and friend of Gessner, who interpreted in this way some fossils from Prussia, kept in the chest of drawers, containing his famous collection of *fossilia* (Fig. 6.11).

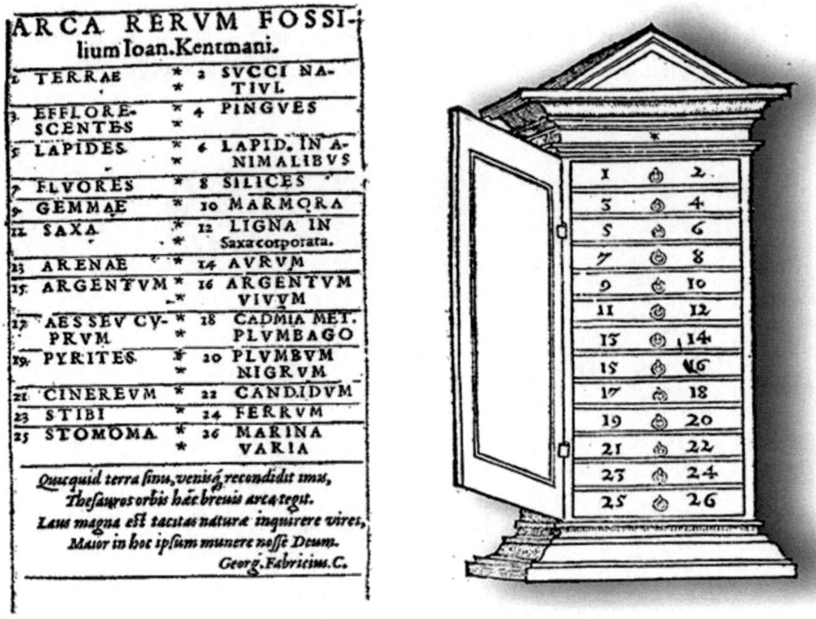

Fig. 6.11 *Ark Rerum Fossilium Ioan. Kentmani*. The first printed illustration of a fossil cabinet: Johannes Kentmann's collection was stored in this cabinet. The contents were arranged according to the division and numbering of the individual drawers. Many of the fossils as we understand them today were included in the category of *Lapides* (5) and *Ligna in saxa corporata* (12) (Gessner 1565) [graphic composition Maurizio Forli, modified from https://doi.org/10.3931/e-rara-4176 / Public Domain Mark]

The Thevet ichthyodontolite, illustrated and described by Gessner as belonging to *Cane carcharia*, is a kind of guide fossil for the oldest illustrations of glossopetres (tonguestones) being repeatedly used as an explanatory image in other texts; for example, in Boëtius de Boodt Anselmus (1636: p. 341), it is reproposed together with the other identical finds in *De omni rerum fossilium*, reversing and inverting the graphic composition.

Aldrovandi in *Musaeum Metallicum, in libros IIII distributum* (1648), printed and published posthumously by Bartolomeo Ambrosini, devotes chapter X of the fourth book to glossopetrae, writing about their synonymy and etymology, description, nature and differences between the various types of forms with which they occur, and various considerations (*Musaeum Metallicum*, pp. 600–603). The find taken up by Gessner is figured at the end of the chapter on page 603. In those same years, however, there were those who still reported the Plinian text with few variations, still considering the tonguestones as mysterious and magical objects:

> *Glosopetra, overo Goloso, è una pietra simile alla lingua humana, da che ha preso il nome. Dicono alcuni, ch'ella non nasce in terra; ma, quando sciema la Luna, cade dal cielo. I*

Magi le danno un picciolo podere nell'arte loro: percioche dicono, che con quella sanno eccitare i moti Lunari (Dolce 1565, 2: pp. 46–47).

Glosopetra, or Goloso, is a stone similar to the human tongue, from which it takes its name. Some say that it is not born on earth, but when the moon rises, it falls from heaven. The Magi give it a small power in their art: for they say that with it they know how to excite the Lunar motions.

Various authors have contributed to the understanding of the real nature of stone tongues and to the identification of fossils with shark teeth, demonstrating their organic origin, more or less independently one from the other in particular we can mention Fabio Colonna, Niccolò Stenone and Agostino Scilla, who succeeded in doing so respectively through experimentation, reasoning, and morphological comparison.

The Neapolitan naturalist Fabio Colonna, among other things, was also interested in fossils, managing to demonstrate the organic origin of tonguestones and their coincidence with the teeth of sharks. In his treatise *Ekphrasis altera* (1616), he includes the pamphlet *De Glossopetris dissertatio* where he describes how these are not bizarre natural objects but fossil shark teeth:

Qua ostenditur, melitenses, Linguas Serpentinas, sive Glossopetrae dictas, non esse Lapideas, ut quisdam asserunt; sed osseas, et Carchariae, Lamiae, sive Canicularum, et similium dentes, Maris aestu olim Terra tenui et lutosa obrutos. (Colonna 1616)

Where we show, that the serpent tongues, or glossopetrae, of the island of Malta, are not of stone, as some assert; but bones and teeth similar to those of Carchariae, Lamiae, or Canicule, which once of the Sea, were buried in the thin muddy soil.

Following the morphological similarity between tonguestones and the corresponding teeth of living sharks, it is reasonable to support their common origin even if, according to the conceptions of that time, it cannot be totally proved that they are parts of animals once alive. To overcome a whole series of possible objections from those who thought that the fossils were real stones, Colonna resorted to experiments consisting in the analysis of the residues obtained burning the Maltese tongues still adhering to the matrix, to distinguish the organic from the inorganic part. As the organic remains carbonize while the inorganic parts calcine, this allowed to that the teeth were originally composed of organic substance, and hence they were not rocks. Supported by this result, arguing that nature operates today as it has always operated and that parts of identical organisms perform the same function, he supported the correspondence of fossil teeth with those of present sharks.

A few years later Niels Stensen (Niccolò Stenone), anatomist and geologist *ante litteram*, in 1667 published *Elementorum myologiae specimen: seu Musculi descriptio geometrica. cui accedunt canis Carchariae dissectum caput, et dissectus piscis ex canum genere* in which he gives an account of the dissection of a head of *Carcharocles carcharias*, made him deliver for this purpose by the Grand Duke of Tuscany Ferdinand II de 'Medici, fished off the port of Livorno. To embellish, complete, and clarify his thought, Steno made use of a figure taken from a copper engraving that was originally intended to illustrate the work of Michele Mercati

(1541–1593), a physician and important naturalist scholar, active at that time at the papal court, then published posthumously in 1717 by Giovanni Maria Lancisi, with the title of *Metallotheca Vaticana*.

It is here that one discovers, on reading what is honestly reported, that about a hundred years before, Mercati had almost solved the mysterious origin of the stone tongues, but, like Gessner and others before him, had not drawn the conclusions from his own observations having already compared the corresponding teeth with those of *Carcharocles*:

> *The difference between the teeth (of the fish) and those glossopetres is quite imperceptible. Generally the glossopetres are larger, the teeth are thinner and less white, so that from the appearance we can make a distinction between the bony and the stony matter of the glossopetres. Moreover, the colour of the teeth is always equally white or yellowish because of the age; the glossopetres are of various colours.* (translation from Marzollo, *Niccolò Stenone. Scientific Works*, 1986: p. 134)

Apparently also Steno seems at first uncertain about the identification of glossopetres with the teeth of sharks; it would seem that he is cautious for what he says instead more decisively at the end of the treatise, leaving however some possibility of error, in reference to the mode of formation of fossil petrifications. This caution, can probably be explained by the fear, common to other scientists more or less contemporary, not to incur in the conservative or ecclesiastical censorship apparatuses:

> *The dispute about the largest glossopetres stones, that is, whether they are the teeth of sharks or stones from the earth, is not yet settled. Some have pretended that the bodies which are extracted from the earth, resembling the parts of animals, are the remains of animals which once lived in those places; others, without calling the animals into question, believe that they were produced in the place itself. I do not yet have such a knowledge of this subject as to be able to express my opinion here, and although my wandering has led me to various places of this kind, I would not dare to guarantee that what I will observe in the remaining part of my journey will be similar to what I have observed so far* [...] *so from what I have observed so far I could produce evidence for which those bodies are attributed to animals, but perhaps at another time I will be able to set out the reasons for a contrary opinion, always waiting for the right opinion from those who know more.* (Marzollo 1986: p. 144)

> *Their form convinces us that they are shark's teeth, since the flat parts, edges and base of the teeth are very similar to those of the shark* [...] *therefore, since those bodies similar to parts of animals which are extracted from the ground can be considered as parts of animals, since the shape of the glossopetres is similar to shark teeth, like an egg to an egg, and since neither the number of them nor the position of the ground convince the contrary, it seems to me that those who claim that the largest glossopetres are shark teeth are not far from the truth.* (Marzollo 1986: p. 154)

However, the question remained quite undefined, as is also testified by the explanation given a few years later by Lorenzo Legati (1677) in the description of the finds of the Cospiano Museum attached to the Aldrovandi Museum. After giving his version on the organic origin of glossopetrae, by which he means the large and triangular teeth of *C. megalodon*, quoting Steno for their mode of formation, he first reports their apotropaic use under the diction *usus in medicina* and then, curiously, describes them together with the other types of teeth of the selachians, under a different meaning, showing a lot of indecision in this regard:

> *Due Denti di Cane Carcaria, de' maggiori, che si trovino, gentilmente legati in argento per commodità di chi se ne servì secondo l'uso del volgo, che è di appenderli al collo de' bambini, stimando c'habbiano virtù contro i loro timori, e giovino ai denti quando spuntano* (Legati 1677: p. 76)

> *Two teeth of Cane carcaria, among the largest that can be found, delicately tied in silver for the convenience of those who used them according to the use of the common people, which is to hang them around the neck of children, estimating that they have virtue against their fears, and help the teeth when they sprout*

> *Denti, e lingue Fossili, ò diciamoli Sassi con figura naturalissima di Lingue, e Denti di varii Bruti, anzi dell'Huomo stesso, se ne fece buon paragone Plinio, e con esso qualche Moderno. Chiamansi dai Naturalisti Glossopetre, cioè Lingue di Pietra, e le ne trova gran copia nell'Isola di Malta. Il volgo le giudica lingue di Serpenti pietrificate. Taluni le credettero, non so con quale fondamento, Denti d'individui inumani della spezie humana, cioè di quelle scelerate Donne, che succiano il sangue a' Bambini, e lamie s'appellano, come appresso il mio gentilissimo Giovanfrancesco Bonomi in que' versi,*
>
> *Dove le Lamie a fieri incanti intente*
> *Studian fatture a tormentar la gente.*
>
> *Altri con più ragione le stimano denti del pesce Lamia [...] A me però sembrano più simili a' Denti del Cane Carcaria [...] col fondamento delle ragioni ivi toccate, e dell'autorità del dottissimo Stenoni, che da me pregiasi grandemente. Che nulla di meno possano essere determinate spezie di sasso nascente nella sua minera [...] e di tutte apertamente l'afferma l'Aldrovandi, ò chi per lui terminò l'Istoria de' Minerali [...] s'osservano composte di vari strati di terra nulla differente da quella di malta, ch'è loro minera. Il che è grande indizio per far conoscere che [...] si generino nella terra, in cui trovansi, e che perciò sia o Pietre di propria spezie, non parti d'animali sepoltevi, ò lasciate dal Mare in qualche inondazione. Mi rimetto però a miglior giudizio.* (Legati 1677: pp. 166–167)

> *Fossil Teeth and Tongues, or let us call them Stones, with the most natural figure of the Tongues and Teeth of various Brutes, or rather of Man himself, Pliny made a good comparison, and with it some Moderns. Naturalists call them Glossopetrae, i.e. Stone Tongues, and a large abundance can be found on the island of Malta. The people consider them to be the tongues of petrified serpents. I do not know on what grounds they were believed by some to be the teeth of inhuman individuals of the human race, that is to say, of those dastardly women who suck the blood of children, and are called lamias, as my dear Giovanfrancesco Bonomi calls them in those verses,*

The Lamias are intent on making ferocious spells; they study them to torment people:

> *Others, with more reason, consider them as teeth of the fish Lamia* [...] *To me, however, they seem more similar to the Teeth of the Cane carcaria* [...] *on the basis of the reasons discussed there, and of the authority of the most learned Steno, whom I greatly esteem. That nothing less can be determined spices of rock rising in its mine* [...] *and of all of them openly affirmed by Aldrovandi, or whoever finished the Istoria de' Minerali* [...] *they are observed to be composed of various layers of earth, nothing different from that of Malta, which is their mine. This is a great indication to show that* [...] they are *generated in the earth, in which they are found, and that therefore they are either Stones of their own species, not parts of buried animals, or left by the Sea in some flood. I remit myself, however, to a better judgment.*

Immediately after, apparently without knowing the work of Steno, was Agostino Scilla who, a keen observer of natural things, who published *La vana speculazione disingannata dal senso* (*Vain Speculation Undeceived by Sense*). In this book he deals not only with ichthyodonts but also with other marine organisms commonly found fossilized, studying them with the use of the *occhialino*, one of the first microscopes, probably similar to the one sent by Galilei to Prince Federico Cesi (1586–1630), founder of the *Accademia Romana dei Lincei*, a cultured environment with which Scilla had relationships:

> *Tutto ciò, unito con le suddette evidenze, m'obbliga, per finirla, a credere, che le Conchiglie, Echini, Istrici, Denti (che Glossopetre si dicono) Vertebre, Coralli, Pori, Granchi, Spatagi, Turbini, e tant'altri innumerevoli corpi, che alcuno ha giudicato essere generazione di puro sasso, ed ischerzo di Natura, siano stati animali, e corpi di quelle spezie non solamente, ma corpi, ed animali propriissimi del mare, arrivati per qualche accidente fra terra, insieme con la materia loro continente* (1670: p. 153)

> *All this, united with the aforesaid evidence, obliges me, in conclusion, to believe that the Shells, Echini, Porcupines, Teeth (which Glossopetres are called), Vertebras, Corals, Pores, Crabs, Spatangoid, Turbins, and so many other innumerable bodies, which some have judged to be the generation of pure stone, and a trick of Nature, have been animals, and bodies of those species not only, but bodies, and animals proper to the sea, arrived by some accident among the land, together with their continental matter*

In the edition written in Latin is added the treatise of Colonna on *glossopetrae* in which, although this last author understood and declared their true nature, he transferred to shark teeth the thaumaturgical properties connected to them:

> *Caniculae dentes alligati repentinos pavores tolli Plinius recitat. Rondeletius refert, dentes ab aurificibus argento includi, quos serpentis dentes vocant: hos e collo puerorum suspendut mulieres, quia dentitionen juvare credentur, ac etiam puerorum pavores arcere; confici etiam dentrificia, nam etiam ignita duritiem servant, & pulvis asperitate dentes dealbat illos detergendo: Bellonius vero: Lamia enim, qui capiunt, ex earum dentibus, ac maxillis magnum quaestum facere solent, quos ajunt adversus venena conferre, quamobrem auro, et argento includere vulgus eos solet.*

(Scilla, *De corporibus marinis lapidescentibus quæ defossa reperiuntur. Addita Dissertatione Fabii Columnæ de glossopetris*, 1759: p. 82)

> *Pliny reports that shark's teeth immediately remove anxiety. Rondeletius reports that goldsmiths include in silver the teeth that resemble those of snakes; mothers hang them around the necks of children, because they believe they help in teething and also to ward off their fears; they also prepare toothpaste, burnt they maintain hardness and their powder cleanses the roughness of the teeth by whitening them. Belon asserts that Lamia's teeth and jaws are used to make great profits, because they are said to help against poisons and the people are wont to include them in gold and silver.*

Even if at the end of 1600 the scholars had understood and proved that the glossopetres, also known as petroglosses, were nothing else than the teeth of ancient sharks, opening the way to the subsequent dissertations on the organic genesis of some *fossilia* due to a whole series of economic interests directly related to their trade, both as real medicines and as amulets, they remained however sought and appreciated as contravening and still considered by the popular wisdom, particular objects, having however special therapeutic and magic properties, without a well precise distinction of the two functions (Fig. 6.12).

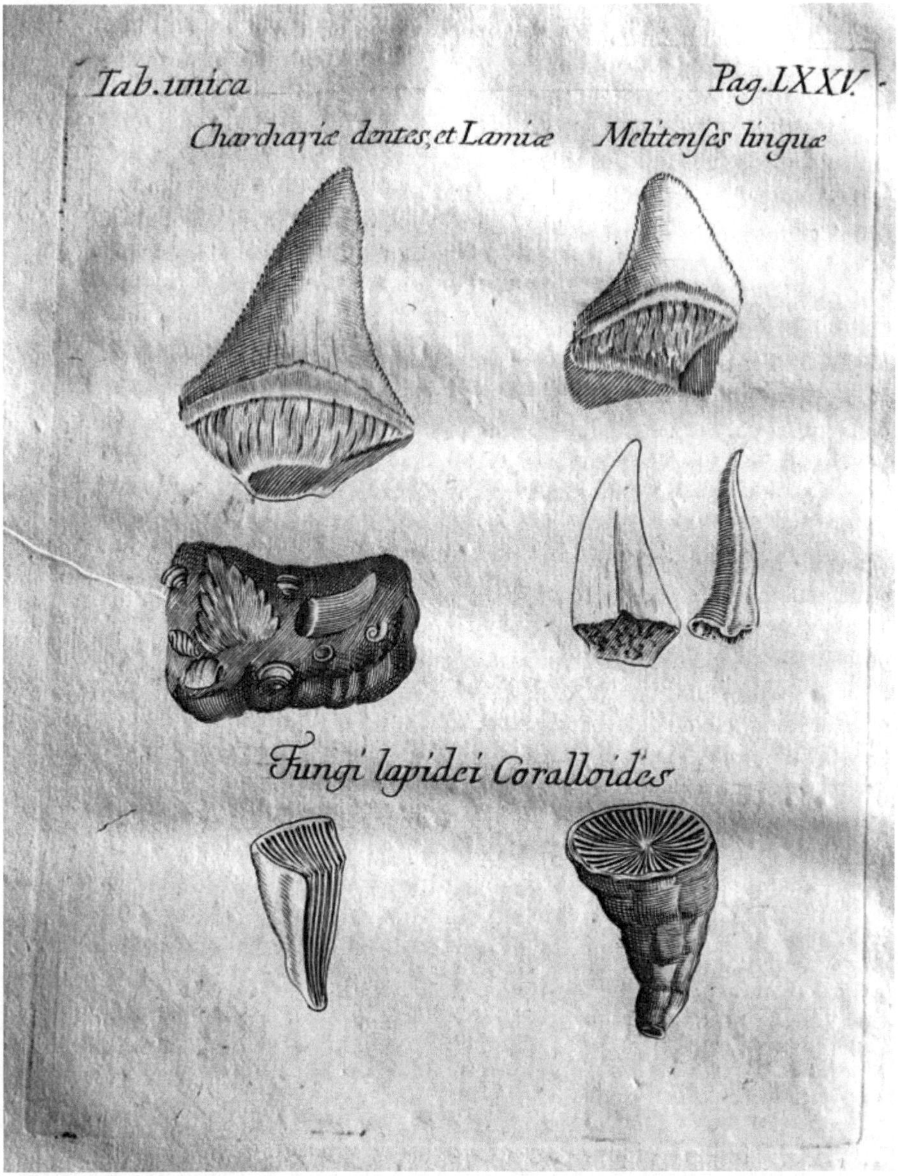

Fig. 6.12 Colonna Fabio (*De Glossopetris dissertatio*, 1759: p. 75). Plate illustrating fossil teeth from Miocene of Malta: *Melitenses lingue, Charcarie Dentes, et Lamie*. Below, two corals indicated as petrified fungi (*Fungi lapidei Coralloides*). After Scilla A (1759) *De corporibus marinis lapidescentibus quæ defossa reperiuntur. Addita Dissertatione Fabii Columnæ de Glossopetris* (personal copy Maurizio Forli)

Superstitions related to snake tongues and their use have their primary center of diffusion in the island of Malta, where these fossils abound, as on the other hand happens, albeit in smaller quantities, in the contemporary Miocene limestones found in Sicily, Calabria, Sardinia, and other Italian and European regions. The origin of these popular beliefs and practices relating to the saint's ability to combat and cancel out the harmful effects of poisoning can be traced back to an apparently marginal episode in his life, which occurred after he was shipwrecked in Malta, during his journey, in captivity, to Rome, around 60 AD:

> *When we were safe, we learned that the island was called Malta. The inhabitants of the island treated us with great humanity, and gathered us together to a fire, which they had kindled, because the rain had come and it was cold. While Paul was gathering up a bundle of dry branches and throwing it on the fire, a viper jumped out of the fire because of the heat, and bit him on the hand. When the people saw the snake hanging from his hand, they said to one another, "Surely this man is a murderer, because although he has escaped from the sea, the goddess of justice has not let him live. But he shook the serpent in the fire and suffered no harm. They expected to see him swell up or fall dead on the spot, but after waiting a long time and seeing that nothing extraordinary happened to him, they changed their minds and said that he was a god.* (2008, Acts of the Apostles, 28, 1–6)

According to tradition, the saint, following this miraculous event and the acceptance of the inhabitants, granted the island the grace of being forever free from the presence of poisonous snakes and, to its soil, powerful therapeutic powers against all types of poison. In addition, he would have annihilated and transformed into stone all the poisonous reptiles, whose rocky remains therefore remain as testimony to the event. In this context, the fossilized shark teeth became snake tongues and the round ones of the sparids, snake eyes, also petrified. Over time, establishing a close relationship between the saint's antiophidic and antivenefic power and the remains and land of St. Paul, this narrative gave rise to a strong relationship between St. Paul, the island of Malta, and the distributors of these objects for profit, up to the creation of legions of his supposed direct descendants, the sanpaolaro, self-proclaimed the true and only dispensers of the miraculous remedies constituted by the fossils and the land itself, the so-called land of St. Paul:

> *The stone or earth of St. Paul, also called the earth of Malta, or Gratie Sancti Paoli* [...] *is one of those clays used in therapy for their absorbent, astringent and sudorific properties, which bear the imprint of a seal as proof of their authenticity. The most famous of these, is the earth of Lemnos, described by Dioscorides. From the Cave of St. Paul, on which stands a church of the Saint, comes the white clay of Malta of which we are talking about* [...] *Michele Mercati* [...] *assures us that it works wonders in cases of fungal poisoning* [...] *The earth of St. Paul was hardened in the oven as it was extracted. Some of the seals whose lozenges were imprinted have been illustrated by various authors.* (Stilon, *Alcuni nostri amuleti melitensi*, 1868)

The earliest attestation of the Pauline tradition connected with Malta, developed from Luke's narration, may be a twelfth-century story by Konrad of Quernfurth in which the Men of St. Paul claimed to have received a special gift from the descendants of the Maltese, whereby they could administer cures for poison and snakebite (Savona-Ventura, *Maltese medical Pauline traditions*, 2010).

As a result of the antiophidic and anti-poisoning abilities attributed to St. Paul, the Apostle, closely related to his cult, developed, with more or less explicit remembrances to the present day, rituals and trade associated with fossilized shark ichthyodontolites. In the past centuries, glossopetrae were much used as antidote against poisons, to stop vomiting and to remove spells. Already in the Middle Ages, they used to soak one of them in the drinks to avoid poisonings, and the first documented evidences of this use, even if sketchy, date back to the half of 1200:

> *La(n)gue de Serpe(n)t est pierre à diverses couleurs aucunes fois sur bla(n)che, ou de couleur de plom(b) noir roux encédre elle resiste au venin / car se on la porte devà(n)t soy elle mue sa couler, elle vault a vices de langue fait bien pler gracieusement et honesteme(n)t* [...].

(*Le Lapidaire en François, composé par messire Jehan de Mandeville, Chevalier*)

Letters in brackets are added to fill in the typographical errors of the original printing.

In the setting of the table in Malta and in the high-ranking residences of Europe, red coral saplings with snake tongues hanging from them were used, always with an anti-poisoning function. Apart from the undoubted aesthetic charm, with these artifacts, combining the magical powers of two powerful thaumaturgical objects, one could protect oneself against poisoning.

In 1682, Pietro Caliari, the Duke of Mantua's jeweler, published his treatise entitled *Il nuovo Lume delle Gioie. Opera vtilissima ad'ogni persona, così a orefici, & speciali, per l'arte pratico nel quale, dissertando della natura e qualità degli occhi e delle lingue di serpente, afferma*:

> *Dette pietre si trovano nell'Isola di Malta, in loco dove S. Paolo vi stette, è maledì li serpenti in quel paese, è voler di Dio morsero tuti, quelli si trovarono, & oggidì vanno cavando da quella terra, e li trovano dentro ochij di serpenti impetriti, e lingue, e denti de medesimi serpenti, & anco di quella terra, la nominano gratia di S. Paolo, e dicano ch'è bona per la febra, e per il veleno, come anco li ochi di serpe sijno boni per il veleno, portandole adosso, che tochi la carne, legati in un anello, & anco le lingue, e denti, della medema spetie, habbia la medema virtù, come già descritta, ma lascio la verità al suo loco.*
>
> *These stones are to be found on the island of Malta, in the place where St. Paul stayed, and cursed the snakes in that country, and by the will of God bitten them all, those found there, & nowadays they are digging from that land, and find them inside the eyes of impetritus snakes, and tongues, and teeth of the same snakes, & also of that land, they name it the grace of St. Paul, and they say that it is good for fevers and for poison, just as snake eyes are good for poison, if you wear them when they touch the flesh, tied in a ring, and also the tongues and teeth of the same snake, they have the same virtue, as has already been described, but I leave the truth to its place.*

and, we would say, good thing that he doubts their efficacy as antidote, seen that by now the tongues and the eyes of snakes or toads, as the case may be, had been identified as teeth of fishes, (selaci and teleosts). Evidently this reality was not yet known or accepted even among the people who, for science or profession, were more in contact with these remains. On the contrary, in the case of the fossils coming from Malta, it was convenient for the many who profited from them to continue to propagate the use of these substances both for medicinal and apotropaic purposes. Following the tradition of the cult of St. Paul, congregations of people had developed who claimed to have been commissioned directly by the saint to make known his glory and his miraculous work. This took place through the diffusion, and sale to the credulous, even of the same earth melitensis, often composed in tablets, itself considered miraculous as the container and custodian of the prodigious remains. Some curious testimonies, of these fraudulent uses and modalities of miraculous administrations, but also of medical persistences linked to the authority of the figure of St. Paul, and maybe not exposing themselves against the facts considered miraculous, because of the supremacy of the ecclesiastical apparatuses that could not agree, are found in a book of the beginning of the 17th century written and published by the physician Scipione Mercurio (*On the Popular Errors of Italy*, 1603), of which we report some excerpts:

> *Il primo di quelli è il mangiar il veleno; il secondo farsi mordere da animali velenosi, come aspidi, vipere e simili. Hora l'inganno stà così, che quando vogliono mangiar il velano, come arsenico, ò risagallo, due hore prima mangiano grandissima quantità di lattuche, con molto oglio, e l'inverno quando non si possono haver lattuche, in loro scambio, mangiano molte trippe di bove ben grosse, ò lasagne, & in tanta quantità che lo stomaco & ivi sepolto non solo non possi corrodere, ò produrre altra mala qualità, che pe sua natura produrrebbe* [...] *fatto ciò prendono quel suo oglio, ò polvere, o elettuario qual vendono contra veleno, & il popolo, che vede, che non muoiono per haver tolto il veleno, crede al sicuro che la causa di ciò sia il medicamento che lodano, e vendeno a tal'effetto, e non le lattughe, ò trippe, ò lasagne, le quali subito gionti alle osterie, ritirandosi in secreto vomitano insieme col veleno tolto: e cosi si burlano d'haver beffato lo sciocco volgo, e vuotatagli la borsa.* [...] *nel mezzo della sua diceria lodando il suo medicamento, come ottimo contra veleno mandano uno di quelli, che gli ascoltano alla spezieria à torre il predetto arsenico, & havendo gli astuti ciarlatani adattati, alcuni pezzetti di zuccaro candido, ne i coperchi di vasetti de' loro medicamenti, con un tantin di cera galantamente, lo pigliano, e lo mostrano in vista* [...] *e cosi mangiando il zuccaro candido per arsenico, mangiano poi il suo lodato medicamento, e restan sani, & allegri, & il popolo corre à strappargli i vasi di mano, & ad empirgli la borsa* [...] *la seconda arte, che usano per farsi mordere* [...] *attizzando detti serpenti à mordere un pezzo di carne, che gli porgono, gli fanno tante volte reiterare, fin che perdono il natio veleno* [...] *si che quando con quest'arte gli hanno cavato il veleno, e le hanno a sua voglia domesticati, si fanno per la piazza mordere, hora la lingua, hora le zinne, e tolto quel rimedio, qual astutamente vogliono vendere ai balordi, gli fanno credere, che il restar illesi dalle morsicature de' serpi, sia per virtù del loro antidoto, e non per la furbissima cautela usata prima.* [...] *ma quel ch'è peggio in queste loro furberie vi vogliono mescolare il nome de Santi, poichè vi sono alcuni cosi sfrontati, e temerarij, che nelle pubbliche piazze ardiscono di farsi della casa di S. Paolo* [...] *non gli basta usar tante bugie, tender tanti inganni, adoprar tanti giuramenti, se anco non s'ingegnano ch'il nome di San Paolo facci il ruffiano alle loro ribalderie, farsi paréti di un Santo* [...] *E pur un verme in forma d'huomo, un vil buffone, un ciarlatano infame s'arroga d'esser suo parente. Nè perciò nego io, che la Terra di Malta non sia buona contro veleni, per lo miracolo occorso a San Paolo* [...] *e per ciò, come si crede quella Terra hà vertu contra gli animali velenosi, & in quell'Isola non si ritrova serpente, ò altro animale che habbia veleno; ma dico, che cotali ciarlatani vendono altra terra simile à quella, e tall'hora un pezzo di calcinaccio con tanto danno del misero Volgo, il quale mentre è morso da qualche serpe, pensandosi d'essere aiutato dalla pietra di San Paolo, l'usano, e non ricevendo aiuto non si provede d'altri rimedij; e cosi in un medesimo tempo restano privi del dinaro, che gettorno via, è della vita stessa.* (1603: pp. 186–187)

The first of those is to eat poison; the second is to be bitten by poisonous animals, such as asps, vipers and the like. The deception is this: when they want to eat poison, such as arsenic or arsenic sulfide (Risagallo), two hours before they eat a great quantity of lettuces, with a lot of oil, and in the winter when they cannot have lettuces, in their exchange, they eat a lot of beef tripe, or lasagna, in such a quantity that the stomach buried by them not only cannot corrode or produce any other bad quality, which by its nature the poison would produce [...] Having done this, they take that oil, or powder, or electuary which they sell against poisoning, and the people who see that they do not die for having annulled the effect of the poison, certainly believe that the cause of this is the medicine that they praise and sell for this purpose, and not the lettuces, or tripe, or lasagne, which immediately upon reaching the taverns, retiring in secret, vomit together with the poison removed: thus they mock having mocked the foolish people and having emptied their purse. [...] while they are speaking, praising their medicine, as excellent against poison, they send one of those who are listening to them to the apothecary's to get the aforementioned arsenic, and having the cunning charlatans adapted some small pieces of white sugar, in the lids of the jars of their medicines, with a little wax, they take it with dexterity and show it to the public [...] and so they eat the arsenic in a way that the people can see [...] and so they eat the white sugar instead of the arsenic, then its praised medicine remaining healthy and cheerful, so the people run to snatch the jars from their hands and to fill their purse [...] the second art that they use is to be bitten [...] luring said snakes to bite a piece of meat that they give them, many times until they lose their poison [...] so that when they have taken away their venom with this art, and have tamed them to their liking, they have their tongues and breasts bitten in the square, and having taken that remedy, which they cunningly want to sell to the beggars, they make them believe that remaining unharmed by the snakes' bites is by virtue of their antidote, and not because of the very cunning caution used before. [...] but what is worse, they want to mix the name of the Saints in their cunning, because there are some so brazen and reckless, that in the public squares they dare to make themselves of the house of St. Paul [...] it is not enough for them to use so many lies, to offer so many deceptions, to make so many oaths, if in addition they do not take it upon themselves that the name of St. Paul is the panderer to their ribaldry, to make themselves related to a Saint [...] And yet a worm in the form of a man, a vile buffoon, an infamous charlatan, claims to be his relative. Nor therefore do I deny that the Land of Malta is not good against poisons, because of the miracle occurred to St. Paul [...] and for this reason, as it is believed that this Land has virtue against poisonous animals, and in that Island one does not find any snake, or other animal that has poison; but I say that these charlatans sell other earth similar to that, and sometimes a piece of rubble to the detriment of the wretched people, who while they are bitten by some snake, thinking to be helped by the stone of St. Paul, use it, and not receiving help do not provide other remedies, and so at the same time they remain deprived of money, which they threw away, and of life itself (Figs. 6.13, 6.14, 6.15, 6.16, 6.17, 6.18, 6.19, 6.20, 6.21, 6.22, 6.23, 6.24, 6.25, 6.26, 6.27).

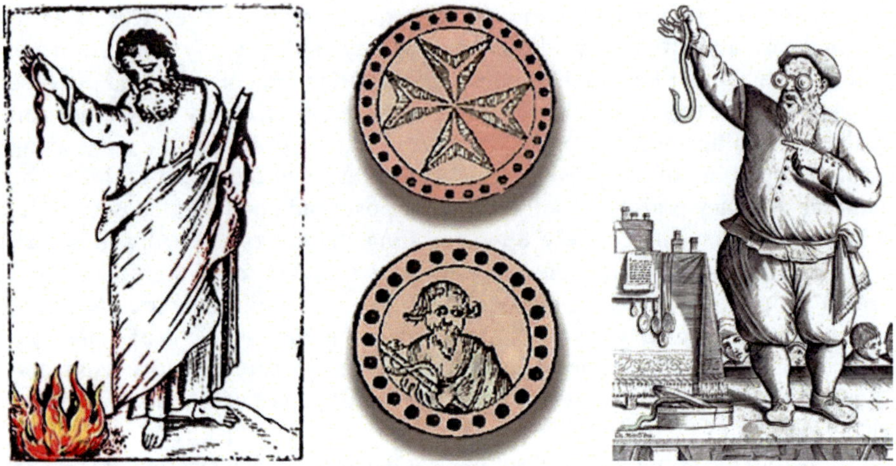

Fig. 6.13 From left to right: holy card with St. Paul, by Abela Francesco G (1647); facsimile of sealed earthen tablet with the effigy of St. Paul and the Maltese Cross, modified by Worm (1655). Charlatan showing a snake in an engraving by Giuseppe Maria Mitelli (*Le Arti per* Via 1660) [modified]. Of course you will notice the similar posture of the characters. [Graphic composition Maurizio Forli. Credit from https://archive.org/details/delladescrittion00abel (not copyright found) https://doi.org/10.5962/bhl.title.63844 (not in copyright). https://archive.org/details/gri_33125010842314/page/n5/mode/2up (not in copyright)]

Fig. 6.14 Malta, the bay where, according to tradition, St. Paul and his companions were shipwrecked. [Modified after Abela Francis (1647), https://archive.org/details/delladescrittion00abel (not copyright found)]

Fig. 6.15 Bag of sanpaularo (quack in the name of Saint Paul) [which became part of the collection around 1884. Coll. Bellucci, MANU (PG). Inv. 338 G]: bag made of kid leather, tanned and dyed violet, probably obtained by reusing the back parts of a pair of female gloves, as can be deduced from the radial stitching on both sides, which was closed with a brass ring depicting a snake. The contents, with the exception of the stellaria stone and the belemnite rostrum, represent an almost complete sample of the therapeutic objects linked to the cult of St. Paul widespread in Abruzzo, particularly in the Aquila area. Lacking any documentation on the subject, it has not been possible to ascertain whether this is the real paraphernalia of an antivenefic and antiophidic operator (sanpaolaro) or whether it is instead an aggregate of objects collected by some collaborator of Bellucci, as a compendium of objects in any case related to the mystical ritual complex referable to St. Paul. The nonfossil materials consist of a small paper wrapping containing some gray clay, probably the matrix of the shark teeth included in the sampler, and a smoothed pebble of ranocchiaia, a green and blackish stone. In addition to the shark teeth, the fossil objects are a limestone slab with colonial hexacorals, polished and honed also in the perimeter except for the longest side, where it is broken. In all probability, an originally heart-shaped stellar stone; two round orange teeth of *Lepidotes*; a dermal plate of skate with a small tooth; a spiral fragment of an internal model of gastropod; a belemnite tip. These finds will be treated later in the chapters devoted to them. (Courtesy of the Ministry for Cultural Heritage and Activities and Tourism, Regional Directorate Museums of Umbria—National Archaeological Museum of Umbria)

Fig. 6.16 Detail of the four fossil shark teeth contained in the sanpaularo bag. (Courtesy of the Ministry for Cultural Heritage and Activities and Tourism, Regional Directorate Museums of Umbria—National Archaeological Museum of Umbria)

TAVOLA I

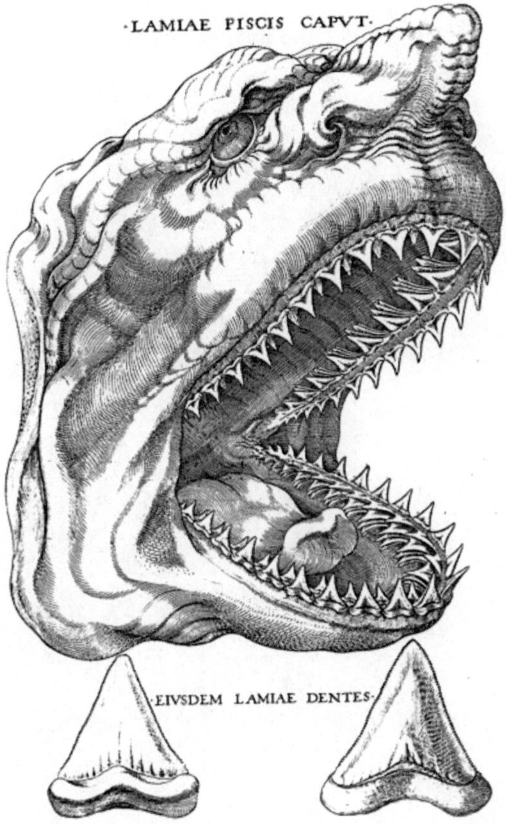

Fig. 6.17 The famous copper engraving, originally intended to illustrate Michele Mercati's *Metallotheca Vaticana* (1717: p. 333), was used by Steno in his treatise *Canis Carchariae Dissectum Caput, Lamiae Piscis Caput* (1667: pl. IV) before it was reproduced by Lancisi in the work of the Samminiatese scholar. [Modified after Marzollo (1986: pl. I)]

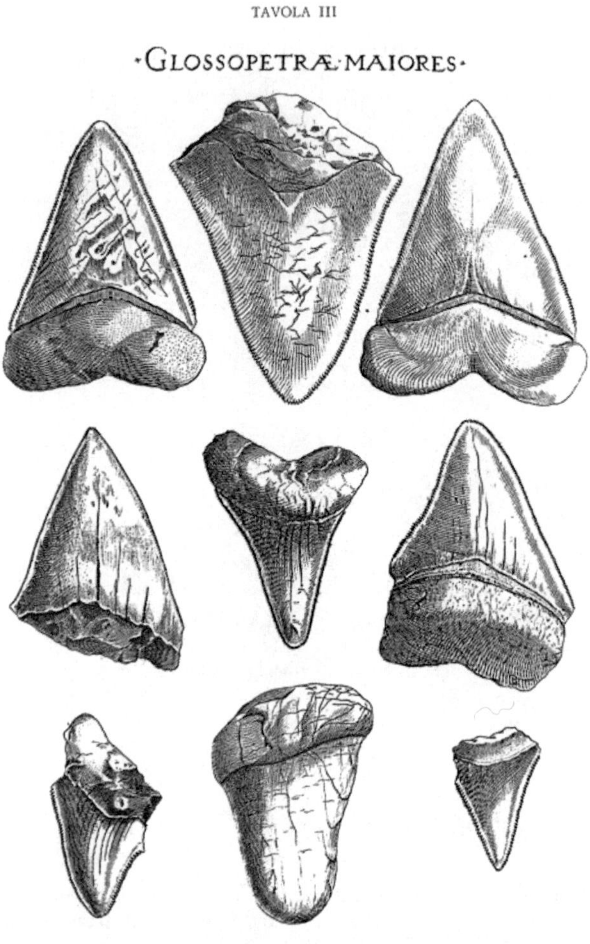

Fig. 6.18 Another copper engraving from the *Metallotheca Vaticana* (1717: pl. 332) used by Steno (1699) in his treatise *Canis Carchariae Dissectum Caput* (1667: pl. VI), *Glossopetrae Maiores*. [Modified after Marzollo (1986: pl. III)]

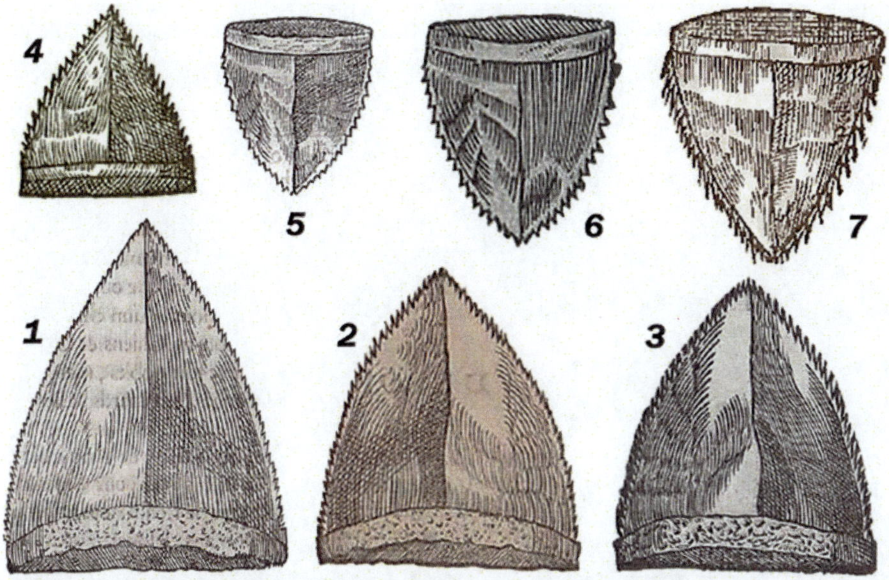

Fig. 6.19 (1) Glossopetra from Thevet (1554 : p. 208), reproposed by many authors, among which we have selected the other images. (2) Gessner (*De omni rerum fossilium genere*, 1565: p. 162). (3) Aldrovandi (*Musaeum Metallicum*, 1648: p. 603). (4) De Boodt (*Gemmarum lapidae Hist*., 1609: p. 171). (5) (*Gemmarum lapidae Hist*., 1636: p. 341). (6) (*Le Parfait Joaillier*, 1644: p. 436). (7) Worm (*Museum Wormianum*, 1655: p. 67) [graphic composition Maurizio Forli. Credit from (1) https://static.lib.virginia.edu/rmds/gordon/gordonimages/ Gordon1554_T54/. (copyright not evaluated) (2) https://doi.org/10.3931/e-rara-4176 / Public Domain Mark (3) https://doi.org/10.3931/e--rara-13017 / Public Domain Mark (4) https://hdl.handle.net/2027/gri.ark:/13960/t8w96041q (Public domain) (5) https://doi.org/10.3931/e-rara-15217 / Public Domain Mark (6) https://doi.org/10.3931/e-rara-12490 / Public Domain Mark (7) https://doi.org/10.5962/bhl.title.63844 (not in copyright)]

Fig. 6.20 Teeth similar to the previous figure, Miocene and Pliocene, maximum height 7 cm. (photo credit Giovanni Todesca)

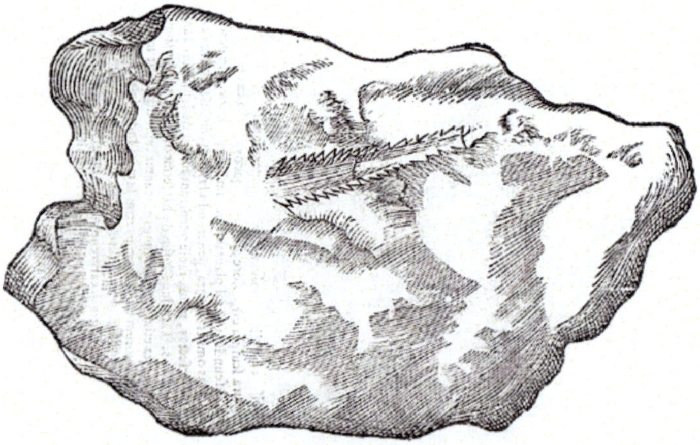

Fig. 6.21 Caudal spine of ray. Aldrovandi is the first to depict this particular fossil, correctly likening it to the corresponding living animal (*Musaeum Metallicum*, 1648: p. 490) [https://doi.org/10.3931/e-rara-13017 / Public Domain Mark]

Fig. 6.22 Detail of an analogous Pliocene sample. Age ~ 4 Ma, province of Siena [photo credit Giovanni Todesca]. [This is an example of how Aldrovandi understood that fossils are ex-vivi (organic genesis). Opposing impressions in reading the *Musaeum Metallicum* derive from (1) not always easy understanding of the Latin phrase, or (2) difficulty in distinguishing Aldrovandi's views from the listing of inorganic views of other authors, or (3) frankly inorganic views introduced by Ambrosini in 1648 (posthumous printing of at least 60–70 years). [G.B. Vai, personal communication, August 2021]

Fig. 6.23 One caudal spine, two dermal plates, and one dental plate of *Myliobatis* sp. [Pliocene (photo credit Giovanni Todesca). The dermal plate in recto-verso view, probably from a *Raja cravatta* (L., 1758), the common spiked ray, is part of the objects contained in the bag of sanpaolaro [Coll. Bellucci, Inv. 338 G. courtesy of the Ministry for Cultural Heritage and Activities and Tourism, Regional Directorate Museums of Umbria—National Archaeological Museum of Umbria]

Fig. 6.24 Stingray toothplate, *Myliobatis* sp., Pliocene, ~4 Ma. (Photo credit Giovanni Todesca)

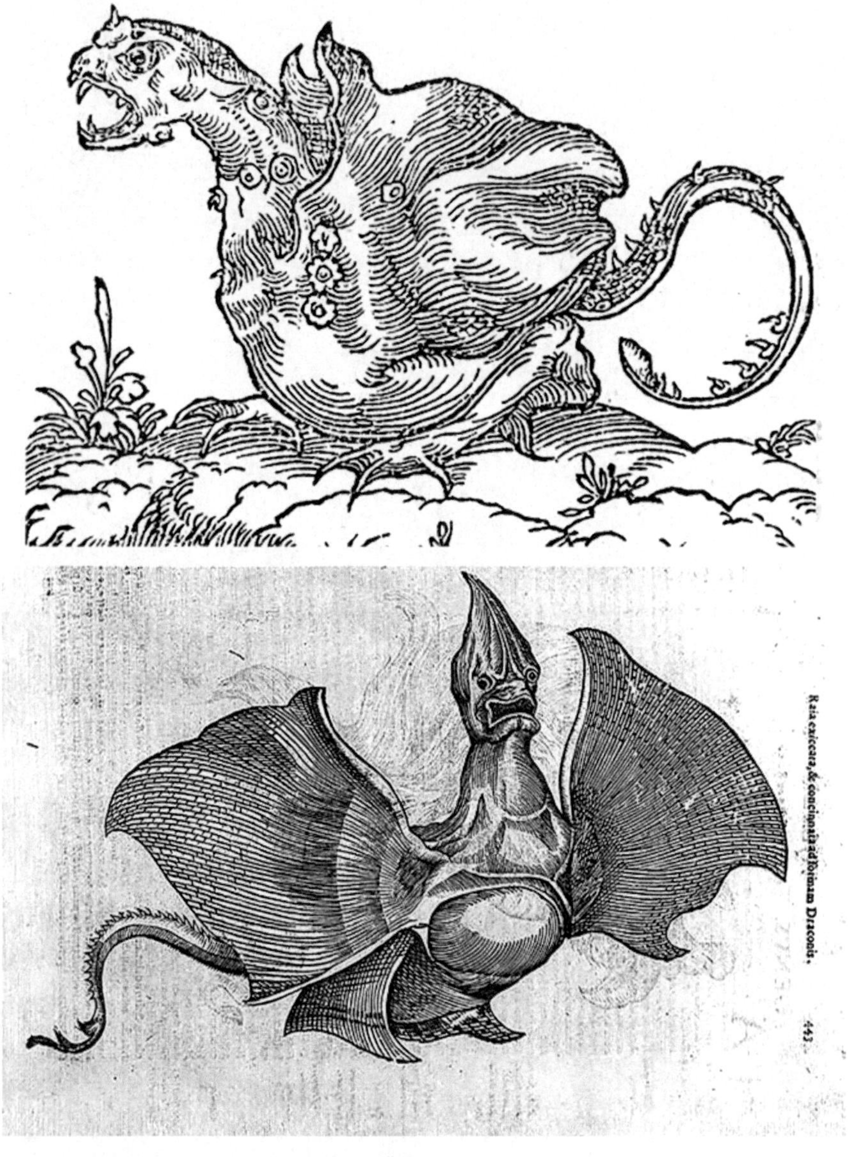

Fig. 6.25 Dried and manipulated dragon-shaped rays. The bodies of these fish lend themselves particularly well to shaping monstrous forms reminiscent of dragons or other sea monsters, and although educated people knew they were forgeries, they made a fine display in numerous "chambers of wonder." (1) Above by Pierre Boaistuau (*Histoires Prodigieuses*, 1561: p. 60) and, below, (2) by Ulisse Aldrovandi (modified after *Musaeum Metallicum*, 1648: p. 443). (Credit: https://archive.org/details/histoiresprodigi00boai (no copyright page found) (2) https://doi.org/10.3931/e-rara-13017 / Public Domain Mark)

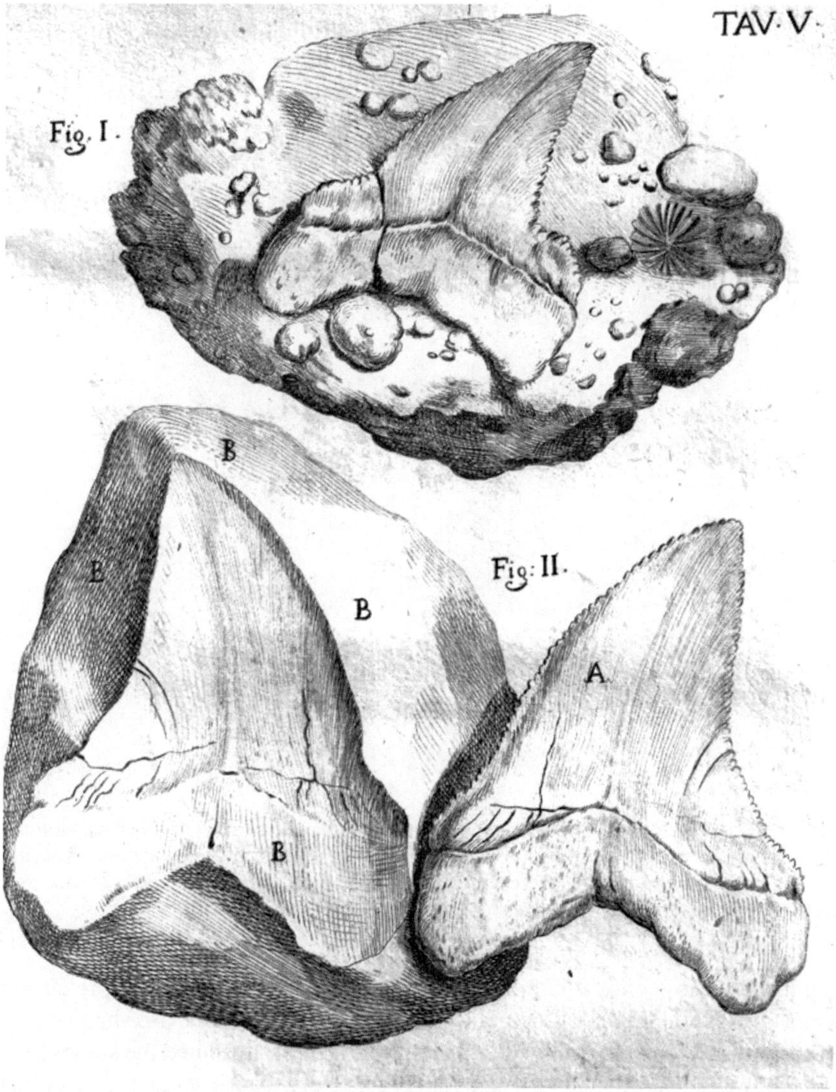

Fig. 6.26 *Petrified Lamia tooth showing its root independent of the strong tuff* [...] *and its imprint left impressed in the marga gentile tuff of Malta*: Scilla (1759: pl. V). With the support of, among others, these images, Scilla argues for the organic origin of glossopetrae, consisting of mineral substances different from those of the matrix containing them (personal copy Maurizio Forli)

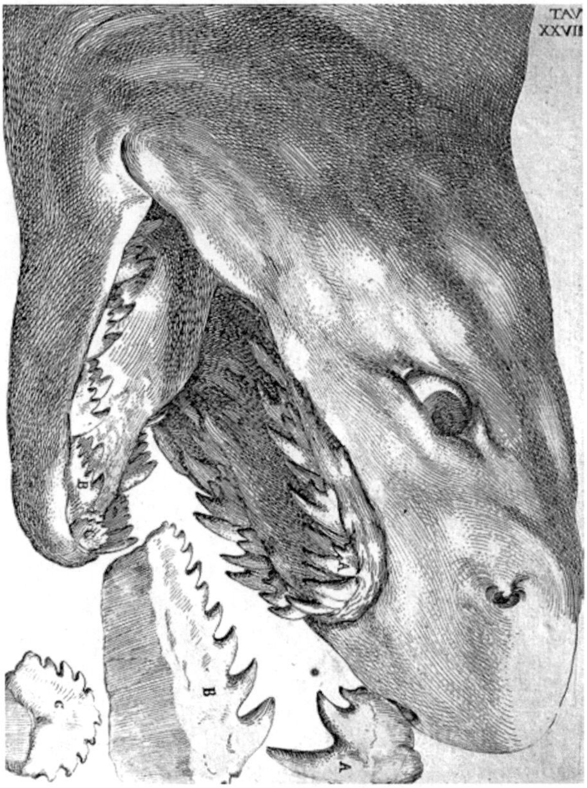

Fig. 6.27 Scilla (1759: pl. XXVII). *Head of the Cowfish portrayed live*. In this plate, Agostino Scilla shows us the head of a *Hexanchus griseus* Bonnaterre, 1788 and its different teeth. This testifies that he had realized that the different shape and size of the glossopetres could be due to the different placement in the mouth of the shark (personal copy Maurizio Forli)

From snake tongues to lightning points, this is another curious interpretation of glossopetrae, associated in this, for their equally triangular and sharp shape, to prehistoric lithic arrowheads. To thunder and lightning were attributed the stones fallen from the sky and in particular two different types of prehistoric tools, the Paleolithic axes or similar objects, as larger and heavier, remains and evidence of thunder, and arrowheads, sagittas, smaller and sharp, as residues and traces of the electric discharge, the thunderbolt. These particular stones, believed to have been generated by lightning and thus to have come from storm clouds, were called ceraunids, from the Greek *ceranyos*, lightning. This interpretation is widespread in most of Europe, for example, in France they are called *coins de foudre*, *pierres de tonnerre*, in England *elf shot* or *thunderstone*, *and* in Germany *strahlstein*:

> *La pietra ceraunio cade dalle nuvole, chi la porta non si può sommergere, ne d'esser percosso dal fulmine, o d'altro accidente, fa l'huomo nelle battaglie, e cause vittorioso, e fa che i sogni vengono dolci, e lieti*. (Bonardo, *La Minera del Mondo*, 1585, 2: p. 22)

The ceraunio stone falls from the clouds, whoever wears it cannot be submerged, nor be struck by lightning, or other accident, makes the man in the battles, and causes victorious, and makes the dreams come sweet, and happy.

Giacinto Gimma, scholar from Bari (1668–1735), writing about the stones rained from the sky, and the formation and consequences of lightning, distinguishing various types as almost all those who dealt with describing them, points out that some were artifacts and could not have been generated in the clouds:

La Ceraunia è da molti annoverata tra le Gemme, facendola spezie di Asteria: e Aldrovando scrisse, che Solino, ed altri, la confondono con la Pietra Ceraunia, di cui qui trattiamo; falsamente persuadendosi, che si trovi ne' luoghi, percossi dal fulmine [...] *che si possano generar pietre nell'aria, o in altra parte del Cielo, l'abbiamo certamente per favola* [...] *Ma che le Ceraunie, che si dicono Saette, e Pietre del Fulmine, si possano generare nell'aria e nelle nuvole, non lo stimiamo affatto impossibile; benchè vera generazione non sia; essendo propria della terra la loro materia* [...] *Vane sono le Gemme, e tutte le altre cose riferite, e da molti credute utili e giovevoli, le quali in niun modo sono valevoli a fugare le saette, e le tempeste*. (Gimma 1730b: pp. 192–202)

Ceraunia is counted by many among the Gems, making it a species of Asteria: and Aldrovando wrote that Solino and others confuse it with the Ceraunia Stone, of which we are speaking here; falsely persuading themselves that it is found in places struck by lightning [...] *that stones can be generated in the air, or in another part of the Sky, we certainly have it as a fable* [...*But that the Ceraunids, that are called Lightnings, and Lightning Stones, can be generated in the air and in the clouds, we don't esteem it at all impossible; although true generation is not; being proper to the earth their matter* [...] *Vain are the Gems, and all the other reported things, and by many believed useful and beneficial, which in no way are valid to escape the lightnings, and the storms*

Ceraunites, ou ceraunia, ou ceraunoides [...] *On désigne par ces noms, plus populaires que philosophiques, par ces noms qui doivent leur origine non à la vérité. mais à l'opinion, diverses sortes de pierres*. (Bertrand 1763)

The teeth of the greater sharks, *Carcharodon*, *Carcharocles*, and similar genera, being triangular and generally shiny on both sides, were much appreciated as amulets against electrocutions. This superstition, as it is obvious, is more diffused in the regions where the fossil shark teeth are of easier finding, in the peninsular Italy in particular from Tuscany and Umbria to Puglia and Calabria, and in the islands, in Sicily and Sardinia, and was especially felt by the rural populations that being for more time to the open air, were more subject to the fears and to the deadly effects of the lightning electrocutions.

In Apulia, in the surroundings of Lecce, the so-called Pietra Leccese is quarried, a light yellow Miocene limestone, commonly used for building constructions, rather rich in ichthyodontolites even of considerable size, whose parts incorporated in the rocky matrix, sometimes sawn with it and then exposed by the consumption caused by the weather, were believed to be the remains of the terminal part of the lightning strike, which had struck there, thus leaving evident traces.

Something of these ancient cults and superstitions, where not explicitly exercised yet, remains more or less perceived and diffused in the cultural substratum of the consolidated traditions and currently little or no longer remembered. Even today, by chance, we have collected two testimonies from friends who live respectively in Gambassi Terme (FI) and in Siena, whose grandparents, still living, were born in the thirties of the twentieth century, continue to call the fossil shark teeth, lightning, identifying them with what remains of the lightning.

Usually it was used to set them to keep them as protective amulets, as it was certain that their presence, worn or kept in the house, preserved from the fall of lightning and prevented them from falling where another had already hit previously. Moreover, they were considered of help in the teething of infants or to cure infestations by parasitic worms. In contemporary times, in addition to fossil teeth, also those of present-day sharks are used as pendants in good luck jewelry, also with the function of exorcising fear toward these animals considered, more or less unconsciously, as sea monsters. A large and fascinating exhibition of these particular therapeutic magic objects, as anticipated, is part of the materials preserved at the National Archaeological Museum of Umbria, in Perugia, from the Giuseppe Bellucci Collection, collected between the end of the nineteenth century and the early twentieth century (Figs. 6.28, 6.29, 6.30, 6.31, 6.32, 6.33, 6.34, and 6.35).

Fig. 6.28 Salt shaker with a branch of red coral and glossopetrae (Natternzungenkredenz, Vienna, Schatzkammer und Museum des Deutschen Ordens, inv. no. K-037)

Fig. 6.29 From snake tongues to lightning stones: this is another curious interpretation of fossilized shark teeth, which, as previously shown also by the testimony of Leonardo da Vinci, were believed to be the remains of the terminal part of the lightning. Free interpretation with a fossil shark tooth and snake exuviae [graphic composition Maurizio Forli. (Photo credit Giovanni Todesca)]

Fig. 6.30 Their preservation and natural exposure in the rocks was a demonstration of the direct fall of the electrocution, which had struck there, thus leaving an evident trace. Another creative image with a Miocene tooth referable to *Carcharocles megalodon* (Agassiz, 1843) [graphic composition Maurizio Forli. (Photo credit Giovanni Todesca)]

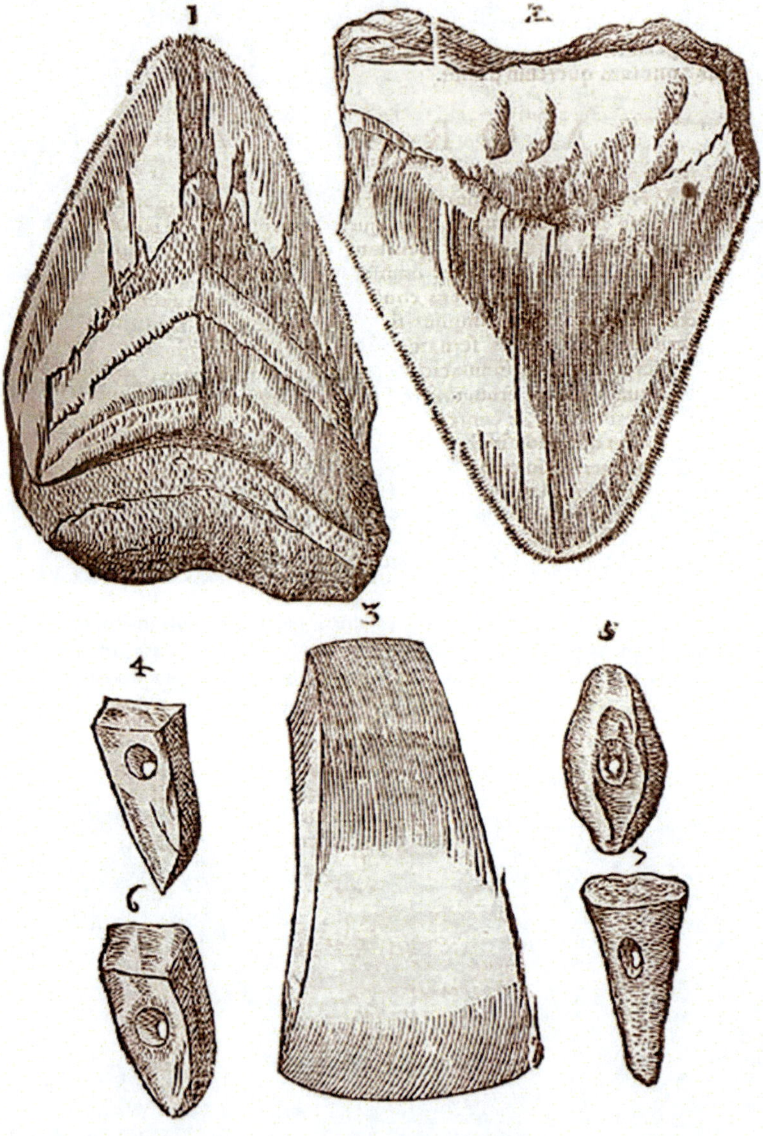

Fig. 6.31 Aldrovandi (*Musaeum Metallicum*, 1648: p. 611) [https://doi.org/10.3931/e-rara-13017 / Public Domain Mark]. *Variae Cerauniarum differentiae* (1–2) Ceraunia in stone, upper part, the same, lower part; (3–7) Ceraunia cuneiform, major, minor, and acuminate, some pierced in the middle. Fossil shark tooth compared with prehistoric polished stone axes. The original appellative, derived from the Greek, was masculine, *Cerauno* or *Ceraunio*, as sometimes it is found in the most ancient texts: *Cerauno is a stone so dinominata in Greek language, and in Latin is appellata lightning; This stone yes falls from the Sky, and is therefore it is found there, where the men are fed from the thunderbolt* (from a seventeenth-century translation reported in various dictionaries among which: Manuzzi Giuseppe, *Vocabolario della Lingua Italiana*, 1833: p. 609). Axes had been proposed earlier by Gessner and de Boodt

Fig. 6.32 Upper part (1–2) *Glossopetrae aliae duae differentiae* (*Musaeum Metallicum*, 1648: p. 604). A tooth of *Carcharocles megalodon* compared with a prehistoric lithic arrowhead. Both objects were believed by the commoners to be the terminal part of the lightning bolt and therefore, when found, were jealously preserved as amulets against lightning (3–4). Down: the tooth of Figure 1 in this plate, still preserved in the Aldrovandi collection displayed in the museum of Palazzo Poggi in Bologna. (Photo credit Alessandro Ceregato, permission Sistema Museale di Ateneo Alma Mater Studiorum—Università di Bologna) [(3–4) photo Maurizio Forli, (1–2) https://doi.org/10.3931/e-rara-13017 / Public Domain Mark]

Fig. 6.33 Amulets from the Bellucci collection, MANU (PG). On the sides, fossil shark tooth (*Lamna appendiculata*) bound in brass; from Spoleto, acquired in 1873 [Inv. 1057 B; mm 19x50x7]. In the center, fossil tooth of *Odontaspis* spp., shaped and bound in silver with a reticulated plate. Foligno, 1886 [Inv. 3136 G; 16 × 41 × 10 mm]. (Reproduced by kind permission of the Ministry for Cultural Heritage and Activities and for Tourism, Regional Directorate for Museums of Umbria—National Archaeological Museum of Umbria)

Fig. 6.34 Amulets from the Bellucci collection, MANU (PG). On the sides, fossil tooth of *Carcharodon carcharias* with silver binding, Scano di Montiferro, Oristano, 1908. [Inv. 6985 G; mm 31x43x5]. In the center, fossil shark tooth shaped like a heart and with a double suspension ring. L'Aquila, late nineteenth century [Inv. 6651 G]. [Reproduced by kind permission of the Ministry for Cultural Heritage and Activities and for Tourism, Regional Directorate for Museums of Umbria—National Archaeological Museum of Umbria]

Fig. 6.35 I remember my early childhood (FM), from my maternal grandmother's house: a heart-shaped case; a cardboard box, lined with hand-embroidered silk, which held small devotional objects such as rosaries, holy cards, necklaces with crosses, and a few briefs, many of which are now lost. Among these objects there were also two small shark's teeth, attached to chains, worn by us children as good luck charms. Procida (Napoli) 1940 circa. (Photo Maurizio Forli, courtesy private collection)

References

AAVV (2008) Acts of the Apostles. The Holy Bible CEI 2008. CEI Italian Episcopal Conference

Abela FG (1647) Della Descrittione di Malta, Isola nel Mare siciliano, con le sue antichita, ed altre notizie. Books four. Paolo Bonacota, Malta. https://archive.org/details/delladescrittion00abel (not copyright found)

Aldrovandi U (1648) Ulyssi Aldrovandi patricii Bononiensis Musaeum metallicum in libros IIII : distributum Bartholomaeus Ambrosinus ... labore, et studio composuit cum indice copisosissimo Arcus Antonius Bernia. Bononiae : typis Io. Baptistae Ferronii, 1648. ETH-Bibliothek Zürich, Rar 1027., https://doi.org/10.3931/e-rara-13017 /Public Domain Mark

Bellon P (1553) De Aquatilibus, Libri duo. C. Stephanum, Parisiis. https://www.biodiversitylibrary.org/bibliography/5765 (Public domain)

Bertrand E (1763) Dictionnaire universel des fossiles propres et des fossiles accidentels. A la Haye : Chez Pierre Gosse Junior, et Daniel Pinet, M. DCC. LXIII [1763]. ETH-Bibliothek Zürich, Rar 5720., https://doi.org/10.3931/e-rara-10463 / Public Domain Mark

Boaistuau P (1561) Histoires prodigieuses, extraictes de plusieurs fameux autheurs Grecz & Latins, sacrez & prophanes: mise en nostre langue par P. Boaistuau, surnomé Launay, natif de Bretaigne: avec les pourtraicts & figures. Pour Jean Longis & Robert le mangnier Libraires, Paris. https://archive.org/details/histoiresprodigi00boai (No copyright page found)

Bonardo GM (1585) La minera del mondo dell'ill. sig. Gio. Maria Bonardo Fratteggiano, conte e cavaliero. Nella qual si tratta delle cose più secrete, e più rare de' corpi semplici nel mondo elementare, e de' corpi composti, inanimati, et animati d'anima vegetativa, sensitiva, e ragionevole. Divisa in quattro libri, & per maggior commodità de' lettori ogni capitolo, c'ha potuto riceverlo, ordinato, con ordine d'alfabeto. Mandato in luce da Luigi Grotto, cieco d'Hadria. Fabio, et Agostin Zoppin Fratelli, Venetia

Boodt A Boëtius de (1636) Gemmarum et lapidum historia. Lugduni Batavorum [Leiden] : ex officina Ioannis Maire, 1636. ETH-Bibliothek Zürich, Rar 1881., https://doi.org/10.3931/e--rara-15217 / Public Domain Mark

Boodt A Boëtius de (1644) Le parfaict Ioaillier ou histoire des pierreries: ou sont amplement descrites leur naissance, iuste prix, moyen de les cognoistre, et se garder des contrefaictes, Facultez medicinales, et proprietez curieuses. Iean-Antoine Huguetan, Lyon. ETH-Bibliothek Zürich, Rar 1010, https://doi.org/10.3931/e-rara-12490 / Public Domain Mark

Caliari P (1682) Il nuovo Lume delle Gioie. Opera utilissima ad ogni persona, così à orefici, et speciali, per l'arte pratico. Iseppo Prodocimo, Venetia

Colonna F (1616) Fabii Columnae Lyncei Minus cognitarum rariorumque nostro coelo orientium stirpium ekphrasis ... https://www.biodiversitylibrary.org/bibliography/723 Public domain

Dolce L (1565) Libri tre ne i quali si tratta delle diverse sorti delle Gemme, che produce la Natura, della qualità, grandezza, bellezza, et virtù loro. Gio Battista, Marchio Sessa, et Fratelli, Venetia

Etter W, Schmidt O (2019) Gessner's fossil crab An icon for the early history of palaeontology. J History Collect. https://doi.org/10.1093/jhc/fhaa003

Etter W, Schmidt O (2020) Gessner's fossil crab An icon for the early history of palaeontology. J History Collect 32(3):411–430. https://doi.org/10.1093/jhc/fhz039

Gessner C (1558) Conradi Gesneri medici Tigurini Historiae animalium liber IIII. qui est de piscium & aquatilium animantium natura - cum iconibus singulorum ad vivum expressis fere omnib. DCCVI.: continentur in hoc volumine Gulielmi Rondeletii quoque, medicinae professoris regii in Schola Monspeliensi, & Petri ... Christoph. Froschoverum, Tiguri. https://doi.org/10.3931/e-rara-16853 / Public Domain Mark

Gessner C (1560) Nomenclator aquatilium animantium : icones animalium aquatilium in mari et dulcibus aquis degentium, plus quam DCC. cum nomenclaturis singulorum Latinis, Grecis, Italicis, Hispanicis, Gallicis, [...]. Tiguri : excudebat Christoph. Froschoverus ..., 1560. ETH-Bibliothek Zürich, Rar 9734: 1, https://doi.org/10.3931/e-rara-48090 / Public Domain Mark

Gessner C (1565) De omni rerum fossilium genere, gemmis, lapidibus, metallis, et huiusmodi, libri aliquot, plerique nunc primum editi. Tiguri.. https://doi.org/10.3931/e-rara-4176 / Public Domain Mark

Gimma G (1730a) Della Storia Naturale delle gemme, delle Pietre, e di tutti i Minerali, ovvero della Fisica Sotterranea. Tomo I, Gennaro Muzio, Napoli

Gimma G (1730b) Della Storia Naturale delle gemme, delle Pietre, e di tutti i Minerali, ovvero della Fisica Sotterranea. Tomo II, Felice Mosca, Napoli

Kentmann J (1565) Nomenclaturae rerum fossilium, que in Misnia praecipue, & in alijs quoque regionibus inveniuntur. Tiguri. In: Gessner C (1565) De omni rerum fossilium genere, gemmis, lapidibus, metallis, et huiusmodi, libri aliquot, plerique nunc primum editi. Tiguri : excudebat Iacobus Gesnerus, anno 1565[–1566] Zentralbibliothek Zürich, NG 1910, https://doi.org/10.3931/e-rara-1340 / Public Domain Mark

Legati L (1677) Museo Cospiano annesso a quello del famoso Ulisse Aldrovandi e donato alla sua Patria dall'Illustrissimo Signor Ferdinando Cospi. Giacomo Monti, Bologna. https://babel.hathitrust.org/cgi/pt?id=gri.ark:/13960/t7dr4608w&view=1up&seq=30&skin=2021 (Public Domain Mark)

Manganelli G, Spadini V (2003) Gli squali del Pliocene senese. Cantagalli Editions, Siena

Manson O (1539) Carta marina et Descriptio septemtrionalium terrarum, ac mirabilium rerum in eis contentarum diligentissime elaborata. Veneciis

Manuzzi G (1833) Vocabolario della Lingua Italiana già compilato dagli accademici della Crusca ed ora novamente corretto ed accresciuto. Tomo Primo Parte prima. Davide Passigli e Soci, Firenze

Marzollo M (1986) Niccolò Stenone. Opere Scientifiche. Traduzione integrale dai testi originali. 2 voll. pp. XIX-304, 296. Giorgi e Gambi Editore, Cassa di Risparmio e Depositi di Prato

Mercati M (1717) Metallotheca Vaticana. Opus posthumum, Autoritate et Munificentia Clementis Undecimi Pontificis Maximi. E tenebris in lucem eductum: Opera autem, & studio Joannis Mariae Lancisii, Archiatri Pontificii illustratum. Ex Officina Jo: Mariae Salvioni Romani, Romae. https://archive.org/details/bub_gb_DC_E6UVWg7EC Public Domain Mark 1.0

Mercurio S (1603) De Gli Errori Popolari d'Italia. Libri sette, divisi in due parti. Parte prima. Gio. Battista Ciotti, Venetia

Mitelli GM (1660) Di Bologna, l'arti per via d'Anibal' Caraci. Gio. Jacomo Rossi le stampe in Roma alla Pace, Roma. https://archive.org/details/gri_33125010842314/page/n5/mode/2up (not in copyright)

Pimiento C, Clements CF (2014) When Did Carcharocles megalodon Become Extinct? a new analysis of the fossil record. PLoS One 10(1):e0117877

Rondelet G (1554) Libri de Piscibus Marinis, in quibus verae Piscium effigies expressae sunt. Matthiam Bonhomme, Lugduni. https://www.biodiversitylibrary.org/bibliography/64229 (not in copyright)

Sansom IJ, Smith MM, Smith MP (2001) The Ordovician radiation of vertebrates. In: Ahlberg PE (ed) Major events in early vertebrate evolution. Taylor & Francis, London, pp 156–171

Savona-Ventura C (2010) Maltese medical Pauline traditions. Hist Perspect 22(2):34–36

Scilla A (1670) La vana speculazione disingannata dal senso. Lettera responsiva Circa i corpi marini, che Petrificati si trovano in varij luoghi terrestri. Andrea Colicchia, Napoli

Scilla A (1759) De corporibus marinis lapidescentibus quæ defossa reperiuntur. Addita Dissertatione Fabii Columnæ de Glossopetris. Joannis Zempel, Romae

Stenone N (1667) Elementorum myologiae specimen: seu Musculi descriptio geometrica. cui accedunt canis Carchariae dissectum caput, et dissectus piscis ex canum genere. Johan. Janssonium, Amstelodami

Stilon A (1868) Alcuni nostri amuleti melitensi. Archivum Melitense, 93–103

Thevet A (1554) Cosmographie de Levant. Ian de Tournes et Guil Gazeu, Lion. https://static.lib.virginia.edu/rmds/gordon/gordonimages/Gordon1554_T54/. (copyright not evaluated)

Worm O (1655) Museum Wormianum, seu Historia rerum rariorum, tum Naturalium, quam Artificialium, tam Domesticarum, quam Exoticarum; quae Hasniae, Danorum in aedibus Authoris Servantur. Iohannem Elsevirium, Lugduni Batavorum. https://doi.org/10.5962/bhl.title.63844 (not in copyright)

Chapter 7
The Understanding of some Ichthyodontolites Through the Time: from Bufonites, Toadstones, and Snake Eyes to Bony Fish Teeth

Abstract Also the teeth of bony fishes have had an important place in the history of the exact definition of fossils and in particular those of hemispherical shape referred both to the extinct genus *Lepidotes* and to those included in the genus *Sparus* and *Diplodus*, fishes still living in the seas. Believed to be small stones located in the head of toads, between the skin and the skull, and for this reason called bufonites, according to tradition they were taken directly or after they had been expelled independently by the animals, just before dying. Considered marvelous objects with magical and therapeutic powers, they were used both as amulets and as medicine, ingested whole or pulverized. They were used against liver disease, tumors, and punctures, but, in particular, it was believed that they were able to cancel poisonings. Among the first news about them we find those handed down, once again, by Pliny the Elder in his *Naturalis Historia*, where the stone is described as *batrachites*. The first depiction of the extraction of batrachites from the head of a toad is given in the *Hortus Sanitatis* by Johann Wonnecke von Kaub, printed in several editions starting from the end of the fifteenth century, while it was Agostino Scilla (La vana speculazione disingannata dal senso. Lettera responsiva Circa i corpi marini, che Petrificati si trovano in varij luoghi terrestri. Andrea Colicchia, Napoli, 1670) who clearly expressed his opinion on their true nature, that of fossil teeth of durophagous fish, comparing the fossils with the corresponding ones of present-day fishes. The Bellucci collection (National Archaeological Museum of Umbria, Perugia) preserves some amulets produced with these fossil ichthyodontoliths, once coveted prodigious objects.

Keywords Fossils · History · *Lepidotes* · *Sparus* · *Diplodus* · Toadstones · Snake eyes

© The Author(s), under exclusive license to Springer Nature Switzerland AG 2022

M. Forli, A. Guerrini, *The History of Fossils Over Centuries*,
https://doi.org/10.1007/978-3-031-04687-2_7

Similar to shark teeth, bony fish teeth have also had a prominent place in the history of exact fossil definition. In particular, the teeth of some durophagous fish, which, before their actual nature was understood, were for centuries believed to be stones derived from toads and considered, as we shall see, marvelous jewels with supernatural properties (Fig. 7.1). The fishes with bony skeleton, Osteichthyes, are distinguished in two great phyletic lines, the Actinopterygii, the group of the ray-finned fishes, and the Sarcopterygii, those with fleshy fins. Within the Actinopterygii, the group of the Teleostei, probably known since the Middle Trias, are the presently most common fishes, characterized, among other things, by having a particular outer covering, except for some bare-skinned forms, formed by bony laminar cutaneous structures contained in the dermis, the *scales*, called *cycloid* and *ctenoid* depending on whether they have a more or less rounded or posteriorly indented margin. Their most ancient representatives date back to the Lower Paleozoic, their systematic treatment and relevant evolutionary history are rather complex and still not totally clarified, and, at the present state of knowledge, the primitive fishes related to some forms of the aforementioned Sarcopterygii, through successive evolutions, even of great importance and diversification, are considered as the progenitors of the tetrapods, the first quadruped terrestrial vertebrates).

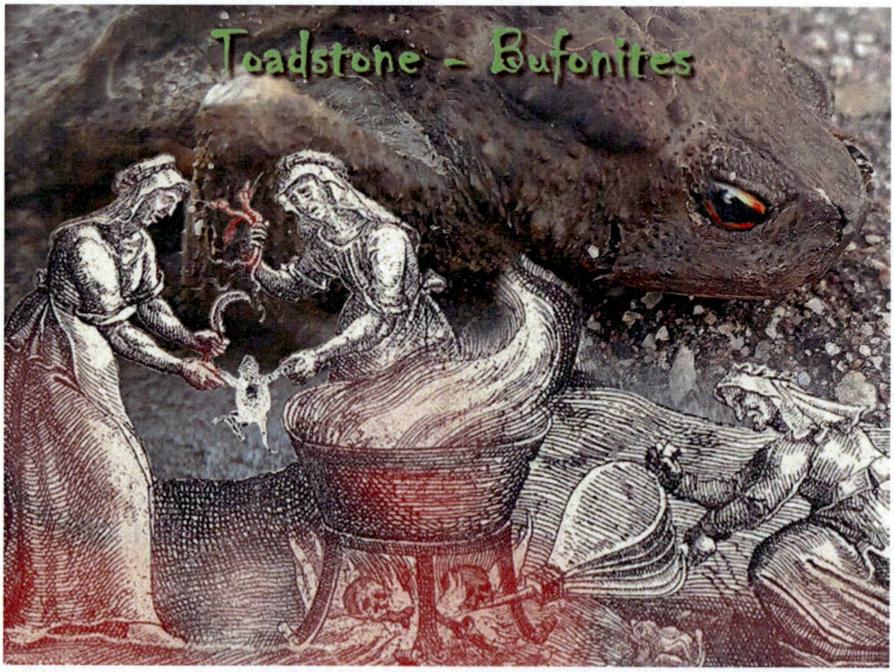

Fig. 7.1 Witches intent on preparing evil potions, using, among other ingredients, toads. (Modified after de Lancre, Tableau de l'Inconstance des mauvais Anges, 1612). [https://digital.library.cornell.edu/catalog/ss:550806 Rights: Presumed to be in the public domain because of creation date). (Photo credit Misa Forli, graphic composition Maurizio Forli)]

The bony fishes that interest us most as subjects of our researches are those referable to two groups of Actinopterygii: the oldest ones belong to the family Semionotidae and to the genus *Lepidotes* Agassiz, 1832 (Lower Jurassic-Lower Cretaceous) and the most recent ones to the family Sparidae (Early Paleogene to Present), to the genus *Sparus* Linnæus, 1758, and to the genus *Diplodus* Rafinesque, 1810, including respectively the present gilthead seabream and breams. Although of lesser importance for our purposes, another group of fossil actinopterygians is represented by the Pycnodontidae (Late Triassic-Middle Eocene, Taverne et al. 2019).

Lepidotes (*Lepidotes* spp.) with heavy *ganoid* scales, often rhomboidal in shape and juxtaposed, formed by a basal layer of bony lamella covered by dentin and above this by a shiny and compact substance similar to enamel, called ganoin, were widespread, with a wide geographic distribution, in the shallow seas and freshwaters of the Lower Jurassic (Toarcian) and Cretaceous. These fishes, massive and not particularly agile, had an average length of around 30–40 cm, even if in some cases they could exceed two meters in length. Representatives of the family Sparidae (Perciformes) occupy a wide range of natural marine environments in the Neritic province and are widespread in all temperate and tropical waters of the world, including the Mediterranean Sea. Some species, such as the White seabream, *Diplodus sargus* (Linnaeus 1758), and the gilthead seabream (*Sparus aurata* Linnæus 1758) are euryhaline, i.e., they tolerate wide variations in salinity, and are therefore also able to live in fresh and brackish waters, such as lagoons or estuaries.

Their finds, consisting mostly of isolated teeth (ichthyodontolites) or carbonate concretions of the inner ear (otoliths), are common and well known in the fossil state. In proportion, in the European Cenozoic and in general, the remains of articulated skeletons are rather scarce even if locally abundant and particularly aesthetically beautiful, such as those coming from the famous Eocene deposits of Bolca in the Veronese area. The abovementioned fishes, which share a similar dental morphology characterized by batteries of grinding teeth, are included from the trophic point of view in the category of the durophagous, that is, those which, being equipped with this particular type of teeth, have a diet consisting of organisms protected by a hard exoskeleton, mostly mollusks but also other benthic organisms. The following is the current traditional classification of two fishes, *Lepidotes maximus* and *Sparus aurata*, whose fossil ichthyodontolites have been known for a long time under the name of *bufonites*, *batrachites*, *borax* or *stones* or *toad eyes*, and *snake eyes* (Figs. 7.2, 7.3, 7.4, 7.5, and 7.6).

Phylum Chordata
Class Actinopterygii
Order Semionotiformes †
Family Semionotidae †
Genus *Lepidotes*
Species *Lepidotes maximus* Wagner, 1863 †

Phylum Chordata
Class Actinopterygii
Order Perciformes
Family Sparidae
Genus *Sparus*
Species *Sparus aurata* Linnæus, 1758

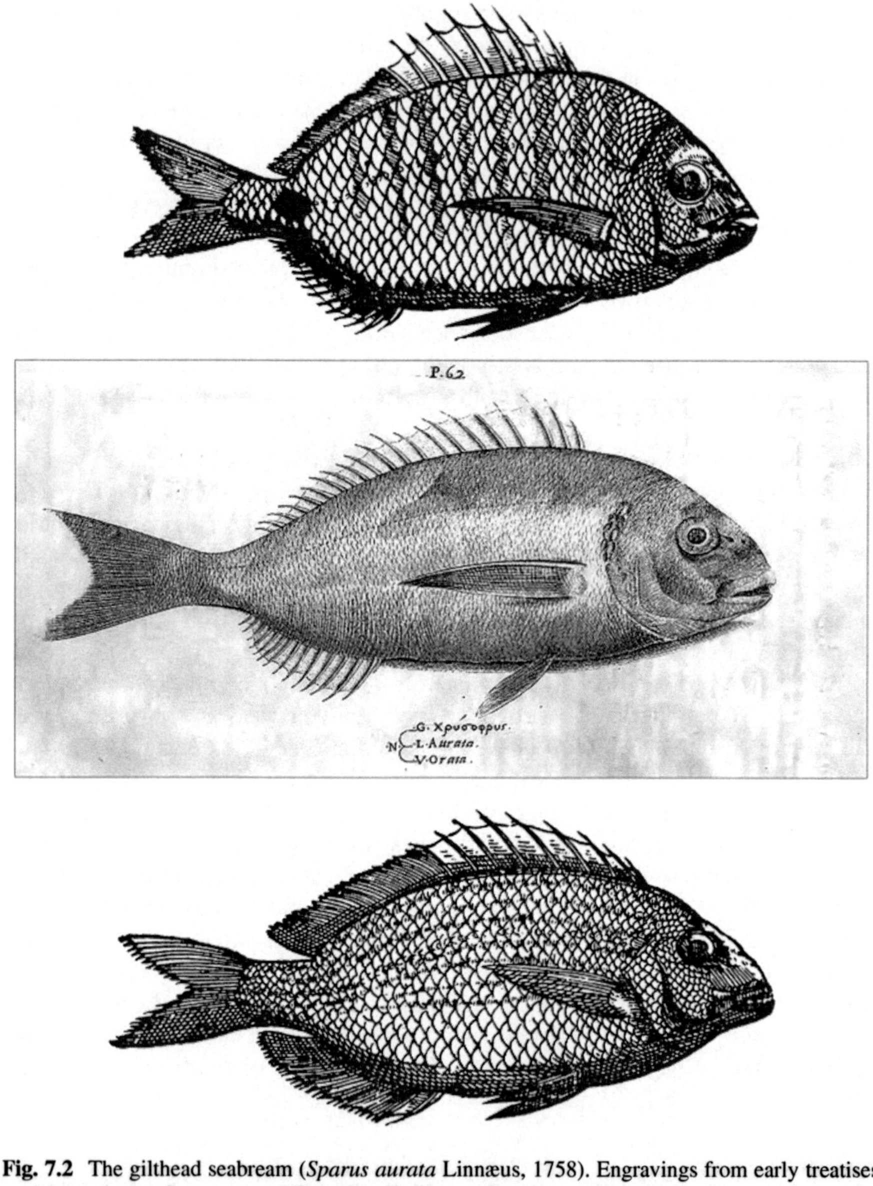

Fig. 7.2 The gilthead seabream (*Sparus aurata* Linnæus, 1758). Engravings from early treatises on ichthyofauna. Bottom, modified after Guillaume Rondelet (*Libri de piscibus marinis* 1554: 115); middle, modified after Salviani Ippolito (*Aquatilium Animalium Historiae* 1554: p. 174). Above, white seabream [*Diplodus sargus* (Linnaeus 1758)] modified after Rondelet (Ibid: p. 122). [Graphic composition Maurizio Forli. Credit from https://www.biodiversitylibrary.org/item/156187#page/5/mode/1up Public domain. https://www.biodiversitylibrary.org/bibliography/64229 (not in copyright)]

Fig. 7.3 Stoup of the church of San Leonardo di Roccasicura (Molise). On the bottom of the basin is carved a toad, representing the defeat of the devil. (Photo credit Franco Valente)

Fig. 7.4 The infernal punishment of gluttons: the souls of the greedy forced by demons to swallow toads. [La Gloutonnerie, Vision of Lazarus, Bellot, ~1497: Le kalendrier des bergiers (https://doi.org/10.3931/e-rara-88383 / Public Domain Mark)]

Fig. 7.5 Pair of toads dressed and adorned with bells, while dancing during the Sabbath. (Modified after de Plancy, *Dictionnaire Infernal*, 1863: p. 189) (Image color Maurizio Forli. Credit https://archive.org/details/bub_gb_DotbAAAAQAAJ Public Domain Mark 1.0)

Fig. 7.6 Exorcism scene with expulsion of a demon with anthropomorphic toad profile. From Pierre Boaistuau, *Histoires prodigieuses et mémorables: extraictes de plusieurs fameux autheurs grecs & latins, sacrez & profanes, divisées en six livres* (1575: p. 109) [https://archive.org/details/histoiresprodigi00boai (no copyright page found)]

The bufonites were considered small stones located in the head of toads, between the skin and the skull, which man could take directly or after they had been expelled independently by the animals, just before dying. They were considered wonderful objects with magical and therapeutic powers, used both as amulets and as medicine, whole or after being pulverized, were ingested in various forms, and were used against liver disease, tumors, and stings; in particular it was believed that they were able to cancel the poisoning. Set in rings they protected from these evils, and it is said that in the presence of poison, they warmed up denouncing its presence. In particular, the toadstones assumed these wide and particular beneficial properties, following the dual scheme common at the time, of similar and opposite functions, as they were produced directly from an animal whose symbolism extremely negative, referred directly to the devil, witches, the most infamous evil. The presence of the toad, inside the diabolic ceremonials, is diffused in different ages and civilizations since the antiquity, and the medieval iconography is strongly influenced by it, attributing to the frogs in general and in particular to the toads, often especially in the most ancient texts indistinctly indicated as batrachians, the most nefarious deadly properties. Their being particularly ugly and repulsive to human eyes made these harmless amphibians laid creatures, symbol of heresy, greed, of taking disgusting pleasure from sexual relations, symbol, and transmutation of diabolic spirits.

They are always depicted as unclean creatures, particularly associated with the sins of gluttony and lust, frequent being the mediaeval representations in which a woman or her sexual attributes are covered or devoured by a toad, and, if a direct personification of the devil, certainly to render him still more repulsive and disagreeable, endowed with bat-like wings and a serpentine tail. What could be better, to arouse loathing and fear, than to show Lucifer, and all infernal demons, as a hideous being born from the association of three of the most diabolical beasts of the medieval imagination, and not only, the toad, the bat, and the snake?

Even today some phrases that are part of the common way of saying derive from these ancient superstitions, and we use them without paying too much attention to their hidden meaning: *you are ugly as a toad*, *swallow the toad*, *spit the toad*, or *kiss the toad*. The reference to the ugliness of the toad is clearly associated with its identification with the devil; swallowing the toad means that we throw down an unpleasant, unwanted thing, and, not by chance in some images of hell, the damned are made to swallow by force, while spitting the toad means exactly the opposite, to free ourselves from an oppression, from something that gave us pain and discomfort, as sometimes this action is represented in depictions of exorcism, in which from the mouth of the possessed or more often possessed, toads come out, often with bat wings, to represent the supreme evil that is driven out and defeated. The same function of primacy of the Christian faith over the devil is also found explicitly in the custom of representing the toad at the bottom of the holy water fonts, in a sort of naturalistic but strongly symbolic representation.

Kissing the toad also has the same ambivalence as it can mean kissing the devil, the worst obscenity, or "Prince Charming," the ideal love. From these brief hints, we can realize how much in the collective imagination, the bufonites, surrounded by a mysterious aura, could be considered *powerful* and therefore sought after and desired (Figs. 7.7, 7.8, 7.9, and 7.10).

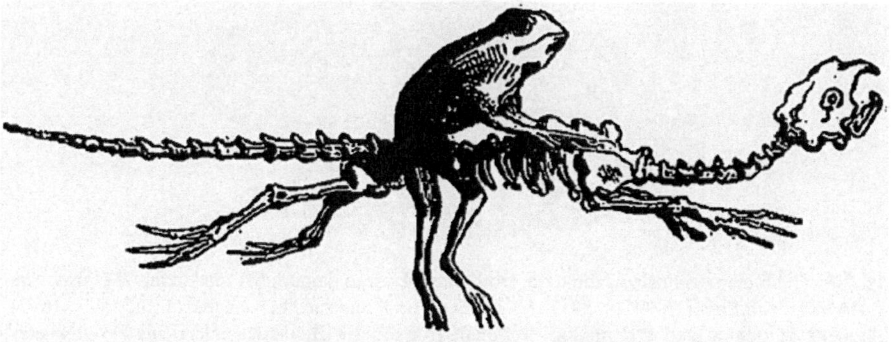

Fig. 7.7 Toad going to the Sabbath riding on the skeleton of a rat. (Modified after de Plancy, *Dictionnaire Infernal*, 1863: p. 189) (https://archive.org/details/bub_gb_DotbAAAAQAAJ Public Domain Mark 1.0)

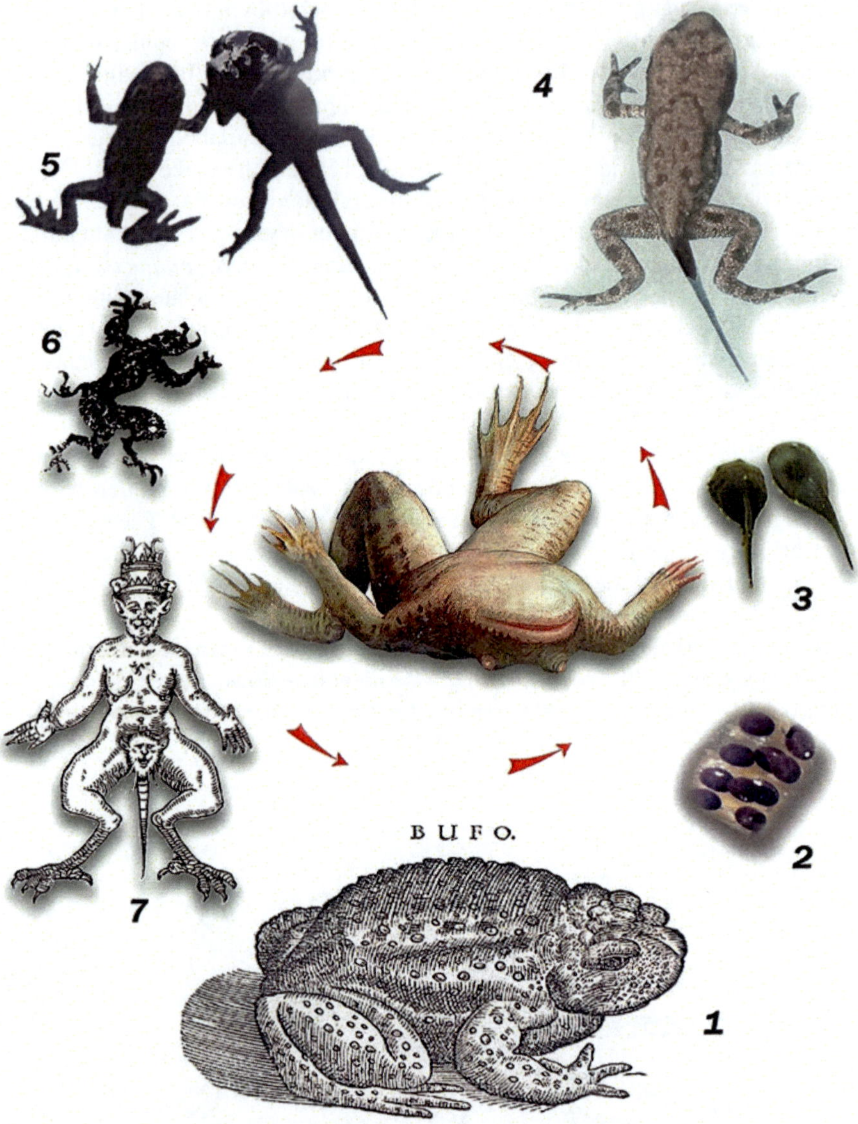

Fig. 7.8 The metamorphosis of the toad, from amphibian to demon: (1) toad, from *Il Dioscoride* by *Andrea Matthioli* (1674: p. 271). (2–5) Eggs and intermediate stages of tadpoles. (6–7) Silhouette of demon and a depiction of Satan; from Pierre Boaistuau, *Histoires Prodigieuses* (1575: p. 109; 1561: p. 1). Is it a coincidence noted by the ancient authors? In some cases, the anthropomorphization of the caudate, juvenile phases of the toads, lends itself more easily to the elaboration of demonic representations. In the center, a toad lying down, symbol of evil defeated and vanquished; detail from a painting by Bramantino, the Madonna of the Towers (1520~), where the batrace lies at the feet of the Madonna and St. Michael. [Graphic composition Maurizio Forli. Credit from (1) https://doi.org/10.3931/e-rara-19400 / Public Domain Mark (2–5) photo Misa Forli (6–7) https://archive.org/details/histoiresprodigi00boai (no copyright page found). In the center: Pinacoteca Ambrosiana from wikipedia.org/wiki/Madonna_delle_Torri (Public Domain)]

Fig. 7.9 Toads perched on the backs of the throne of the Queen of the Sabbath, and another witch, on either side of the Demon. (Detail, modified after de Lancre, *Tableau de l'Inconstance des mauvais Anges* 1612). (https://digital.library.cornell.edu/catalog/ss:550806 Rights: Presumed to be in the public domain because of creation date)

Fig. 7.10 Cornelius Saftleven, 1640: *Le sabbat des sorcières*. Detail with demonic toads [modified (artic.edu/artworks/53495/a-witches-sabbath) Public Domain]

It is again from Pliny the Elder, in book XXXVII of the already mentioned *Naturalis Historia*, that we draw news about the stone described as *Batrachites*:

> *Coptos et Batrachitas mittit: unam ranae similem colore: alteram ebeni: tertiam rubentis e nigro. Baptes, mollis alioqui, odore excellens. Beli oculus albicans pupillam cingit nigram, e medio aureo fulgore lucentem. Haec propter speciem sacratissimo Assyriorum deo dicatur* (Pliny, XXXVII)

The batrachites also come from Coptos: one has a color like that of a frog, a second is similar and also has veins, while a third is red mixed with black. The *baptes* smells excellent, but is otherwise an ordinary soft stone. The "Eye of Baal" has a whitish ground surrounding a dark eye which sends out a golden gleam from its midst. Because of its appearance, the stone is consecrated to the holiest god of the Assyrians.

The descriptions and illustrations in the first printed editions derived from the ancient lapidaries are almost always more or less similar to each other and often are only repropositions of the previous texts and figures. This appears evident in the works of the sixteenth-century authors, known or not, constituted almost always from repetitive compendiums; we must not therefore be surprised if the descriptions also of the toadstones are similar or are slavishly brought back identical from volume to volume of different authors, also to distance of decades the one from the other, often repeating even the errors of interpretation. An example is taken from the Agricola, in which are described under the same valence of toadstone also other lapidaries, borrowed from Pliny, instead sometimes considered distinct, called chelonites or balanites:

> *Le Chelonitidi furono cosi dette per havere con le Testudini somiglianza, perché di dentro sono concave: I Germani chiamano questa gemma di una voce composta di pietra, e di quella rana verde, che chiamano i Latini Rubeta: per che si tengono di certo, che ella ne la testa di questo animale velenoso nasca: alcune ne sono negre; alcune brune, e in qualche parte schizzate di bianco: le negre ne la convexità concava hanno à le volte goccie sanguigne e bianche; à le volte goccie auree: Tutte sono gonfiette in guisa d'uno occhio, ma non tutte concave. Et essendo d'una spetie istessa tutte, di rado sono maggiori di occhi di lupo, e spesso sono minori [...] una sola voce, significherà diverse spetie; come aviene ne le due Batrachite; che essendo amendue simile à l'una ne la figura; l'una ne ha il colore de l'una acerba; l'altra ne è negra. e come aviene ancho de le due Balanite: che l'una ne è alquanto verde, l'altra ha il colore del rame Corintho. La prima ci vien di Copto, la seconda da la contrada de' Trogloditi* [...] *l'una di colore ruffetto, che fregandosi gocciola fuori un sudore purpureo: l'altra è bianca, e ha vista di polve. Questo stesso si vede avenire ancho ne le tre Batrachite: la prima a colore di una ranonchia, onde il suo nome: la seconda l'ha di avorio: la terza è rossa pendente al negro: e queste due ultime ben che non habbiano il colore de la ranonchia, nondimeno per che pure in Copto nascono, dove nasce la prima: sono state chiamate Batrachite.* (De la Natura de le cose Fossili, e che sotto la terra si cavano. Lib. X. Traduzione in volgare dell'opera di Giorgio Agricola a cura di Michele Tramezzino 1550: pp. 307–310)

> *The Chelonites were so called because of their resemblance to the Testudines, because they are concave inside: the Germans call this gem a compound voice of stone, and of that green frog, which the Latins call Rubeta: so they are certain that it is born in the head of this poisonous animal: Some are black, some brown, and in some parts sprinkled with white: the black ones in the concave convexity have sometimes bloody and white drops; sometimes golden drops: all are swollen like an eye, but not all concave. And since they are all of the same sort they are rarely larger than wolf's eyes, and often smaller* [...] *a single voice will signify different sort; as happens with the two Batrachites; which, being both similar to one another in shape, one has the color of a unripe one, the other is black. And as also happens with the two*

Balanites: one is somewhat green, the other has the color of Corinthian copper. The first comes to us from Coptic, the second from the area of the Troglodytes [...] one is of reddish colour, which when rubbed drips out a purple sweat: the other is white, and has a powdery appearance. The same is also seen in the three Batrachites: the first has the colour of a frog, hence its name: the second has ivory: the third is red hanging down to black: and these last two, even if they do not have the colour of a frog, nevertheless because they are born in Coptic, where the first one is born: they have been called Batrachites.

The first representation of bufonite, while it is extracted from the head of a toad, is given in the *Hortus Sanitatis* of Johann Wonnecke von Kaub, otherwise known as Johannes de Cuba. The work, a herbarium devoted to medicinal plants, initially published in German from 1485 under the title *Gart der Gesundheit*, was later translated into Latin and other languages, enlarged and enriched by various parts concerning animals and precious stones. Among these, as in other more or less contemporary writings, said stone is called *Borax*. We reproduce the text from a French version (Cube 1501):

La vipere borax selon albert est ainsi nomee de aucun gére de crapault et est ainsi dicte e appellee pource qu'il la porte en sa teste e en est deux géres. L'une blache aucun pou noire. Et lautre si esse est extraicte du crapault estàt encores en vie et palpiant elle a au milleu vng oeil de couleur bleue. Et en noz teps en fut extraicte dug crapault une petite verte. Et aussi avons deu aucue crapaulx q avoiet en eusx les pierres depaintes qu'estoiet dictes de ce genre. Et vulgairement sont dictes crapauldines.

Camillo Leonardi (1502) in *Speculum Lapidum* describes the toadstones and the medicinal properties attributed to them (Fig. 7.11):

Fig. 7.11 The extraction of bufonites from the toad's head: Johann Wonnecke von Kaub, *Hortus Sanitatis* 1498 and 1501 (modified). [Credit on the left: https://commons.wikimedia.org/wiki/File:Borax-Kr%C3%B6tenstein1.jpg (Public Domain). On the right: https://doi.org/10.3931/e-rara-7477 / Public Domain Mark

Borax: Nosa. Crapondinus synonima sunt eiusde lapis qa buffone extrahit cuius duae sunt spés albus & hic melior ac raro reptus. [...] *In veneno mirabile virtute* [...]

The same thing is done by Lodovico Dolce, who a few decades later takes possession of the entire text of Leonardi, translating it into the vernacular (*Libri tre nei quali si tratta delle diverse sorti delle gemme che produce la Natura*, 1565: pp. 34–35):

Borace, Nosa, Crapondino, sono tutti nomi della stessa pietra, la quale si cava della botta. E sono di due sorti. Bianco, e questo è migliore, e si trova più di rado. L'altro è non nero fosco con color ceruleo, havendo nel mezo somiglianza di occhio: e dee esser cavato dalla botta, quando ella è morta, mentre che ancora è palpitante. E questi sono migliori di quelli, che si cavano dopo lunga dimora sotto la terra. Hanno mirabile virtù nel veleno. percio che se alcuno ha preso il veleno, inghiotta questo: il quale subito cinge l'intestini; et estingue la velenosa qualità impressa in essi intestini, et esce per le parti di dietro, e si conserva. Giova al morso delle biscie, toglie via la febbre: sana le male disposizioni dello stomaco e delle rene, se si piglia l'acqua, ove esso sia stato. Alcuni lo chiamano Sigesteto.

Borax, Nosa, Crapondino, are all names of the same stone, which is hollowed out of the toad. And they are of two kinds. White, and this is better, and is found more rarely. The other one is not black with a cerulean color, having in the middle the resemblance of an eye: and it must be extracted from the toad, when it is dead, while it is still throbbing. And these are better than those which are taken out after long residence under the earth. They have admirable virtue in poison. Therefore if anyone has taken poison, he swallows this: which immediately encircles the intestines; and extinguishes the poisonous quality impressed in the intestines, and comes out through the back parts, and is preserved. It helps at the bite of the snakes, it takes away the fever: it heals the bad dispositions of the stomach and of the kidneys, if you take the water, where it has been. Some call it Sigesteto.

Even in another pharmacological indication of Bonardo, a little later (*La Minera del Mondo*, 1585, 2: p. 21), the uses and effects of the *Pietra della Botta* were taken for granted:

Binding the stone of the Botta (toadstone) in a golden ring so that it touches the flesh of the finger, immediately, when the poison appears to her, it induces in that finger of the hand so much warmth that it seems really a burning coal, which burns it.

In Conrad Gessner (1565): *Batrachitae vel Crapodinae vulgò dicti lapis, à Bufonibus, quorum capitibus eos contineri persuasum est.*

Ordinary people call the stone Batrachite or Crapodina, which they believe is contained by the head of the Toad.

In addition to giving this concise definition, Gessner provides a brief list of the names used to describe the morphology in which these stones may occur, and illustrates four different types of which the elongated toadstone, corresponding to a tooth of *Coelodus*, a genus belonging to the Pycnodontidae, marked with the number four, is also called snake stone:

Quartus maior & oblongior est. Caeteros ferè Krottenstein, id est, Bufonum lapides nostri vocat. Quartum, qui rarior est, audio Schlangenstein oder grosser Krottenstein vocari, id est, Serpentium lapidem, tanquam Ophiten dicas.

The sense of the discourse is that the fourth elongated stone, besides being a toadstone, is by some called a snakestone like the ophiolites which are also called serpentine. *Krötenstein* and *Schlangenstein* are the words taken from the German language, corresponding to toadstone and snakestone, respectively. Today's interpretation of ophiolites is of course different and broader than Gessner's, but, as evidence of the use of calling some stones serpentine, the term serpentinites is still used to indicate a particular type of green metamorphic rocks, which, due to their graphic representation and color variations, resemble the skin of some snakes. Ferrante Imperato (*Dell'Historia Naturale*, 1599) also discusses toadstone, listing it among the exhibits in his museum. In order to verify the truthfulness about its origin, as he did not believe that it was growing between the skin and the skull of the batrachians, he did experiments keeping prisoners many toads, in order to see if some of them were really emitting, as everybody said, the magic stone (Accordi 1981). His interpretation is that they were however particular objects, with shape and color of small fungi, which were born in the rocks. In the writing of the text, he keeps however a certain balance between his opinion and the commonly considered one; in fact, even if openly manifesting his incredulity about the origin of the bufonites, he does not disdain to note their presumed medicinal properties:

> *La pietra di rospo, cosi detta dalla volgare opinione, che essa si ritroui nel capo del Rospo, animal che è spezie di Rana, nella figura è simile a testa di chiodo, couessa dalla parte soprana, e concua dalla sottana, di onde anco mostra il vestigio del suo picciuolo. L'istessa nella parte sua conuessa è di color leporino ammelato, e nella concaua liscia, e bianca. Sono communemente di figura circolare, ma se ne veggono anco in figura ouale, quantunque nel resto delle altre cose tutte simili alle circolari, nascono negli sassi à modo di piccoli fungi, à quali non solo di figura, ma di colore anco somigliano. è stimata detta pietra antidoto contra veleni, & à romper le pietre de reni, e scaccia il rigor delle febri lunghe*. (1599: p. 661)

> *The toadstone, so called by the vulgar opinion that it is found in the head of the Toad, an animal which is a sort of the Frog, is similar in shape to the head of a nail, with the upper part of the head and the lower part of the underbelly, from which it also shows the vestige of its petiole. The same in its convex part is of a softened hare colour, and in the concave part smooth and white. They are commonly circular in shape, but they are also seen in the shape of a circle, although in the rest of the other things they are all similar to the circular ones, they are born in the stones in the manner of small fungi, to which not only in shape, but also in colour they resemble. It is esteemed that this stone is an antidote to poisons, and to break the stones of the kidneys, and chases away the rigor of long fevers.*

His son Francesco published in 1628 a work to comment on his father's one, giving news and further descriptions of the objects in the family museum, entitled *Discorsi intorno a diverse cose naturali*, where the fourth chapter is dedicated to the observations and explanations on the *toadstone*. In it he contests the authors, contemporary to his father and to himself, who continue to spread the belief that these stones are extracted from the head of the toad in the most bizarre ways and times, concluding with a statement, accurate from the point of view of scientific method, which seems to us also funny:

La pietra volgarmente detta dal Rospo, che dentro la terra si ritrova, è stata da mio Padre nella sua Historia naturale depinta, e brevemente descritta; ove con ragione si è dimostrato alieno dall'opinione d'alcuni che reputan detta pietra ritrovandosi nel capo del Rospo; [...] *che il Rospo debbia esser vecchio* [...] *& anco doversi battere; che in questo modo vomita la pietra; il che appare chiaramente esser falso e fabuloso;* [...] *doversi detta pietra cavar nel mese d'Agosto; altri à Luna mancante, altri nel giorno canicolare; altri che il Rospo doppo essere ammazzato, si debbia riporre in luoghi, dove son formiche, che al fin frà le ossa si ritrovarà anco la pietra; nè ripugna all'opinion di mio Padre la diversità di forme, e colori di dette pietre; atteso ne habbiamo molte fossili dell'istessi colori, & forme, da gli Autori falsamente attribuite à quelle, che dicono cavarsi dalli capi de Rospi, anzi quì vi son stati più volte venditori di gemme, e pietre, che di quelle ne han condotto migliaia, e le vendevano à vil prezzo; onde per la detta lor quantità non si possono riputare esser cavate dalli capi de Rospi, mentre gli Autori assegnano tante difficoltà, e circostanze nel cavarsi.* [...] *hà mio Padre preso ogni fatica, & usata esattissima diligenza per ritrovare e cavar detta pietra dalli Rospi; e con osservar la verità de tempi, e staggioni, moti lunari, diversità di Rospi, e di luoghi, ove quelli si ritrovano, non hà giamai potuto giongere al suo intento; per il che mi bisogna concludere, ò che li Rospi che si ritrovano in questo regno di Napoli non generano pietre, o che son fabulose.*

The stone vulgarly called the Toad stone, which is found inside the earth, has been described by my Father in his Historia naturale; where he has rightly shown himself to be opposed to the opinion of some who believe that this stone is found in the head of the Toad; [...] *that the Toad must be old* [...] *& also must be beaten; that in this way it vomits the stone; which clearly appears to be false and fraudulent;* [...] *that the said stone must be quarried in the month of August; others when the moon is absent, others on the cyclical day; others that after being killed, the Toad must be placed in places where they are formed.*[...] *the said stone must be quarried in the month of August; others in the absence of the Moon, others on the canicular day; others that the Toad, after being killed, must be placed in places where there are ants, so that at the end, between the bones, the stone will also be found; nor does the diversity of forms and colours of the said stones repel the opinion of my Father; since we have many fossils of the same colours and forms, attributed by the Authors falsely, to the same Toad. shapes, falsely attributed by the Authors to those, which they say to be extracted from the heads of the Toads, indeed here there have been many times gems and stones sellers, who have brought thousands of them, and sold them at a low price; so that for the said quantity they cannot be considered to be extracted from the heads of the Toads, while the Authors attribute so many difficulties, and circumstances in extracting them.* [...] *my Father has taken every effort, & used the most exact diligence to find & extract the said stone from the Toads; & by observing the truth of the times, & seasons, & lunar motions, & diversity of the Toads, & of the places, where they are found, he has never been able to reach his goal; for which I must conclude, or that the Toads found in this kingdom of Naples do not generate stones, or these are fabulous stories..*

In Aldrovandi's *Musaeum Metallicum*, there are two references and several illustrations: *Bufonites, seu Batrachites* (1648: p. 814; figs 1–3, 7) corresponding to three teeth of *Coelodus* sp. (figs 1–3) and one of *Lepidotus gigas* Agassiz, 1832 (fig 7). The teeth of *Coelodus were* worked as if they were gems, being partly flattened and faceted (Sarti 2003). The same disbelief on the origin of the toadstone is clearly manifested also by the nurse friar Felice Passera (1688) who, in his book *Il nuovo tesoro degli arcani farmacologici* (*The new treasure of the pharmacological arcana*), expresses himself in this way about the Toadstone or Zatto:

It is truly a great and universal fable, not only among the opinions of men, but also among authors, that the Toad has this stone in its body, which, placing itself in a cage with a red cloth underneath, vomits it out, and then returns it, when it can, to be swallowed, and I have seen this experience with many and many large Toads, and it was not true. An author, who treats this matter, says that he has seen it by chance in the same living Toad at the top of his head covered with green skin, which is hollowed out and polished, and is clearly different from the skin described below, called Vertigo, since this appears to be especially below bone with its veins, and is full, and on the surface has certain porous spots, like the stellar; and which are similar to the spots on the skin of the Animal.

At the same time, however, like other authors, it reports the alleged medicinal virtues, not excluding a priori its effectiveness:

Many Authors, both ancient and modern, attribute to it virtue against all poisons, carrying it in the left arm, tied up, touching the flesh, in the presence of the poison it cooks, & burns, as Michel Mercato says, and having taken the poison immediately swallow, so whole, this Stone; because it exceeds the strength of the poison and then it is rendered by digestion, and is recovered, and acquires new strength. [...] *with the contact, and lightly rubbed on it, dissolves the inflation, and inflammation made by venomous Beasts, and discards all malignity. Therefore, if Mouse, Spider, Wasp, Beetle, Gira, or other poisoned animal offends any part, when this stone is placed on it, it mitigates the pain and stops all inflammation.* [...] *They report that in the presence of a poisoned Bichier, this stone changes colour and sweats.*

It is evident from a comparison with the Leonardi text, which is almost 200 years more antique, how, although the same convictions persist, they are no longer accepted uncritically. On the contrary one can say that they are subjected to some experimental verification that makes one doubt, for some, their validity. Also Passera as Francesco Imperato gives account of the wide commerce of *false* toadstones, in truth very abundant on the contrary of the original and "unobtainable" Toad Stones, calling them however *Stones of the Vertigo*, attributing to them some medical powers, and giving us news on the modalities of their collection and on the places where they were found:

Da pochi anni in qua si sono sparse, quasi per tutta l'Europa; mà più per l'Italia, e per la Francia, certe Pietre volgarmente chiamate del Rospo. Quantunque non sia il vero, le quali portansi contra la Vertigine, & il mal di fianco, & contra li accidenti di cuore. Questa è di varij colori: mà per lo più del color della Nespola [...] *Altre poco; mà del tutto nere. Altre biancheggiano, e formano come un'occhio nel mezzo. Alle volte è cinto d'un cerchio nero, o bianco sottilissimo, & altre volte rosseggiante. La forma sua è come una meza nocciuola perfettamente tonda, come meza balla d'archibugio. Altre volte è come un lupino.* [...] *La sua sostanza è come di vetro, durissima rilucente, & imbrunita dalla natura* [...] *nasce negli scogli d'alcuni Mari di Francia, di Sardegna, e di Maiorica, e forse altroue ch'io non hò saputo. In quella parte dello scoglio, o ripa, doue percote il Mare con l'onde si ritroua un sasso bianco, come calcinaccio, & assai duro; ma fragile, o frangibile in cui percotendo li huomini pratichi con mazze ferrate, & martelli grossi, né schiacciano, e cavano gran pezzi, quali minuzzando, & rompendo in pezzetti ritrovano dentro queste Pietre. Appunto come si ritrovano l'ostiche, chiamate dattili, dentro à sassi delli scogli, nel fondo del Mare; E sono come nodi nelle tavole, più duri, e più lustri, o come stelle nel Cielo.*

Since a few years ago, certain stones, vulgarly called Toad Stones, have been scattered almost all over Europe, but more so in Italy and France. Although it is not true, they are used against Vertigo, and the pain of the flank (colic of kidneys), and against the accidents of the heart. This one is of various colours: but for the most part of the colour of Medlar [...]

Others are little; but entirely black. Others are white, and form an eye in the middle. Sometimes it is surrounded by a black circle, or a very thin white one, and sometimes reddish. Its shape is like a perfectly round half-hazel, like a half-bale of an arquebus. At other times it is like a lupine. [...] *Its substance is like glass, very hard, shining, and darkened by nature* [...] *it is born in the rocks of some Seas of France, Sardinia, and Majorica, and perhaps other which I have not known. In that part of the cliff, or bank, where it strikes the Sea with the waves, a white stone is found, like rubble, very hard; but fragile, or breakable, in which, when men strike it with mallets, & large hammers, they crush it, & hollow out great pieces, which, by breaking it into small pieces, they find inside these Stones. Just as the oysters ones, called dactyls, are found in the stones of the rocks at the bottom of the sea: and they are like knots in the boards, harder and shinier, or like stars in the sky.*

As it can be deduced from the text, the author involuntarily makes us aware of the methods of search of these fossils, of the places of origin, and of the rocks that contain them, allowing us to understand in the light of the actual knowledge that the *Stones of Vertigo*, that is, the fossil ichthyodontolites, were and are still contained in white calcareous rocks and that today we know of Miocene age, are overlooked or come out directly from the sea because together with the fossils, the *dactyls* were found, that is, the mollusks *Lithophaga lithophaga* (Linnæus, 1758) that exactly live embedded in the rock, from the level of the sea up to about 20 meters of depth, as it happens for instance in Sardinia. Nowadays, the collection of the fossils as well as of the sea dates is forbidden; in particular, for what these last ones are concerned, seen the seriousness of the effects caused by the destruction of the rocks where they nest, which cause serious damages to the marine ecosystem, it is forbidden, besides the fishing, also the only commercialization.

Another imaginative way to indicate the circular teeth of fossil sparids, in addition to the more common *toad's eyes* and *snake's eyes*, is the name of *cat's eyes*, usually used to specify instead of agate or jasper stones, reported, for example, in the pages of the Cospiano Museum:

fossils, such as those, which are dense in different pieces of white Maltese soil, which has the surface from place to place sandy [...] *among which are distinguished many heterogeneous parts* [...] *shells, and round stones, similar to those that are called Cat's Eyes, for the similarity, which have with light of eyes of that Animal.* (Legati 1677: p. 76)

These presumed toadstones, mostly consisting of the fossilized teeth of White seabream and gilthead seabream, being smaller in size and much more abundant than the "authentic" toadstones, mostly teeth of *Lepidotus*, were offered and sold at lower prices than those of the larger stones which were the prerogative of the richer social classes or the nobility. As for the glossopetrae, those coming from the Maltese islands, where these fossil remains are very common, were particularly known and appreciated. Geologically speaking, the Maltese archipelago is made up almost exclusively of white Miocene limestone, rich in fossil remains, in particular echinids, shark teeth, and bony fish, the latter two being closely linked to the mythical-religious rituals linked to the cult of St. Paul the Apostle and to his antivenom power, in particular against snakes but also against poisoning in general. Agostino Scilla, the painter, sculptor, and naturalist from Messina, was responsible for the recognition and explanation, particularly precise and documented, of their exact nature:

Sono i sassolini, chiamati occhi di serpe, apertissimi denti di pesci. [...] *Dirò prima, che di simili pietre, insieme con Glossopietre, se ne trova grandissima quantità nella Sicilia, e in particolare in Corleone. Da ciò si deve comprendere, che gli occhi di serpe non sono in Malta solamente generati per miracolo, o per ispeziale virtù del terreno puro, ed alessifarmaco. Dico questo per cagione, ch'io ne conservo appresso di me di quei di Corleone molti, e molti, che in un tufo forte, ma renoso, ed impuro, anzi noioso d'odore sono involti, secondochè il caso l'accompagnò, con alcune Glossopietre, e con molte sporcizie; onde non sono eglino, benchè similissimi di corpo, compariscenti, e coloriti, a guisa di quelli di Malta, ma cinericci, neri, e spesso macchiati. La diversità de' colori poco monta; sono così questi, come codesti, denti de' pesci, Sarco, Orata, Dentici ed infiniti altri simili, le spezie de' quali, dopo de' primi denti, che terminano có l'estremità della bocca, furono dotate dalla Natura d'una copia grandissima di denti con bell'ordine schierati, e diffusi dentro di essa bocca, così nella parte di sopra, come nella di sotto.* [...] *Non mancherò per tanto di ridurre a veduta alcune loro ganasce spolpate .a. per paragonare i loro denti con gl'impietrati, che pur disegnerò più sotto in quella maggiore, e varia quantità, che m'è stato possibile raccogliere, per via di molti amici, da Malta; e questo non solamente per espressione della mia osservazione volentieri farò, ma anche per dare campo co' disegni ad ogn'occhio di riscontrarne l'istessità delle parti, e per poterne far conoscere, cioè, che di quelle pietre, volgarmente dette occhi di serpe, furono un tempo denti, e parti della bocca de' sarchi, Dentici, Orate, e simili, che in molto numero, e varietà per tutti i mari nuotano e si pescano.* (*La vana speculazione disingannata dal senso*. Scilla 1670: pp. 66–68)

These are the pebbles, called snake eyes, very open fish teeth. [...] *I will say first that of such stones, together with glossopetres, there is a great quantity in Sicily, and in particular in Corleone. From this it must be understood that the snake eyes are not only generated in Malta by a miracle, or by special virtue of the pure soil, and counterpoisons. I say this for the reason that I have many of those of Corleone, and many, which in a strong tuff, but sandy, and impure, even boring in smell, are wrapped, according to what the case accompanied it, with some Glossopetres, and with a lot of dirt; so that they are not they, although very similar in body, compariscent, and coloured, like those of Malta, but ashy, black, and often stained. The diversity of colors does not matter much; they are so these, as well as these, teeth of fishes, Sarcus, Gilthead, Dentex and countless other similar ones, the species of which, after the first teeth, which end with the extremity of the mouth, were endowed by Nature with a very large number of teeth with a beautiful order, arranged and spread inside the mouth, so in the upper part, as well as in the lower part.* [...] *I will not fail, therefore, to reduce to view some of their stripped jaws ... in order to compare their teeth with the petrified ones, which I will draw below in the greater and more varied quantity that I have been able to collect, through many friends, from Malta; and this not only as an expression of my observation I will gladly do, but also in order to give scope with the drawings to each eye to ascertain the same parts, and to be able to make known, that is, that of those stones, vulgarly called snake eyes, were once teeth, and parts of the mouth of the Sea breams, dentexes, giltheads, and the like, which in great number and variety throughout the seas swim and are fished.* (*Vain Speculation Undeceived by Sense*. Scilla 1670: pp. 66–68)

The work of Scilla, whose first edition in the vernacular, was not very well known, was reprinted in Latin in 1747, thus reaching a wider circulation among the European scholars of that time, who often continued to pass on, albeit more and more dubiously, the hypothetical medicinal properties of those stones (Figs. 7.12, 7.13, 7.14, 7.15, 7.16, 7.17, 7.18, 7.19, and 7.20).

Fig. 7.12 A craftsman intent on mounting toadstones on rings [*Hortus Sanitatis* 1501 (modified from https://doi.org/10.3931/e-rara-7477 / Public Domain Mark)]. Amulet from the Bellucci collection, MANU (PG). (Inv. 7815 B). Brass ring with a notched bezel fixing a palatal tooth of *Lepidotes*. Entered in the collection in 1916, it comes from the town of Gubbio (Perugia). Such rings were used to protect from the effects of poisonous substances (snake's eye, 20 × 22 × 10). (Graphic composition Maurizio Forli)

Fig. 7.13 Ichthyodontolites from the Jurassic of the central Apennines, probably teeth of *Lepidotes* sp. (average diameter ~ 10 mm). (Photo Maurizio Forli)

Fig. 7.14 Agostino Scilla (1759, *De corporibus marinis lapidescentibus quæ defossa reperiuntur. Addita Dissertatione Fabii Columnæ*: pl. II). (I, III), Jaws of fishes called by the people Sarco (I, III), *Diplodus sargus* (Linnaeus 1758), Orata (II] *Sparus aurata* Linnæus, 1758), and Dentato [(IIII) *Pagrus pagrus* (Linnaeus 1758)]; V)

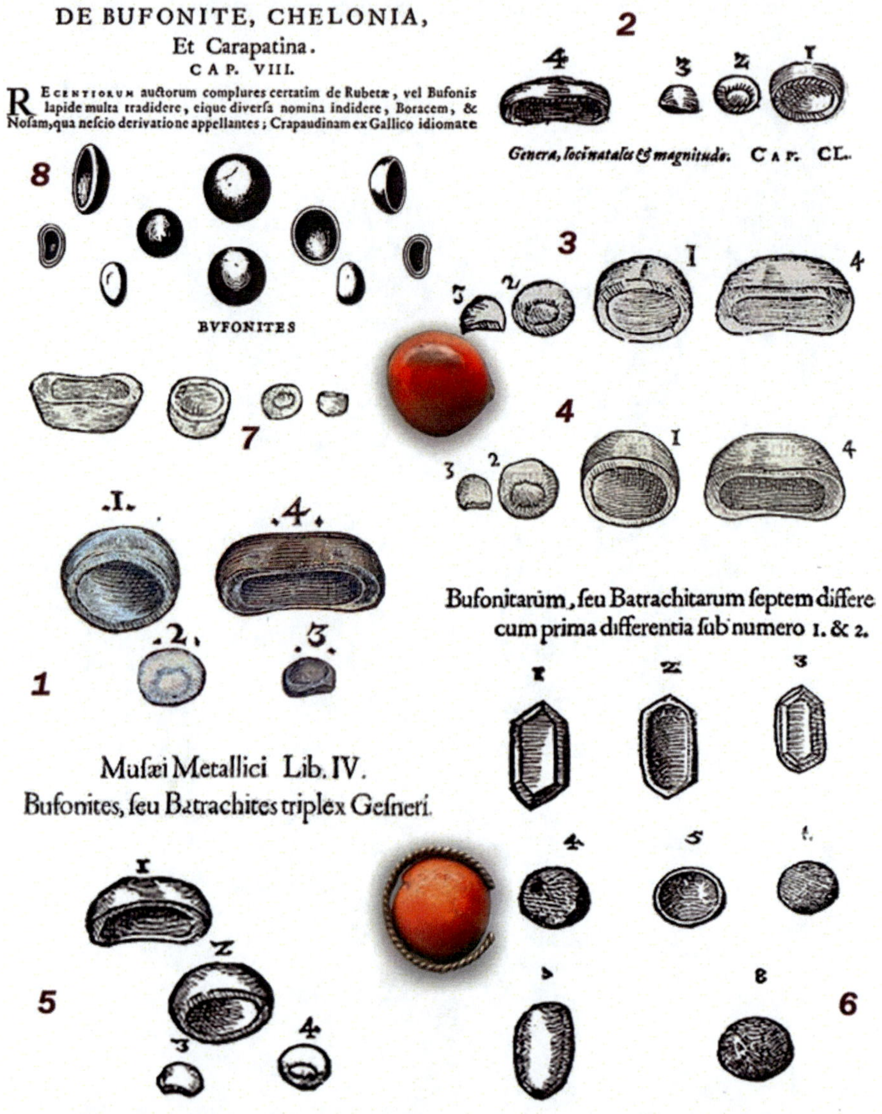

Fig. 7.15 Depictions of bufonites, batrachites, chelonites, and carapatinae: (1) Gessner (*De Omni rerum fossilium*, 1565: p. 161). (2–3) De Boodt (*Gemmarum lapidae Hist.*, 1609: 152; 1636: 301). (4) (*Le Parfaict Joaillier*, 1644: p. 386). (5–6) Aldrovandi (*Musaeum Metallicum*, 1648: pp. 813–814). (7) Worm (*Museum Wormianum*, 1655: p. 107). (8) Mercati (*Metallotheca Vaticana*, 1717: p. 184). Among the so-called Carapatine, distinguished by the Mercati, are also included other teeth, pointed cylindrical, similar to human nails, from the Balearic Islands and Sardinia (*Metallotheca Vaticana*, 1717: pp. 336–338). In the center, fossil teeth of *Lepidotes* forming part of the contents of the bag of a sanpaularo (Coll. Bellucci, Inv. 338 G). [Graphic composition Maurizio Forli. Credit from (1) https://doi.org/10.3931/e-rara-4176 / Public Domain Mark (2–3) https://hdl.handle.net/2027/gri.ark:/13960/t8w96041q (Public domain) (4) https://doi.org/10.3931/e-rara-12490 / Public Domain Mark (5–6) https://doi.org/10.3931/e-rara-13017 / Public Domain Mark (7) https://doi.org/10.5962/bhl.title.63844 (not in copyright) (8) https://archive.org/details/bub_gb_DC_E6UVWg7EC (Public Domain Mark 1.0) Coll. Bellucci, reproduced by kind permission of the Ministry of Cultural Heritage and Activities and Tourism, Regional Directorate Museums of Umbria – National Archeological Museum of Umbria]

Fig. 7.16 Toadstone, ichthyodontolites embedded in the rock matrix and loose. Ferrante Imperato (***Historia Naturale***, 1672: p. **577**). [https://www.biodiversitylibrary.org/item/183177#page/5/mode/1up **(not in copyright)]**

Fig. 7.17 Wooden matrix used for the printing of the relevant page of the *Musaeum Metallicum* (1648: p. 814), with some original findings, still preserved in the Museum of Palazzo Poggi in Bologna. (Photo Maurizio Forli, permission *Sistema Museale di Ateneo Alma Mater Studiorum*—Università di Bologna)

Fig. 7.18 Fake toad fossils. Along with many others, these finds were carved and then hidden in the places where Dr. Beringer, Professor of Natural History at the University of Würzburg, used to take his students for research. [Modified after Beringer, *Lithographia Wirceburgensis*, 1767: p l. XII. (https://www.biodiversitylibrary.org/bibliography/150131)]

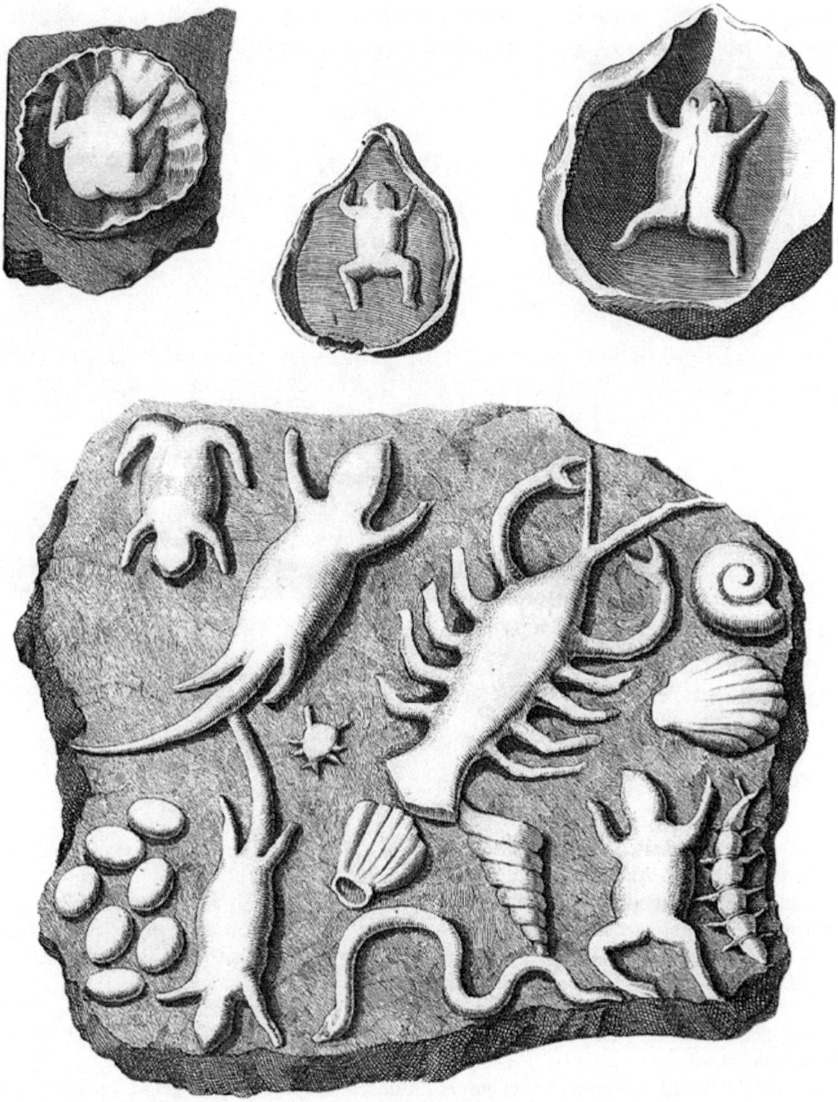

Fig. 7.19 An advocate of the inorganic origin of fossils, which he considered to be freaks of nature, Beringer, misled by carved rocks buried in the place where he was carrying out research, illustrated and described these forgeries in *Lithographia Wirceburgensis* (1726). Of course the diabolical joke, set up and conducted by two of his students, came to light and caused, of course, resentment and court cases. In the meantime, however, the book, for its worth, circulated around Europe and, despite the author tried to withdraw as many copies as possible, it was also reprinted in a second edition, from which we have taken the illustrations (1767: pls I–V (modified). [https://www.biodiversitylibrary.org/bibliography/150131 (Public Domain)]

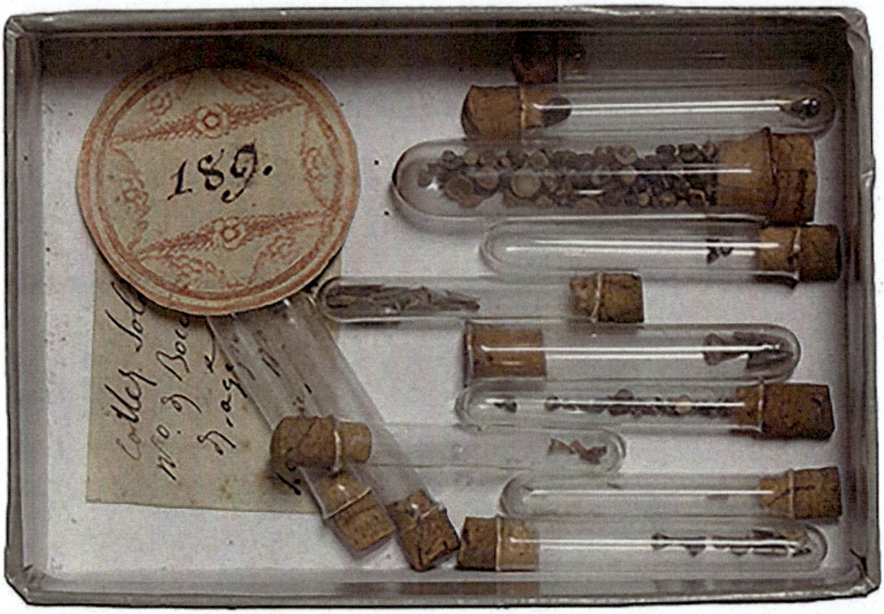

Fig. 7.20 Test tubes with fossil teeth from material in the Soldani collection discussed in his work (1780) (preserved in the Natural History Museum of University of Florence, Section of Geology and Paleontology). (Photo credit Stefano Dominici)

References

Accordi B (1981) Ferrante Imperato (Napoli, 1550–1625) e il suo contributo alla storia della geologia. Geol Romana 20:43–56

Aldrovandi U (1648) Ulyssi Aldrovandi patricii Bononiensis Musaeum metallicum in libros IIII : distributum Bartholomaeus Ambrosinus ... labore, et studio composuit cum indice copisosissimo Arcus Antonius Bernia. typis Io. Baptistae Ferronii, Bononiae. ETH-Bibliothek Zürich, Rar 1027, https://doi.org/10.3931/e-rara-13017 / Public Domain Mark

Bellot J (1497) Le kalendrier des bergiers. Imprime a genesue : [Jean Bellot], [nicht vor 1497]. Universitätsbibliothek Bern, MUE Inc I 84: 1, https://doi.org/10.3931/e-rara-88383 / Public Domain Mark

Beringer JBA (1726) Lithographiæ Wirceburgensis, ducentis lapidum figuratorum, a potiori insectiformium, prodigiosis, imaginibus exornatae. Marci Antonii Engmann, Würzburg

Beringer JBA (1767) Lithographiæ Wirceburgensis, ducentis lapidum figuratorum, a potiori insectiformium, prodigiosis, imaginibus exornata. Tobiam Goebhardt, Francofurti et Leipzig. https://www.biodiversitylibrary.org/bibliography/150131 (Public Domain)

Boaistuau P (1575) Histoires prodigieuses, extraictes de plusieurs fameux autheurs Grecz & Latins, sacrez & prophanes: mise en nostre langue par P. Boaistuau, surnomé Launay, natif de Bretaigne: avec les pourtraicts & figures. Chez Charles Macé à l'enseigne de la Pyramide, Paris

Boodt A Boëtius de (1609) Gemmarum et lapidum Historia. Claudiam Marnium, Hanoviae. https://hdl.handle.net/2027/gri.ark:/13960/t8w96041q (Public domain)

Boodt A Boëtius de (1636) Gemmarum et lapidum historia. ex officina Ioannis Maire, Lugduni Batavorum [Leiden]. ETH-Bibliothek Zürich, Rar 1881, https://doi.org/10.3931/e-rara-15217 / Public Domain Mark

Boodt A Boëtius de (1644) Le parfaict Ioaillier ou histoire des pierreries: ou sont amplement descrites leur naissance, iuste prix, moyen de les cognoistre, et se garder des contrefaictes, Facultez medicinales, et proprietez curieuses. Iean-Antoine Huguetan, Lyon. ETH-Bibliothek Zürich, Rar 1010, https://doi.org/10.3931/e-rara-12490 / Public Domain Mark

Bonardo GM (1585) La minera del mondo dell'ill. sig. Gio. Maria Bonardo Fratteggiano, conte e cavaliero. Nella qual si tratta delle cose più secrete, e più rare de' corpi semplici nel mondo elementare, e de' corpi composti, inanimati, et animati d'anima vegetativa, sensitiva, e ragionevole. Divisa in quattro libri, & per maggior commodità de' lettori ogni capitolo, c'ha potuto riceverlo, ordinato, con ordine d'alfabeto. Mandato in luce da Luigi Grotto, cieco d'Hadria. Venetia, Fabio, et Agostin Zoppin Fratelli

Camilli L (1502) Speculum Lapidum et D. Petri Arlensis de scudalupis, presbyteri hierosolymitani, Sympathia, septem metallorum ac septem selectorum lapidum ad planetas accredit Magia Astrologica Petri Constantii Albinii villanovensis. Apud Christianum Liebezeit, Hamburgi

Cube J von (1501) Ortus sanitatis. Nouvelleme[n]t imprimé à Paris : pour Anthoine Vérard, [vers 1501?]. BPU Neuchâtel, A.F. B 10, https://doi.org/10.3931/e-rara-7477 / Public Domain Mark

de Lancre P (1612) Tableau de l'inconstance des mauvais anges et démons, où il est amplement traicté des sorciers et de la sorcellerie... Chez Jean Berjon, Paris. http://numelyo.bm-lyon.fr/BML:BML_00GOO0100137001100737043

de Plancy JASC (1863) Dictionnaire infernal: répertoire universel des êtres, des personnages, des livres... qui tiennent aux esprits, aux demons, aux sorciers, au commerce de l'enfer, aux divinations, aux maléfices, a la cabale et aux autres sciences occultes, aux prodigies, aux impostures, aux superstitions diverses et aux prognostics, aux fait actuels du spiritisme, et généralement a toutes les fausses croyances merveilleuses, surprenantes, mysterieuses et surnaturelles, 6e éd. par J. Collin de Plancy. https://archive.org/details/bub_gb_DotbAAAAQAAJ Public Domain Mark 1.0

Dolce L (1565) Libri tre ne i quali si tratta delle diverse sorti delle Gemme, che produce la Natura, della qualità, grandezza, bellezza, et virtù loro. Venetia, Gio Battista, Marchio Sessa, et Fratelli

Gessner C (1565) De omni rerum fossilium genere, gemmis, lapidibus, metallis, et huiusmodi, libri aliquot, plerique nunc primum editi. Tiguri. https://doi.org/10.3931/e-rara-4176 / Public Domain Mark

Imperato F (1599) Dell'Historia Naturale di Ferrante Imperato napolitano. Libri XXVIII. Nella quale ordinatamente di tratta della diversa condition di miniere, e pietre. Con alcune historie di Piante, et Animali fin'ora non date in luce. Costantino Vitale, Napoli. https://www.biodiversitylibrary.org/item/183177#page/5/mode/1up (not in copyright)

Imperato F (1672) Historia naturale di Ferrante Imperato napolitano. Nella quale ordinatamente si tratta della diversa condition di minere, pietre pretiose, & altre curiosità. [...]. Venetia : presso Combi & La Noù, 1672. ETH-Bibliothek Zürich, Rar 607, https://doi.org/10.3931/e--rara-11175 / Public Domain Mark

Imperato F (1628) Discorsi intorno a diverse cose naturali. Opera non meno curiosa, che utile, e necessaria à Professori della natural Filosofia. Egidio Longo, Napoli

Legati L (1677) Museo Cospiano annesso a quello del famoso Ulisse Aldrovandi e donato alla sua Patria dall'Illustrissimo Signor Ferdinando Cospi. Giacomo Monti, Bologna. https://babel.hathitrust.org/cgi/pt?id=gri.ark:/13960/t7dr4608w&view=1up&seq=30&skin=2021 (Public Domain Mark)

Mattioli MP (1674) Opera quae extant omnia, hoc est, commentarii in VI. libros Pedacii Dioscoridis Anazarbei de medica materia, adjectis in margine [...]. sumptibus Ioannis König, Basileae. ETH-Bibliothek Zürich, Rar 3536, https://doi.org/10.3931/e-rara-19400 / Public Domain Mark

Passera F (1688) Il Nuovo Tesoro degli Arcani Farmacologici, Galenici. & Chimici, ò Spargirici. Giovanni Parè, Venetia

Rondelet G (1554) Libri de Piscibus Marinis, in quibus verae Piscium effigies expressae sunt. Matthiam Bonhomme, Lugduni. https://www.biodiversitylibrary.org/bibliography/64229 (not in copyright)

Salviani I (1554) Aquatilium animalium historiae, liber primus : cum eorumdem formis, aere excusis. https://www.biodiversitylibrary.org/item/156187#page/5/mode/1up Public domain

Sarti C (2003) Le Collezioni di geologia nel Museo di Ulisse Aldrovandi. In: Vai GB, Cavazza W (eds) Quadricentenario della parola Geologia. Ulisse Aldrovandi 1603 Bologna, Minerva, p 152–167

Scilla A (1670) La vana speculazione disingannata dal senso. Lettera responsiva Circa i corpi marini, che Petrificati si trovano in varij luoghi terrestri. Andrea Colicchia, Napoli

Scilla A (1759) De corporibus marinis lapidescentibus quæ defossa reperiuntur. Addita Dissertatione Fabii Columnæ de Glossopetris. Joannis Zempel, Romae

Soldani A (1780) Saggio Orittografico ovvero osservazioni sopra le terre nautilitiche ed ammonitiche della Toscana. Con Appendice o Indice Latino Ragionato de' piccoli Testacei, e d'altri fossili d'origin marina per schiarimento dell'Opera. Stamperia di Vincenzo Pazzini Carli e Figli, Siena. https://doi.org/10.3931/e-rara-14788 / Public Domain Mark

Taverne L, Capasso L, Del Re M (2019) The pycnodont fishes from the Lower Cretaceous of the Capo d'Orlando, near Castellammare di Stabia (Napoli, Campania, southern Italy), with the description of the new genus Costapycnodus. Geo-Eco-Trop 43(1):53–74

Tramezzino M (1550) Di Giorgio Agricola De la generatione de le cose che sotto la terra sono e de le cause de' loro effetti e nature, lib. V; De la natura di quelle cose che dalla terra scorrono, lib. IIII; De la natura de le cose fossili e che sotto la terra si cauano, lib. X; De le minere antiche e moderne, lib. II; Il Bermanno, o De le cose metallice, dialogo / recato tutto hora dal latino in buona lingua volgare. In: Vinegia : per Michele Tramezzino. https://archive.org/details/hinwel-all-00002012-001 (not copyright found)

Worm O (1655) Museum Wormianum, seu Historia rerum rariorum, tum Naturalium, quam Artificialium, tam Domesticarum, quam Exoticarum; quae Hasniae, Danorum in aedibus Authoris Servantur. Iohannem Elsevirium, Lugduni Batavorum. https://doi.org/10.5962/bhl.title.63844

Chapter 8
Stellaria or Stregonia Stone. About Some Corals in Folklore and Science

Abstract With the broad term corals, we commonly mean exclusively marine organisms with a calcareous skeleton, more or less arborescent, usually red or white, or more or less rounded masses of cells vaguely resembling wasp nests, globules, balls, and whips, belonging to the phylum Cnidaria or Coelenterata. Mediterranean red coral has been famous since Greco-Roman antiquity. Referring to the myth of Perseus and Andromeda, it was believed that it was born from the contact of some algae with the severed head of Medusa, transferring the petrifying property of her eyes to the coral itself. Very famous in the mid-1600s, a human skull with placed a sprig of coral on top is believed to have been recovered in the deep sea, part of the Medici collections. The first depictions of fossil coral aggregates, called *asterias* or *astroites*, worked as ornaments, are found in Gessner (*De omni rerum fossilium genere, gemmis, lapidibus, metallis, et huiusmodi, libri aliquot, plerique nunc primum editi*. Tiguri. https://doi.org/10.3931/e-rara-4176 / Public Domain Mark, 1565). These objects were believed to be bearers of benefits, used as medicines but above all as amulets. A tradition in this sense still persists in some Italian regions.

Keywords Fossils · Corals · Medusa · Asteria · Astroite · Stregonia

Corallite, Stellaria and Astroite: Corals, Cnidaria or Coelenterata Anthozoa (Zoantharia)

With the broad term *corals*, we commonly mean exclusively marine organisms with calcareous skeleton, more or less arborescent usually red or white, or more or less rounded masses of cells vaguely resembling wasp nests, globules, balls, and whips, belonging to the phylum Cnidaria or Coelenterata. Ancientlt coral has been considered a vegetable, from this belief derives its classification in the class of Anthozoa, Greek word meaning *animal flowers*. Throughout the centuries, both fossil and living, corals have been used attributing to them magical, apotropaic and therapeutic

© The Author(s), under exclusive license to Springer Nature Switzerland AG 2022

M. Forli, A. Guerrini, *The History of Fossils Over Centuries*,
https://doi.org/10.1007/978-3-031-04687-2_8

powers (Fig. 8.1). The vital part is made up of tiny organisms called polyps that live in colonies, even very large ones, or in a solitary state; fixed in their cells, contiguous and welded together, they are responsible for the most varied forms anchored to the seabed. They can have semirigid, rigid, or soft structures; distributed from a few centimeters to over 4000 m deep, they are of considerable importance in the global marine ecosystem.

Fig. 8.1 Stones against witches (stregonia). Witches had to flee at dawn light, unable to perform evil deeds, "forced" to count, without succeeding, the countless holes and rays on the surface of the stones themselves. Three witches during a night spell. (Modified after Francesco Maria Guaccio, *Compendium Maleficarum*, 1626: p. 278) (Photo and graphic composition Maurizio Forli. Credit: https://archive.org/details/image4995MiscellaneaOpal/page/n293/mode/2up. Under CC/publicdomain/zero/1.0)

Their body is sac-shaped with radial symmetry; the central gelatinous part, the *mesoglea*, is surrounded inside by the *gastroderm* and outside by the *ectoderm*. It is from the latter that the calcium carbonate is secreted, from which the skeleton is derived. The lower extremity is characterized by the presence of the basal disc that is the point with which the polyp adheres to the rest of the colony or to the substratum; above it the oral disc opens, surrounded by a crown of tentacles, in which there is an orifice that serves both as the mouth and anus. From this opening a very short pharynx departs, which joins the mouth to the internal gastrovascular cavity, subdivided into chambers, by means of vertical walls (the mesenteries), where the digestive cells are located. A particular structure formed by urticant cells, called cnids or cnidocytes, is typical of them and gives the name to the group (Cnidaria). Depending on the number and arrangement of mesenteries, corals are divided into two categories: octocorals and hexacorals. Simplifying, we can say that polyps are small organisms, about 2 mm long, with retractable tentacles that float in the water, collecting plankton on which they feed. In order to better intercept food, they are always arranged transversally with respect to the main sea current. The great availability of food is decisive for the appearance of coral reefs, agglomerations of different species, even hundreds of kilometers long, placed at low depths.

Coral reproduction can be both sexual and asexual. In the first case, fertilization takes place in the water, with the eggs and spermatozoa fusing to form a larva. This will wander till when it will find the suitable bottom for anchoring. The asexual reproduction, on the contrary, takes place by scission of a polyp, from which will originate two distinct individuals.

Being a large group of marine animals similar to others close to them, the classification of these organisms is not yet well defined, and there are still discrepancies and disagreements among scientists on their actual systematic placement, in a single *phylum* or in more *phyla*.

Their earliest representatives date back to the Paleozoic, over 500 million years ago. The class Anthozoa, subclass Hexacorallia, is the one that has the greatest relevance from the point of view of their importance as fossils. The various orders included in Hexacorallia are classically distinguished according to the arrangement of the mesenteries and their respective septa as well as on the morphology of the skeleton when present and, recently, also on a genetic basis. The orders are Rugosa or Tetracorallia, solitary or colonial, and Tabulata, only colonial, extinct in the Permian.

In the Triassic there are forms attributable to the order Scleractinia (cycles in multiples of six, hexacorals) also present today. Colonial bioconstructing forms of reefs are also termed Madrepores or Madreporaria; the latter include solitary or colonial species with calcareous, with rare exceptions, aragonitic exoskeletons, with sclerosects appearing and being ordered in cycles of six and multiples of six; hexacorals are known from the Middle Trias of Germany, the southern Alps and Sicily. From the Upper Trias to the Lower Lias (~ 220 to ~180 Ma), they have a worldwide distribution extending between 60° North latitude and 10° South latitude. In the Middle Jurassic, there was a further evolutionary radiation with widespread reefs in the seas of the Tethys, while, at the end of the Mesozoic, the Cretaceous is characterized by alternating phases of development and regression. In the Cenozoic and in particular in the Neogene (Miocene), the differentiation of coral reefs into the current two faunal provinces begins: Caribbean and Indo-Pacific.

The Mediterranean Sea was also characterized until the end of the Middle Miocene (Upper Tortonian–Lower Messinian, ~ 7.2 Ma) by a complex system of reefs that reached, as a northern limit in Italy, up to present-day Livorno. They all disappeared during the subsequent Upper Messinian (~ 6 Ma) during the so-called *Salinity Crisis*, which involved the more or less complete evaporation of our sea. Since the Pliocene the reefs distribution has not changed.

Numerous species of fossil corals are present in the fossil state in Italian Pliocene soils; particularly in Tuscany, during the Pliocene, there were numerous representatives of solitary corals, e.g., *Flabellum*. In the Lower Pleistocene, the cushion madrepora *Cladocora caespitosa* (Linnæus 1767), an endemic species of the Mediterranean, still widespread, became common.

Phylum Cnidaria
Class Anthozoa
Subclass Hexacorallia
Family Scleractinia
Genus *Cladocora*
Species *Cladocora caespitosa* (Linnæus 1767)
(Figs. 8.2, 8.3, and 8.4)

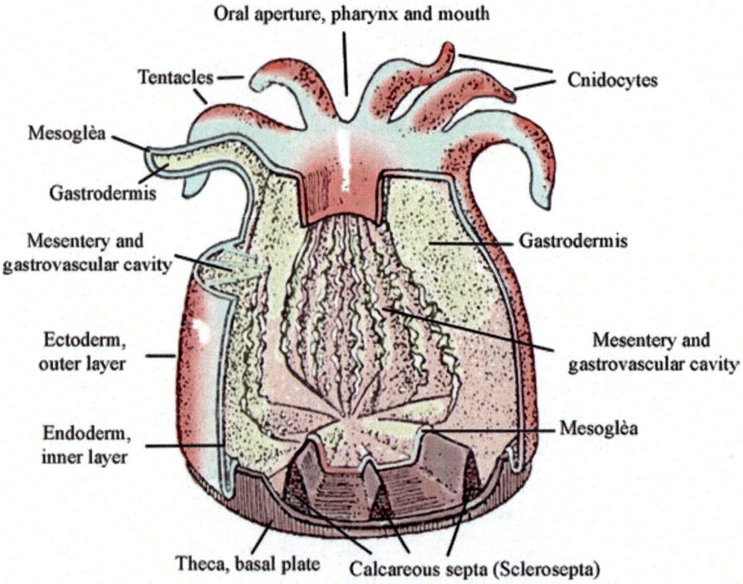

Fig. 8.2 Simplified morphological scheme of an isolated coral. The body is sac-shaped with radial symmetry; the central gelatinous part, the mesoglea, is surrounded inside by the gastroderm and outside by the ectoderm, from which the calcium carbonate is secreted, from which the skeleton is derived. At the lower end, there is the basal disc that is the point with which the polyp adheres to the rest of the colony or to the substratum; above there is the oral disc, surrounded by a crown of tentacles, in which there is an orifice serving both as the mouth and anus. From this opening departs a very short pharynx that joins the mouth to the internal gastrovascular cavity, subdivided into chambers, by vertical walls (the mesenteries), in which are located the digestive cells. A particular structure constituted by urticant cells, called cnids or cnidocytes, is typical of them and gives the name to the group (Cnidaria). [Drawing Maurizio Forli modified after Vialli (1974)]

Fig. 8.3 In overlay, ***Pseudocoralium album fungosum***, illustrated by Ulysses Aldrovandi in ***Musaeum Metallicum*** (1648: p. 292) (modified). In the background the corresponding coral, a present colony of *Cladocora caespitosa* (Linnæus 1767). (Photo and graphic composition Maurizio Forli. Credit: https://doi.org/10.3931/e-rara-13017 / Public Domain Mark)

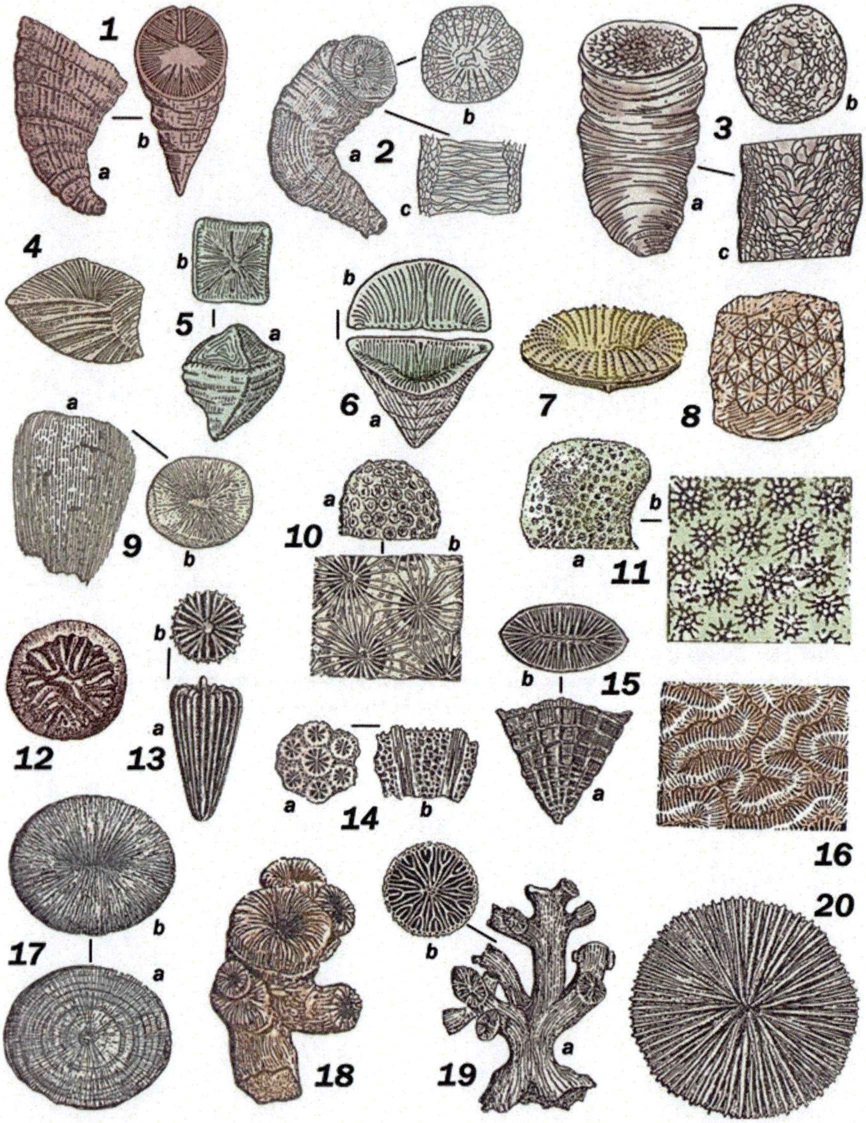

Fig. 8.4 Examples of solitary and colonial fossil tetracorals: (1) *Zaphrentia*, Devonian. (2) *Caninia*, Carboniferous-Permian. (3) *Cystiphyllum*, Silurian. (4) *Goniophyllum*, Silurian. (5) *Goniophyllum*, complete specimen, Silurian. (6) *Calceola*, Devonian. (7) *Porpites*, Silurian-Devonian. (8) *Isastrea heliantoides*, Cretaceous. (9) *Montlivaltia ellipsocentra*, Cretaceous. (10) *Heliastrea conoidea*, Miocene. (11) *Porites pellegrinii*, Eocene. (12) *Aspidiscus cristatus*, Cretaceous. (13) *Turbinolia bowerbanki*, Eocene. (14) *Madrepora*, from Eocene to present-day. (15) *Flabellum*, from Eocene to present-day. (16) *Meandrina*, from Eocene to present-day. (17) *Cyclolites elliptica*, Cretaceous. (18) *Thecosmilia trichotoma*, Cretaceous. (19) *Dendrophyllia elegans*, Oligocene. (20) *Fungia*, from Miocene to present-day. [Color and graphic composition Maurizio Forli modified after Vialli (1974)]

Mediterranean red coral (*Corallium rubrum*, belonging to the Octocorallia, another subclass of Anthozoa) is considered the most valuable. It has a treelike structure and an endoskeleton made of calcium carbonate, usually bright red pigmented, and prefers poorly illuminated marine environments such as steep walls, crevices, and caves, from the coastal area to hundreds of meters deep, without ever being too common, neither beached nor fossilized. From this it is evident that the supply for intensive use had to be made, as it happens also in our days, by particular ways of fishing (Marsili 1725: pp. 111–112; figs 107–111). Its use for ornamental purposes in grains, more or less regular, for necklaces and bracelets, or as encrustations in the ornamentation of objects, is testified in Europe from the Neolithic.

In Greek and Roman antiquity, it was believed to be created by the contact of some seaweed with the severed head of Medusa, transferring the petrifying property of her gaze into the coral itself. Below is the passage of the Latin poet Ovid that tells the story (Metamorphoses, IV: pp. 740–752):

> *Ipse manus hausta victrices abluit unda, anguiferumque caput dura ne laedat harena, mollit humum foliis natasque sub aequore virgas sternit et inponit Phorcynidos ora Medusae.*
> *Virga recens bibulaque etiamnum viva medulla vim rapuit monstri tactuque induruit huius*
> *percepitque novum ramis et fronde rigorem. At pelagi nymphae factum mirabile temptant pluribus in virgis et idem contingere gaudent, seminaque ex illis iterant iactata per undas:*
> *Nunc quoque curaliis eadem natura remansit, duritiam tacto capiant ut ab aere quodque vimen in aequore erat, fiat super aequora saxum.*

> *In the meantime the hero draws water and washes his victorious hands; then, so that the rough sand does not damage the snake-bristling head of Forco's daughter, he softens it with leaves, covers it with aquatic twigs and places Medusa's face on it. The fresh, still-alive twigs absorb the force of the snake in their marrow and in contact with the monster they harden, assuming an unprecedented rigidity in their arms and leaves. The sea nymphs try again with many other twigs and are amused to see that the miracle is repeated; so they make them multiply by throwing their seeds into the sea. Even today, corals retain unchanged the property of hardening in contact with air, so that what in the water was wicker, sprouting out of it is petrified.* – (Text and translation from Ovidio, Metamorfosi. Ramous M., Garzanti, Milano, 1995)

Below we report instead the version of the same passage of the *Metamorphoses*, adapted in rhyme by Giovanni Andrea dell'Anguillara, who for some art scholars influenced Giorgio Vasari in the ideation of his famous painting *Perseus frees Andromeda, birth of the Coral*, executed for the *Studiolo of Francesco I de' Medici*, in Palazzo Vecchio in Florence. The composition instead, with Andromeda tied to a vertical rocky outcrop, placed in the center of the work, in our opinion, finds a strong similarity with an engraved plate of the volume *La Vita et Metamorfoseo d'Ovidio* by M. Gabriello Symeoni (1559: p. 73) (Figs. 8.5 and 8.6):

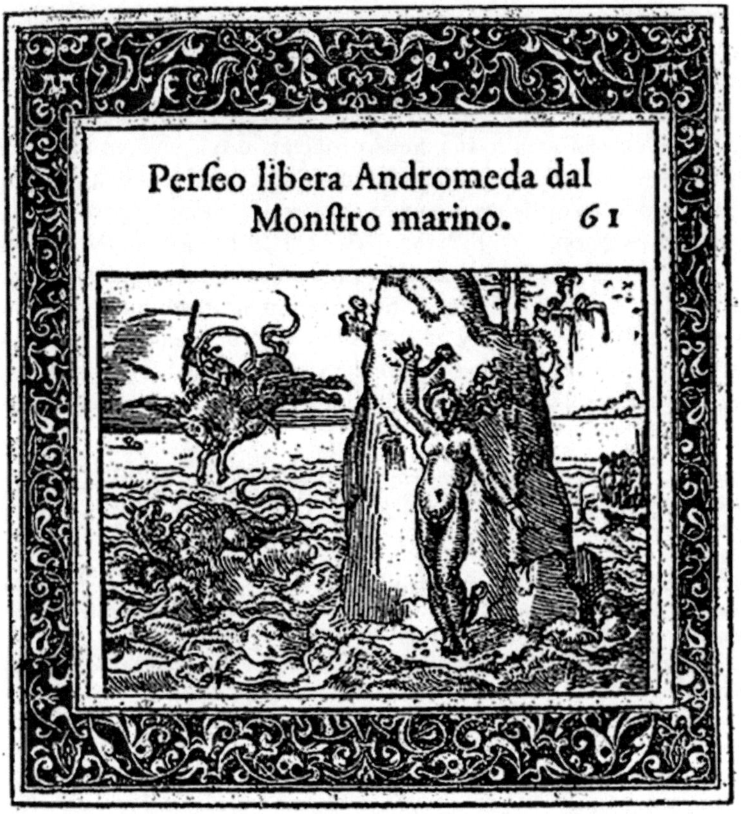

Fig. 8.5 Perseus frees Andromeda from the sea monster. M. Gabriello Symeoni (1559: p. 73) (modified). For the composition of his work, Vasari may have been inspired by this engraving, with Andromeda chained to a vertical rock in the foreground. His painting, if one can say so, seems to be a snapshot taken shortly after the killing of the monster, which in Vasari is seen pulled to shore, as the hero frees Andromeda, after laying Medusa's severed head just to the side. In the context of our book, we have not delved much into the subject of Vasari's sources of inspiration regarding this subject, and although we are aware of other hypotheses, our observation seems pertinent. Perhaps others before us have arrived at the same determination. This same compositional model is also found used by other artists to illustrate one of the salient episodes of Orlando Furioso, the liberation of Angelica by Ruggero. [https://numelyo.bm-lyon.fr/f_view/BML:BML_00GOO0100137001105317585?pid=BML:BML_00GOO0100137001105317585&pg_titre=(Domain Public)]

Fig. 8.6 Giorgio Vasari (1571), Perseus frees Andromeda, Birth of the Coral. Oil on slate, height 117 cm, width 100 cm. Bottom right Medusa's head and the birth of coral, although the latter does not arise from direct contact of Medusa's head with seaweed as in the sources, but from the petrification of the Gorgon's own blood on contact with water. [https://it.wikipedia.org/wiki/File:Vasari,_perseo_e_andromeda,_studiolo.jpg (Public domain)]

E, perche in terra offeso non restasse
Il volto, che fe sasso la balena,
Certe ramose verghe del mar trasse,
E gli fe un letto in su la trita arena.
Io non credo, ch'à pena le toccasse,
Che la scorza di fuor, dentro la vena,
Alterar si sentì la sua natura,

E farsi pietra pretiosa, e dura.
Ma le Nereide, ch' immortali, e dive
Non han punto à temer di quella testa,
Con altre verghe assai bagnate, e vive
Voller toccar la serpentina cresta.
Vistole poi restar del legno prive,
Ne fer con l'altre Ninfe una gran festa.
Co'l seme anchor la vennero à toccare,
E quel poi seminar per tutto il mare.
Cosi nacque il corallo, e anchor ritiene
Simil natura, che nel mar più basso
È tenero virgulto, e come viene
A l'aria s'indurisce, e si fa sasso. [...]

Giovanni Andrea dell'Anguillara, *Le Metamorfosi d'Ovidio ridotte in ottava rima* (1563: p. 71v)

The general interest in the representation of the severed head of the Gorgon, between the second half of the sixteenth century and the first half of the seventeenth century, is expressed in numerous works, among which the shield painted by Caravaggio, now in the Uffizi Gallery, stands out. It was also owned by Francesco I de' Medici, a family that, from the Renaissance onward, acquired many works related to the myth of Medusa (Fig. 8.7).

Fig. 8.7 The famous artifact from the Medici collection (Courtesy Museum of Natural History, University of Pisa, photo credit Stefano Puzzuoli), below its figure from Athanasius Kircher, *Mundus Subterraneus* (1664: p. 158). (https://doi.org/10.3931/e-rara-14326 / Public Domain Mark)

In this regard, we linger briefly on a typical "chamber of wonders" object that has particularly impressed us since the first time we saw it years ago: a human skull surmounted by a sprig of red coral. We had asked ourselves the question whether it could be an authentic natural composition or a manipulation, not being able to explain how a sprig of coral could have grown in that position on a human skull, without any encrustations and other marine organisms. The find is known since 1600 and is depicted in a famous still life of the Baroque period, painted by Domenico Remps, called the Scarabattolo, which is a small closet, where was kept a small collection of natural and artificial things, in practice a mini *Wunderkammer*, also owned by the Medici family, and in the *Mundus Subterraneus* by Athanasius Kircher (1664). A brief annotation also appears in Scarabelli (1667: p. 60): [...] *pensando egli in tal guisa puoter essere naturale la produzione del corallo, che sovra un Cranio humano nel Museo Pisano del Sereniss. Gran'Duca di Toscana conservato si mira.*

> [...] *thinking that in such a way the production of coral can be natural, that over a human skull in the Pisan Museum of the Sereniss. Grand Duke of Tuscany is preserved.*

Annotation taken up by Lorenzo Legati in his description of the Cospiano Museum, describing the coral "seedlings": *Ragione, che se qui sussiste, può insieme persuadere come naturalmente habbiano potuto generarsi que' coralli* [...] *che radicato sopra un Cranio humano, con essa conservasi nel Museo Pisano del Serenissimo Granduca di Toscana.* (1677: p. 130).

> *Reason, that if it subsists here, it can together persuade how naturally has been able to generate those corals* [...] *that rooted above a human skull, with it conserved in the Pisan Museum of the Most Serene Grand Duke of Tuscany.* (1677: p. 130)

We now know that, although Savi (1828) passes on to us the news that

> *This skull was fished in the sea near Sardinia* [...] *by a certain Francesco fisherman of corals,*

it is an artifice, as we had already been led to believe by the bone surface almost entirely covered by a greyish lamina of calcite, which seems to be more produced by the stay in some cave or travertine thermal spring than in the depths of the sea. We may reasonably suppose that the mind that devised this ingenious and symbolic deception was also inspired by the petrifying face and blood of Medusa, a myth so popular in the Medici household.

The fascination it exudes remains intact and, for those who want to see it, the reference to reflections on vanitas.

The red coral used the past in various ways as a medicine, for the many symbolic meanings of which it is invested, was and still is widely used in the mounting of apotropaic jewelry. In the late Middle Ages and later Renaissance, it was considered so powerful that it was even depicted around the neck of the Baby Jesus, reinterpreted in a Christian key as a symbol of the Passion. In many *Madonnas with Child* of that period, Jesus is elegantly depicted with red coral necklaces and amulets, testifying its common use, by virtue of the superstitious and protective powers against evil and evil influences in general, which were unanimously recognized. Still widespread is the use of giving to newborns of the small pendants or bracelets of coral for protection against evil eye. For the vastness of the correlated information to this particular type of coral, being insignificant its presence like fossil, we send back to other authors the eventual curiosity of the readers on the matter (Figs. 8.8, 8.9, 8.10, 8.11, and 8.12).

Fig. 8.8 Domenico Remps, Still Life in Deception (Scarabattolo), second half of the seventeenth century, with coral-encrusted skull at top right [https://commons.wikimedia.org/wiki/File:Cabinet_of_Curiosities_1690s_Domenico_Remps.jpg (public domain)]

Fig. 8.9 The present location of the find in a showcase of the Hall of Wonders. (Courtesy Museum of Natural History of the University of Pisa, photo credit Stefano Puzzuoli)

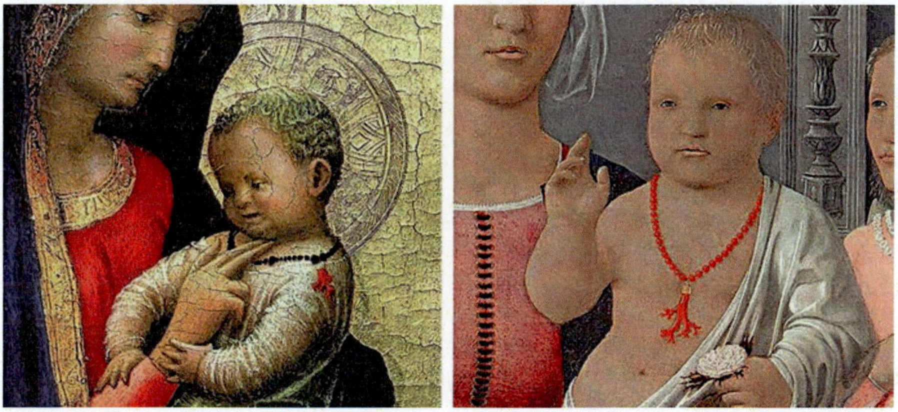

Fig. 8.10 Details of two famous paintings where the Child is portrayed wearing a necklace with a coral pendant. On the left the so-called Madonna del Solletico, attributed to Masaccio, painted around 1426–1427, and on the right the Madonna di Senigallia, by Piero della Francesca, painted in 1470 [https://commons.wikimedia.org/w/index.php?curid=6597840 & https://commons.wikimedia.org/wiki/File:Madonna_di_Senigallia.jpg **(Public domain)**]

Fig. 8.11 From Luigi Ferdinando Marsili, *Histoire physique de la mer* (1725, pl. XXII, fig. 107, pl. XXIII, figs 108–109, pl. XXIV, figs 110–111, p, 112, tools and techniques for fishing; coral sprig, pl. XXV, fig. 114). (Color and graphic composition Maurizio Forli. credit from https://doi.org/10.3931/e-rara-15280 / Public Domain Mark)

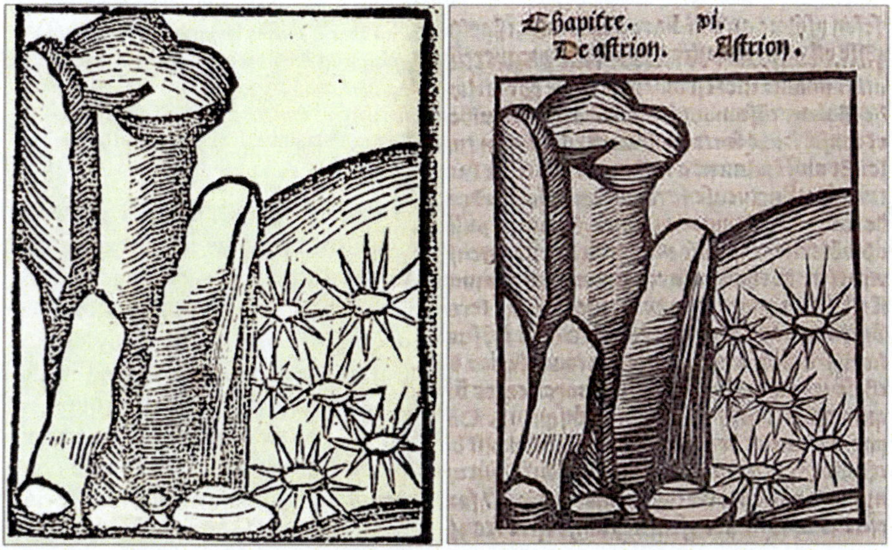

Fig. 8.12 *Astrion et Asterite*, crystal-like gems, shining with their own light: Johann Wonnecke von Kaub (von Cube), *Hortus Sanitatis*, 1498 and 1501 (https://doi.org/10.3931/e-rara-7477 / Public Domain Mark)

Numerous were the stones whose names contained the prefix *Aster*, practically all those which recalled any starry form. Among these are placed, owing to their peculiar radiated internal structure, some coral aggregates which appear in section as if they were small stars on the polished surfaces of the rocks containing them.

The first representation of this type of fossils appears to us to be the one related to two samples, called *astroites* or *asterias*, precisely because of their similarity to the starry sky, described by Gessner (1565: p. 35) in the chapter devoted to the figured stones, reporting as usual at the time the opinions of authors who had dealt with them previously such as Pliny and Agricola. The image on the left is the one that corresponds to the typical astroites; the other on the right, deriving from a meandering coral, is the one that was also commonly described as *Scolopendrite*, as the signs highlighted by the polished sections of these stones resembled the body, and its legs, of the scolopendre.

The same icons are found in Boetius (1636: p. 298; 1644: p. 382) with information also drawn from Gerolamo Cardano's *De Subtilitate* (1560).

A beautiful engraved plate, with various types of these figured stones, comes from the treatise *De Fossilibus Opusculum* by Francesco Imperato, Ferrante's eldest son, to whom his father had left the management of his museum collection. In it, with the exception of the first one on the upper left, various shapes of coral aggregates are depicted (1610; fig. 3):

> *Finally, the asterites or astrites or astroites according to some, which is numbered among the porous stones. It shows on every side many stars, both on the surface and in the interior*

to the centre. This stone by another name is called Victory, as they believe that he who wears it defeats his enemies. It possesses internal spaces, so that it moves spontaneously within the vinegar and lemon juice, and offers a pleasant vision to the observer. In these liquids both asterites and all kinds with porosity move; the latter are very numerous, though they show very different figures, such as mountains, woods, waves, spots, animals, and the like. – (translation from Fr. Imperato edited by Mariantonietta Paladini 2015: pp. 162–164 and notes, pl. 3)

An excellently reproduced coral among the finds in the Calceolari museum (Cerutus and Chioccus 1622: 417) makes it easy to recognize a tropical representative of the genus *Fungia* as well as the one in the plate on page 187 of Lodovico Moscardo's work (1656). Given the differences in form and substance of organisms understood as corals in the broad sense, which still today do not allow to have an unambiguous classification, it is not uncommon to find in the texts of the seventeenth century their representations in various groups of objects, sometimes assembled by shape or color. For example, in Aldrovandi, apart from the pages dedicated to them as "corals" properly so called and considered, one finds actual and fossil ones scattered among the various categories described by the author. So it is for a figurative fragment among the *Stelechites*, i.e., stones simulating trunks or twigs, an individual fossil coral of the genus *Flabellum* and a concave madrepore still among the *Fungites*, another flabelliform among the *Lapis figura conis Laricis*, i.e., stones shaped like larch cones, still one after the belemnites, among the stones called Lincurium, *Lincurij legitimi icon*, while some polished rocks with star aggregates are obviously included, similarly to Gessner's opinion, among the Astroites (*Musaeum Metallicum*, 1648: pp. 304, 491, 496, 508, 622, 877–878). In the *Metallotheca Vaticana*, the asteriae were kept in the ninth cupboard with the other idiomorphic stones, i.e., rocks of various materials with indeterminate figures; among those figured, two *cabochons* are particularly representative of the type being completely formed by the radiated coral cells (1717: p. 235; figs 3–4). Two other images of stellar stones can be found on page 124 of the same work … *Quare vulgus petram stellariam vocat* … named respectively *Spongia corallites* and *Lapis corallites*, at the end of the description of corals, among other marine things in the sixth cabinet.

These particular stones were considered real gems extremely rare and beautiful:

Questa rarissima pietra, stellaria nominata per la similitudine che ha con le stelle, ma picciole di color bianco – ma il fondo di detta pietra è nericcio -, non è pietra matrosa, piglia buon lustro, ma non si vede se non pezzi piccioli come cipolle stiacciate rosse. Dicono alcuni che si trovano nella Magna. Io ne ho viste due di queste pietre in Firenze, una nella galleria segreta del signor cavalier Gaddi con altre pietre bellissime ed un'altra a un antiquario forestiero. (Agostino del Riccio, Istoria delle pietre 1597)

This very rare stone, named stellaria because of its resemblance to stars, is small and white – but the bottom of the stone is black -, it is not a pitted stone, it has a good brilliance, but you cannot see it except for small pieces like crushed red onions Some say that they are found in Magna. I have seen two of these stones in Florence, one in the secret gallery of

Signor Cavalier Gaddi with other beautiful stones and another in the possession of a foreign antique dealer

Pliny describes several stones with similar names: *asteria*, *astrios*, and *astroite*. It is the latter that is identified with that formed by fossilized coral calyxes:

> *Astroites. Celebrant et astroiten, mirasque laudes ejus in magicis artibus Zoroastrem cecinisse, qui circa eas diligentes sunt, produnt.* (Pliny XXXVII: p. 49)

> *About Astroite. The astroite joy is still much celebrated, and those who have written about it with diligence say that Zoroaster attributes to it marvellous virtues in the magical art.* (Domenichi 1844: p. 1422)

> *The Astroite gem is that which is vulgarly called Stellaria; which is found in Saxony, as Agricola writes, it is white, or ashy, and full of stars, from which it takes its name. Placed in vinegar on a table, or of marble, or of another material, it moves from one place to another: and is called Victory by others: because he who wears it gains victory: Pliny relates that Zoroaster marvellously praises it in the art of magic.* (Moscardo, ib.: p. 134)

Agostino Scilla's treatise discusses and represents numerous fossil coral remains, both single calyxes and branched specimens, from around Messina or the island of Malta. Some individual corals are found scattered here and there in the plates as components of various fossiliferous aggregates, but those specifically devoted to them are numbered XX–XXII:

> *Io ne trovo molte branche ben ramificate nelle nostre colline, imbrogliate insieme con gli Echini, e conchiglie & c. ed ho osservato qualche parte de' detti coralli calcinata, e rotta* [...]. *Con tutto ciò tra tanta rotta quantità, che se ne vede, ne ho cavato una rama non affatto intera, ma ben conservata in riguardo del tempo. Le nostre colline non sono contente di farci vedere coralli comuni mezzi calcinati e rotti, ma anche degli altri fistolosi in buon numero, sì bene più maltrattati de' primi* [...]. *Ad ogni maniera avendo usato qualche studio, ne ho esatto da un solo pezzo di tufo quattro branche, che prima d'aver patito, una sola rama di corallo fistoloso certamente formavano.* (1670: pp. 141–142)

> *I find many branches of it well branched in our hills, entangled together with the Echini, and shells & c. and I have observed some part of the said corals calcined, and broken* [...]. *With all of this, among such a broken quantity, which can be seen, I dug out a branch which is not at all whole, but well preserved with respect to time. Our hills are not content to show us common corals half calcined and broken, but also other fistulose corals in good numbers, although more ill-treated than the first* [...]. *In any case, having used some study, I extracted from a single piece of tufa four branches, which before having suffered, a single branch of fistulous coral certainly formed.* (1670: pp. 141–142)

Like Scilla, other authors of the second half of the seventeenth century understand the true nature of coral fossil remains by comparing them to their living counterparts, thus abandoning the undefined nomenclature with which they were described. For example, in Boccone we still find the description of the rock astroite or stellaria but is given the correct genesis:

> [...] *I do not doubt, that this Stellaria Stone is produced by means of an aggregate of many coralloid and porous torches, in the middle of which are star-shaped and striated in the*

surface, & that they are of the Nature of those Productions, which Imperato calls Millepora, and Madrepora, in the Cavities of which aggregate, adding Clay, Earth, or Arenas, by their iuxta position then harden, and then fix all the compound in such a way, that no longer Millepora, and Madrepora, but Stellaria Stone, or Astrites appears with spots. (1697: p. 286, pl. IX; figs 2–4, 7)

In the *Dictionnaire universel des fossiles…* (Bertrand 1763: pp. 60–63), among the various names mentioned, astroites are distinguished in two classes, both formed by masses of coral tubules, but one with the surface decorated by circular stars and the other by angular (hexagonal) stars. A list of prodigious virtues attributed to the stellaria is found in Giacinto Gimma (1668–1735) who, despite skepticism on the subject, reports diligently as in other cases the opinion of his predecessors adding notes on the marketing of the same:

[...] *per la peste, e per ammazzare i vermi si loda la polvere della Stellaria* [...] *ed alcuni la portano come amuleto. Che faccia fuggire i ragni dice esser falso Olao Vormio. I venditori di quelle pietre, che dicono medicinali, molte virtù mostrano della Stellaria, e dispensano colla pietra la sua Ricetta stampata.* [...] *Vogliono che portata addosso dia fortezza, guardi l'Uomo dal male della goccia. Portata al collo, toglie via il tremore de' membri, conforta la vista, fa capelli mediocri, morbidi e sottili, e difende dalla peste, e da qualsivoglia febbre pestifera* [...] *Scaccia la paura, e le cattive ombre notturne* [...] *libera dal mal caduco, fa star l'uomo vigilante, e che non possa essere offeso da animali selvaggi: che conservi gli spiriti, e rallegri il cuore* [...] *che stagni il sangue* [...] *che giovi al male di fianco, e che sia mirabile al male del granchio* [...] *conforti il polmone, e 'l fegato, corrobori il sangue, e vaglia contro le vertigini. Queste, ed altre virtù celebrano nelle loro Ricette, perchè si muovono le persone a comprarle, a vil prezzo ancora concedendole.* (1730, I: pp. 288–289)

[...] *for the plague, and to kill worms, the powder of Stellaria is praised* [...] *and some wear it as an amulet. That it makes the spiders flee is said to be false by Olao Vormio. The sellers of those stones, which they say are medicinal, show many virtues of Stellaria, and dispense with the stone its printed recipe.* [...] *They want it to give strength when worn, to protect Man from the illness of the drop. Worn around the neck, it takes away the trembling of the members, comforts the sight, makes the hair medium, soft and thin, and defends from the plague, and from any pestiferous fever* [...] *It drives away the fear, and the bad night shadows* [...] *frees from the illness of death, makes the man vigilant, and that he cannot be offended by wild animals: that it preserves the spirits, and rejoices the heart* [...] *that it stagnates the blood* [...] *that it helps the illness of the flank, and that it is admirable to the illness of the crab* [...] *it comforts the lung, and the liver, corroborates the blood, and is good against vertigo. These, and other virtues they celebrate in their Recipes, because they move people to buy them, at a low price still granting them.*

Another appellation attributed to them by popular imagination is that of Stone Witch or against witches (stregonia):

With the vulgar designation of witch stones are qualified some minerals, represented by fossil madreporites or polyporites, which show on their surfaces numerous spots, irregularly circular, of dark colour, formed by the radiated sections of the single madreporic animals, already living in society, and cemented by a calcareous paste of grey colour. The popular imagination has seen in these radiated sections the form of small stars and has therefore designated the whole of the fossils, also with the name of stellar stones or simply astroites. – (Bellucci 1907: p. 100)

The amulets made with these stone against witches, being mostly intended to protect children, as a sign of affection toward those who wore them, were often heart-shaped and set in notched silver frames. Witches had to flee at dawn light, unable to perform nocturnal evil deeds because they were "forced" to count, without succeeding, the countless holes and rays on the surface of the stones themselves. Substituting and superimposing the "pagan" beliefs with those of the Christian religion, for an even more effective performance, thus multiplying the power against evil influences, the amulet could be engraved or carved with sacred symbols and images:

> *So the faith for ancient tradition placed in the* stones against witch (stregonia), *comes for the Christian believer to invigorate even more, through the sacred figures carved in them; and the wretched witches, already made impossible to bring evil spells for the presence of the enemy stone, from the innumerable and innumerable holes, remain completely terrified and fled in the presence of the sacred images most venerated and, according to the popular concept, most feared.* (Bellucci 1907: pp. 102–104; figs 55–56).

Of course, the commercialization of pendants is very active even in our days; the undoubted aesthetic qualities of such objects mean that they are still appreciated as jewelry despite the fact that the awareness of their apotropaic function has probably been lost, which, it seems to us, continues more by tradition than by conviction (Figs. 8.13, 8.14, 8.15, 8.16, 8.17, 8.18, 8.19, 8.20, 8.21, 8.22, and 8.23).

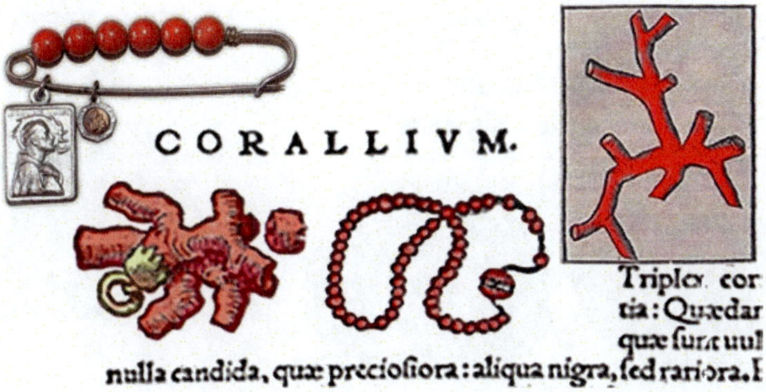

Fig. 8.13 Antique prints of coral elements, in the background a pendant and a necklace (1) Lonicerus (1551: 323 v), on the right (2) a rough sprig from Johann Wonnecke von Kaub, also known as Johann von Cube, (1501: CCIII). Above left (3) a brooch with coral beads and votive medals depicting St. Francis and the Madonna of Montenero (Livorno, Tuscany, ~ 1960, courtesy private collection). (Photo (3) and graphic composition Maurizio Forli. Credit from (1) http://mateo.uni-mannheim.de/camenaref/lonitzer/loni1/jpg/s323b.html (2) https://doi.org/10.3931/e-rara-7477 / Public Domain Mark)

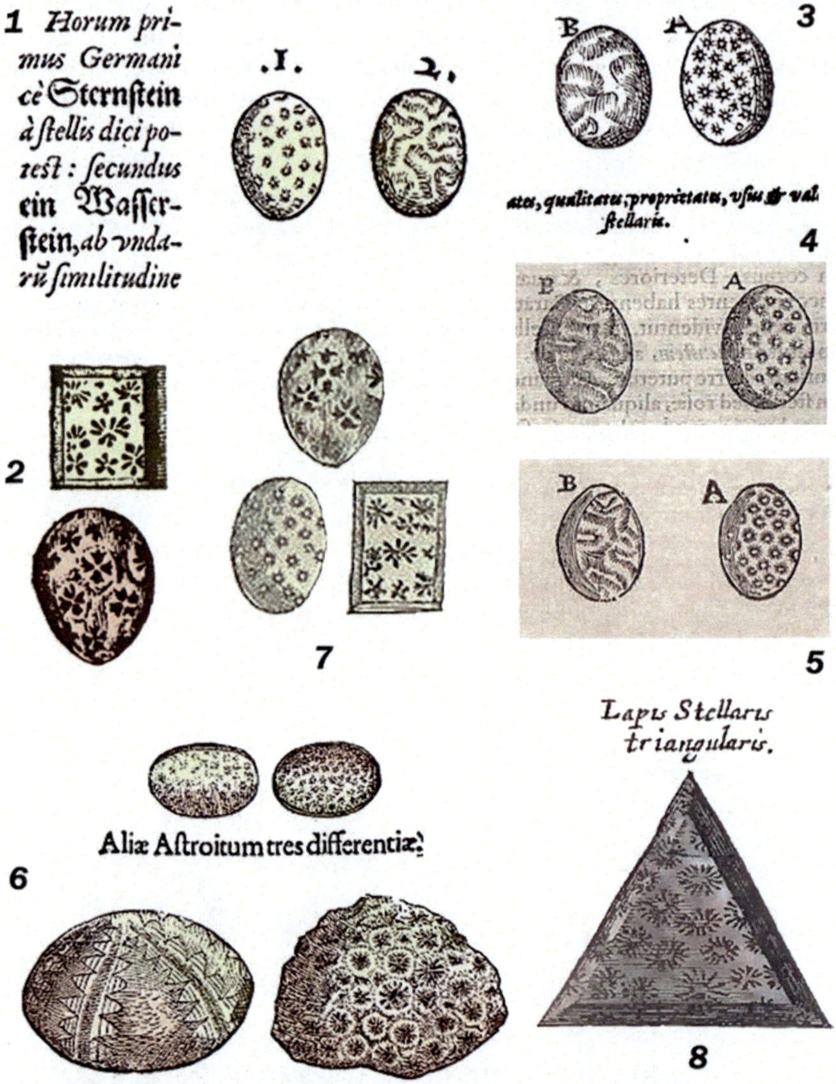

Fig. 8.14 Astroite, asteria, or scolopendrite gems, for their resemblance to the starry sky and to the figure of the scolopendra. (1) Gessner (*De Omni rerum fossilium*, 1565: p. 35), as far as we know this image concerning two cabochons in semiprecious stones, obtained from fossil corals, is the first of its kind. (2) De Laet (*De Gemmis et Lapidibus*, 1647: p. 97). (3–4) De Boodt (*Gemmarum lapidae Hist*. 1609: p. 151) (1636: p. 301). (5) (*Le Parfaict Joaillier*, 1644: p. 382). (6) Aldrovandi (*Musaeum Metallicum*, 1648: p. 878), in this category of starlike radiated stones, the Bolognese author inserts many other lapids among which also the echinid figured in the lower left. (7) Worm (*Museum Wormianum*, 1655: p. 68). (8) Sachs (*Gammarologia*, 1665: p. 158), *Lapis Stellaris Triangularis*, an unusual trigonal cut. [Color and graphic composition Maurizio Forli. Credit from (1) https://doi.org/10.3931/e-rara-4176 / Public Domain Mark (2) https://digital.clarkart.edu/digital/collection/p16245coll5/id/106300 (Public domain) (3) https://doi.org/10.3931/e-rara-15217 / Public Domain Mark (4) https://doi.org/10.3931/e-rara-12490 / Public Domain Mark (6) https://doi.org/10.3931/e-rara-13017 / Public Domain Mark (7) https://doi.org/10.5962/bhl.title.63844 (not in copyright) (8) biodiversitylibrary.org/item/209611#page/9/mode/1up]

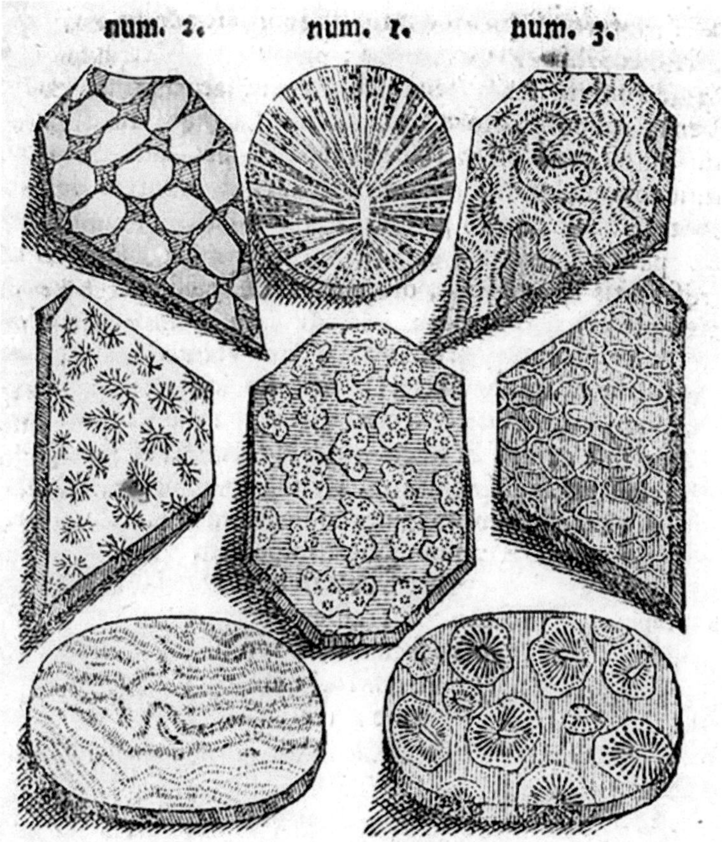

Fig. 8.15 Stellar stones. A plate from the *De Fossilibus Opusculum* by Francesco Imperato, Ferrante's eldest son, to whom his father had left the management of his museum collection. In it, except for the first in the upper left, which is a fragment of an echinid, are depicted different shapes of coral aggregates. The ones in the upper right and lower left represent typical *scolopendrites* (1610: fig. 3) (modified). [https://archive.org/details/bub_gb_yPMwLxv1n4kC (Public Domain Mark 1.0)]

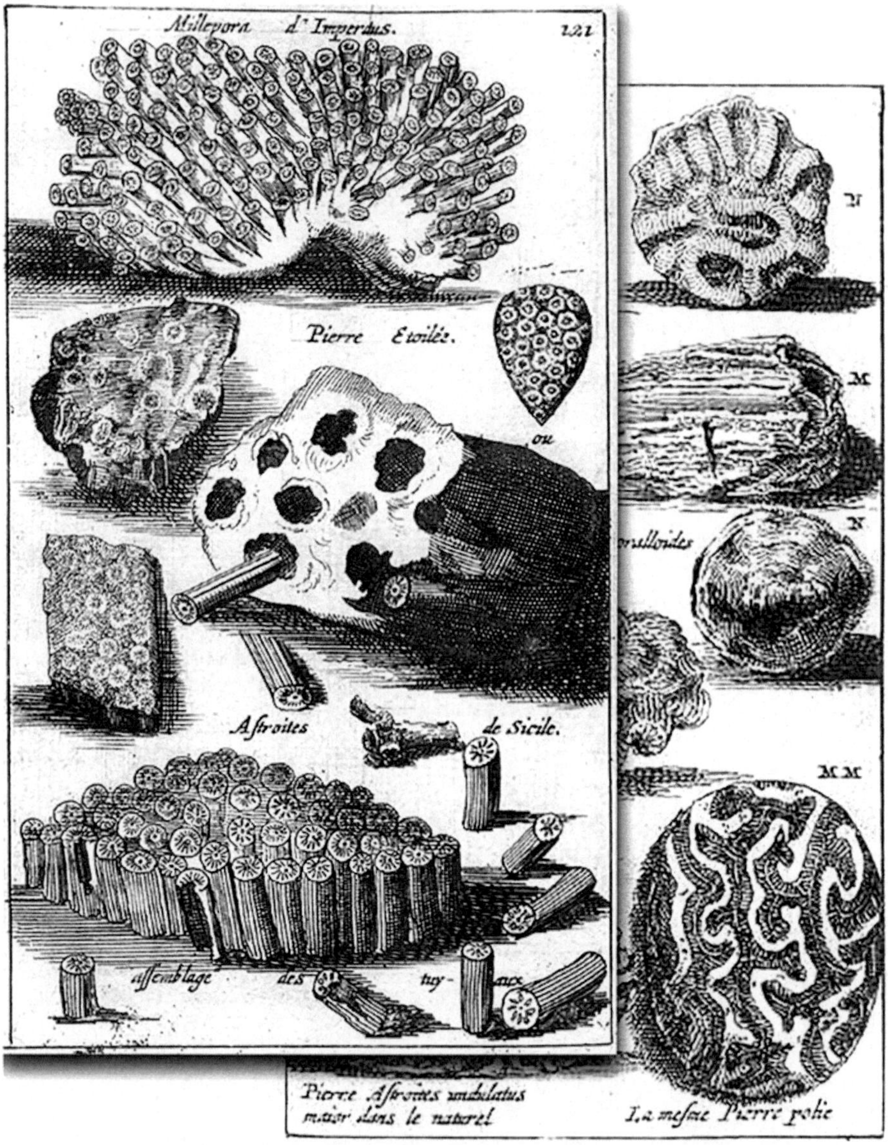

Fig. 8.16 Two plates from the work of the Palermo botanist Paolo Silvio Boccone, *Recherches et observations naturelles touchant le corail, la pierre étoilée…* in the upper one there are depicted some fossil corals [*…I don't doubt that this Stellaria Stone is produced by means of an aggregate of many coralloid and porous torches, in the middle stellate and in the surface striated, &… that they are of the Nature of those Productions, which Imperato calls Millepora, and Madrepora, in the Cavities of which aggregate overlapping Clay, Earth, or Arenas, for their iuxta position then harden, and then fix all the compound in such a way, that no more Millepora, and Madrepora, but Stellaria Stone, or Astrites appears of spots*. In the underlying of the corals coming from the tropical seas. (1674: pp. 121, 142) (Graphic composition Maurizio Forli. Credit from https://doi.org/10.3931/e-rara-21877 / Public Domain Mark)

Fig. 8.17 Two Plio-Pleistocene fossil coral branches from the hills around Messina. (Agostino Scilla 1759. *De corporibus marinis lapidescentibus quæ defossa reperiuntur. Addita Dissertatione Fabii Columnæ*, plate XX: I, pp. 141–142). (I) *Coralium simplex durissimum, sed decoloratum.* (II) *Coralium fistulosum, quod copiosum in collibus Messanensibus conspicitur.* [...] *Our hills are not content to show us common corals half calcined and broken, but also other fistulose corals in good number, so much more ill-treated than the first* [...] *you will observe that everything also corresponds to the corals of the sea; which concludes that there were once these too.* (Personal copy Maurizio Forli)

Fig. 8.18 *Stellaria* or *stregonia stone*, recto and verso, part of the "ingredients" contained in the bag of a sanpaolaro (Coll. Bellucci, MANU (PG). Inv. 338). (Reproduced by kind permission of the Ministry for Cultural Heritage and Activities and for Tourism, Regional Directorate Museums of Umbria—National Archaeological Museum of Umbria). In the background, *Lapides Idiomorphi, Astroiti*. (Mercati, *Metallotheca Vaticana*, 1717: p. 235). [Graphic composition Maurizio Forli. Credit from https://archive.org/details/bub_gb_DC_E6UVWg7EC (Public Domain Mark 1.0)]

Fig. 8.19 On the left a colony of *Porites* sp. (Miocene), in the center and on the right colonial fossil corals, worked in cabochons, the so-called stellar stones or sorcerous stones (*stellaria* and *stregonia*). (Photo and graphic composition Maurizio Forli)

Fig. 8.20 *The witch stone is a much sought-after and common amulet in a considerable area of Italy, whose center of diffusion is in the region of L'Aquila, where the Cretaceous and Eocene deposits are rich in madreporites and coral fossils. The mutations occurred in the shape of this amulet, [...] starting from the natural one of a discoidal pebble, is assuming a more and more regular outline, that of a heart shape, at first simply perforated at the base, and finally elegantly outlined in silver, with a fixed ring, to hang it on the person.* (Bellucci 1907: pp. 100–101). Heart-shaped limestone slab, in colonial hexacorals, set in silver, with floral-patterned dentils, and hanger. At the bottom of the front, at the tip of the heart, the three sacred nails of the Passion are over-engraved. Collection G. Bellucci, MANU (PG). (Inv. 329 G). (Reproduced by kind permission of the Ministry for Cultural Heritage and Activities and for Tourism, Regional Directorate for Museums of Umbria—National Archaeological Museum of Umbria)

Fig. 8.21 *They are very interesting from the point of view of the adaptation of the Christian doctrines to the primitive pagan fetishes* [...] On the *witch stone the figure of the Redeemer is carved on the right side, and on the reverse side the image of Christ crucified. So the faith for ancient tradition reposed in the witch stones, comes for the Christian believer to invigorate itself more, through the sacred figures carved in them; and the miserable witches, already made in the impossibility to bring evil spells for the presence of the enemy stone, from the innumerable and innumerable holes, remain absolutely terrified and escaped in the presence of the sacred images more venerated and, according to the popular concept, more feared* (Bellucci 1907: pp. 103–104). Bellucci Collection, M.A.N.U. (PG). (Inv. 720959). (Reproduced by kind permission of the Ministry for Cultural Heritage and Activities and for Tourism, Regional Directorate for Museums of Umbria—National Archaeological Museum of Umbria)

Fig. 8.22 Composite amulet, recto-verso view. Heart-shaped amulet stone with silver binding, fitted with a suspension ring in which, by means of a brown cloth ribbon, it is joined to a current shark otolith, also with silver binding, with a double ring at the sides. Castelnuovo di San Pio delle Camere (AQ). Late nineteenth century (Coll. Bellucci. MANU (PG) Inv. 4521 G). Otolith protects from ear pain. (Reproduced by kind permission of the Ministry for Cultural Assets and Activities and for Tourism, Regional Directorate Museums of Umbria—National Archaeological Museum of Umbria)

Fig. 8.23 Necklace and small branches of coral, Procida (Napoli), 1950 circa. (Photo Maurizio Forli, courtesy, private collection)

References

Aldrovandi U (1648) Ulyssi Aldrovandi patricii Bononiensis Musaeum metallicum in libros IIII : distributum Bartholomaeus Ambrosinus ... labore, et studio composuit cum indice copisosissimo Arcus Antonius Bernia. typis Io. Baptistae Ferronii, Bononiae. ETH-Bibliothek Zürich, Rar 1027, https://doi.org/10.3931/e-rara-13017 / Public Domain Mark

Bellucci G (1907) Il feticismo primitivo in Italia e le sue forme di adattamento. Unione Tipografica Coop. Editrice, Perugia

Bertrand E (1763) Dictionnaire universel des fossiles propres et des fossiles accidentels. A la Haye : Chez Pierre Gosse Junior, et Daniel Pinet, M. DCC. LXIII [1763]. ETH-Bibliothek Zürich, Rar 5720, https://doi.org/10.3931/e-rara-10463 / Public Domain Mark

Boccone PS (1674) Recherches et observations naturelles touchant le corail, la pierre étoilée, les pierres de figure de coquilles, le corne d'ammon, l'astroïte Undulatus, [...]: examinées à diverses fois dans l'Assemblée de Messieurs de Societé Royale de Londres, & dans les conferences de Monsieur l'Abbé Bourdelot à Paris. chez Jean Jansson à Waesberge, A Amsterdam. ETH-Bibliothek Zürich, Rar 5705, https://doi.org/10.3931/e-rara-21877 / Public Domain Mark

Boccone PS (1697) Museo di fisica e di esperienze variato, e decorato di osservazioni naturali, note medicinali, e ragionamenti secondo i principij de' moderni. Giacomo Combi, Venetia

Boodt A Boëtius de (1636) Gemmarum et lapidum historia. Lugduni Batavorum. ex officina Ioannis Maire, Leiden. ETH-Bibliothek Zürich, Rar 1881, https://doi.org/10.3931/e-rara-15217 / Public Domain Mark

Boodt A Boëtius de (1644) Le parfaict Ioaillier ou histoire des pierreries: ou sont amplement descrites leur naissance, iuste prix, moyen de les cognoistre, et se garder des contrefaictes, Facultez medicinales, et proprietez curieuses. Iean-Antoine Huguetan, Lyon. ETH-Bibliothek Zürich, Rar 1010. https://doi.org/10.3931/e-rara-12490 / Public Domain Mark

Cardano G (1560) Hieronymi Cardani Mediolanensis medici de subtilitate libri XXI, Basilaea

Cerutus B, Chioccus A (1622) Musaeum Franc. Calceolari iun Veronensis. Apud Angelum Tamum, Veronae. https://doi.org/10.3931/e-rara-50568. (Public Domain Mark)

De Laet J (1647) De Gemmis et Lapidibus, libri Duo. Quibus praemittur Theophrasti, liber de Lapidìbus Graece & latine cum Brevibus Annotationibus. Ioannis Maire, Lugduni Batavorum. https://doi.org/10.3931/e-rara-35487 / Public Domain Mark

Del Riccio A. (1597) Istoria delle pietre. Riproduzione anastatica del MS 230 della Biblioteca Riccardiana (a cura di Barocchi P.), SPES, Firenze. 1979; 82.

dell' Anguillara GA (1563) Le Metamorfosi di Ovidio ridotte da Giovanni Andrea dell'Anguillara in ottava rima, Al Christianissimo Re di Francia Henrico Secondo, di nuovo dal proprio auttore rivedute, & corrette; Con le annotationi di M. Gioseppe Horologgi. Francesco de' Franceschi senese, Venetia

Domenichi ML (1844) Della Storia Naturale di C. Plinio Secondo, Libri XXXVII. Traduzione di M. Lodovico Domenichi emendata per la prima volta secondo il testo latino con l'aggiunta di un nuovo indice generale, vol Secondo. Giuseppe Antonelli Ed., Venezia

Gessner C (1565) De omni rerum fossilium genere, gemmis, lapidibus, metallis, et huiusmodi, libri aliquot, plerique nunc primum editi. Tiguri. https://doi.org/10.3931/e-rara-4176 / Public Domain Mark

Gimma G (1730) Della Storia Naturale delle gemme, delle Pietre, e di tutti i Minerali, ovvero della Fisica Sotterranea. Tomo II, Felice Mosca, Napoli

Guaccio FM (1626) Compendium Maleficarum. Ex Collegi Ambrosiani Typographia, Mediolani. https://archive.org/details/image4995MiscellaneaOpal/page/n293/mode/2up. licensed under CC/publicdomain/zero/1.0

Imperato F (1610) De Fossilibus Opusculum. In quo miro ordine continentur Naturalis disciplinae scitu dignissima, eiusque professoribus omnino necessaria; ab alijs minime excogitata. Accesserunt. Dominici Roncalioli, Neapoli. https://archive.org/details/bub_gb_yPMwLx-v1n4kC. Public Domain Mark 1.0

Kircher A (1664) Athanasii Kircheri Mundus subterraneus: in XII libros digestus. apud Ioannem Ianssonium & Elizeum Weyerstraten, Amstelodami. ETH-Bibliothek Zürich, Rar 1365 GF, https://doi.org/10.3931/e-rara-14326 / Public Domain Mark

Legati L (1677) Museo Cospiano annesso a quello del famoso Ulisse Aldrovandi e donato alla sua Patria dall'Illustrissimo Signor Ferdinando Cospi. Giacomo Monti, Bologna. https://babel.hathitrust.org/cgi/pt?id=gri.ark:/13960/t7dr4608w&view=1up&seq=30&skin=2021 (Public Domain Mark)

Lonitzer A (1551) Naturalis historiae opus novum: in quo tractatur de natura et viribus arborum, fruticum, herbarum, animantiumque terrestrium, volatilium & aquatilium; item, gemmarum, metallorum, succorumque concretorum. apud Chr. Egenolphum, Francofurti. http://mateo.uni-mannheim.de/camenaref/lonitzer/loni1/jpg/s323b.html

Marsili F (1725) Histoire physique de la mer. aux dépens de la Compagnie, Amsterdam. ETH-Bibliothek Zürich, Rar 2263. https://doi.org/10.3931/e-rara-15280 / Public Domain Mark

Mercati M (1717) Metallotheca Vaticana. Opus posthumum, Autoritate et Munificentia Clementis Undecimi Pontificis Maximi. E tenebris in lucem eductum: Opera autem, & studio Joannis Mariae Lancisii, Archiatri Pontificii illustratum. Ex Officina Jo: Mariae Salvioni Romani, Romae. https://archive.org/details/bub_gb_DC_E6UVWg7EC Public Domain Mark 1.0

Moscardo L (1656) Note overo memorie del museo di Lodovico Moscardo nobile veronese. Academico Filarmonico, dal medesimo descritte, Et in Tre Libri distinte. Per Paolo Frambotto, Padoa. https://www.biodiversitylibrary.org/bibliography/157058 (not in copyright)

Paladini M (trans: it. ed) (2015) De Fossilibus Opusculum, 1610. Quad Accad Pontiniana 61:121–210

Ramous M (trans: it. ed) (1995) Ovidio, Metamorfosi. Garzanti, Milano

Sachs PJ (1665) Gammarologia sive Gammarorum vulgo Cancrorum consideratio Physico, Philologico, Historico, Medico, Chimica, Frankfurt & Leipzig. biodiversitylibrary.org/item/209611#page/9/mode/1up

Savi G (1828) Notizie per servire alla storia del Giardino e Museo della I. e R. Università di Pisa. Tipografia Nistri, Pisa

Scarabelli PF (1667) Museo ò Galeria Adunata dal Sapere, e dallo Studio del Sig. Canonico Manfredo Settala nobile milanese. descritta in Latino dal Sig. Dott. Fis. Coll. Paolo Maria Terzago Et hora in Italiano dal Sig. Pietro Francesco Scarabelli, Fott. Fis. di Voghera. E dal medesimo accresciuta. Per li Figliuoli del qd. Eliseo Viola, Tortona. https://archive.org/details/museoogaleriaadu00terz_0 no copyright page found

Scilla A (1670) La vana speculazione disingannata dal senso. Lettera responsiva Circa i corpi marini, che Petrificati si trovano in varij luoghi terrestri. Andrea Colicchia, Napoli

Scilla A (1759) De corporibus marinis lapidescentibus quæ defossa reperiuntur. Addita Dissertatione Fabii Columnæ de Glossopetris. Joannis Zempel, Romae

Symeoni MG (1559) La vita et Metamorfoseo d'Ovidio, figured & abbreviated in the form of epigrams by M. Gabriello Symeoni. John of Tornes, Lyon. https://numelyo.bm-lyon.fr/f_view/BML:BML_00GOO0100137001105317585?pid=BML:BML_00GOO0100137001105317585&pg_titre= (Domain Public)

Vialli V (1974) Appunti di Paleontologia. Pitagora Editrice, Bologna

Cube J von (1501) Ortus sanitatis. Nouvelleme[n]t imprimé à Paris: pour Anthoine Vérard, [vers 1501?]. BPU Neuchâtel, A.F. B 10, https://doi.org/10.3931/e-rara-7477 / Public Domain Mark

Worm O (1655) Museum Wormianum, seu Historia rerum rariorum, tum Naturalium, quam Artificialium, tam Domesticarum, quam Exoticarum; quae Hasniae, Danorum in aedibus Authoris Servantur. Iohannem Elsevirium, Lugduni Batavorum. https://doi.org/10.5962/bhl.title.63844. (not in copyright)

Chapter 9
Nummulites: Are they Stone Lentils, Frumentaries Stone or Devil's Coins? No, are Giant Foraminifers

Abstract The nummulites, which derive their name from the flattened discoidal shape of their shell, similar to that of a stone coin, were and are unicellular marine macroforaminifera, whose size varies from a few millimeters to about 12 cm in maximum diameter for some fossil species. Their abundance in certain rocks, now called nummulitic limestones, had already been noticed even by the ancient Egyptians and later Greeks and Romans. They were believed to be petrified grains, and under this appellation are described in the earliest printed books beginning with Agricola and Gessner. Their true origin remained disputed throughout the eighteenth century. The real nummulites or similar blackish discs of marcassite, according to popular tradition, were traces of the passage or work of the devil, who had lost them or had thus petrified the coins paid to commit a crime or a spell, making it in the end fruitless for the perpetrator.

Keywords Foraminifera · Nummulite · Frument stones · Devil's coins

Stone Lentils and Stone Frumentaries: Nummulites

These giant foraminifera were believed for centuries to be petrified grains, and under this appellation are described in the earliest printed books. The finding of these discs was interpreted as presence of the evil one by the folk belief (Fig. 9.1). In fact, the *Foraminifera* d'Orbigny, 1828, are unicellular protozoa provided of shell, mainly marine but partially present also in brackish and fresh waters, which form a very complex systematic group. Briefly, they can be distinguished, due to their living environment, in planktonic foraminifera, that is, floating marine organisms which, being unable to move actively, are transported by the currents and by the wave motion, or in benthic foraminifera, that is, strictly bound to the seabed, often adhering to a solid substratum. Their size is rather small, mostly under 1 mm, even

© The Author(s), under exclusive license to Springer Nature Switzerland AG 2022

M. Forli, A. Guerrini, *The History of Fossils Over Centuries*,
https://doi.org/10.1007/978-3-031-04687-2_9

if some benthic species can exceed one centimeter. Foraminifera are currently widespread in the oceans, including about 5000 species, with shells of various shapes, sometimes very complex, living in the most diverse environmental conditions, varying in different marine habitats. Their first fossil evidence dates back to the Precambrian, since then they spread unceasingly in the course of time, even if entire groups suffered extinctions like other prehistoric organisms. In particular, the mass extinction occurred at the Cretaceous-Tertiary transition, a moment dated at 66 Ma, generally known as the K-T limit, where K stands for the Cretaceous Period and T for the subsequent Tertiary Era, now sometimes replaced by the equivalent indication K-Pg (Cretaceous-Paleogene), has created the conditions for the spread of large nummulites characteristic of the Eocene terrain (~56 to ~34 Ma). In palaeontology many foraminifers are of extreme importance finding applications in biostratigraphic analyses. They are usually studied in equatorial and axial section).

Fig. 9.1 In the Middle Ages, the finding of these discs was interpreted as a trace of the passage or the work of the devil, who had lost them or had thus petrified the coins paid to commit a crime or a curse, making it in the end fruitless for the perpetrator. (Photo and graphic composition Maurizio Forli)

The nummulites, from the Latin *nummus* moneta + *lithos* pietra, which derive their name from the flattened discoidal shape of the shell, similar to that of a stone coin, were and are unicellular marine macroforaminifera, whose size varies from a few millimeters to about 12 cm in maximum diameter for some fossil species. Precisely because of their large average size, they can easily be observed in their entirety without the aid of lenses or microscopes, contrary to most other foraminifers which, as we said above, are instead small or very small. The oldest tertiary nummulite species appear in the Upper Paleocene (~58 to ~56 Ma) sediments of North Africa (Trevisan and Giglia 2005) and, although with a reduced number of species, are still widespread in present-day seas. They are used stratigraphically as *guiding fossils* of the Eocene, and their remains are sometimes so concentrated in the sediments of that period that they have given rise to particular rocks that take the name of *nummulite limestones*. Their shell has bilateral symmetry and is formed by the spiral winding of a V-shaped calcareous lamina, which gradually covers the previous turns. The canal is divided by septa which separate intercommunicating spaces; the communication between the inside and the outside is done through the pores of the shell and a reduced opening at the base of the terminal septum. The sexual dimorphism is diffused.

Phylum Foraminifera
Class Globothalamea
Superfamily Nummulitoidea
Family Nummulitidae
Genus *Nummulites*
Species *Nummulites laevigata* (Bruguiere 1792) †
(Fig. 9.2)

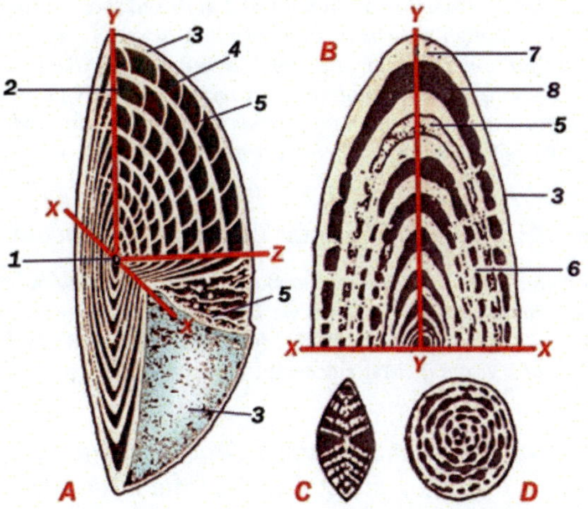

Fig. 9.2 Simplified structural scheme of nummulite, internal cavities are dark colored: (A) general schematic structure; (B) axial section, details; (C, D) axial and equatorial sections. (*x*–*x*) Winding axis; (*y*–*y*) equatorial plane; (*z*) cross section; (1) initial loggia; (2) oral opening; (3) spiral lamina or wall, exterior and thickness; (4) loggias; (5) septa; (6) pillars; (7) basal cord; (8) spiral canal. [Drawing Maurizio Forli modified after Vialli (1974)]

To the best of our knowledge, the most ancient presence of a nummulite in a prehistoric context, as a nonutilitarian object, perhaps to be interpreted as an ancient amulet, is to be referred to a specimen of *Nummulites perforatus* (Montfort 1808), bearing a cruciform sign, from the locality of Tata, Hungary, with finds dated between 130 and 70,000 years ago (Moncel et al. 2012).

In the Middle Ages, the finding of these discs was interpreted as a trace of the passage or work of the devil, who had lost them or had thus petrified the coins paid to commit a crime or a curse, making it in the end fruitless for the perpetrator. A beautiful woodcut of a single nummulite, and its description, comes from the aforementioned *De Fossilibus Opusculum* by Francesco Imperato:

> *La pietra piatta, come viene detta per il fatto che ha una forma schiacciata, ha la grandezza di una moneta, e si scompone, anche se non facilmente, in sottilissimi strati. Conserva un colore bianco che va nell'ocra, ma per alcuni elementi prominenti è inciso come le monete, grazie alla rottura dello strato superiore: la parte rimanente rimane leggermente a rilievo. Due tipi di pietra come questa ce li mandò il famosissimo Carolus Clusius, aggiungendovi queste parole: moneta di Dracola; moneta di D. Cratone; i poeti raccontano che la pietra piatta era nel novero di quelle che un tempo furono monete* (Imperato Fr 1610)

e, nella nota 339 della traduttrice, riferendosi alla definizione di *Dracolae nummus*:

> *La frase non è chiara. Forse si tratta di esemplari di nummuliti che per dimensione ed ornamentazione esterna ricordano o permettono di supporre che un tempo fossero state vere monete. Dracola dovrebbe riferirsi a Vlad II (1390–1447), voivoda* di Valacchia, che coniò monete con l'effige di un drago* (Mariantonietta Paladini 2015: pp. 173–174 e note, * principe)

> *The flat stone, as it is called from the fact that it has a flattened shape, is about the size of a coin, and breaks down, though not easily, into very thin layers. It retains a white colour going into ochre, but for some prominent features it is engraved like coins, owing to the breaking up of the upper layer: the remaining part remains slightly in relief. Two types of stone like this one were sent to us by the very famous Carolus Clusius, adding these words: coin of Dracola; coin of D. Cratone; the poets say that the flat stone was among those that were once coins*

and, in translator's note 339, referring to the definition of *Dracolae nummus*:

> *The phrase is not clear. Perhaps these are nummulite specimens that in size and external ornamentation are reminiscent of or allow one to assume that they had once been real coins. Dracola should refer to Vlad II (1390–1447), voivode* of Wallachia, who minted coins with the effigy of a dragon* (Mariantonietta Paladini 2015: pp. 173–174 and notes, * prince)

or, in our opinion, *Dracolae nummus*, since *Dracul*, can mean both *dragon* and *devil*, is still the classic definition of devil coins, from *Drackuluj* or "of the devil." (Fig. 9.3)

Fig. 9.3 (1) Nummulite from Francesco Imperato (*De Fossilibus Opusculum*, 1610: p. 97). (2) Coin of the prince of Wallachia, Vlad II, minted between 1445 and 1446, which in the effigy recalls the specimen figured and remembered by Imperato as *Dracola coin*. (3) *Nummulites* sp., similar to that illustrated by the Neapolitan scholar; 25 mm diameter, Eocene, Veneto. [Credit: (1) https://archive.org/details/bub_gb_yPMwLxv1n4kC Public Domain Mark 1.0 (2) modified after cng-coins.com/Coin.aspx?CoinID=247640 (3) photo and graphic composition Maurizio Forli)]

Aldrovandi illustrates and describes the *devil's coins* from the surroundings of Siena, near Vescovado di Murlo (*Musaeum Metallicum*, 1648: pp. 120–121; fig. 1) as blackish discs produced naturally but believed by the illiterate people to be devilish coins minted in hell. Since only Pliocene soils are found in that area, the attribution of these finds to nummulites, characteristic of other, more ancient soils, cannot be supported. In fact, these are metallic discs produced by the mineralization of marcasite with clay, which appear as blackish and eroded coins. The *diabolic nummas* illustrated in the *Metallotheca Vaticana* are also the same blackish discoidal formations, pseudometallic, and also coming from the area of the Crete Senesi, San Quirico d'Orcia, Montalceto, which is in the present-day municipality of Asciano, and other similar places:

> *Eruuntur è terra cretacea in ditione Senensi ad oppida Sancti Quirici & Montem Alcetum Senis ab exortu solis propinquiùs, & aliis ibidem loci* (Mercati, Metallotheca Vaticana, 1717: p. 240).

San Quirico, as a place where devil coins are dispensed, is also mentioned by Targioni-Tozzetti (1754: p. 233) in the *Report of the journey made in 1733 by Pier Antonio Micheli through various places of the Sienese State* [...] and by Giovanni Battista Brocchi (1814: p. 75):

> *Niuna sostanza metallica v' ha ne colli marnosi, se eccettuare si vogliano le piriti di ferro che si mostrano per lo più in grani e in cristalli isolati. Le così dette monete del diavolo di San Quirico e di Montalceto sono piriti discoidee e compresse che hanno esternamente una tinta bruna*

> *There is no metallic substance in marly hills, except if you prefer the iron pyrites which appear mostly in grains and in isolated crystals. The so-called coins of the devil of San Quirico and Montalceto are discoid and compressed pyrites that have a brown tint externally.*

while Montalceto is still quoted to describe the similar formations from the hills of Bacedasco in Parma (Cortesi 1819: p. 2): *These grains, however, in the hills of Bacedasco, are compressed and have a discoid shape, like the so-called coins of the Devil of S. Quirico and Montalceto, so it seems to be there a kind of lenticular penetrated by the iron* (Figs. 9.4 and 9.5)

Fig. 9.4 *Moneta Diaboli*, by Aldrovandi (1648: p. 120), and *Nummi Diabolici*, by Mercati (1717: p. 240) compared to those still found in some places in the Sienese territory. [Photo credit Massimo Cresti, graphic composition Maurizio Forli. Credit on the left: https://doi.org/10.3931/e-rara-13017 / Public Domain Mark. On the right: https://archive.org/details/bub_gb_DC_E6UVWg7EC (Public Domain Mark 1.0)]

Fig. 9.5 The same "coins" as they appeared naturally arranged in the Pliocene clays around Asciano (Siena). Largest size 23 mm. (Photo credit Massimo Cresti)

The nummulites and other types of similar foraminifera have been described also with different names, sometimes rather fanciful, for example, *Stone naturally carved in figures of wheat, and seeds of legumes* (Imperato, 1599: 664), or just as *wheat stones*, that is, as lapidary stones constituted by rocky aggregates of various seeds, mostly similar to wheat. So it is for the *Triticites molybdoides*, and the *Umbilicorum petrificatorum congerie* (*Musaeum Metallicum*, 1648: pp. 170, 843, tab. XXII), or *the Lapis frumentarius helvethicus* or the *Salicites helvethicus niger foliolis candidis* of the Swiss Karl Nikolaus Lang (*Historia lapidum figuratorum Helvetiae*, 1708: p. 69, pl. 18).

This similarity of the nummulites with various grains and seeds has originated, since the times of the old Greeks, the belief that those present in the blocks with which the Egyptian pyramids are built, a nummulitic limestone dating back to the middle Eocene, were nothing else than petrified lentils, remains of the meals of the old builders:

> *The oldest of the Greek historical writers Herodotus mentioned their abundance, and the way they were stratified in various parts of Africa. Something more Strabo left us as he said:*
> *Mirum est quod penes Pyramides ingentes lapides extent, plura minuta lenticularis figurae corpora recipientes* (Michelotti 1841: pp. 253–254). It is surprising (strange) that at the pyramids are visible so many stones, containing many small lentil-like bodies.

The nummulites of this particular area are also mentioned in Agricola [1546, *De Natura fossilium...* (see also 1558 edition)] *lentis, arena Africae; & lapilli ad pyramides in Aegypto* who mentions them among the fossil things with similarities to the present ones, without identifying them as remains of ancient marine organisms.

In the northeast of France, in an area where nummulitic limestones outcrop, there is a legend characteristic of this type of superstitious credulity. A farmer was sowing lentils while a traveler passing by asked him what he was doing. The farmer, just for fun, told him that he was sowing small stones, to which the other, who of course was the devil, replied that then he would have collected small stones. And so from that moment on, the farmers of the area could no longer sow and harvest lentils (Mozzani, *Légendes et mystères des régions de France*, 2015). The work of the devil that makes the use of these objects, both lentils and coins, futile is a typical explanation to justify their presence as something diabolical and nefarious.

Numerous localities in the Veronese area where nummulitic limestones outcrop, such as Roncà, famous for the abundance of its fossils, are enumerated by Spada (*Corporum lapidefactorum agri Veronensis Catalogus*, 1744 : pp. 49–51) who, in disagreement with his predecessors, who considered the "seeding" of the stones to be the work of the wind, points out that the said seeds are mixed together with the remains of shells and marine organisms and that therefore their origin is to be found in the waters of the sea (Vai and Cavazza 2006: 411, p. 53, fig. 6). In the Tavole Acquarellate of Aldrovandi (Ceregato, *Natura Picta*, 2007: p. 605, fig. 489), MS reproduced and plundered for two centuries all over Europe, a sample of nummulite limestone from the Museum of Aldrovandi is represented (Vai, "personal communication", August, 2021) coming from the Eocene of the Vicentino.

In Bertrand (1763: p. 73) in addition to the specification of the etymology according to their shape and origin, it is explained how they can be opened and observed internally, by throwing them into water after red hot on coals:

> *Pour découvrir leur forme intérieure, il n'y a qu'à échaufer une de ces pierres sur un carbon. Jettez-la toute chaude dans l'eau froide, vous la verrez se lever par couches minces, comme les Besoards. Vous observerez alors dans la coupe du milieu sur les deux surface planes correspondantes, deux spirales, dans d'autres des cercles concentrique & correspondans.*

To discover their internal shape, simply heat one of these stones over charcoal. Throw it completely hot into the cold water; you will see it rise a little from the bottom to the top, like the Bezoards*. You will then observe in the middle cut, on the two corresponding flat surfaces, two spirals, in others concentric and corresponding circles (Fig. 9.6).

Fig. 9.6 Our replication of a simple experiment: nummulite from the Eocene of Veneto, opened by boiling and subsequent rapid immersion in cold water. On the fourth attempt, the shell cracked and separated, revealing the interior. (Photo Maurizio Forli)

*(Bezoard: stony concretions or calculi, formed in the body of certain animals, believed to possess miraculous powers, both in medicine and magic)

This particular structure has contributed not a little to generate confusion, leaving many doubts on their real essence, until relatively recent times, being believed from time to time, bivalve shells, opercula of gastropods or ammonites. Curiously, also other small foraminifera, like tiny ammonites, have been mistaken for *horns of Ammon*, that is, real ammonites in today's sense, as having a flattened and spiral shape, were similar to the circumvolutions ammonoid then not well understood (Figs. 9.7 and 9.8).

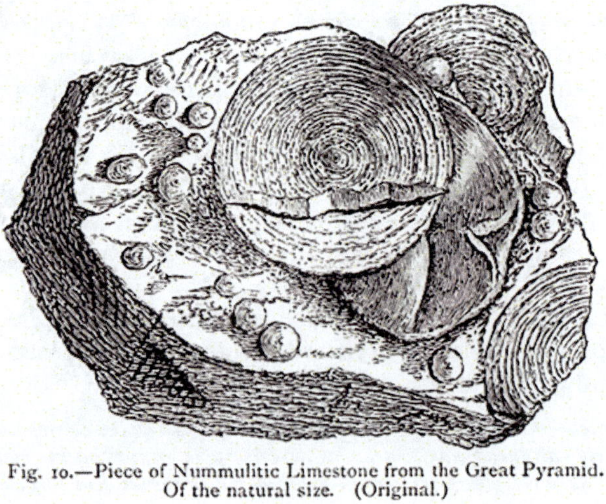

Fig. 9.7 Nummulitic limestone from the Pyramid of Cheops. (Modified after Nicholson, *The ancient life-history of the earth*, 1877: p. 25) (https://www.biodiversitylibrary.org/bibliography/13890 Public Domain)

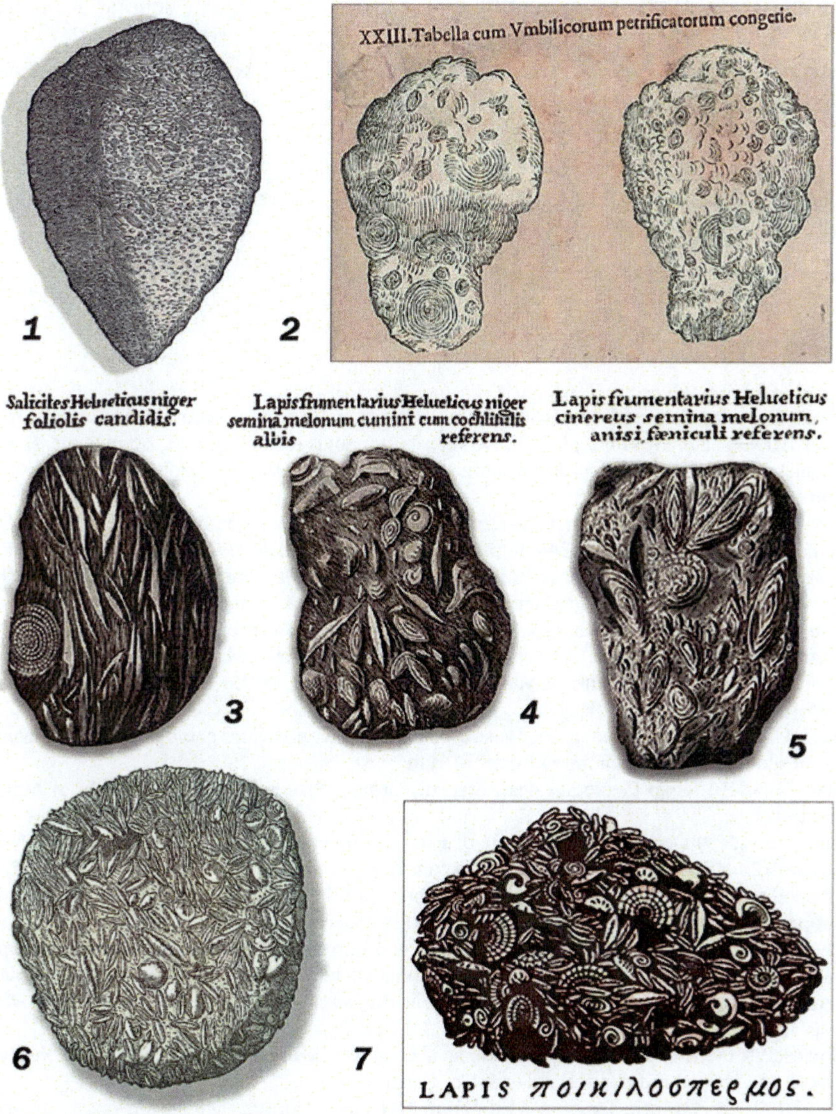

Fig. 9.8 Samples of nummulite rocks: (1) Lead-colored rock with petrified wheat grains: *Triticites molybdoides* (Aldrovandi 1648: p. 170); (2) confused cluster of umbilicus-shaped *petrifications*: *Umbilicorum petrificatorum congerie* (Aldrovandi, 1648: p. 843); (3–5) figurative stones: dark Helvetic rock, representing candid willow leaves, *Salicites helvethicus niger foliolis candidis*; frumentary stones with various types of seeds, cumin, anise, melon, etc. (Lang 1708: p. 69, pl. 18); (6) stone naturally carved in figure of wheat, and seeds of legumes (Imperato 1599: p. 664). (7) Scattered seeds, of various categories, petrified in the rock: *Poecilospermos* (Mercati 1717: p. 285). [Color and graphic composition Maurizio Forli. Credit from (1–2) https://doi.org/10.3931/e-rara-13017 / Public Domain Mark (3–5) https://doi.org/10.3931/e-rara-10398 / Public Domain Mark (6) https://www.biodiversitylibrary.org/item/183177#page/5/mode/1up (not in copyright) (7) https://archive.org/details/bub_gb_DC_E6UVWg7EC (Public Domain Mark 1.0)]

In the early 1900s to nummulites in general, the English scholar Randolph Kirkpatrick devoted a scientific work published in a book entitled *Nummulosphere: An Account of the Organic Origin of the So-Called Igneous Rocks and Abyssal Red Clays* (1913), in which he proposed that the Earth was originally covered by seas in which nummulite foraminifera thrived; these, with their remains, would have given rise to a global sedimentary layer called the *Nummulosphere* from which all other rocks on the planet would then derive. An ingenious theory but, in this case, absolutely unfounded.

References

Agricola G (1546) De ortu & causis subterraneorum Lib. V. De natura eorum quae effluent ex terra Lib. IIII. De natura fossilium Lib. X. De veteribus & novis metallis Lib. II. Bermannus, sive de re metallica dialogus. Interpretatio Germanica vocum rei metallicae, addito Indice foecundissimo. Basel: Hieronymus Froben

Agricola G (1558) De ortu et causis, subterraneorum lib. V. Basileae : [Froben], M D LVIII. [1558]. ETH-Bibliothek Zürich, Rar 244, https://doi.org/10.3931/e-rara-9796 / Public Domain Mark

Aldrovandi U (1648) Ulyssi Aldrovandi patricii Bononiensis Musaeum metallicum in libros IIII: Distributum Bartholomaeus Ambrosinus ... labore, et studio composuit cum indice copisosissimo Arcus Antonius Bernia. typis Io. Baptistae Ferronii, Bononiae. ETH-Bibliothek Zürich, Rar 1027, https://doi.org/10.3931/e-rara-13017 / Public Domain Mark

Bertrand E (1763) Dictionnaire universel des fossiles propres et des fossils accidentels. Pierre Gosse et Daniel Pinet, La Haye

Brocchi GB (1814) Conchiologia fossile subapennina con osservazioni geologiche sugli Apennini e sul suolo adiacente. Stamperia Reale, Milano

Cortesi G (1819) Saggi Geologici degli Stati di Parma e Piacenza. From the Torchj of Maino, Piacenza

Imperato F (1599) Dell'Historia Naturale di Ferrante Imperato napolitano. Libri XXVIII. Nella quale ordinatamente di tratta della diversa condition di miniere, e pietre. Con alcune historie di Piante, et Animali fin'ora non date in luce. Costantino Vitale, Napoli. https://www.biodiversitylibrary.org/item/183177#page/5/mode/1up (not in copyright)

Imperato F (1610) De Fossilibus Opusculum. In: quo miro ordine continentur Naturalis disciplinae scitu dignissima, eiusque professoribus omnino necessaria; ab alijs minime excogitata. Accesserunt. Dominici Roncalioli, Neapoli. https://archive.org/details/bub_gb_yPMwLxv1n4kC. (Public Domain Mark 1.0)

Kirkpatrick R (1913) The nummulosphere: an account of the organic origin of so-called igneous rocks and abyssal red clays. Lamley & Co, London

Lang KN (1708) Historia lapidum figuratorum Helvetiae, eiusque viciniae: in qua non solum enarrantur omnia eorum genera, species et vires aeneisque tabulis repraesentantur, sed insuper adducuntur eorum loca nativa, in quibus [...]. sumptibus authoris, typis Jacobi Tomasini, Venetiis. ETH-Bibliothek Zürich, Rar 5680, https://doi.org/10.3931/e-rara-10398 / Public Domain Mark

Mercati M (1717) Metallotheca Vaticana. Opus posthumum, Autoritate et Munificentia Clementis Undecimi Pontificis Maximi. E tenebris in lucem eductum: Opera autem, & studio Joannis Mariae Lancisii, Archiatri Pontificii illustratum. Ex Officina Jo, Mariae Salvioni Romani, Romae. https://archive.org/details/bub_gb_DC_E6UVWg7EC (Public Domain Mark 1.0)

Michelotti G (1841) Saggio storico dei Rizopodi caratteristici dei terreni sopracretacei. Memorie Mat Fis Soc Ital Sci 22(2):253–302

Moncel MH, Chiotti L, Gaillard C, Onoratini G, Pleurdeau D (2012) Non-utilitarian lithic objects from the European paleolithic. Archaeol Ethnol Anthropol Eurasia 40(1):24–40

Mozzani E (2015) Légendes et mystères des régions de France. Éditions Robert Laffont, Paris

Nicholson HA (1877) The ancient life-history of the earth: a comprehensive outline of the principles and leading facts of palæontological science. https://www.biodiversitylibrary.org/bibliography/13890 Public Domain

Paladini M (trans: it. ed.) (2015) De Fossilibus Opusculum, 1610. Quad Accad Pontiniana 61:121–210

Spada G (1744) Corporum lapidefactorum agri Veronensis catalogus. Typis Dionysii Ramanzinii : Bibliopolæ apud S. Thomam, Veronæ. Zentralbibliothek Zürich, NG 58 I G, https://doi.org/10.3931/e-rara-50868 / Public Domain Mark

Targioni-Tozzetti G (1754) Relazione del viaggio fatto nel 1733 da Pier Antonio Micheli per diversi luoghi dello Stato Senese. In: Relazioni d'alcuni Viaggi Fatti in diversi Parti della Toscana, per osservare le Produzioni Naturali, e gli Antichi Monumenti di essa. Stamperia Imperiale, Firenze

Trevisan L, Giglia G (2005) Introduzione alla Geologia, seconda edizione. Pacini Editore, Pisa. ISBN 88-7781-482-9

Vialli V (1974) Appunti di Paleontologia. Pitagora Editrice, Bologna

Chapter 10
Mollusca

Abstract The classes included in the phylum Mollusca are treated: Polyplacophora, Bivalvia, Gastropoda, Scaphopoda, and Cephalopoda. After each introduction, concerning information on the anatomy and geological distribution of the members of the individual classes, the same are discussed from the historical and scientific point of view, with notes on some folkloristic aspects related to them.

Keywords *Fossilia* · Fossils · History · Folklore · Mollusca

Apart from the very first images borrowed from the miniatures of medieval codices, already in the oldest printed works, the mollusks become part of the iconography more or less naturalistic in the modern sense, with various species of gastropods, bivalves, and cephalopods illustrated with a wealth of realistic details, especially when it comes to tropical shells before unknown in Europe. Especially for some cephalopods, reconstructions and consequent frightening representations, in the form of monstrous beings while attacking and destroying vessels persist for long time. These medieval reminiscences disappeared as the knowledge of seas and fauna increased. One has sometimes the impression that these images are used more as an imaginative aesthetic corollary to adorn the works, for a greater satisfaction of the reader, or to discourage the daredevils willing to go to the farthest corners of the world and compete in the exploitation and search for new riches, than for the conviction of the writers of their actual existence (Fig. 10.1). The classes included in the phylum Mollusca are treated: Polyplacophora, Bivalvia, Gastropoda, Scaphopoda, and Cephalopoda. Their classification is in a constant state of revision. For simplicity of exposition, we have preferred for the hierarchically broader taxonomic categories to continue using the long-established ones.

© The Author(s), under exclusive license to Springer Nature Switzerland AG 2022

M. Forli, A. Guerrini, *The History of Fossils Over Centuries*,
https://doi.org/10.1007/978-3-031-04687-2_10

Fig. 10.1 Summary plate of the most common classes belonging to the phylum Mollusca. From top left, bottom right: (1) Bivalvia, represented by a *Pecten jacobaeus* Bellon (1553: p. 409), (2) Gastropoda (land snail) from Moscardo (1656: p. 180), (3) Scaphopoda (*Dentalium elephantinum*) from Buonanni (1681: pl. II, figs 8), (4) Cephalopoda (*Octopus vulgaris*) from Bellon (1553: p. 332), (5) Polyplacophora (*Chiton* sp.) from Rondelet (1554: 529). In the background, detail from the frontispiece of the famous work by the Jesuit monk Filippo Buonanni: *Ricreatione dell'occhio e della mente nell'osservation' delle chiocciole* (1681). [Graphic composition Maurizio Forli. Credit from: (1, 4) https://www.biodiversitylibrary.org/bibliography/7124 (Public domain) (2) https://www.biodiversitylibrary.org/bibliography/157058 (not in copyright) (3) https://doi.org/10.3931/e-rara-39893 / Public Domain Mark (5) www.biodiversitylibrary.org/bibliography/64229 (not in copyright)]

References

Bellon P (1553) De Aquatilibus, Libri duo. C. Stephanum, Parisiis. https://www.biodiversitylibrary.org/bibliography/5765 (Public domain)

Buonanni F (1681) Ricreatione dell'occhio e della mente nell'osservation' delle chiocciole : proposta a' curiosi delle opere della natura. In Roma : per il Varese : a spese di Felice Cesaretti ..., Zentralbibliothek Zürich, NNN 713 | G, https://doi.org/10.3931/e-rara-39893 / Public Domain Mark

Moscardo L (1656) Note overo memorie del museo di Lodovico Moscardo nobile veronese. Academico Filarmonico, dal medesimo descritte, Et in Tre Libri distinte. Per Paolo Frambotto, Padoa. https://www.biodiversitylibrary.org/bibliography/157058 (not in copyright)

Rondelet G (1554) Libri de Piscibus Marinis, in quibus verae Piscium effigies expressae sunt. Matthiam Bonhomme, Lugduni. https://www.biodiversitylibrary.org/bibliography/64229 (not in copyright)

Chapter 11
Polyplacophora: from *Physsalus*, Oscabrion, to Chiton

Abstract The Polyplacophores, commonly called chitons, are strictly benthic mollusks, exclusively marine, characterized by a flattened body with bilateral symmetry, currently covered by a shell formed by eight superimposed dorsal plates, imbricated and articulated. Their presence in the fossil state was recognized only in the nineteenth century, while the first printed representations were made by Rondelet (*Libri de Piscibus Marinis, in quibus verae Piscium effigies expressae sunt*. Matthiam Bonhomme, Lugduni. https://www.biodiversitylibrary.org/bibliography/64229 (not in copyright), 1554; *Universae aquatilium Historiae pars altera, cum veris ipsorum Imaginibus*. Matthiam Bonhomme, Lugduni. https://archive.org/details/ARes071032/page/n13/mode/2up (Public Domain 1.0), 1555), *Gessner (Historiae Animalium Liber IIII. qui est de Piscium & quatilium animantium natura*. Christoph. Froschoverum, Tiguri. https://doi.org/10.3931/e-rara-16853 (Public Domain Mark), 1558), and Aldrovandi (*De Reliquis Animalibus exanguibus libri quatuor, post mortem eius editi: nemp' De Mollibus, Crustaceis, Testaceis, et Zoophytis*. Apud Io: Baptistam Bellagambam, Bononie. https://www.biodiversitylibrary.org/bibliography/119999 / Public Domain, 1606). It is our belief that the representation and description of the fish *Physsalus*, which accompanies the description made by Rondelet (*Libri de Piscibus Marinis, in quibus verae Piscium effigies expressae sunt*. Matthiam Bonhomme, Lugduni, p 428. https://www.biodiversitylibrary.org/bibliography/64229 (not in copyright), 1554), is to be attributed to a chiton.

Keywords Polyplacophora · Chitoni · *Physsalus* · *Lepas parva* · Oscabrion

The Polyplacophora, commonly called chitons, are strictly benthic mollusks, exclusively marine, characterized by a flattened body with bilateral symmetry, presently covered by a shell formed by eight superimposed dorsal plates, imbricate and articulated. The shell is surrounded by a thick fold of the mantle, like a belt, called perinotum, in its turn covered by several calcareous or horny spines, scales, or bristles and also by sensory organs. Ventrally, the mollusk has an elongated fleshy foot,

© The Author(s), under exclusive license to Springer Nature Switzerland AG 2022
M. Forli, A. Guerrini, *The History of Fossils Over Centuries*,
https://doi.org/10.1007/978-3-031-04687-2_11

surrounded by the pallial groove where the gills are contained; they do not have eyes, cephalic tentacles, and statocysts. The sense organs are represented by the subradular organ, probably gustatory, and by the aesthetes. These last ones, typical of this group, are constituted by mantle cells lodged inside tiny vertical canals of the superior tegmentum. The channels containing the sensory terminations reach a cap situated beneath the surface of the shell plates. Their density is extremely high and their form and arrangement is used in systematics as a distinguishing character between one species and another. The precise function of these peculiar organs is not clear, even though it seems that they are used to sense the presence of a food source or conspecifics. In the Chitonidae, some of the aesthetes are modified and function as single eyes or ocelli, concentrated in particular on the anterior plates. The surface of the mantle, which extends to variously cover the plates of the shell, has numerous tactile and photoreceptor cells even though spines and aculei are present. In the pallial groove, various areas of the wall are covered by sensory epithelium considered similar to the olfactory sense organ of the other mollusks. They have separate sexes but the fecundation is external, in the water medium, or it may take place inside the pallial cavity of the female. In this case, this is also the place where eggs incubation takes place. The process of fecundation is favored by their gregarious habits. Generally, the ontogenesis takes place by typical spiral segmentation, which leads to the formation of a planktonic larva called trocophora From this freely swimming larva, a young chiton forms directly, without passing through the stage of veliger typical of other acquatic mollusks. At least in one case, *Calloplax vivipara* (Plate, 1899), it has been observed that the young come out from the pallial cavity of the mother already completely formed (Dell'Angelo and Smriglio 1999). These animals are generally active at night when they generally feed on algae. The largest species *Cryptochiton stelleri* (Middendorff, 1847), widespread in the North Pacific from Alaska to Japan, reach 30 cm length. They live mainly in rocky or shallow intertidal marine environments, adhering tenaciously to the rocks with the foot functioning as a sucker, but smaller species are also found in very diverse environments, even at a few thousand meters depth, in particular *Lepidopleurus vitjazi* Sirenko, 1977, was dredged off the Solomon Islands between 6920 and 7657 meters depth (Dell'Angelo and Smriglio 1999). The articulation of the dorsal plates, accompanied by the presence of a powerful longitudinal muscle, allows them to roll up on the ventral side to protect themselves from predators. Among these, we can count octopuses, several starfishes, and some fishes like cod and sole (Fig. 11.1).

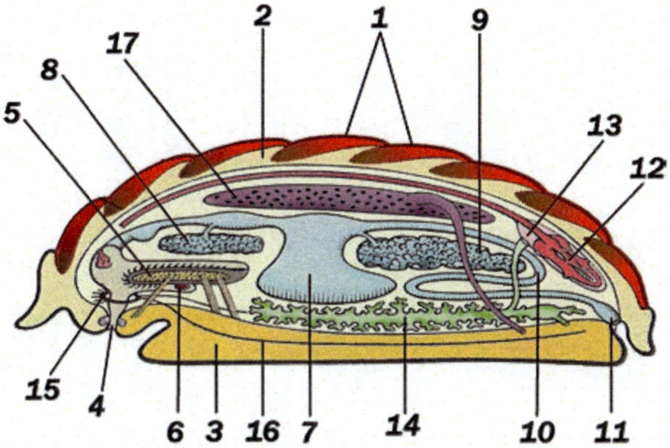

Fig. 11.1 Schematic longitudinal section of the body of a chiton to show its anatomy. (1) Dorsal valves or valves of the shell. (2) Mantle. (3) Foot. (4) Oral aperture. (5) Radula. (6) Subradular gland. (7) Stomach. (8) Anterior digestive gland. (9) Posterior digestive gland. (10) Intestine. (11) Anal aperture. (12) Heart. (13) Pericardium cavity. (14) Excretory organ. (15) Nerve ring. (16) Nerve termination of foot. (17) Gonad. (Modified drawing Maurizio Forli, courtesy Società Italiana di Malacologia)

The most ancient fossil evidence dates back to the Devonian, even if there are other uncertain, even more ancient ones. Their evolution develops during the Mesozoic, but it is with the following Cenozoic that the shape and organization of the body is completed, reaching the present model. The fossil species known at the moment are little less than five hundred and the present ones are about 950. Their systematic position within the Phylum Mollusca has been controversial since the nineteenth century. Their place in the hierarchical scale of invertebrates was then still under discussion, not having first well understood whether they were crustaceans, insects, or mollusks. The class that contains them within the Phylum Mollusca, still considered valid today, was proposed by De Blainville in 1816 as Polyplaxiphores corrected in Polyplacophora by Gray (1821) a few years later. A revision of the chiton systematics, currently followed, was proposed just over a decade ago by Sirenko (2006), based not only on conch morphology but also on other anatomical characters. In polyplacophores, sometimes for a series of anomalies, the plates can be in number different from eight. If they are in excess, we speak of hypermeria, if instead they are in smaller quantity, we speak of hypomeria. Between the two, the hypomeria seems to be more frequent, and several nineteenth-century authors reported of teratological chitons with a lower number of plates, from five to seven, but the cases of hypermeria are quite rare. The best known representation of a specimen with nine plates is in Scheuchzer, also like Gessner, a physician and naturalist from Zurich (1733: pl. 554, fig. B), although it is probably it is an error of the designer and not a real anomalous chiton (Dell'Angelo and Schwabe 2010) (Figs. 11.2, 11.3, 11.4, 11.5, 11.6, 11.7, 11.8, and 11.9).

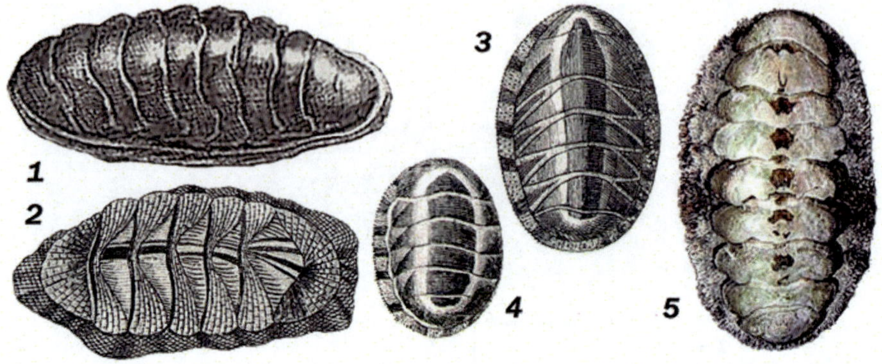

Fig. 11.2 Cases of hypermeria, nine-valved specimens: (1) from Scheuchzer (1733: pl. 554, fig. B). (5) *Acanthopleura gemmata* (Dell'Angelo and Schwabe 2010). (Courtesy Società Italiana di Malacologia). Cases of hypomeria, seven- and six-valved specimens. From Bruguière (1791): (2) pl. 162, fig. 10. (3–4) pl. 163, figs. 4–5. (Personal copy Maurizio Forli)

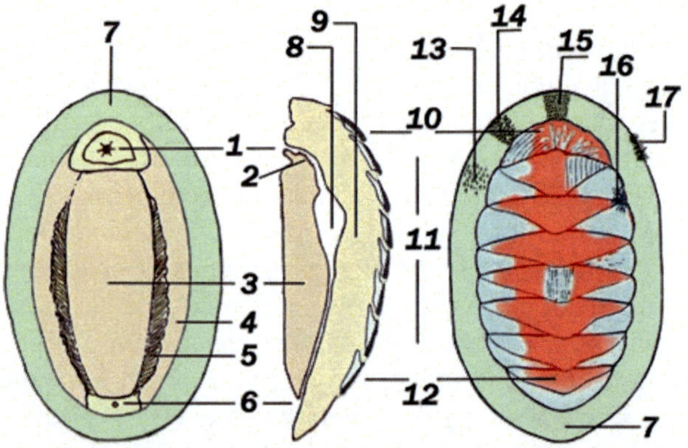

Fig. 11.3 Schematic structure of polyplacophora in basal view, lateral and dorsal sections. (1) Mouth. (2) Radula. (3) Foot. (4) Pallial fold. (5) Gills. (6) Anus. (7) Perinotum. (8) Digestive tract. (9) Visceral mass. (10) Anterior shell plate. (11) Intermediate shell plates. (12) Posterior shell plate. (13) Spicole. (14) Granules. (15) Scales. (16) Bristles. (17) Circles. [Drawing Maurizio Forli modified after Dell'Angelo and Smriglio (1999)]

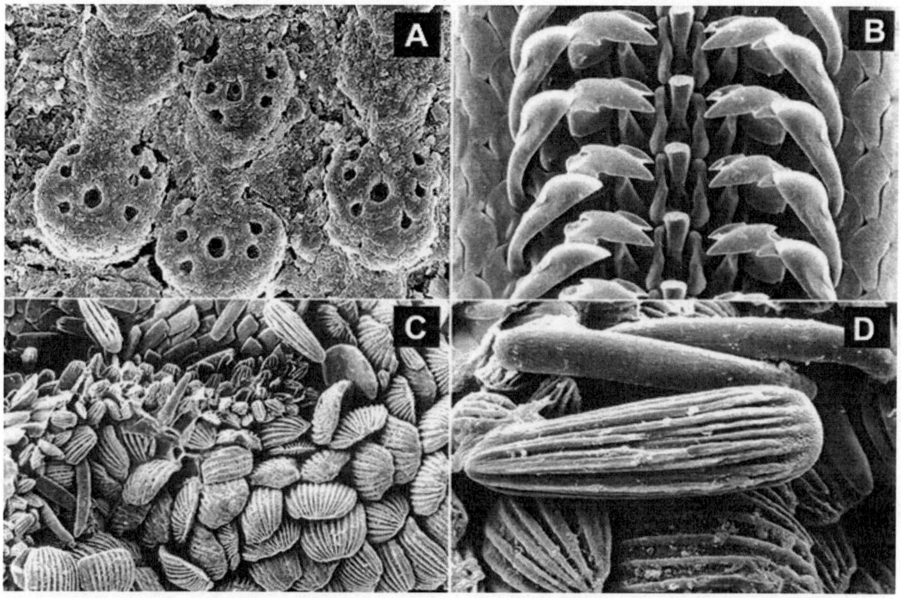

Fig. 11.4 Electron microscopic images with some details of the body of a polyplacophora: (**A**) Tubercles with macro- and micro-aesthetes (40×). (**B**) Radula (320×). (**C**) Dorsal scales and marginal spicules of perinotum (160×). (**D**) Spicole (640×). [Graphic composition Maurizio Forli modified after Dell'Angelo and Smriglio (1999)]

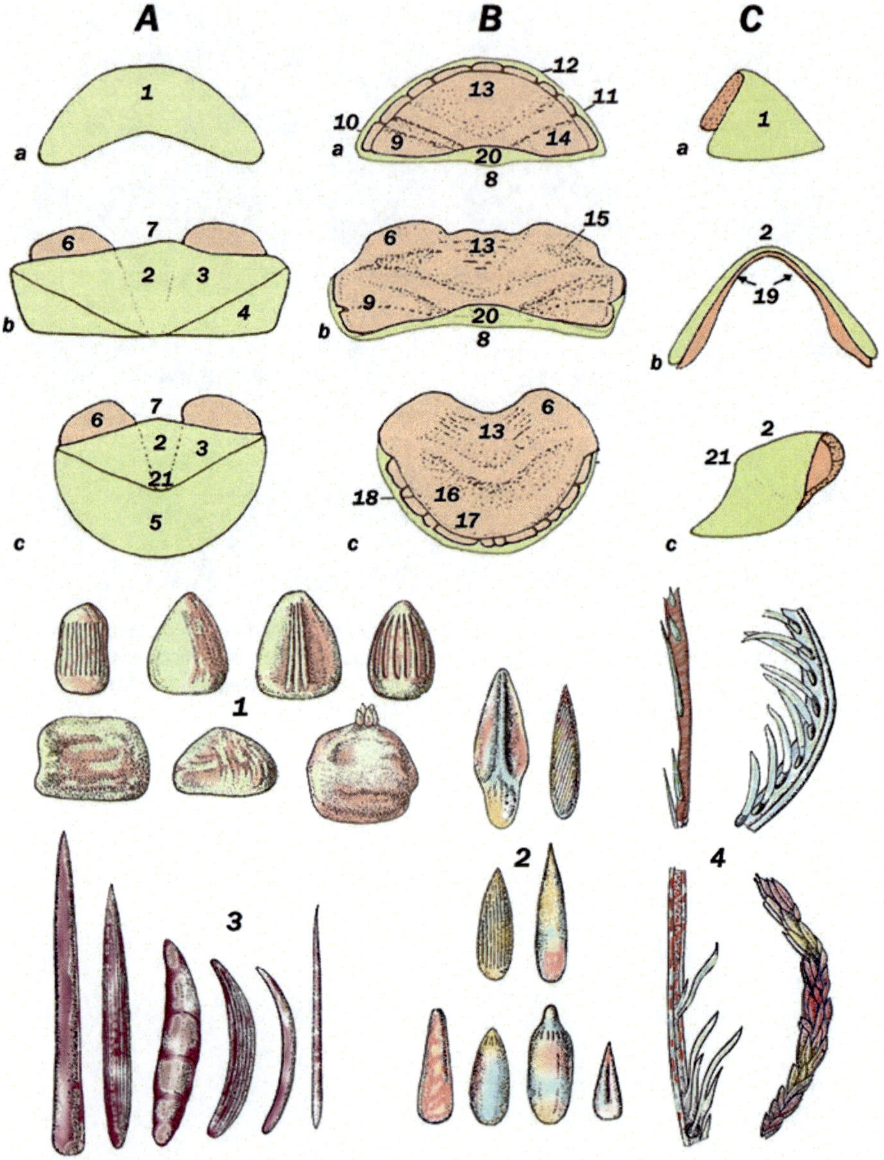

Fig. 11.5 Above, morphology of the plates' types. (**A–C**) Dorsal, ventral, and lateral sides: (a–c) Cephalic valva; intermediate and caudal valva. (1) Anterior area. (2) Jugal area. (3) Pleural area. (2 + 3) Central area. (4) Lateral area. (5) Posterior area. (6) Apophysis or sutural laminae. (7) Jugal sinus. (8) Apex. (9) Radial groove. (10) Insertion lamina. (11) Incision. (12) Margin of tegmentum. (13–14) Median and lateral fossa. (15) Muscle impression. (16) Posterior fossa. (17) Callus. (18) Periostracal sulcus. (19) Jugal angle. (20) Apical area. (21) Mucron. Bottom, perinotum formations: (1) Scales. (2) Spicules. (3) Thorns, (4) Bristles. [Graphic composition Maurizio Forli modified after Dell'Angelo and Smriglio (1999) (colorations plausible but fanciful)]

Fig. 11.6 Three current chitons: (1) *Chaetopleura chelazziana* (69 mm). (2) *Lepidochitona caprearum* (10 mm). (3) *Ischnochiton rissoi* (23 mm). (Photo Stefano Bartolini, credit Bruno Dell'Angelo)

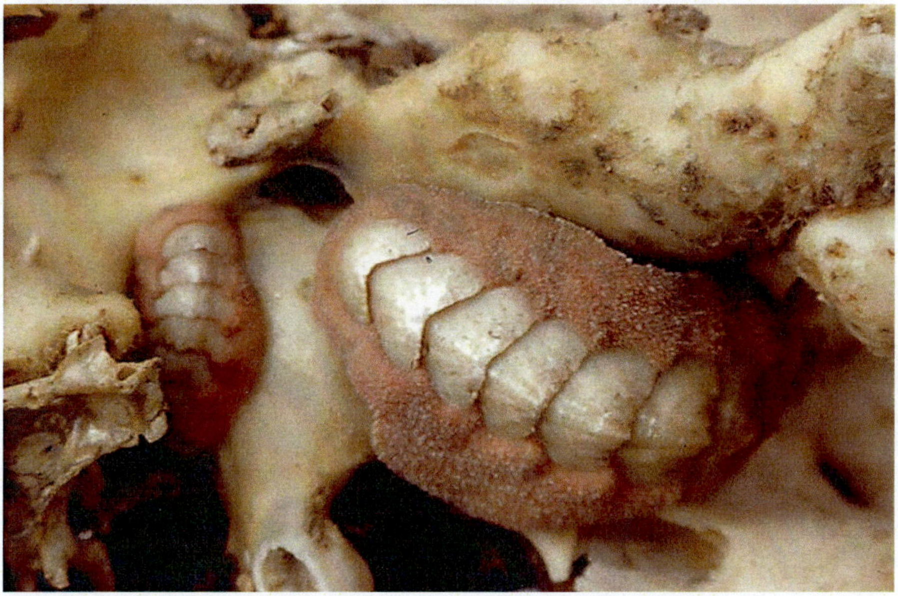

Fig. 11.7 A pair of polyplacophora in their natural environment: *Hanleya nagelfar* on white corals. (Courtesy Società Italiana di Malacologia)

Fig. 11.8 *Cryptochiton stelleri* (Middendorff, 1847), the largest species, widespread in the North Pacific from Alaska to Japan. Dried specimen, length 200 mm. (Photo and graphic composition Maurizio Forli)

Fig. 11.9 *Chiton saeniensis* Laghi, 1984, compiled specimen, Serre di Rapolano (Siena, Tuscany), Pliocene. Top left *Proleptochiton* sp., (11.5 mm) articulated chiton, Carboniferous, Republic of Czechia [(Mikuláš et al. 2017) courtesy of Società Paleontologica Italiana]. (Graphic composition Maurizio Forli, photo credit Massimo Cresti)

Apparently, there is no trace of these marine animals in the ancient classics that have come down to us. Before they were universally known under the name of chitons, with the genus *Chiton* (from the ancient Greek *χιτών, khitōn*, tunic or armor) proposed by Linnaeus in his *Systema Naturae* (1758: 667, number 266), they had been referred to by various names including *Oscabrion*. This last term, borrowed from an analogous precedent referred to another marine organism deemed to be similar, *Oscabiorn*, from the Icelandic language, had a favorable reception especially among the French authors, and it is just under this name that the chitons are described, for instance, in the *Dictionnaire des Sciences Naturelles* by de Blainville (1825: 519–555). The latter, after recalling those who had dealt in some way with them, writes of them giving an extensive anatomical description comparing them with two groups of apparently similar gastropods, the Limpets and the Nudibranchs (Phyllidies). The first known printed figure is the one reported in Rondelet (1554: 529), which shows a block of rock in which, among other organisms typical of the marine biocoenosis of the rocky mesolittoral, we can easily recognize a chiton in dorsal view. This chiton has supernumerary plates, eleven instead of the normal eight, and in the text, as also indicated by the nineteenth-century authors, there are no notes or descriptions about it. In the following work by Rondelet *Universae aquatilium Historiae pars altera* (1555: 3–5), two pages after the image of the usual piece of rock, also appears the description of this marine organism, *Lepas parva*, in truth a bit 'confused', although the author takes care to warn us of the accuracy of his studies even for small things:

> Cum *marinas res etiam minutissimas quasque diligenti indagatione persequerer, lepadem parvam saxo haerentem deprehendimus, quam a saxo evulsam repraesentamus in superiore pictura.*

Even with marine things of very small dimensions we try to make accurate investigations, we have taken the small lepade adhering to the rock, which avulsed from the stone we represent in the figure above.

Also this iconography is copied in the works of later authors, sometimes several times in the same text, illustrating the different organisms that were described separately from time to time, for example, in Gessner (1558: 463, 806) where the chiton is reported with the letter E and the name of *Lepas parva*, or in Aldrovandi (1606: 410, 567), reported practically without any comment (Figs. 11.10, 11.11, and 11.12).

Fig. 11.10 The first report of a neogenic fossil chiton, with the plates still in anatomical connection. *Craspedochiton altavillensis* (Seguenza, 1876) still adherent in the inner part of the giant pectinid *Macrochlamys latissima* (Brocchi, 1814) from the Tuscan Pliocene (Dell'Angelo et al. 2003). (Redrawn and graphic composition Maurizio Forli. Courtesy Bruno Dell'Angelo)

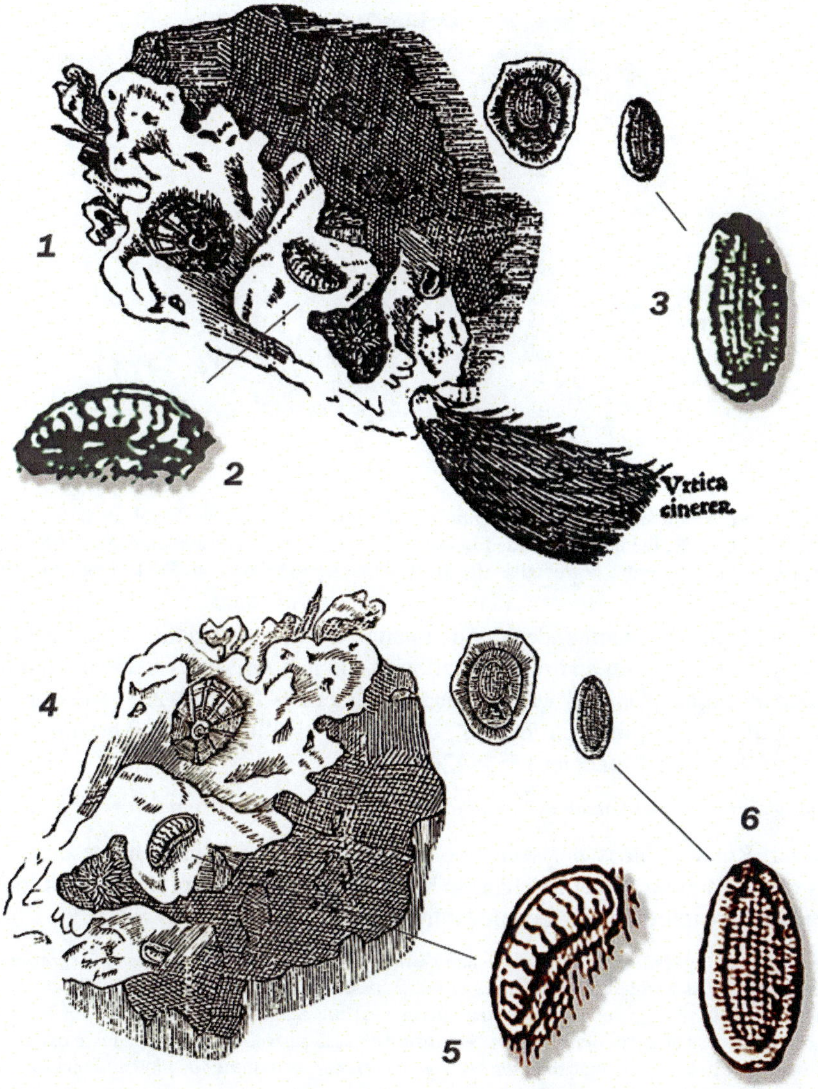

Fig. 11.11 From Rondelet (*Libri de Piscibus marinis, in quibus verae Piscium effigies expressae sunt*, 1554: pp. 529, 582). These are the first representations of chiton in a printed work; it appears with other invertebrates on a rocky block but is neither named nor described. (1) Image from the paragraph on *Urtica cinerea*, an actinia of Anthozoa class. (2–3) Enlarged details of chiton in dorsal and ventral view. (4) The same small piece of rock, without *Urtica cinerea*, used in the paragraph dedicated to the description *De Echinorum fifth genus*. (5–6) Enlargements in dorsal and ventral view. [Graphic composition Maurizio Forli. Credit from: https://www.biodiversitylibrary.org/bibliography/64229 (not in copyright)]

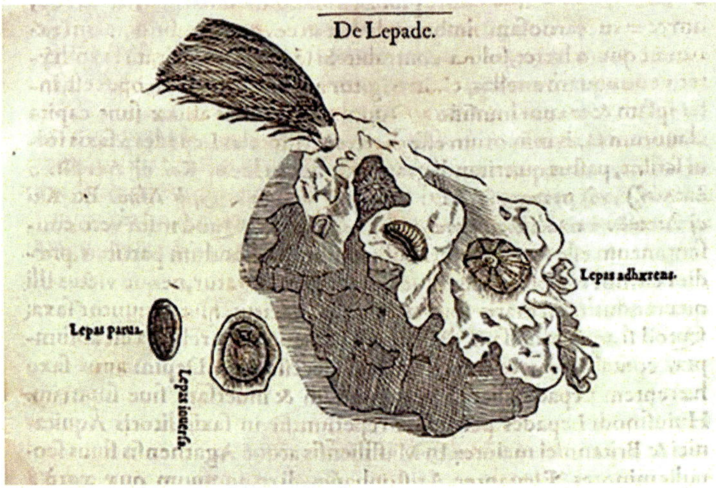

Fig. 11.12 From Rondelet (*Universae aquatilium Historiae pars altera*, 1555: pp. 3–5). In the paragraph *De Lepade*, the chiton included in the figure used above is described and referred to as *Lepas parva*. [https://archive.org/details/ARes071032/page/n13/mode/2up (Public Domain 1.0)]

It is our belief that another depiction of chiton is concealed in the French author's text: it is the description and representation of the fish *Physsalus* made by quoting the Roman philosopher and writer Claudius Elianus (~165/170–235) who, apparently, dealt with a spiny "puffer fish" capable of swelling and swimming at the surface of the water (Rondelet 1554: 428).

> *Inter pisces maris rubri Aelianus Physsalum ex Leonide recenset et describit his verbis.* [...]

However, Rondelet does not seem too convinced that the "fish" he describes is really corresponding to that of Elian, so he tells us that, if it is not the same, it looks very much like it, otherwise it could be traced to some kind of sea worm.

> *Sed si physsalus non est, ei certe non valde dissimilis iure dicetur, vel erucarum marinorum generi rectius subiicietur. Ore et oculis caret, in medio latior est, extrema gracilescunt et incurvantur, in ventre sive in supina parte rugosus est, pubendi muliebris species referens, in prona parte sive in dorso tumores parvi eminent, verrucas piscatores nostri vocant, è quibus pili virides existunt. At contactum intumescit et supernatat, ut Aelianus de physsalo scripsit.*

The description of above passage, more or less freely translated, reads as follows:

Mouth and eyes are missing, it is wide in the middle, thinned and curved at the ends, in the belly or supine part, it is wrinkled, similar to a woman's genital organ, in the prone part small protuberances stand out, which our fishermen call warts, and here there are green hairs. On contact it swells and swims to the surface as Elian wrote of the physsalus.

This representation is still nowadays suitable for two marine invertebrates similar but completely different from the zoological point of view. The first one is presently called *Aphrodita aculeata* L., 1758 and the connection to the Rondelet figure we found in *Tractatus de Phisalo*, a small chapter in the larger collection of the works of the Dutch naturalist Ian Swammerdam's (1637–1680) *Biblia naturae* published posthumously in two parts, in 1737 and 1738, by Herman Boerhaave (Swammerdam

1738: 902–906, pl. X, figs. VIII–XVI) and in Francesco Redi's treatise *Intorno agli animali viventi che si trovano negli animali viventi* (1684: 187, pl. 25, figs. 1–2):

> *Mi fu portato in Livorno un'Animaletto marino della figura, e grandezza perappunto disegnata nella Tav. ventesimaquinta, Fig. prima, e seconda, a cui piacemi di dar nome di spinoso marino, o d'Istrice marino. Questo avea il ventre per di sotto di color bianco liscio, non peloso, ma con rughe rette trasversali, e rilevate, come tanti cordoni, Dal capo, fino alla coda nell'uno e nell'altro de' lati del ventre era tutto circondato come da tanti pennellini di setole. Ne contai ventisei per ogni banda; sicche in tutto furono cinquantadue.* [...] *Queste setole sembravano tutte di color nericcio, e filigginoso, eccetto quella del mezzo, la quale essendo sempre la più lunga, e la più grossa, ella è parimente sempre di color oro lustrante, che alcuna volta rassembra velato di verde, secondo gli sbattimenti della luce* [...] *Dall'una delle estremità si apre il forame della bocca, all'intorno della quale pendono due antenne, o cornetti carnosi flosci, e bianchi.* [...] *L'intestino senza avvolgimenti va diritto alla volta del podice; ma dal piloro fin quasi per tutta la lunghezza dell'intestino pendono due ordini paralleli d'intestini ciechi* [...] *Quindi tutti essi intestini ciechi trapassano con la loro cieca estremità, ed entrano in altrettante guaine, le quali guaine non istanno nella cavità del ventre, ma sfondano, e riescono in un'altra cavità, che occupa per di sopra tutto il dorso, e la schiena dell'animale da capo a piedi, e queste tali guaine circondate intorno intorno da una espansione membranosa, formano la figura di quaranta ventarole col manico; e tale espansione membranosa è doppia, e internamente scanalata, e tra una membrana, e l'altra vi corre un fluido limpidissimo, che talvolta ne gonfia il lembo.*

> *They brought me in Livorno a small sea animal of the figure and size drawn in plate twenty-fifth, Fig. first and second, which I like to call sea spiny, or sea porcupine. This one had a smooth white underbelly, not hairy, but with straight transversal wrinkles, like so many cords, and from the head to the tail on either side of the belly it was surrounded as if by many small bristle brushes. I counted twenty-six on each side, so that there were fifty-two in all.* [...] *These bristles all appeared to be blackish and filigreed, except for the middle one, which, being always the longest and thickest, is also always of a lustrous gold colour, which sometimes appears veiled in green, according to the fluttering of the light* [...] *From one of the extremities opens the foramen of the mouth, around which hang two antennae, or fleshy, floppy, white horns.* [...] *The intestine, without coils, goes straight to anus; but from the pylorus to almost the entire length of the intestine hang two parallel orders of blind intestines* [...]. *So all these blind intestines pass through with their blind end, and enter into as many sheaths, which sheaths do not stay in the cavity of the belly, but break through, and succeed in another cavity, which occupies above all the back, and the back of the animal from head to foot, and these sheaths surrounded around by a membranous expansion, form the figure of forty fans with a handle; and this membranous expansion is double, and internally grooved, and between one membrane and the other there runs a very clear fluid, which sometimes swells its edge.*

The second invertebrate we propose as an alternative to *A. aculeata* is a polyplacophore of the genus *Acanthochitona*, common in the European seas with the common species *A. fascicularis* (Linnæus, 1767) and *A. crinita* (Pennant, 1777). This last one, in our opinion, could correspond to Physalus, due to sevaral of its characters. These include the disposition, number, and shape of the tufts of bristles in eight pairs of warts distributed on the sides of the body, in correspondence of what might be the dorsal plates, the back at times covered by the expansions of the lateral areas, the poor visibility of the mouth and the anus which make them seem absent, the vulva shape of the foot, and its capacity to "swell" when touched. Certainly it does not swim floating, but it can be found at water surface on the sea rocks. Moreover, it has no whiskers, antennae, at the oral opening, unlike the *A. aculeata,* and the latter has many more bristles, in tufts or not, than those represented in Rondelet's drawing (Figs. 11.13 and 11.14). * These observations about *Physallus* were made

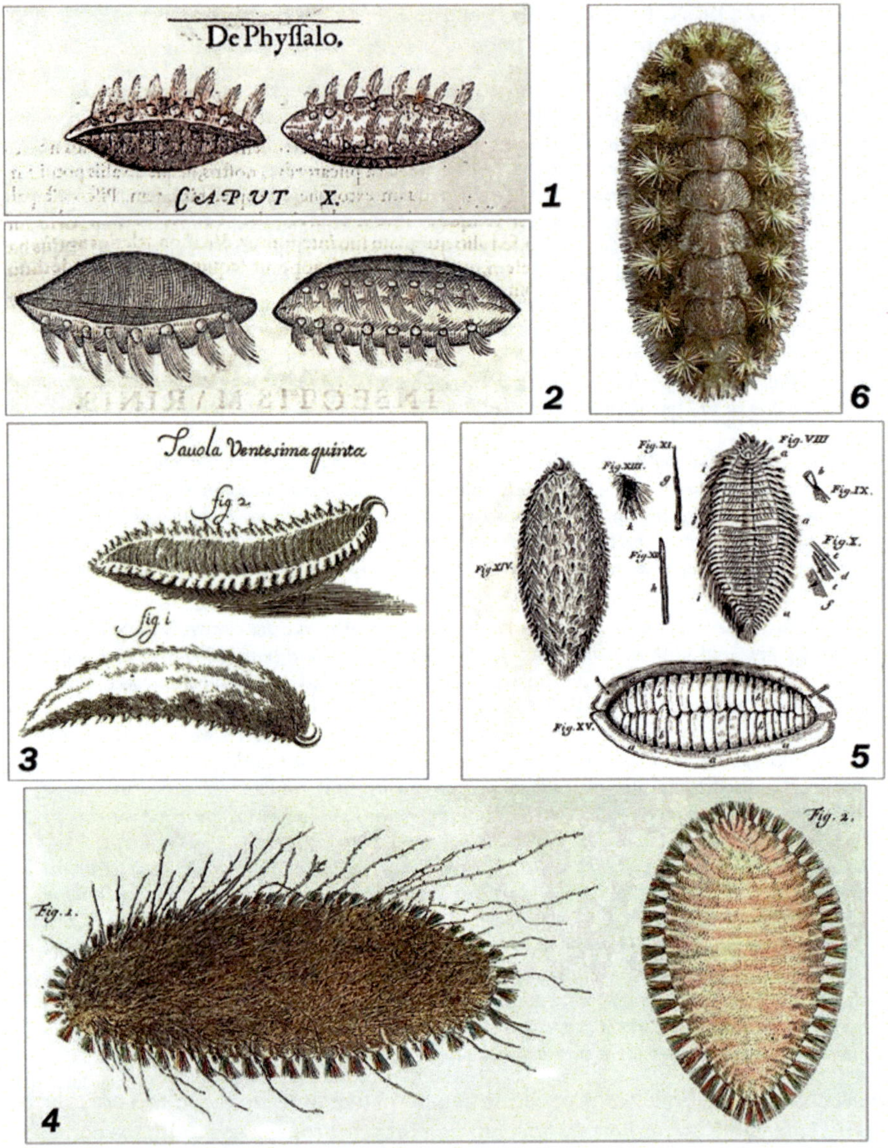

Fig. 11.13 (1) The strange "swelling fish" described by Rondelet in the special paragraph called *De Physsalo* (*Libri de Piscibus marinis*, 1554: p. 428). (2) The same organism reproposed a few years later by Gessner, placing it among marine insects (*Nomenclator aquatilium Animantium*, 1560: p. 268). *Aphrodita aculeata* L., 1758, from: (3) Redi (*Intorno agli animali viventi che si trovano negli animali viventi*, 1684: p. 187, pl. 25). (4) "Chenille de Mer", Seba (*Locupletissimi rerum naturalium thesauri accurata descriptio,* 1734: pp. 141–142, pl. XC). (5) Swammerdam (*Biblia naturae*, 1738: pp. 902–906, pl. X). (6) *Acanthochitona crinita* (Pennant, 1777), length 14 mm. [Graphic composition Maurizio Forli. Credit from: (1) https://www.biodiversitylibrary.org/bibliography/64229 (not in copyright). (2) https://doi.org/10.3931/e-rara-48090 / Public Domain Mark. (3) https://archive.org/details/bub_gb_oaW7gCi6R5gC/page/n293/mode/2up (Public Domain Mark 1.0). (4) https://www.biodiversitylibrary.org/bibliography/62760 (not in copyright). (5) https://www.biodiversitylibrary.org/bibliography/119987 (Public Domain). (6) Photo credit Gofas S., Benalmádena, Spain marinespecies.org/aphia.php?p=image&tid=138675&pic=70143 (modified)]

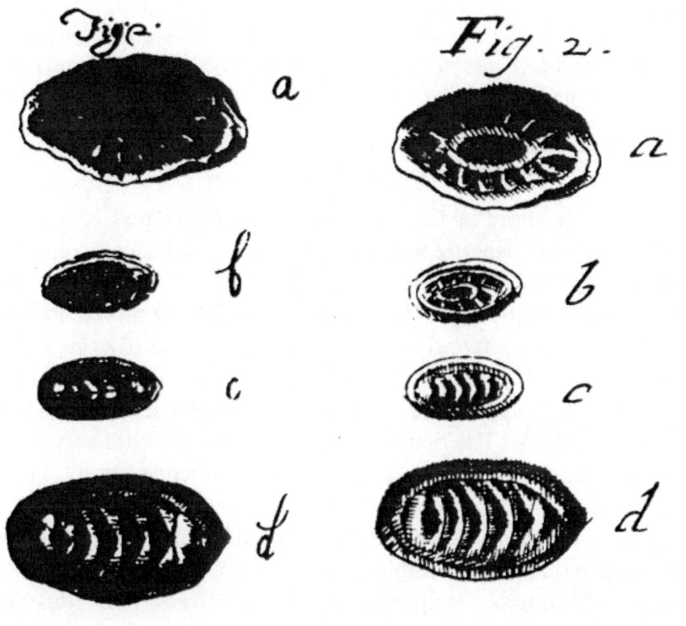

Fig. 11.14 On the left, illustrations from Vallisneri (*Raccolta di varj trattati del Sig. Antonio Vallisneri, Accresciuti con Annotazioni, e Giunte, e con Figure in Rame*, 1715: pl. XVI, figs a–d); on the right, from Vallisneri A (son) (*Opere Fisico-Mediche stampate e manoscritte del Kavalier Antonio Vallisneri raccolte da Antonio suo Figliuolo*, 1733: pp. 95–96, pl. XVII). The "Sea-bug" illustrated by Antonio Vallisneri, and a few years later by his son, in the reprinting of his father's works: the first true print of the body of a polyplacophorus. [https://archive.org/details/bub_gb_u74q8xvussmc (Public Domain Mark 1.0), https://archive.org/details/bub_gb_AFZ9V57bxbkC/page/n37/mode/2up (Public Domain Mark 1.0)]

independently by us in the course of our research. In intellectual honesty a few days before going to press, we found that part of our observations were made before by Vallot (1848) who identified Rondelet's *Lepas parva*, with *Chiton fascicularis*, now *Acanthochitona fascicularis* (Linnaeus, 1767).

In any case, as De Blainville reminds us [op. cit.], it is Antonio Vallisneri, Italian physician, scientist, naturalist, and biologist (1661–1730), who, while dealing with plant parasites, in correspondence with the Livornese Diacinto Cestoni, illustrates with certainty the chitons, describing them, however, as sea bugs, assimilating them to some Isopod crustaceans, better known as fish fleas, and to limpets:

> *Appiccati strettamente alle pinne, o ad altri crostacei di mare ha trovato sovente il nostro Autore certi insetti similissimi alle cimici degli agrumi, dette da altri patelle* [...]. *Anche quelle di mare non si muovono dal luogo, dove prima si piantano, sono ermafrodite, o piantanimali, e fanno i feti nella stessa maniera, osservati dal nostro Autore. Si vegga la figura seconda. b. c. Figure della cimice marina grande al naturale, la prima delle quali moftra la parte di sotto, la seconda quella di sopra. Le altre due figure, a d. la mostrano alquanto ingrandita con una lente, dimostrando la prima la inferior parte, e la seconda la superiore* (1715: p. 247, pl. XVI, figs a–d).

> *Clinging tightly to the fins, or other crustaceans of the sea our Author often found certain insects similar to citrus bugs, called by others limpets* [...]. *Even those of the sea do not move from the place, where they first plant themselves, they are hermaphrodite, or plant-animals, and make the fetuses in the same way, observed by our Author. See the second figure.* b. c. *Figures of the large natural sea-bug, the first of which exhibit the lower part, the second the upper. The other two figures* on the right *show it somewhat enlarged with a lens, the first showing the lower part, and the second the upper part.*

It is rather rare to find more or less intact polyplacophore fossils with articulated plates. A recent scientific paper reports the exceptional finding of a chiton, from the Carboniferous of the Czech Republic (Mikuláš et al. 2017), at the end of its locomotion track. For Italy, the only report to our knowledge of a complete individual, in a living position, is the one referred to a Pliocene specimen of *Craspedochiton altavillensis* (Seguenza, 1876), made by Dell'Angelo et al. (2003), adhering to the inside of a valva of a giant pectinid. Usually, they are the separate plates found in the fossiliferous sediments. The first reported fossil species, with valves of few millimeters of size, was the *Chiton grinionensis*, later amended as *C. grignonensis*, described by Lamarck in 1802, from the Lutetian sediments (Middle Eocene) of Grignon, locality of the Paris basin. The plates, in sufficient number to be assembled to reconstruct a specimen in a plausible manner, had been found by another naturalist contemporary of Lamarck, Jacques Louis Marin Defrance. His illustration dates from more than twenty years later and is found in Deshayes G. P., *Description des coquilles fossiles des environs de Paris*, t.2, *Atlas*, 1824: pl. I, figs. 1–7 (Fig. 11.15).

A paleontological and bibliographical summary of discoveries related to fossil species is given by the French zoologist (1836–1912) Alphonse Trémeau de Rochebrune:

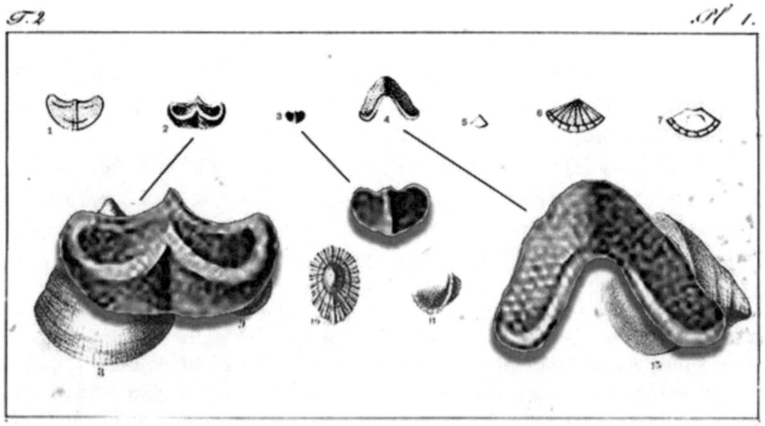

Fig. 11.15 *Chiton grinionensis* Lamarck, 1802. (Figures from Deshayes, 1824, pl. I, figs 1–7 [Figs 2–4 enlarged and superimposed on the original plate, reduced and modified, missing the lower part referring to other invertebrates]. Personal copy Maurizio Forli)

> *La première découverte paléontologique relative aux Chitons, remonte à l'annèe 1802* [...] *plus tard, en 1834, Conrad en signalait deux espèces dans les dépots tertiaries de l'Alabama, et vers 1836, Puzos et Duchastel, rencontraient dans les argiles Carboniferes de Tournay, plusieurs fragments à l'aide desquels, Munster reconstituait, en 1839, son classique* C. priscus *(1839)* [...] *dès la fin de 1840, Sandberger annonçait leur présence dans le calcaire Devonien de Willmar, en décrivait quelques-uns des mêmes gisements, de 1842 à 1850* [...].

The French author continuing his account, after a few pages, concludes with considerations on the number of fossil species discovered after *C. grignonensis*, discarding from the resulting total of 82, 39 attributions and consequent dubious determinations, concluding with a total of 43 certain species, which he will describe in the continuation of his fundamental work on fossil chitons (*Monographie des especes fossiles appartenant à la classe des Polyplaxiphores*, 1883: pp. 2–5, pls 1–3).

Rudolph Amandus Philippi, German zoologist and botanist, naturalized Chilean, was the first to report a fossil chiton in our country. In his work *Enumeratio Molluscorum Siciliae* (1836: 108), he describes the finding and subsequent immediate loss of a single plate in the fossiliferous sediments, probably Pleistocene, of Capo Milazzo:

> *Singulam valvulam* Chitonis squamosi *prope Melazzo una* cum *reliquis testaceis* fossilem *inveni, sed in itinere perdidi.*

In Tuscany, the findings of fossil polyplacophores are quite frequent in soils of Plio-Pleistocene age. In particular, in the province of Siena, from a Pliocene locality known in literature as Terre Rosse, come thousands of plates of the species *Chiton saeniensis* Laghi, 1984 while, from the sediments of the Calabrian (Lower Pleistocene) in the surroundings of Riparbella, come, among others, the remains of another chiton described as a new species about 20 years ago, *Chiton etruscus* Dell'Angelo and Forli, 1995.

As there is no news from the antiquity, referred to the chitons, there are no evidences about beliefs or their possible utilization in medicine. A superstition indirectly related to them through the definition of *Oscabiorn*, a similar animal, but described with legs and eyes, probably an isopod crustacean, nowadays known as *Cymothoa*, parasite of fishes, commonly known also as sea flea or tongue-eating flea, concerns the belief that whoever had found in the body of these a small stone would have seen his wishes granted. A curious interpretation about the identity of the chitons is that reported by Georg Friedrich Franck von Franckenau (1727: pp. 63–64, *Calva Serpentis americana diademata*), who compares a specimen of chiton to diadem-shaped expansions of the head of an American serpent, similar to those of the basilisk described in Aldrovandi.

Occasionally, today are sold set on rings or as ornamental pendants for necklaces (Figs. 11.16, 11.17, 11.18, 11.19, and 11.20).

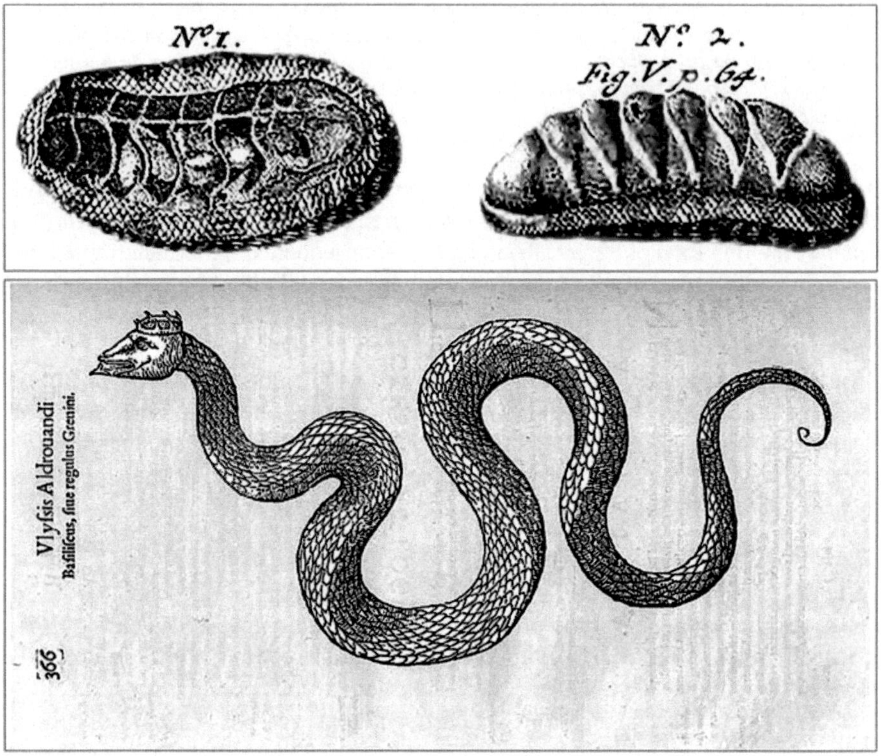

Fig. 11.16 (1) The chiton illustrated by von Franckenau (1727), internally free of viscera (left) and in lateral view (right), with the scaly perinotum in evidence, which may recall the skin of a snake. (2) Below the basilisk in Aldrovandi's interpretation, deduced from classical texts, of the *King of Snakes*, marked by the crown and the arrow coming out of its mouth, symbolizing its supremacy over all other serpentine animals and the power of its venom ((Bartolomeo Ambrosino) Aldrovandi, *Serpentum et Draconum Historiae*, Libri Duo, 1640). (Graphic composition Maurizio Forli. Credit from: (1) https://archive.org/details/actaphysicomedic11727acad No copyright page found (2) https://doi.org/10.3931/e-rara-5287 / Public Domain Mark)

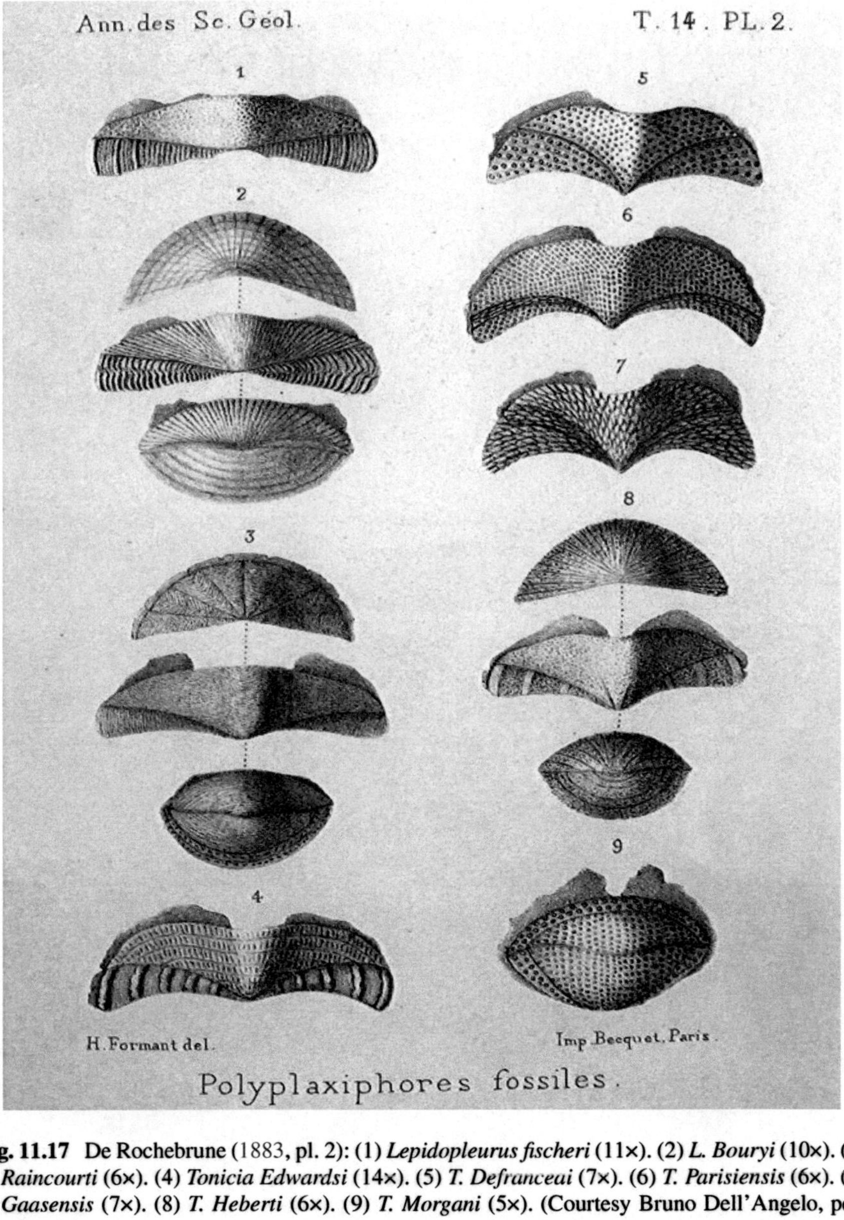

Fig. 11.17 De Rochebrune (1883, pl. 2): (1) *Lepidopleurus fischeri* (11×). (2) *L. Bouryi* (10×). (3) *L. Raincourti* (6×). (4) *Tonicia Edwardsi* (14×). (5) *T. Defranceai* (7×). (6) *T. Parisiensis* (6×). (7) *T. Gaasensis* (7×). (8) *T. Heberti* (6×). (9) *T. Morgani* (5×). (Courtesy Bruno Dell'Angelo, personal copy)

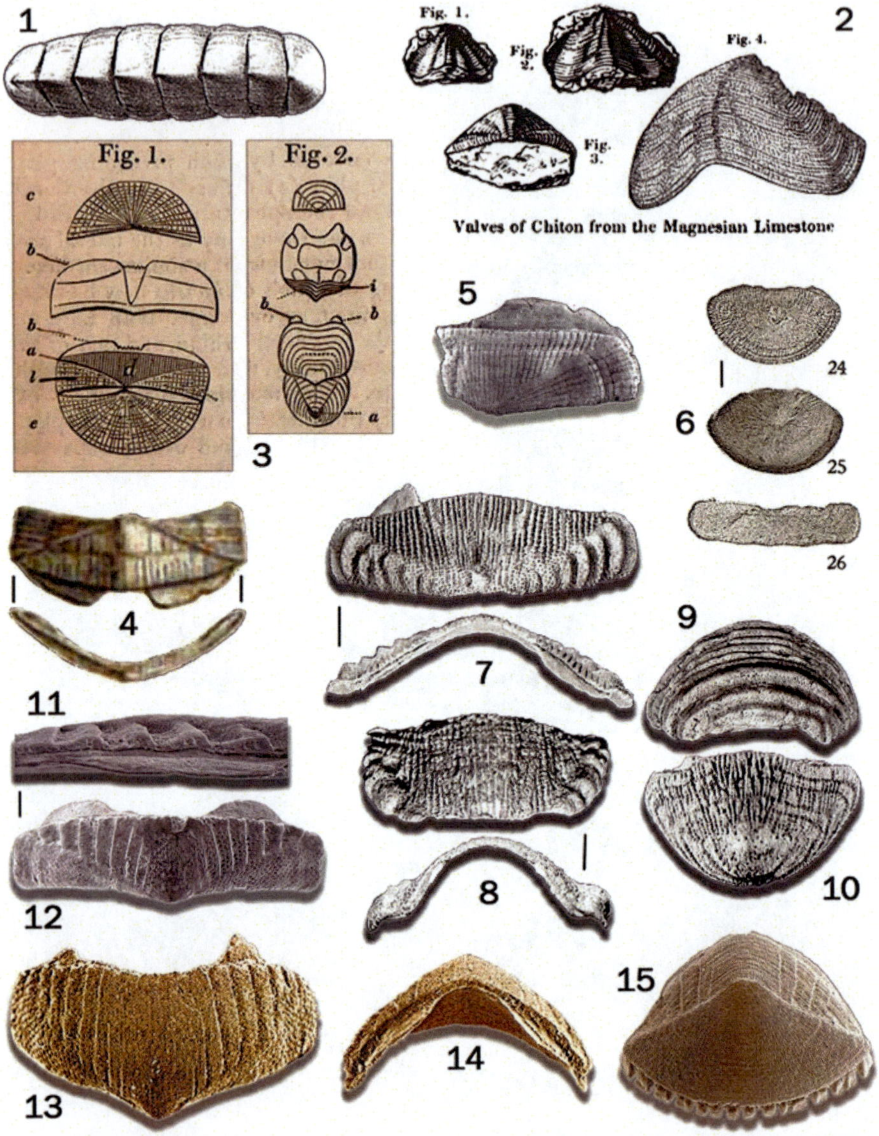

Fig. 11.18 Fossil chitons: (1) *Chiton priscus* (Münster 1839: p. 38, pl. XIII, Fig. 4). (2) Valve (King 1846: p. 12). (3) Fig. 1 *Chiton cumungii*; Fig. 2 *Helminthochiton priscus* (Salter 1847: p. 49). (4) *Chiton miocenicus* (Michelotti 1847: p. 132, pl. XVI, fig 7) (5) *C. miocenicus* (6.5 mm) (Forli et al. 1999: p. 113, pl. 1, fig 5). (6) *Chiton benoisti* (12 mm) (Cossmann and Peyrot 1918: p. 33, pl. II, figs 24–26). (7–10) *Lepidopleurus cajetanus* (Dell'Angelo et al. 2013: p. 68, pl. 1, figs A–F). (11–12) *Chiton olivaceus*, middle plate (4.4 mm) and detail, in lateral view, of the longitudinal ribs, (Dell'Angelo et al. 2013: p. 85, pl. 5, figs T–U). (13–14) *Chiton etruscus* (2.8 mm) (Dell'Angelo and Forli 1995: p. 233, figs 4–5). (15) *Callochiton calcatus* (2.7 mm) (Dell'Angelo and Palazzi 1994). (Graphic composition Maurizio Forli, courtesy Società Italiana di Malacologia)

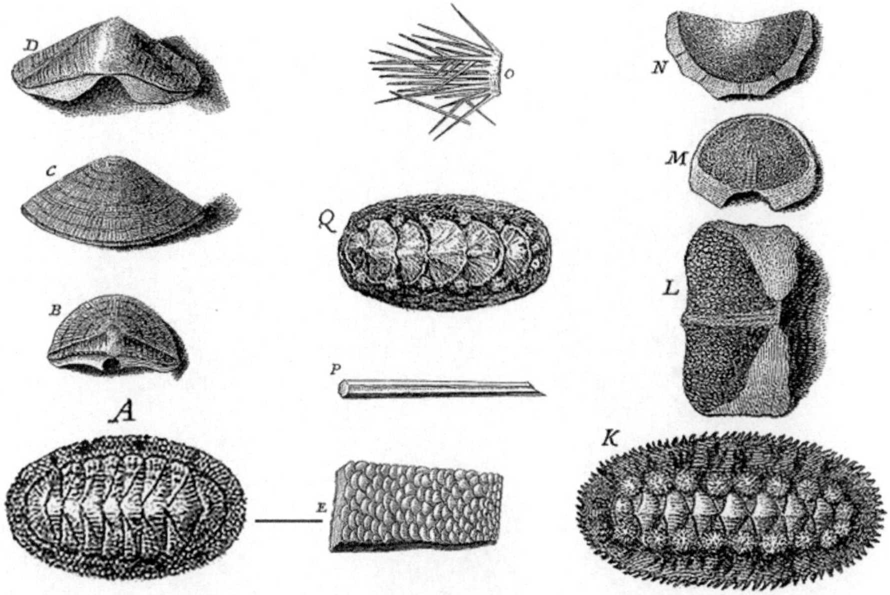

Fig. 11.19 *Articulated limpets. Among the many marine life, which are fished in the Port of Bua... deserve particular description two species of long limpets, articulated, called Babusche, by those fishermen... Although the articulated limpet is never found so close to the Lido, that the low tide can leave it out of the water, it loves the air, and shows it with singular precision ...* [Fortis, *Viaggio in Dalmatia,* 1774: 2: pp. 20–21, pl. IX (modified)]. The specimen on the left (A) should be *Lepidochitona cinerea* (Linnaeus, 1767), while the one on the right (K), an *Acanthochitona fascicularis* (Linnaeus, 1767). In the middle (Q) an anomalous specimen with only six plates. (https://doi.org/10.3931/e-rara-40493 / Public Domain Mark)

Fig. 11.20 Discarded specimens of chitons, often from Southeast Asian countries, are currently offered set on the rings or as ornamental pendants for necklaces. (Photo and graphic composition Maurizio Forli, courtesy private collection)

References

Aldrovandi U (1606) De Reliquis Animalibus exanguibus libri quatuor, post mortem eius editi: nemp' De Mollibus, Crustaceis, Testaceis, et Zoophytis. Apud Io: Baptistam Bellagambam, Bononie. https://www.biodiversitylibrary.org/bibliography/119999 / Public Domain

Aldrovandi U (1640) Serpentum et draconum historiae libri duo. Bononiae: apud Clementem Ferronium. Stiftung der Werke von C.G.Jung, Zürich, Online. https://doi.org/10.3931/e-rara-5287 / Public Domain Mark

Bruguière M (1791) Tableau encyclopédique et méthodique des trois règnes de la nature contenant l'helminthologie, ou les vers infusoires, les vers intestins, les vers mollusques, &c. Septième Livraison. Panckoucke, Paris

Cossmann MM, Peyrot A (1918) Conchologie Néogénique de l'Aquitaine. Actes de la Société Linnéenne de Bordeaux 70:5–180

de Blainville H-MD (1825) Dictionnaire des Sciences Naturelles dans lequel on traite méthodiquement des différens êtres de la nature, considérés soit en eux-mêmes, d'après l'état actuel de nos connoissances, soit relativement a l'utilité qu'en peuvent retirer la médecine, l'agriculture, le commerce et les arts. Tome Trente-sixième, Levrault, Paris de Rochebrune AT (1883) Monographie des espèces fossiles appartenant à la classe des Polyplaxiphores. Ann Sci Géol 14(1):1–74, pls 1-3

Dell'Angelo B, Forli M (1995) The Polyplacophora of the lower Pleistocene of Riparbella (Pisa), with list of molluscs found. Malacol Bull 30(9–12):221–252

Dell'Angelo B, Palazzi S (1994) Callochiton calcatus n. sp. with notes on Callochiton septemvalvis (Montagu, 1803). Shell 273:15–23

Dell'Angelo B, Schwabe E (2010) Teratology in chitons (Mollusca, Polyplacophora): a brief summary. Malacol Bull 46(1):9–15

Dell'Angelo B, Smriglio C (1999) Chitoni viventi del Mediterraneo. Edizioni Evolver s.r.l., Roma

Dell'Angelo B, Gallorini H, Taviani M (2003) First Neogene record of an articulated Polyplacophoran. J Paleontol. 77(6):1193–1194

Dell'Angelo B, Sosso M, Prudenza M, Bonfitto A (2013) Notes on fossil chitons. 5. Polyplacophora from the Pliocene of Western Liguria, Northwest Italy. Riv Ital Paleontol Stratigr 119(1):65–107

Deshayes GP (1824) Description des coquilles fossiles des environs de Paris. Atlas. L'auteur, chez Bechet jeune, Paris

Forli M, Dell'Angelo B, Taviani M (1999) Molluscs of the Tuscan Pliocene: the Montenero section (Grosseto). Malacol Bull 34(9–12):109–122

Fortis A (1774) Viaggio in Dalmazia. In Venezia: presso Alvise Milocco ..., 1774. Zentralbibliothek Zürich, NR 842. https://doi.org/10.3931/e-rara-40493 / Public Domain Mark

Gessner C (1558) Historiae Animalium Liber IIII. qui est de Piscium & quatilium animantium natura. Christoph. Froschoverum, Tiguri. https://doi.org/10.3931/e-rara-16853 (Public Domain Mark)

Gessner C (1560) Nomenclator aquatilium animantium: icones animalium aquatilium in mari et dulcibus aquis degentium, plus quam DCC. cum nomenclaturis singulorum Latinis, Grecis, Italicis, Hispanicis, Gallicis, [...]. Tiguri: excudebat Christoph. Froschoverus ..., 1560. ETH-Bibliothek Zürich, Rar 9734: 1, https://doi.org/10.3931/e-rara-48090 / Public Domain Mark

Gray JE (1821) Natural arrangement of Mollusca according to their internal structure. Lond Med Repos 15:234.

Linnæus C (1758) Systema Naturae per Regna Tria Naturae, secumdum Classes, Ordines, Genera, species, cum characteribus differentiis, synonimis, locis. Tomus I. Editio decima, Reformata. Laurentii Salvii, Holmiae. https://www.biodiversitylibrary.org/bibliography/542 / Public domain

Michelotti G (1847) Déscription des Fossiles des terrains miocènes de l'Italie septentrionale. Naturkundige verhandelingen V. D. Hollandsche Maatschappij der Wetenschappen te Haarlem:1–407

Mikuláš R, Rindsberg AK, Santos A, Pavela M (2017) A carboniferous chiton (Mollusca, Polyplacophora) at the end of its trail: A unique find from the Czech Republic. Boll Soc Paleontol Ital 56(2):171–179. https://doi.org/10.4435/BSPI.2017.15

Münster GG (1839) Beiträge zur petrefacten-kunde mit XVIII nach der Natur gezeichneten Tafeln. In Commision der Buchner'schen Buchhandlung, Bayreuth

Philippi RA (1836) Enumeratio Molluscorum Siciliae, cum viventium tum in tellure tertiaria fossilium quae in itinere suo observavit. Berolini

Redi F (1684) Osservazioni di Francesco Redi, Accademico della Crusca, Intorno Agli Animali Viventi che si trovano Negli Animali Viventi. Piero Matini, Florence. https://archive.org/details/bub_gb_oaW7gCi6R5gC/page/n293/mode/2up (Public Domain Mark 1.0)

Rondelet G (1554) Libri de Piscibus Marinis, in quibus verae Piscium effigies expressae sunt. Matthiam Bonhomme, Lugduni. https://www.biodiversitylibrary.org/bibliography/64229 (not in copyright)

Rondelet G (1555) Universae aquatilium Historiae pars altera, cum veris ipsorum Imaginibus. Matthiam Bonhomme, Lugduni. https://archive.org/details/ARes071032/page/n13/mode/2up (Public Domain 1.0)

Scheuchzer JJ (1733) Phisica Sacra : iconibus aeneis illustrata. Augustae Vindelicorum ; Ulmae, 1732-1735. Bibliothèque des Pasteurs, BPU Neuchâtel, PA P 49.7.17, https://doi.org/10.3931/e-rara-1706 / Public Domain Mark

Seba A (1734) Locupletissimi rerum naturalium thesauri accurata descriptio, et iconibus artificiosissimis expressio, per universam physices historiam: opus, cui, in hoc rerum genere, nullum par exstitit. https://www.biodiversitylibrary.org/bibliography/62760 not in copyright

Swammerdam I (1738) Biblia Naturae; sive Historia Insectorum, in classes certa redacta, nec non exemplis, et anatomico variorum animalculorum examine, aeneisque tabulis illustrata. Insertis numerosis rariorum naturae observationibus. Isaacum Severinum, Balduin Vander Aa, Petrum Vander Aa, Leydae. https://www.biodiversitylibrary.org/bibliography/119987 / Public Domain

Vallisneri A (1715) Raccolta di varj trattati del Sig. Antonio Vallisneri, Accresciuti con Annotazioni, e Giunte, e con Figure in Rame. Gio. Gabbriello, Venice. https://archive.org/details/bub_gb_u74q8xVUsSMC (Public Domain Mark 1.0)

Vallisneri A (son) (1733) Opere Fisico-Mediche stampate e manoscritte del Kavalier Antonio Vallisneri raccolte da Antonio suo Figliuolo, Corredate d'una prefazione in genere sopra tutte, e d'una in particolare sopra il Vocabolario della Storia Naturale. Sebastiano Coleti, Venezia. https://archive.org/details/bub_gb_AFZ9V57bxbkC/page/n37/mode/2up (Public Domain Mark 1.0)

Vallot DM (1848) Détermination de plusieurs animaux aquatiques mentionnés par Rondelet, et négligés par les naturalistes modernes. Observation d'Histoire Naturelle. Mémoire de l'Académie des Sciences, arts et Belles-lettres de Dijon, 179–218

von Franckenau GFF (1727) Observatio XXI. Calva Serpentis Americani diademata. In: Acta Physico-Medica Academiae Caesareae Leopoldino-Carolinae, Naturae Curiosorum exibentia Ephemerides sive Observationes Historias et experimenta. Volumen Primum cun Appendice. Officina W. M. Endteriana, Nurembergae, pp 63–64. https://archive.org/details/actaphysicomedic11727acad No copyright page found

Chapter 12
Bivalvia: Devil's Nails, Reflections Between Superstition and Science

Abstract The bivalves, with a shell usually formed by two valves, are a class of mollusks diffused in the aquatic environments of the whole world. Most of them are infaunal and filter feeders; that is, they live buried in the marine and dulcicolous sediments in the intertidal and sublittoral marine zones, often immediately under the level of the bottom, feeding on the organic particles suspended in the water. They are fundamental fossils in paleoecology and some species may be utilized in biostratigraphy at regional level. And it is precisely of a bivalve, sometimes believed to be the valva of a Cardium, the first explanatory image of a fossil, printed in Christophorus Encelius (*De re metallica: hoc est, de origine varietate, & natura corporum metallicorum, lapidum, gemmarum, atque aliarum, quae ex fodinis eruuntur, rerum, ad medicinae deservientium*, [...]. Franc. [Frankfurt]: apud Chr. Egenolphum, [15--]. ETH-Bibliothek Zürich, Rar 2259, p 227. https://doi.org/10.3931/e-rara-29357 / Public Domain Mark, 1551). Our interpretation in this regard is discordant as we believe that the shell is that of a nonfossil *Pecten maximus.* In ancient times, all fossilia, which had a hemispherical shape, were called *chelonites*, in similarity to the shape of the turtle shell, and it is often under this name that their illustrations are found in the earliest printed texts. Some particular shapes evoked various subjects to our predecessors; for example, the hooked shells of the Mesozoic *Gryphaea arcuata* were estimated to be among other things, devil's nails; the sections of the heart-shaped Triassic Megalodonts, demonic footprints; the bivalve Glossus, or others with similar shapes, were believed to be petrified oxhearts. As pendants, the pierced shells of mollusks were often used in the manufacture of amulets against evil eye, individually or accompanied by other objects believed to be apotropaic.

Keywords Bivalvia · Fossils · Encelium · Chelonite · Gryphaea arcuata · Devil's nails

© The Author(s), under exclusive license to Springer Nature Switzerland AG 2022

M. Forli, A. Guerrini, *The History of Fossils Over Centuries*,
https://doi.org/10.1007/978-3-031-04687-2_12

Bivalves constitute a class of invertebrates so widespread in aquatic environments all over the world, that it can certainly be said that they are cosmopolitan and ubiquitous. Most are infaunal and filter feeders; that is, they live buried in marine and freshwater sediments, often immediately below the bottom level, feeding on organic particles that are suspended in the water medium. A large number of bivalves are settled in the marine intertidal and sublittoral zones, but some can also survive in environments considered extreme, such as under polar ice, abyssal depths, clustered around hydrothermal vents, or in the supralittoral plane, that is, the upper part of terrestrial areas generally exposed only to sea spray).

The shells vary in size from a few millimeters on the longest side to a maximum of just over 1 meter in *Tridacna gigas* (Linnæus, 1758), which is also the largest bivalve mollusk, reaching two quintals in weight, while some others equipped with a long siphonal canal, for instance, some species of *Panopea*, can exceed one and a half meter in total length. Their lifespan is generally not too long, but some of them easily 100 years of age. The record of longevity is attributed to an *Arctica islandica* (Linnæus, 1767) coming from the Icelandic marine waters, whose age, based on the count of the annual growth lines visible in the thickness of the shell, has exceeded the 500 years of existence. Their paleontological classification, especially for the most ancient species, is based exclusively on the morphological characters of the shell, except in the rare cases in which we can make use of impressions of the soft parts of the body. For the more recent fossil species, with more or less close relatives to the present species, or still present, we utilize also biological and genetic criteria. The generic and specific determination is relatively easy, while that of the higher systematic categories is continuously under revision (Fig. 12.1).

More in detail, the shell of bivalves is constituted by two valves, one right and one left, symmetrical with respect to a plane passing between the valves, called commissure. Generally, they are equivalent and inequilateral, that is, asymmetrical with respect to the geometric center of the valve, even if there are some inequivalent exceptions (e.g., *Pecten* and *Ostrea*) in which one valve is more developed than the other, up to extreme cases in the Rudiste. In the dorsal part of the shell are recognizable two protuberances, called umbones, that correspond to the first part of the shell formed from the larval shell called prodissoconca. The umbones are generally turned toward the front (prosogyrate), more rarely toward the back (opisthogyrate), or toward each other (orthogyrate). In order to orient a bivalve, let us imagine to take it in our hands with the umbones facing forward and the commissure plane vertical: the valva held by the right hand will be the right valva, the one held by the left hand, the left valva. In general, the following criteria are useful:

- In most bivalves, the umbones are turned anteriorly.
- The pallial sinus is always posterior.
- If one of the adductor muscles is reduced or absent, it is almost always the anterior.

If external, the ligament is generally posterior (Fig. 12.2).

The *ligament*, or *ligamen*, is the structure, which has the function of joining the two valves dorsally, is secreted by the outer fold of the margin of the mantle, and, from the

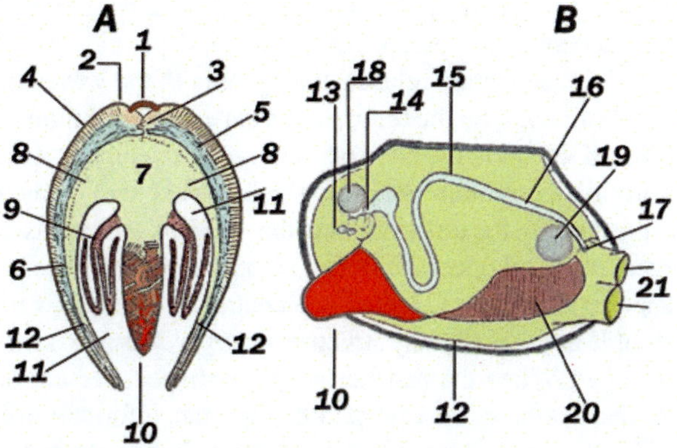

Fig. 12.1 Simplified general organization of lamellibranch: (**A**) specimen with joined valves in cross section; (**B**) internal lateral view. (1) *Ligament* or *Ligamen*. (2) Umbone. (3) Hinge. (4) Periostracum, organic covering of the external part of the shell. (5) Mesostracum, intermediate layer of the shell constituted by calcium carbonate. (6) Endostracum, internal layer of the shell, constituted by Aragonite, often mother of pearl. (7) Body of the mollusk. (8) Right and left lobes of the body, covering it, called Mantle or Pallium. (9) Gills. (10) Foot. (11) Pallial cavity, produced by the space left between the body of the mollusk and the mantle. (12) Pallial line, the imprint left by the body inside the shell. (13) Labial palpi, appendages connected to the filtering apparatus (*Ctenidium*), which direct the food particles to the mouth. (14) Oral aperture. (15) Digestive tube. (16) Intestine. (17) Anal aperture. (18) Anterior adductor muscle. (19) Posterior adductor muscle. (20) Gonads. (21) Siphons. In the bivalves that have them, there are two, one inhaling, which sucks in the water and its contents, and one exhaling, which expels it after it has passed through the pallial cavity, the gills, and the filtering apparatus. [Drawing Maurizio Forli modified after Vialli (1974)]

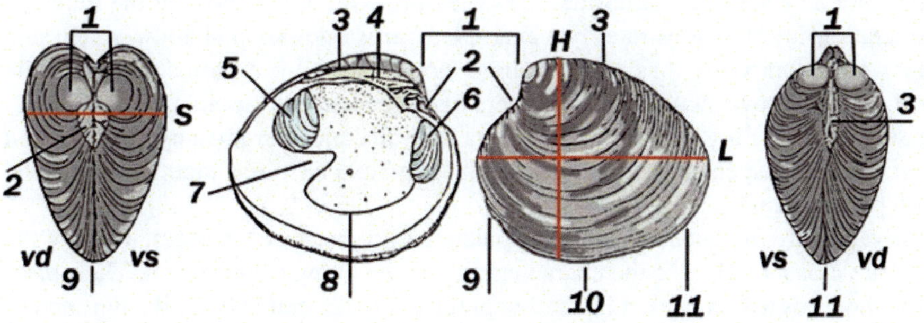

Fig. 12.2 Simplified description of the shell of a generic lamellibranch: in the center a left valva seen internally and externally, on the left a complete shell in anterior view, on the right in posterior view. (vd-vs) Right valva and left valva. (H) Height of valva. (L) Width. (S) Thickness of shell. (1) Umbones, in the umbonal zone are included the prodissoconch, the larval shell, from which the adult shell develops. (2) Lunula. (3) Ligamentary dimple. (4) Hinge. (5) Posterior adductor muscle. (6) Anterior adductor muscle. (7) Pallial sinus. (8) Pallial line. (9) Anterior margin. (10) Inferior margin. (11) Posterior margin. [Drawing Maurizio Forli modified after Vialli (1974)]

structural point of view, is formed by elastic, lamellar, or fibrous conchiolin. When the shell is closed, the ligament is compressed or stretched, depending on its position and, as it is elastic, when the musculature relaxes, the opening of the valves takes place, taking advantage of its elasticity, with the return to the state of equilibrium. Other morphological characteristics of the shell, useful in taxonomy, are the lunula, a flat heart-shaped area placed anteriorly under the umbones; the shield, where the ligament originates, a flat elongated area posterior to the umbones; and the *resilium*, which has a function similar to that of the ligament. More specifically, this structure, also referred to as the internal ligament, is a triangular-shaped pad located in a central dimple below the cardinal and dorsal lines. The dimple or generally any structure that serves as an insertion for the ligament is called a *resilifer*. When the resilifer is spoon-shaped, present only on the left valve or on both valves, it is called chondrophore. Important information can be deduced inside the valves. These ones open pivoting on a flat area in dorsal position, the cardinal area, on which is placed the hinge, a structure formed by teeth alternated to dimples, corresponding respectively to dimples and teeth on the other valve. The hinge, in addition to regulating the closing of the valves allowing a perfect joint, prevents the lateral slipping, allowing the mollusk to dig without losing the alignment between the two parts of the shell. The number and disposition of teeth and dimples have an important taxonomic meaning in paleontology. As previously mentioned, inside the valves, it is also possible to recognize the impressions of insertion of the adductor muscles and the pallial line, which may connect the two impressions, parallel to the edge of the shell, possibly incised by the pallial sinus. The external ornamentation of the shell is characterized by striae, ribs, grooves, wrinkles, tubercles, and spines, which develop both concentrically and radially with respect to the umbones. The growth striae are easily recognizable on the external surface of the valva and correspond to periods of stasis during the growth, marking the previous positions of the margins of the valva.

In present-day fossorial bivalves, the siphons, inhaling and exhaling, are of considerable ecological importance, as they provide information on the depth of penetration into the substrate. For example, *Cerastoderma*, a shallow-water suspensivorous burrower, has short siphons not covered by a periostracum, while *Tellina*, a shallow, depositional digger, has independent siphons. *Loripes* possesse a rear exhaling siphon that cements its walls and prevents any collapse and *Mya*, a deep suspensivorous digger, with long siphons fused together and lined with periostracum.

Bivalves are fundamental fossils in paleontology, because the peculiarities of the shell are closely linked to the characteristics of the living environment. Their presence allows us to reconstruct in a rather precise way the main physical environmental parameters, such as the type of substrate, temperature, salinity, oxygenation, depth, and speed of the current. In addition to their paleoecological importance, some species can be used in biostratigraphy at a regional level. Among the most important forms are *Posidonia becheri* Bronn, 1828 (guide species of the Lower Carboniferous), *Megalodon gümbeli* Stoppani, 1865 (Upper Triassic), *Claraia clarai* Emmrich, 1844, and *Daonella lommeli* Wissmann, 1841, guide fossils in the Triassic of the Dolomites, *Arctica islandica* (Linnæus, 1767) marker of the Lower Pleistocene in the Mediterranean area (Figs. 12.3 and 12.4).

Fig. 12.3 *Arctica islandica* (Linnaeus, 1767). In the background, the two valves of a present specimen from the Lofoten Islands (Norway), in the foreground a fossil valva from the Lower Pleistocene in the province of Livorno (Tuscany). Width ~90 mm. The species is usually well represented in sandy and clayey soils dating from that part of the Lower Pleistocene called the Calabrian Stage dated between 1.806 Ma and 781.000 years ago. (Photo Maurizio Forli)

Like Gastropods, bivalves were classified by Thiele (1929–35) from an earlier work by Cossmann and Peyrot (1909). Later the classification was revised by Newell (1965) based mainly on the type of hinge and/or cardinal area. There are different classification methods that take into account other criteria, for example, the one proposed by Pelseneer (structure and evolution of gills), as well as Douvillé's classification (phylogenetic basis) (Allasinaz 1992), up to the most recent one proposed by Bouchet and Rocroi (2010).

The origin of the present European Atlantic marine fauna is to be found in the ancient Miocene Cenozoic sediments and in those at the Miocene-Pliocene transition for the Mediterranean one. In Italy, the marine sediments of these geological periods are widespread and the fossiliferous mollusk associations, quite common and extensive, are the source of the specimens described on several occasions by Italian scholars who have given their contribution, often fundamental, to the definition, growth, and consecration of paleomalacological studies.

Below is the classification of two species, which we will discuss a little further on, the first being related to superstitious manifestations, the second as the first figure of a fossil bivalve illustrated by Imperato (1599).

Class Bivalvia Linnæus, 1758

Family	Gryphaeidae	Family	Glossidae
Genus	*Gryphaea*	Genus	*Glossus*
Species	*Gryphaea arcuata* Lamarck, 1801†	Species	*Glossus humanus* (Linnæus, 1758)

Fig. 12.4 Some types of hinges with examples of bivalves possessing them. (1) Taxodont: (a) *Glycymeris* (Bellon 1553), (b) *Arca* (Colonna 1616). (2) Dysodont: (a) *Pecten maximus* (Bellon 1553), (b) *Aequipecten* (Aldrovandi 1648); (3) Desmodont: (a) *Mya* (Scilla 1759), (b) *Mya* (Sowerby 1812). (4) Schizodont: (a) *Trigonia* (Lhuyd 1699). (5) Isodont: (a) *Spondylus* (Aldrovandi 1648); (b) *Spondylus* (Bellon 1553). (6) Heterodont: (a) *Venus verrucosa* (Bellon 1553); (b) *Cardium* (Colonna, 1616). [Color and graphic composition Maurizio Forli. Credit: (1) (a), (2) (a), (5) (b), (6) (a) https://www.biodiversitylibrary.org/bibliography/5765 (Public domain). (2) (b); (5) (a) https://doi.org/10.3931/e-rara-13017 / Public Domain Mark. (3) (a) personal copy. (3) (b) https://www.biodiversitylibrary.org/bibliography/14408 (not in copyright). (1) (b), (6) (b) https://www.biodiversitylibrary.org/bibliography/723 (Public domain)]

Bivalves have been used since Paleolithic times, both as food and to make ornaments or tools for daily use. The study of the numerous archaeological testimonies left by our ancestors allows today's researchers to identify them through their dietary, aesthetic and cultural choices, sometimes recreating distribution networks that contribute to the understanding of the socioeconomic activities of our predecessors. Even today, by reading the labels on the packages of edible mollusks sold in supermarkets, and perhaps then being able to recognize the various species offered cooked at home or in delicatessens, we can realize the importance of trade and food habits related to their circulation. In our markets, for example, the Chilean or Thai mussels (*Mytilus chilensis* & *Perna viridis*) or Vietnamese clams (*Meretrix lyrata*), are more easily found than the corresponding native Mediterranean species. Furthermore, the shells assume the most varied functions, from the simple ornament to the container, formed by the concave part of the valve, for example, transformed into a banal ashtray, to the elaborate ornaments in refined and precious jewels of mother-of-pearl and natural pearls. Currently, the oldest use of bivalves for ornamental purposes in European prehistory is confirmed for Spanish sites of the Middle Paleolithic, dated around 50,000 years ago. These populations of Neanderthal people used shells of *Acanthocardia tuberculata* and *Glycymeris nummaria* (*Glycymeris insubrica* in the original text, from Zilhão et al. 2010), perforated in the umbonal region, to make pendants. Often the shell wall of these species is consumed autonomously giving rise to natural perforations in that area as a prominent point of the shell, subjected to the passive rolling induced by the waves, as it is easy to see even today by collecting the same types of shells, piled up along the shoreline by the wave motion. Our attentive ancestors will not have missed this fact, which will have allowed them to easily adapt the shells to the desired use. Other species, *Pecten* and *Spondylus*, have been found, in coeval sediments of other Iberian localities, with traces of yellow or reddish coloration, which induce to think that they have been intentionally painted or used as containers of pigments.

And it is precisely of a bivalve that is believed to be the first explanatory image of a fossil, described as such even if under the name of chelonite, which is found in *De re metallica* ... by Cristoforo Encelio (1551: p. 227). Chelonites, a name borrowed from Pliny, were those remains or stone objects, often fossils, whose shape recalled the shell of turtles. Although in the text it is written right next to the image what Encelio refers to, this shell has been and continues to be believed and cited as *Cardium* (see, e.g., Lanza 1984: p. 117, fig. 2).

> *... a me inventa est chelonitis, quae forma omnino repraesentabat cochleas marinas, wie ein Iacobs muschel, oder meer muschel ... wie ein Iacobs muschel*

... as a *James' shell,* that is, since the place where it was found was near the mouth of the Elbe, in the present North Sea, nothing else than a local *Pecten maximus* called, like the similar Mediterranean *Pecten jacobaeus*, *scallop* or *St. James' shell.* Among other things, the representation, in its schematic nature, is quite faithful to the actual shape of this species of pectinid, showing the two auricles on the front side and backside of the umbo, absent in *Cardium*, the wider ribs and fewer than in *Cardium*, and an appropriate hint of coloration represented by the two transverse

Fig. 12.5 A chelonite, a stone object, often a fossil, shaped like a turtle shell. In this case, a septaria, that is, a mineral concretion, of a calcareous-clayey nature, crossed by a series of cracks that cross each other. (Photo Maurizio Forli, courtesy Savino Montanari)

bands. This last feature leaves us doubtful about its actual fossil *status*, because, although it is not uncommon to find remnants of the original coloration in the shells of fossil mollusks, it is not so easy that the chromatic ornamentation is preserved in scallops in general. The same image ex Encelius, reported printed vice versa, is in Gessner (*De omni rerum fossilium genere*, 1565: p. 165) who in the same page, after a few lines, referring to an excellent figuration of a *Pecten jacobaeus*, proposed earlier by Rondelet (*Universae aquatilium Historiae pars altera*, 1555: p. 15) describes one with higher probability of being a fossilized find:

Pecten lapideus, substantia intus est pallida, albicans, ceu luti in lapidem indurati.

Pecten petrified, the substance within is pale, whitish, like hardened clay in stone.

It is not clear to us if the proposition of the images on the same page by Gessner joins the identification with the specimen of Encelius or not (Figs. 12.5 and 12.6).

Numerous present-day bivalves are reproduced with remarkable verisimilitude, already in the printed texts of the mid-sixteenth century, Bellon (*De Aquatilibus, Libri duo*, 1553), Rondelet (1555) or figured as fossils in the form of internal molds, sometimes called by their proper name Bucardie (Imperato 1599: p. 666).

Bucardia. La Bucardia pietra così detta dalla figura, e grandezza, che imita di cor di bue, è di superficie liscia, simile dall'una, e dall'altra parte: onde dimostra con densamento fatto dentro concavità di conca di ambe valve convesse...

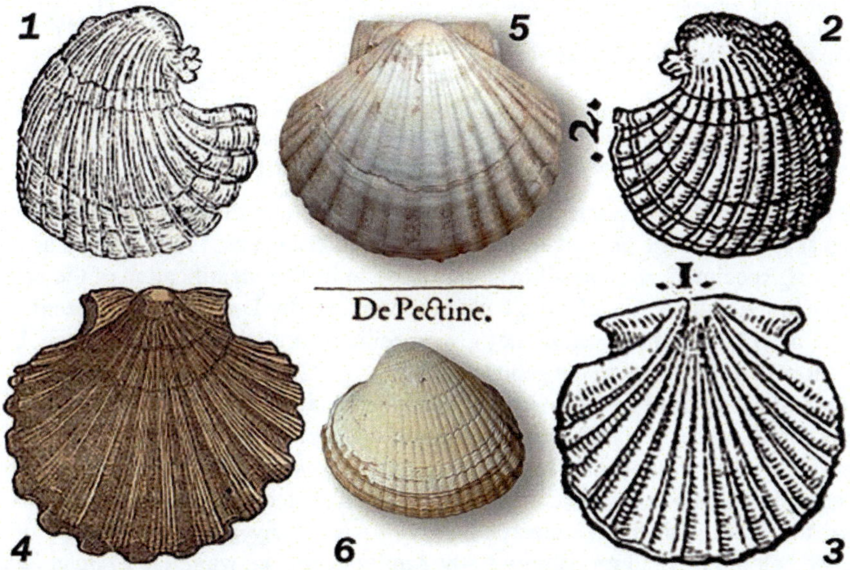

Fig. 12.6 (1) From Encelius (1551: p. 227). The representation of *Pecten maximus* (Linnaeus, 1758), which is believed to be the first explanatory image of a fossil. (2) The same one, reproposed a few years later by Gessner (1565: p. 165). (3) Another fossil, the same species drawn from the left side, taken again from Gessner (1565: p. 165) but inspired by the following figure by Rondelet. (4) A complete specimen of living *P. maximus*, (Rondelet 1555: p. 15). (5) A present specimen of *P. maximus* from Saint Valery en Caux (Normandy), width 130 mm. (6) *Cardium edule* Linnaeus, 1758, from Camaret- sur-Mer (Brittany), width 35 mm. [Photo and graphic composition Maurizio Forli. Credit from: (1) https://doi.org/10.3931/e-rara-29357 / Public Domain Mark (2–3) https://doi.org/10.3931/e-rara-4176 / Public Domain Mark (4) https://archive.org/details/ARes071032/page/n13/mode/2up (Public Domain 1.0)]

> *Bucardia. The Bucardia stone, so called because of its shape and size, which imitates the heart of an ox, has a smooth surface, similar on one side and on the other: so that it demonstrates with a densification made inside the concavity of the basin of both convex valves...*

In the case of the so-called bucardiae, these are equivalent and equilateral inner shell molds with curved umbones, curled, facing forward, attributable almost exclusively, in the case of Pliocene-Pleistocene finds, to the species described as *Glossus humanus* or *Isocardia cor.* For the specimen depicted by Ferrante Imperato, it is assumed, with reasonable certainty, to be an internal model referable to a specimen of lower Pleistocene age, since Pliocene specimens sometimes attributed to the same species have smaller and less pronounced umbones.

Even in some works of lesser size, but not less important, of the other Neapolitan naturalist Fabio Colonna, student and friend of Imperato, are illustrated bivalves fossils (Colonna, *Aquatilium et terrestrium aliquot animaliu, aliarumquae naturalium rerum observationes*, 1616). The first images are included in the plate on page lii (52) where at the top are scallops and at the bottom other petrified testaceans.

> *Rispetto ai testacei fossili, egli si trattiene segnatamente su quelli delle colline di Andria nella Puglia e de' contorni di Campochiaro* [...]. *Ebbe l'avvertenza il Colonna di distinguere i gusci dai semplici nuclei e dalle impressioni* [...] (Brocchi 1814: XI).

> *With regard to testaceous fossils, he focuses in particular on those of the hills of Andria in Apulia and of the surroundings of Campochiaro* [...]. *Colonna was careful to distinguish the shells from simple cores and impressions* [...] (Brocchi 1814: XI).

The first one, named *Pectunculus lapideus*, is of uncertain determination; its elongated shape recalls the present *Manupecten pesfelis* (Linnæus, 1758), as well as the following specimen, not a fossil, which is close to it. The identification of the present specimen with *M. pesfelis* has been proposed also by Dollfus and Dautzenberg (1932: p. 313) who, however, do not give specific information about the fossil, but only note that it belongs to the genus *Chlamys*.

In the lower part of the same plate on page lii there are the other specimens definitely fossil, respectively, named *Concha lapidea recurva*, *Concha lapidea nautili effigie*, and *Concha lapidea gibbosa*. The *Concha lapidea recurva*, given the shape of the umbonal region folded, can be reasonably attributed to a specimen of *G. humanus*, the other with the features of a *Nautilus* is clearly an internal model of a bivalve belonging in all probability to the family Cardidae, while the *Concha lapidea gibbosa* can be interpreted as a bivalve of the family Chamidae, for example, *Chama placentina* (Defrance, 1817), or a brachiopod, as we will see later when dealing with these invertebrates. Another internal model, partially filled with calcite crystals, called *Concha fasciata gemmea concretione repleta*, illustrated in two different poses, could also probably belong to a *Cardium* (Colonna, *Purpura*, 1616: pp. 22, 25, 26).

We owe to Ulisse Aldrovandi the first extensive publication on fossil bivalves, even if not exclusively dedicated to them, mostly coming from the Pliocene terrain of Bologna or from the Eocene terrain of Vicenza. Specimens that can be recognized for a correct specific attribution are illustrated under various names. Giovanni Battista Brocchi in his work *Conchiologia fossile subapennina* (1814), still fundamental for the geological and paleontological knowledge and for the specific attributions of the discussed shells, reports a good part of them in the synonyms that accompany the descriptions of the testacei. Lodovico Foresti (1887) also attempted to identify these and other fossils belonging to the Aldrovandi collection, leaving out some of those of more uncertain recognition and consequent attribution. More than a decade ago, a treatment of the finds still presents in the original collection, preserved and displayed in the museum of Palazzo Poggi in Bologna, on the occasion of the quadricentenary of Aldrovandi's death, was edited by Sarti (2003).

Among the easily determinable species, one of the most obvious is the one called *Argyroconchites* (*Musaeum Metallicum*, 1648: p. 87), that is, silver shell-like. This Pliocene species, exclusively fossil, lived in littoral marine waters, at a depth of 10–15 meters, in a subtropical climate warmer than the present one, forming extensive shoals, both in power and length. The shell, foliate nacreous, has an oval profile with a beak-like protrusion at the anterior margin of the hinge, consisting of long, parallel teeth and dimples. The current name is *Isognomon maxillatus* Lamarck 1819.

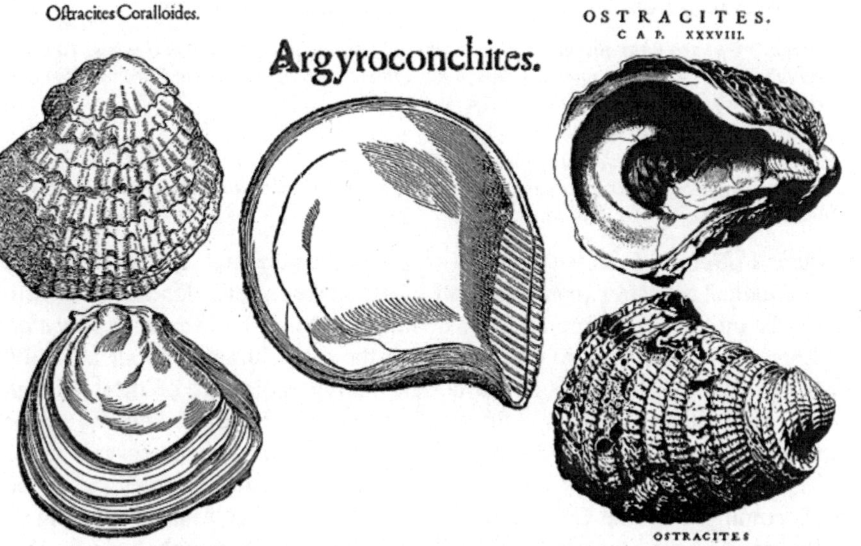

Fig. 12.7 *Ostracites*: Pliocene fossils of *Ostrea* sp. figured respectively, on the left, in Aldrovandi's *Musaeum Metallicum*, drawn from a xylographic matrix in pear wood (1648: p. 463) and, on the right, from a copper engraving, in the *Metallotheca Vaticana* (Mercati 1717: p. 293). In the center, the *Argyroconchites*, from Aldrovandi (1648: p. 87). [Graphic composition Maurizio Forli. Credit from: https://doi.org/10.3931/e-rara-13017 / Public Domain Mark - https://archive.org/details/bub_gb_DC_E6UVWg7EC (Public Domain Mark 1.0)]

More or less at the same time as Aldrovandi, Michele Mercati, describing and cataloguing the collections of the Vatican collection, deals with a wide range of objects, including fossils related to lithic models of bivalves. The illustrations, drawn from copper engravings, are more precise and detailed than those of his Bolognese contemporary, derived from wooden plates, and are used for the numerous specimens mostly from the area around San Miniato, where Mercati was originally from (Fig. 12.7).

In particular, among those called *Conchites*, that is, shell-shaped objects, we recognize internal models of *Glycymeris*, *Acanthocardia*, a venerid (*Metallotheca Vaticana*, 1717: p. 290) and a valva, excellently reproduced on both sides, of *Ostrea lamellosa* Brocchi, 1814, usually referred to as *Ostracites*. These were recognized as oysters or objects simulating them, and many different definitions were given. In Bertrand, in order to summarize their morphological variations and consequent different names, even within the group "Conchites", six pages of the dictionary are dedicated to them (1763: pp. 92–97).

> *Ostracites lapis est in similitudine Testae ostrea, a qua nomen sumpsit, i usu pumicis ad cutem levigandum habetur: sistere sanguinem virtus ejus est si in potu tribuatus. Melleque tritus mamillarum dolores compescit* (Leonardi 1502: p. 107).

The above, taken from Dolce (1565: p. 57), is translated as follows:

> *Ostracite è pietra fatta alla guisa d'un teschio d'ostrica, da cui ha preso il nome. Tiensi a uso di pomice per polir la pelle. È la sua virtù di stagnare il sangue, se è data nel bere. E trita cō melle sana il dolor delle mammelle.*
>
> *Ostracite is a stone made in the guise of an oyster skull, from which it took its name. It is used as a pumice to polish the skin. Its virtue is to stagnate the blood, if it is given to drink. And it heals the pain of the udders.*

In Mercati (*Metallotheca Vaticana*, 1717: p. 292), a particular species of bivalve is the one called *conchites lunatus*, a half-moon shaped shell, also characterized by having the umbonal area curved into a hook, what is now known as *Gryphaea arcuata* Lamarck, 1801. A fossil dating back to the early Jurassic, about 200 million years ago, with distribution in the corresponding sediments of most of central Europe, particularly abundant in France, Germany, and England. These same specimens are cited by Brocchi (1814: p. 472) under the name of *Anomia gryphus* L., that is, the true Mesozoic *Gryphaea arcuata,* as specified in the footnotes, described as fossils coming from the Crete Senesi, that is, Pliocene, evidently confusing them with other species of the genus *Ostrea.* Due to its peculiar morphology, frequently accompanied by its blackish color, it has often been considered as the remains of hooked nails of birds of prey or similar birds. The most common denomination, which has originated also the scientific name, is that of *Gryphitae*:

> *Gryphitae vocantus quia unguibus aduncis gryphorum similes sunt* (Spada 1744: p. 40).
>
> Or (Bertrand 1763: p. 255) *Elle a un rapport bien éloigné avec une griffe d'oiseau, ce qui cependant lui a fait donner le nom de gryphite.*

Other names always winking at the hooked form are *Crow-stone*, *Crow-cups* (Merret, *Pinax rerum naturalium Britannicarum*, 1667: p. 216) (Plot, *The Natural History of Oxfordshire*, 1677: p. 75), *Ague Shell* (Llwyd, *Lithophylacii Britannici ichnographia*, 1699: p. 27, no. 480), *Hawk's-Bill* (Thomas, *Philosophical Transactions*, 1708: p. 78). It is also known as the Devil's nail or *Devil's toe-nail* (Smith 1877) (Duffin and Davidson 2011).

The finding of the shell convinced that the evil had passed through there and the places were feared and avoided. Other supposed evidence of the devil's passage is due to the dissected impressions of the bucardia, this time understood as Mesozoic bivalves belonging to the genera *Conchodon* and *Megalodon*. These fossils are characteristic of the first period of the Mesozoic era, the Triassic. They lived in dense shoals within lagoons with high salinity waters of the ancient seas and the two most common species, both considered to be guide fossils of the Upper Triassic, are *Megalodon gumbeli* Stoppani, 1865, and *Conchodon infraliasicus* Stoppani, 1865.

What we believe to be a megalodontidae is shown by Gessner described as *Ctenitae*, petrified shell, similar to the mouth of a whale:

> *Oris ceti* [...] *figuram repraesentans* (1565: p. 165 versus, fig. B).

In the *Musaeum Metallicum* (p. 479) three bucardiae, *Bucardites triplex*, are illustrated, all of which seem to us to be produced from internal models of megalodontids; in particular, the one indicated at number 1 of the plate, seems to correspond to the specimen reported by Sarti (2003: p. 163) *Lapide simulante patellam seu molam genu* [*Musaeum Metallicum*, p. 483(sic) (fig. 5.13)]:

> *two internal models of Megalodon sp. belonging to two different specimens. The larger valve, which is the one shown on the MM, has an ancient inscription in china "horidi* (?) *humani lapis f.3".*

In section in the exposed rocks, worn and eroded by atmospheric agents, their heart shape has been attributed to footprints of transhumant herds of ungulates and also, and especially, to the disturbing footprints of the forked hooves of the devil's foot. In the light of our knowledge, the extravagance of the interpretations is sometimes so accentuated that the beliefs related to the devilish walks, in consideration of the historical period in which they originated, are almost normal compared to the idea proposed by Jean-Baptiste-René Robinet, a French naturalist philosopher, considered among the precursors of the theory of evolution, who expresses himself about them as follows:

> *Lithocardites & Bucardites. Plusieurs Naturalistes parlent des Lithocardites & Bucardites. On en connoît un grand nombre d'espèces.* [...] *Il n'est pas étonnant, que la Nature ait multiplié, avec tant de profusion, les exemplaires ébauchés d'une sî noble partie animale, le siége du mouvement vital. En comparant les differentes sortes de Lithocardites que produit chaque pays, on remarque entre elles une gradation de ressémblance qui plaît à l' imagination* (1768: p. 19, pl. 1, fig. 2).

In the *Metallotheca Vaticana* (1717: p. 330) are indicated under the denomination of *Onychites*, twenty-two fragments of shells of recent bivalves, *Glycymeris* and *Acanthocardia*, reduced to flakes by the natural wear and tear of the beaching, but voluntarily pierced by some of our ancient similar, to form small pendants, so that their structure lends itself to make a jewel or an ornament (Fig. 12.8).

> *Lapides ad humani unguis vel integri, vel dimidiati similitudinem alludentes Onychitas appellare libuit.*
>
> Stones, both whole and halved, alluding to the human fingernail, liked to call them *Onychitas.*

Mercati wonders how nature could have produced such perforated objects with such precision and skill, doubting that it is the work of some population. *Singuli in extremitate sui, foramen, quasi terebra confossum habent. Hoc nescio cujus solertiae vulgus ascribit.*

Isolated at their end, a hole, as if a drill had pierced it. I know of no people to whom I can attribute this skill. The finds come from the Sienese lands, in the locality of Cerreto Alto, not far from Monte Ilcineo, today's Montalcino. Other names given to similar fragments of worn shells, of a color and consistency reminiscent of onyx, have been called *Unguis lapideus* and *Pierre d'ongle* (Fig. 12.9).

As pendants, the pierced shells of mollusks were often used in the making of amulets, singly or more frequently accompanied by other objects believed to be apotropaic (Fig. 12.10). It is by resorting again to the Bellucci collection that we transcribe the description of an anti-malocchio amulet with these characteristics:

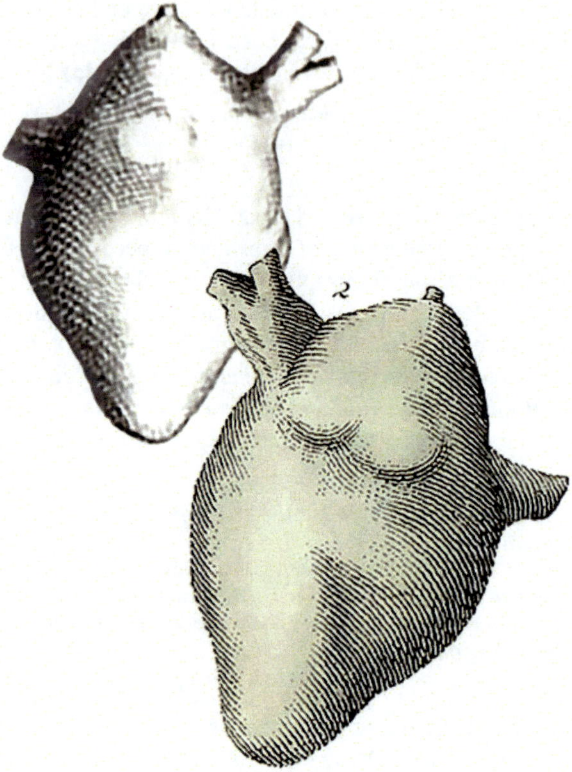

Fig. 12.8 *Anthropocardithes*: stone in the shape of a human heart; in the background by Plot (1677: p. 130, pl. VIII, fig. 1) and in the foreground by Robinet (1768: p. 19, pl. 1, fig. 2). [Graphic composition Maurizio Forli. Credit from https://www.biodiversitylibrary.org/bibliography/23488 (not in copyright); https://wellcomecollection.org/works/udrvrm8b (Public Domain Mark)]

> From Colle Armena, Aquila. *Perforated* Cardita *wormhole, with a ribbon ending in a knot for hanging. In the concave part of the valve there is a tuft of yew hair* (Meles taxus L.), *which greatly increases the effectiveness of the* Cardita *valve against the evil eye. This amulet is especially used by women, who are used to adapt it to the left shoulder, hanging it externally to the band, which passing on the shoulder, supports the bust. Sometimes you can see a group of these valves of* Cardita, *up to 10 or 12, adapted to the belt in the left side, or hanging from the* pincer, *placed on the chest, in its upper left side. The* pincer *is a tool made of silver or bone, which contains a small hook, through which you pass the wire, which is twisted from the distaff to the spindle, so that it does not get caught in other parts* (Bellucci 1898: pp. 60–61).

The apotropaic use of internal bivalve models, of fossil shells, is known to us from Sardinia. In the Sassari area, it was customary to carry such a fossil in one's pocket for superstitious purposes (2007, personal communication Mario Doria, Sassari), while in the Apennine part of Emilia Romagna, their use is documented in some rituals related to popular medicine. In both cases, the shape and color of these fossils is similar, being models composed of grey calcareous marly sediments of Miocene age, even

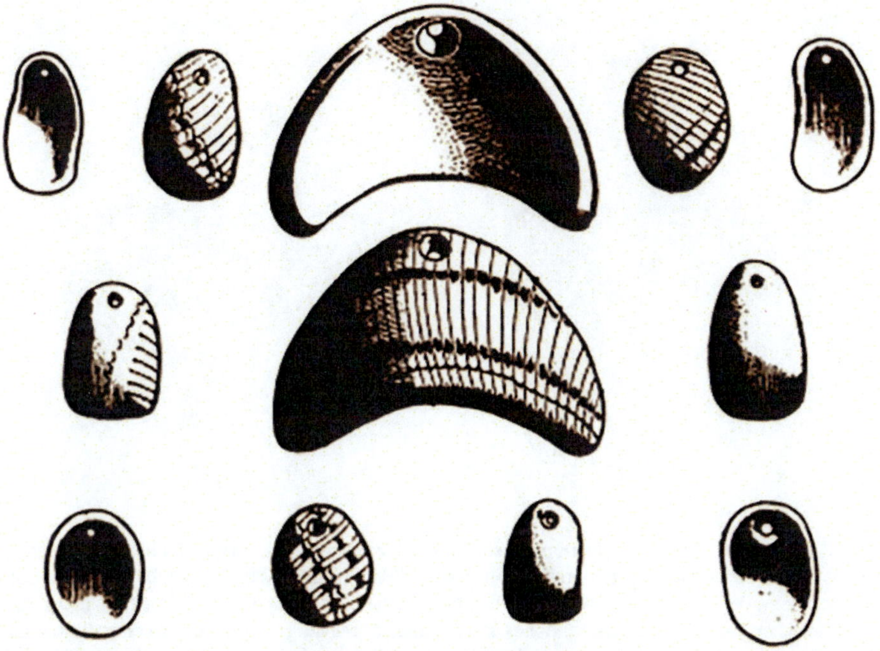

Fig. 12.9 Fragments of the shells of recent bivalves, *Glycymeris* spp. and *Acanthocardia* spp., reduced to flakes by the natural wear and tear of beaching, but voluntarily pierced by some of our ancient peers, to form small pendants (Mercati, *Metallotheca Vaticana*, 1717: p. 330) (https://archive.org/details/bub_gb_DC_E6UVWg7EC Public Domain Mark 1.0)

if they belong to different species. In the case of the Apennine finds in the Modena area, the possible connection between the places with toponyms recalling the term "Sasso delle Streghe" and the presence of Miocene Lucinid fossils seems interesting and, if necessary, a reason for further investigation. These fossils are bivalves, shells and their models almost always still connected and in a position of life, with the hinge facing upward, which constitute with their remains, large banks that characterize some sediments of limited extension called "Calcare a Lucina", distributed along the Apennine margin. These rocks are the evidence of ancient areas of deep sea, where these particular mollusks, which still today feed through the exploitation of chemosynthetic bacteria capable of obtaining organic matter from the gaseous exhalations, developed near the methane gas or hydrogen sulfide leaks from the bottom. These, or similar models, could be the fossils used within some rituals whose testimonies have been collected and published by Maddalena De Bernardi (2016). In a strange (?) coincidence, the use of fossils is documented for a practitioner from Sardinia, but living in Bologna at the time of the interview, who assured however that she practiced her arts also in Emilia Romagna (Figs. 12.11, Fig. 12.12, 12.13, 12.14, 12.15, 12.16, 12.17, 12.18, 12.19, 12.20, 12.21, and 12.22).

Fig. 12.10 From the Bellucci collection. Presumed fossil echinus, in the shape of a heart, in which a hole has been made at the top, crossed by a rough string used to suspend the object. In our opinion, it is an internal model of a worn bivalve. It is in fact evident the bilateral symmetry of the find, in which, even if worn, we can recognize the umbones and the trace of the hinge, which is not compatible with the structure of an echinid. For a comparison, see the following figures related to these particular bivalves. To be interpreted therefore as a *Bucardia*. From Roccamorice (Pescara), hanging in the stable was used to prevent epidemic diseases of animals. (Coll. Bellucci. M.A.N.U. (PG) Inv. 6530. Reproduced by kind permission of the Ministry for Cultural Heritage and Activities and for Tourism, Regional Directorate for Museums of Umbria - National Archeological Museum of Umbria)

Fig. 12.11 In the foreground, specimens of *Lucina* sp. from the surroundings of Castagno d'Andrea (FI), (average ø 100 mm). In the background *Lucina pomum* Dujardin, from Gioli 1887, pl. XV, figs 1, 6. (Photo and graphic composition Maurizio Forli)

Fig. 12.12 Devil's nails, fanciful reconstruction of the so-called Devil's nails. Demon from Sebastian P. Münster (*Cosmographia universale Lib. VI*, 1575: p. 1166). [Graphic composition Maurizio Forli. Credit from: https://archive.org/details/cosmographiavniu00mnst/page/1166/mode/2up (no copyright)]

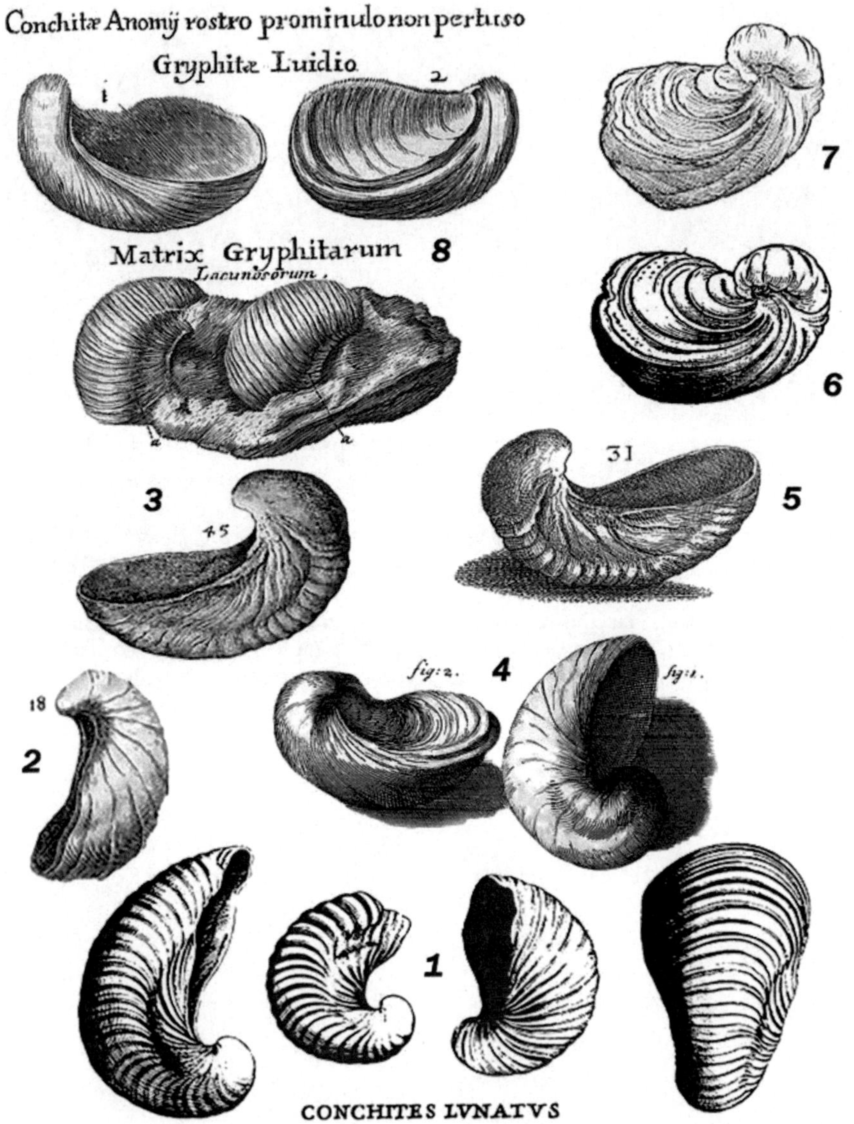

Fig. 12.13 "Devil's nails": various representations and designations of *Griphaea arcuata* Lamarck, 1801. (1) *Conchites lunatus* (Mercati 1717: p. 292). (2) *Crow-Stone* (Plot 1677: p. 105, pl. IIII, fig. 18). (3) *Conchites anomius rugosus* (Lister 1678: p. 238, pl. 8, fig. 45) reproposed from Plot (1677). (4) ... *a sort of shell* ... (Waller 1705: p. 285, pl. IV, figs 1–2). (5) *shell of the English Sea capricious in form* (Buonanni 1681: p. 158, fig. 31) reproposed from Lister. (6) *Conchites* (De Laet 1647: p. 192). (7) *Conchites* (Worm 1655: p. 80) from De Laet. (8) *Conchitae Anomij rostro prominulo non pertuso* [Lang 1708: p. 152, pl. 48(2)]. [(Graphic composition Maurizio Forli. Credit from: (1) https://archive.org/details/bub_gb_DC_E6UVWg7EC (Public Domain Mark 1.0). (2) https://www.biodiversitylibrary.org/bibliography/23488 (not in copyright). (3) https://www.biodiversitylibrary.org/bibliography/65746 (Public domain). (4) https://wellcomecollection.org/works/xxggvyj3 (Public Domain Mark). (5) https://doi.org/10.3931/e-rara-39893 / Public Domain Mark. (6) https://doi.org/10.3931/e-rara-35487 / Public Domain Mark. (7) https://doi.org/10.5962/bhl.title.63844 (not in copyright). (8) https://doi.org/10.3931/e-rara-16400 / Public Domain Mark)]

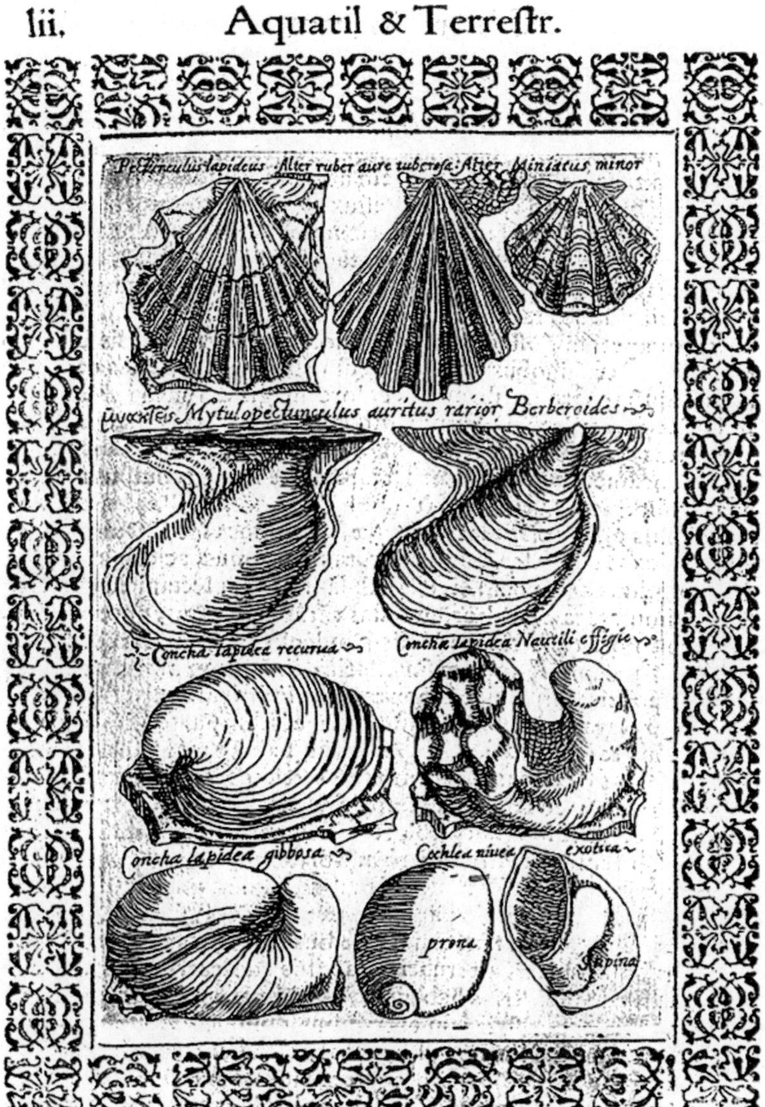

Fig. 12.14 Some shells from Fabio Colonna (1616): top left *Manupecten pesfelis* (L., 1758) fossil and a similar living one, next right probably *Flexopecten flexuosus* (Poli, 1795). After two *Pteria hirundo* (L., 1758) there are three fossils, the *Concha lapidea recurva* is referable to *Glossus humanus* (L., 1758), the *Concha lapidea Nautili effigie* is an indeterminable partial remnant, plausibly attributable to the family Cardidae, while the *Concha lapidea gibbosa* can be interpreted as a bivalve of the family Chamidae, for example, the exclusively fossil *Chama placentina* (Defrance, 1817) or the similar *Chama gryphoides* L., 1758, both fossil and living species. Bottom right, an exotic naticid that we identify as *Polinices mammilla* (L., 1758), widespread in the Indian Ocean (https://www.biodiversitylibrary.org/bibliography/723 Public domain)

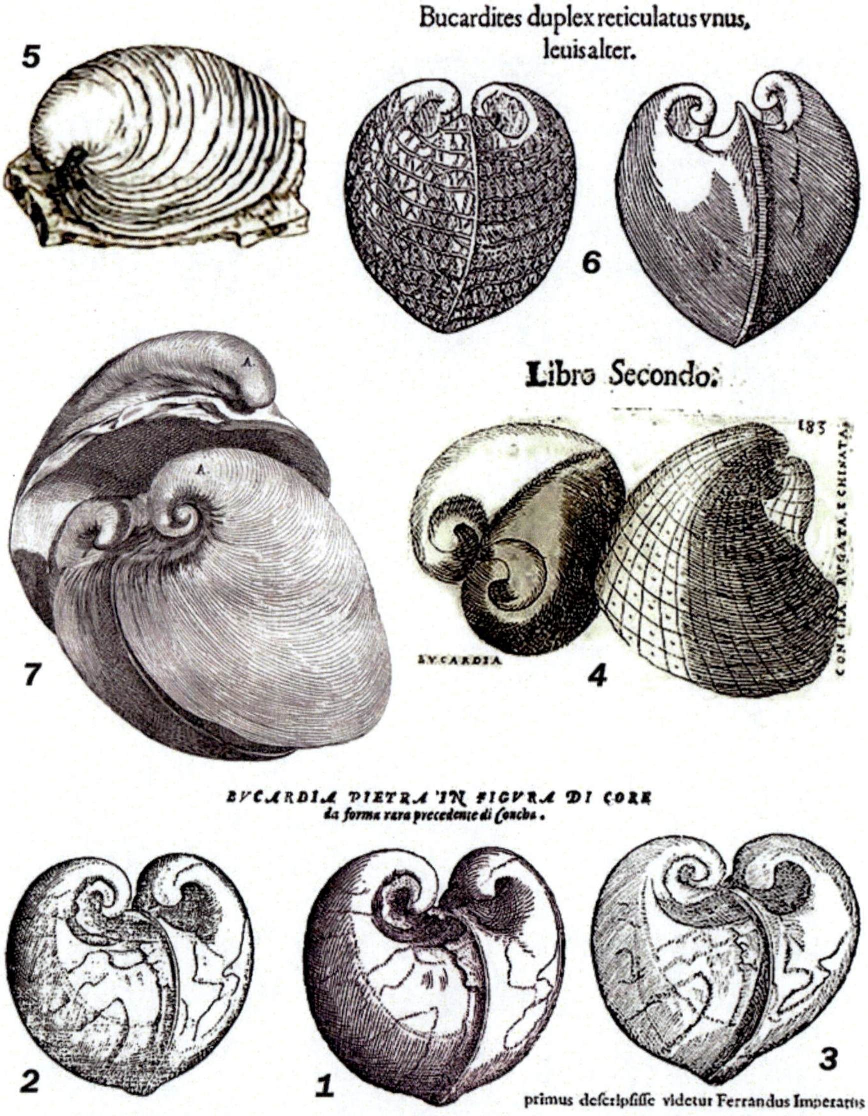

Fig. 12.15 Bucardia heart-shaped stones derived from the internal pattern of fossil shells often referable to the bivalve *Glossus humanus* (Linnaeus, 1758) ex *Isocardia cor* (L., 1767) or *Megalodon* sp.: (1) from Ferrante Imperato (1599: p. 666) *Bucardia pietra in figura di core da forma rara precedente di concha*. (2–3) De Laet (1647: p. 210) and Worm (1655: p. 80); both taken from Imperato's figure. (4) Moscardo (1656: p. 183) Bucardia next to a recent specimen probably of *Acanthocardia* (5) *Concha lapidea recurva* from Colonna (1616: lii). (6) *Bucardites duplex reticulatus unus levis alter* fossils from the environs of Bologna vulgarly called "Tortelli," Aldrovandi (*Musaeum Metallicum* 1648: p. 479). (7) *Rarissima cŏca* .A. *detta Bugardia* Monti di Calabria left valva and complete specimen from Scilla (1759: pl. XVI, fig. A). [Graphic composition Maurizio Forli. Credit from: (1) https://www.biodiversitylibrary.org/item/183177#page/5/mode/1up (not in copyright) (2) https://doi.org/10.3931/e-rara-35487 / Public Domain Mark (3) https://doi.org/10.5962/bhl.title.63844 (not in copyright) (4) https://www.biodiversitylibrary.org/bibliography/157058 (not in copyright) (5) https://www.biodiversitylibrary.org/bibliography/723 (Public domain) (6) https://doi.org/10.3931/e-rara-13017 / Public Domain Mark (7) personal copy Maurizio Forli]

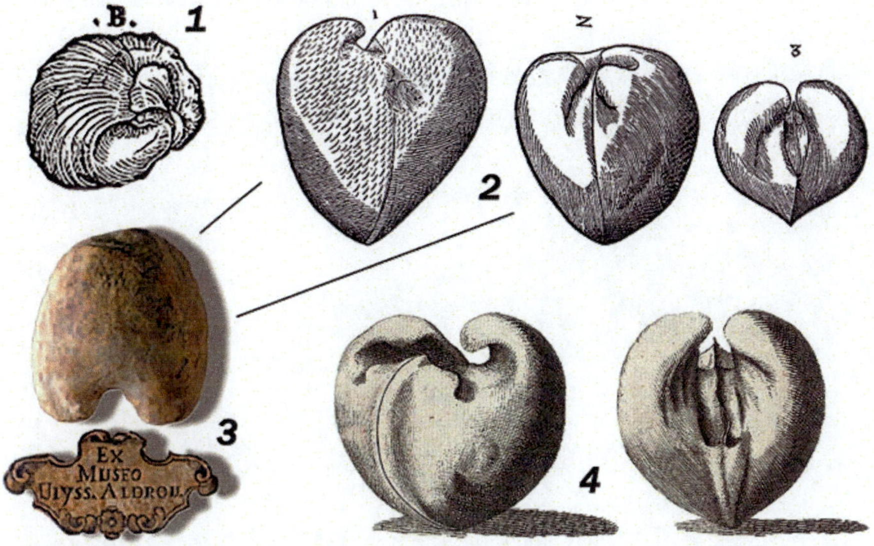

Fig. 12.16 Fossil remains of megalodontids. (1) *Ctenitae*, from Gessner (1565: p. 165). (2–3) *Bucardites triplex* (*Musaeum Metallicum*, 1648: p. 480); it is one of these that corresponds to the specimen cited by Sarti (2003) with the eighteenth-century tag that identifies it as part of the Aldrovandi collection (height ~100 mm). (4) Two models, well preserved, from Baier (1708: pl. IV, figs. 20 and 21). [Graphic composition Maurizio Forli. Credit from: (1) https://doi.org/10.3931/e-rara-4176 / Public Domain Mark (2–3) https://doi.org/10.3931/e-rara-13017 / Public Domain Mark (4) https://doi.org/10.3931/e-rara-18618 / Public Domain Mark]

Fig. 12.17 A typical bucardia constituted by the internal mould of a complete specimen of *Megalodon* sp. In particular, these remains were considered "ox heart" shaped *fossils* that were sometimes distinguished from those formed by the models of bivalves, approached for their different shape and smaller size to the human heart, therefore calling them *Anthropocardithes*. (Photo and graphic composition Maurizio Forli)

Fig. 12.18 Devil's hoof or Devil's footprints, a natural section of the shell of a megalodontidae from the surroundings of Madonna di Campiglio (Trento). (Photo credit Massimo Cresti)

Fig. 12.19 Details from the spectacular cluster of shells of *Megalodon* sp., found in the locality Orti della Regina, in the Dolomites, near Madonna di Campiglio. (Photo credit Massimo Cresti)

Fig. 12.20 Other details of *Megalodon* shells from the same rocky block. Dolomites, Orti della Regina: from Madonna di Campiglio. (Photo credit Massimo Cresti)

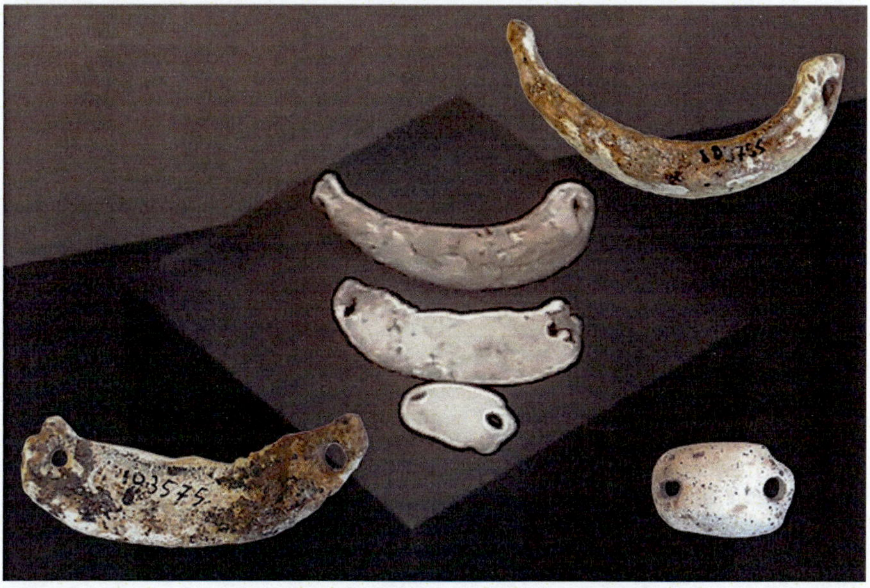

Fig. 12.21 Ornaments made from bivalve shells. From the Lion's Cave in Agnano, on Monte Pisano. [Courtesy Museum of Natural History, University of Pisa (photo credit Lucia Angeli) photo in background, credit Maria Teresa Tommasi, graphic composition Maurizio Forli]

Fig. 12.22 Lower valve of *Pecten maximus* (Linnaeus, 1758), the shell of St. James, symbol and evidence of the pilgrimage to Santiago de Compostela and the visit to the Saint's tomb (photo credit Massimo Cresti). On the right, a silver necklace with an enamelled pendant in the shape of a *Cardium* valve, worn as a good luck charm (Procida (Napoli) 1970 (courtesy private collection). (Graphic composition and photo on the right Maurizio Forli)

References

Aldrovandi U (1648) Ulyssi Aldrovandi patricii Bononiensis Musaeum metallicum in libros IIII: distributum Bartholomaeus Ambrosinus ... labore, et studio composuit cum indice copisosissimo Arcus Antonius Bernia. Bononiae: typis Io. Baptistae Ferronii, 1648. ETH-Bibliothek Zürich, Rar 1027. https://doi.org/10.3931/e-rara-13017 / Public Domain Mark

Allasinaz A (1992) Paleontologia generale e sistematica degli invertebrati. Edizioni Culturali Internazionali Genova

Baier JJ (1708) Oryktographia Norica sive rerum fossilium et ad minerale regnum pertinentium, in territorio Norimbergensi eiusque vicinia observatarum [...]. Norimbergae: impensis Wolfgangi Michahellis, 1708. ETH-Bibliothek Zürich, Rar 3707. https://doi.org/10.3931/e-rara-18618 / Public Domain Mark

Bellon P (1553) De Aquatilibus, Libri duo. C. Stephanum, Parisiis. https://www.biodiversitylibrary.org/bibliography/5765 (Public domain)

Bellucci G (1898) Amuleti italiani contemporanei. Catalogo descrittivo della collezione inviata all'Esposizione Nazionale di Torino. Unione Tipografica Cooperativa, Perugia

Bertrand E (1763) Dictionnaire universel des fossiles propres et des fossiles accidentels. A la Haye: Chez Pierre Gosse Junior, et Daniel Pinet, M. DCC. LXIII [1763]. ETH-Bibliothek Zürich, Rar 5720. https://doi.org/10.3931/e-rara-10463 / Public Domain Mark

Bouchet P, Rocroi J-P (2010) Nomenclator of bivalve families. Malacologia 52(2):1–184

Brocchi GB (1814) Conchiologia fossile subapennina con osservazioni geologiche sugli Apennini e sul suolo adiacente. Stamperia Reale, Milano

Buonanni F (1681) Ricreatione dell'occhio e della mente nell'osservation' delle chiocciole: proposta a' curiosi delle opere della natura. In Roma: per il Varese: a spese di Felice Cesaretti ..., Zentralbibliothek Zürich, NNN 713 | G. https://doi.org/10.3931/e-rara-39893 / Public Domain Mark

Colonna F (1616) Fabii Columnae Lyncei Minus cognitarum rariorumque nostro coelo orientium stirpium ekphrasis … https://www.biodiversitylibrary.org/bibliography/723 Public domain

Cossmann MM, Peyrot A (1909) Conchologie Néogénique de l'Aquitaine. Actes de la Société Linnéenne de Bordeaux 63:73–293

De Bernardi M (2016) Marking healing: ethnosemiotics of a magical act: semiotic reading of the practice of “signatures” in the transcultural perspective relative to the Apennine area of the Italian region of Emilia Romagna. Social Anthropology and ethnology. Université Toulouse le Mirail - Toulouse II, 2015. English. < NNT: 2015TOU20039>. <tel-01326386>

De Laet J (1647) De Gemmis et Lapidibus, libri Duo. Quibus praemittur Theophrasti, liber de Lapidìbus Graece & latine cum Brevibus Annotationibus. Ioannis Maire, Lugduni Batavorum. https://doi.org/10.3931/e-rara-35487 / Public Domain Mark

Dolce L (1565) Libri tre ne i quali si tratta delle diverse sorti delle Gemme, che produce la Natura, della qualità, grandezza, bellezza, et virtù loro. Gio Battista, Marchio Sessa, et Fratelli, Venetia

Dollfus G, Dautzenberg Ph (1932) Les Mollusques de Fabius Columna. Journal de Conchyliologie, LXXVI(4): 283-333 https://www.vliz.be/en/search-persons?module=ref&refid=15733

Duffin C, Davidson J (2011) Geology and the dark side. Proc Geol Assoc 122(1):7–15

Encelius C (1551) De re metallica: hoc est, de origine varietate, & natura corporum metallicorum, lapidum, gemmarum, atque aliarum, quae ex fodinis eruuntur, rerum, ad medicinae deservientium, [...]. Franc. [Frankfurt]: apud Chr. Egenolphum, [15--]. ETH-Bibliothek Zürich, Rar 2259. https://doi.org/10.3931/e-rara-29357 / Public Domain Mark

Foresti L (1887) Sopra alcuni fossili illustrati e descritti nel Musaeum Metallicum di Ulisse Aldrovandi. Boll Soc Geol Ital 6:81–121

Gessner C (1565) De omni rerum fossilium genere, gemmis, lapidibus, metallis, et huiusmodi, libri aliquot, plerique nunc primum editi. Tiguri. https://doi.org/10.3931/e-rara-4176 / Public Domain Mark

Gioli G (1887) La Lucina pomum Duj. Atti della Società Toscana di Scienze Naturali 8:301–313

Imperato F (1599) Dell'Historia Naturale di Ferrante Imperato napolitano. Libri XXVIII. Nella quale ordinatamente di tratta della diversa condition di miniere, e pietre. Con alcune historie di Piante, et Animali fin'ora non date in luce. Costantino Vitale, Napoli. https://www.biodiversitylibrary.org/item/183177#page/5/mode/1up (not in copyright)

Lang KN (1708) Tractatus de origine lapidum figuratorum in quo diffuse disseritur, utrum nimirum sint corpora marina a diluvio ad montes translata, [...]. Lucernae [Luzern]: sumptibus authoris, typis Annae Felicitatis Hautt, 1709. ETH-Bibliothek Zürich, Rar 2773. https://doi.org/10.3931/e-rara-16400 / Public Domain Mark

Lanza V (1984) Il “De Re Metallica…” di Christophorus Encelius (1571-1583). Le prime illustrazioni esplicative dei fossili. Geol Romana 23:111–120

Leonardi C (1502) Speculum Lapidum et D. Petri Arlensis de scudalupis, presbyteri hierosolymitani, Sympathia, septem metallorum ac septem selectorum lapidum ad planetas accredit Magia Astrologica Petri Constantii Albinii villanovensis. Apud Christianum Liebezeit, Hamburgi

Lhuyd E (1699) Eduardi Luidii Lithophylacii Britannici ichnographia: sive lapidum aliorumque fossilium Britannicorum singulari figura insignium quotquot hactenus vel ipse invenit vel ab amicis accepit, distributio classica, [...]. Londini: ex officina M.C ; Lipsiae: sumt. Joh. Ludw. Gleditsch & Weidmann, 1699. Zentralbibliothek Zürich, NG 1883. https://doi.org/10.3931/e--rara-30541 / Public Domain Mark

Lister M (1678) Angliae Historiæ animalium Angliæ, tres tractatus. Unus de araneis. Alter de cochleis tum terrestribus, tum fluviatilibus. Tertius de cochleis marinis. Quibus adjectus est quartus de lapidibus ejusdem insulæ ad cochlearum quandam imaginem figuratis. Joh. Martyn, Londini. https://www.biodiversitylibrary.org/bibliography/65746 Public domain

Mercati M (1717) Metallotheca Vaticana. Opus posthumum, Autoritate et Munificentia Clementis Undecimi Pontificis Maximi. E tenebris in lucem eductum: Opera autem, & studio Joannis Mariae Lancisii, Archiatri Pontificii illustratum. Ex Officina Jo: Mariae Salvioni Romani, Romae. https://archive.org/details/bub_gb_DC_E6UVWg7EC (Public Domain Mark 1.0)

Merret C (1667) Pinax rerum naturalium Britannicarum, continens Vegetabilia, Animalia et Fossilia. In hac Insula reperta inchoatus. Typis T. Royerost, Impensis Cave Pulleyn, Londini

Moscardo L (1656) Note overo memorie del museo di Lodovico Moscardo nobile veronese. Academico Filarmonico, dal medesimo descritte, Et in Tre Libri distinte. Per Paolo Frambotto, Padoa. https://www.biodiversitylibrary.org/bibliography/157058 (not in copyright)

Münster SP (1575) Cosmographia vniuersale, nella quale secondo che n'hanno parlato: i piu veraci scrittori, son designati i siti di tutti gli paësi. I siti de tutte le parti del mondo habitabile, & le proprie doti: le tauole topographice delle regioni ... https://archive.org/details/cosmographiavniu00mnst/page/1166/mode/2up (no copyright found)

Newell ND (1965) Classification of the Bivalvia. Am Mus Novit 2206:1–25

Plot R (1677) The natural history of Oxfordshire, being an essay toward the natural history of England. Printed at the Theater, Oxford. https://www.biodiversitylibrary.org/bibliography/23488 (not in copyright)

Robinet JBR (1768) Consideration Philosophiques de la Gradation Naturelle des forms de l'etre, ou les essais de la Nature qui apprend a faire l'homme. Charles Saillant, Paris. https://wellcomecollection.org/works/udrvrm8b Public Domain Mark

Rondelet G (1555) Universae aquatilium Historiae pars altera, cum veris ipsorum Imaginibus. Matthiam Bonhomme, Lugduni. https://archive.org/details/ARes071032/page/n13/mode/2up (Public Domain 1.0)

Sarti C (2003) Le Collezioni di geologia nel Museo di Ulisse Aldrovandi. In: Vai GB, Cavazza W (eds) Quadricentenario della parola Geologia. Ulisse Aldrovandi 1603 Bologna, Minerva, pp 152–167

Scilla A (1759) De corporibus marinis lapidescentibus quæ defossa reperiuntur. Addita Dissertatione Fabii Columnæ de Glossopetris. Joannis Zempel, Romae

Smith WJB (1877) The common Lias fossil. Notes Queries VII(5):15

Sowerby J (1812) The Mineral Conchology of Great Britain; or coloured figures and descriptions of those remains of testaceous animals or shells, which have been preserved at various times and depths in the Earth. Vol. I. Benjamin Meredith, London. https://www.biodiversitylibrary.org/bibliography/14408 (not in copyright)

Spada G (1744) Corporum lapidefactorum agri Veronensis catalogus. Veronæ: Typis Dionysii Ramanzinii: Bibliopolæ apud S. Thomam, 1744. Zentralbibliothek Zürich, NG 58 I G. https://doi.org/10.3931/e-rara-50868 / Public Domain Mark

Thiele J (1929–35) Handbuch der Systematischen Weichtierkunde. Voll. 1-4. Gustav Fischer, Jena

Thomas A (1708) Advertissement. Philosophical Transactions, 26: 77-80 https://doi.org/10.1098/rstl.1708.0012

Vialli V (1974) Appunti di Paleontologia. Pitagora Editrice, Bologna

Waller R (1705) The posthumous works of Robert Hooke ... containing his Cutlerian lectures, and other discourses, read at the meetings of the illustrious Royal Society ... Illustrated with sculptures. To these discourses is prefixt the author's life, giving an account of his studies and employments, with an enumeration of the many experiments, instruments, contrivances and inventions, by him made and produc'd as Curator of Experiments to the Royal Society. Printed by Sam. Smith and Benj. Walford (Printers to the Royal Society) at the Prices Arms in St. Paul's Church-yard, London. https://wellcomecollection.org/works/xxggvyj3 Public Domain Mark

Worm O (1655) Museum Wormianum, seu Historia rerum rariorum, tum Naturalium, quam Artificialium, tam Domesticarum, quam Exoticarum; quae Hasniae, Danorum in aedibus Authoris Servantur. Iohannem Elsevirium, Lugduni Batavorum. https://doi.org/10.5962/bhl.title.63844 (not in copyright)

Zilhão J, Angelucci DE, Badal-Garcia E, d'Errico F, Daniel F, Dayet L, Douka K, Higham TFG, Martínez-Sánchez MJ, Montes-Bernárdez R, Murcia-Mascarós S, Pérez-Sirvent C, Roldán-García C, Vanhaeren M, Villaverde V, Wood R, Zapata J (2010) Symbolic use of marine shells and mineral pigments by Iberian Neandertals. Proc Natl Acad Sci 107(3):1023–1028

Chapter 13
Curiosity: The Strange Mussels of Mr. Van Aken (Maurizio Forli)

Abstract Shells, from prehistoric meal scraps, have become the main object of attention and have taken on symbolic meanings that, though varying over time, almost always refer to concepts of beauty and rebirth or death and the transience of life. Particularly in figurative artistic representations, they are often associated with pearls, elements of water, symbolizing purity, and motherhood and as such have been depicted in many paintings, including famous ones. Thus, often in these paintings, it is the female sexual organ that is masked and hidden behind the image of a shell. It is in this context that some representations of *Mytilus edulis* are part of two works by Hieronymus Bosch, the *Garden of Earthly Delights* and the *Last Judgement*, the first shell highlighted as a heavenly place and the second as a receptacle of damnation.

Keywords Hieronymus Bosch · *Garden of Earthly Delights* · *Last Judgement* · *Mytilus edulis*

Shells have always exerted an undeniable fascination on us. From meal scraps they have become the main object of attention and have taken on symbolic meanings that, though variable over time, almost always refer to concepts of beauty and rebirth or death and transience of life

In artistic representations and in particular in those of the figurative arts, shells are often associated with the pearl, elements of water, symbolize purity, and motherhood and as such have been depicted in numerous paintings, including famous ones, such as Botticelli's Birth of Venus. Thus, often in these paintings, it is the female sexual organ that is masked and hidden behind the image of a shell. This type of representation can be considered "joyful," referring precisely to concepts of purity, beauty, and pleasure, and in contradiction with the other way of depicting the shells, "macabre," which wants them only empty shells, cadaverous remains similar to skulls, which remind man of the emptiness and vanity of life, deprived of the comfort of faith. A certainly more current and modern way of interpreting shells is that of a more or less attractive natural object, and as such received, as if it were not part of a living being but a natural object in itself. To summarize, we say that, depending

© The Author(s), under exclusive license to Springer Nature Switzerland AG 2022

M. Forli, A. Guerrini, *The History of Fossils Over Centuries*,
https://doi.org/10.1007/978-3-031-04687-2_13

on the point of view, a shell is a natural production, often beautiful, that can symbolize concepts related to motherhood, birth, the magnificence of life, or the skeleton, with all the unpleasantness associated with this definition, of a being of which there is no longer any trace or memory.

These more or less latent concepts can be traced in the way of depicting them but also in the way of collecting and ultimately owning them. Moreover, if we add the fact that with the use of current scientific determination, each of us can link his or her name to a shell, as the author of a new species or as a person to whom one has been dedicated that was never described before, and for this reason remembered in the future, *immortalized* in this way even after his or her earthly disappearance, the shell symbology becomes an even more manifest expression of the dualism of life–death.

It is in this context that I place two illustrations of mussels unusually loaded with meaning.

Their depiction already appears to be uncommon, as usually in painting specimens of *Pecten* or "exuberant" gastropods, such as muricids and tritons, were depicted. In this, too, Mr. Jeroen Van Aken stands out from his contemporaries. This Dutch gentleman, better known as Hieronymus Bosch, who lived more or less from 1453 to 1516, depicted in at least two of his most famous paintings, the *Triptych of the Garden of Delights*, now in Madrid, and the *Triptych of the Last Judgment*, in Bruges, two mussels with half-closed valves with apparently obscure meanings.

Then, why two mussels?

The starting point for these reflections came from reading the booklet devoted to Bosch, in the series "I classici dell'arte" published by the Rizzoli Corriere della Sera group in 2004. In the preface taken from a writing by Dino Buzzati, the conclusion is reached that Bosch, albeit with a very personal visionary manner, did not represent anything other than what surrounded him. Agreeing with this thought, I cannot help in observing that the naturalistic representations of fishes, insects, amphibians, birds, and sometimes mixed together in an absurd way, giving origin to more or less fantastic bodies, at that time must not have been too much different from the "monstrous" anatomical preparations, such as the basilisk, the sea monkeys, or the sirens, handcrafted from the union of parts of more animals. The mussels, very common on the coasts of the Channel, and still widely used as gastronomic delicacies, with their obvious figurative reference to the vagina, were not at all unknown to Bosch, so here is a plausible reason to use the shell as an alcove in the triptych of the *Garden of Earthly Delights*. The shell of the mussel, black/dark brown on the outside and pearly on the inside, because of the shape of the mollusk's body, which is orange-red when open, raw or cooked, cannot but be charged with erotic meanings. It is in this guise that in my opinion she is depicted in both paintings. Among other things, in this representation, appear around the characters lying down, some translucent pearls sometimes interpreted as drops of semen. Leaving aside their allegorical meaning, I will dwell instead on the fact that Bosch must surely have known these mollusks very well because, in order to place pearls there, he must have known that mussels produce them, often very small and translucent ones, sometimes becoming, because of this fact, inedible (Fig. 13.1).

Fig. 13.1 Hieronymus Bosch, *Garden of Earthly Delights,* central panel, detail, (1480–1490) Prado Museum, Madrid (Spain) (https://commons.wikimedia.org/wiki/File:El_jard%C3%ADn_de_las_Delicias,_de_El_Bosco.jpg Public domain)

In the *Garden of Earthly Delights* the shell serves as a safe haven for lovers, protected and hidden by the shell, abandoned to the pleasures of the senses, while in the triptych of the *Last Judgement of Bruges*, the shell appears as a place of torment for the sin committed. The position of the two shells is no longer horizontal, like a cradle, a bed, and an alcove, but vertical, with the shells dilated by a knife, similar to the one still used today to open oysters, which has become a phallic symbol, which rapes the shell with its penetration and at the same time punishes a man (or a woman, as it could be by observing what looks like blond hair stretched out on his back), lying and forced ventrally on the blade, tormented, and torn to pieces by a demon. In short, ecstasy and torment, forgiveness and punishment are other ways of expressing the union of opposites (Figs. 13.2 and 13.3).

Fig. 13.2 Hieronymus Bosch, *The Last Judgement*, detail, (c. 1486); Groeningemuseum, Bruges (Belgium) (https://it.wikipedia.org/wiki/Trittico_del_Giudizio_di_Bruges#/media/File:Bosch_laatste_oordeel_drieluik.jpg Public domain).

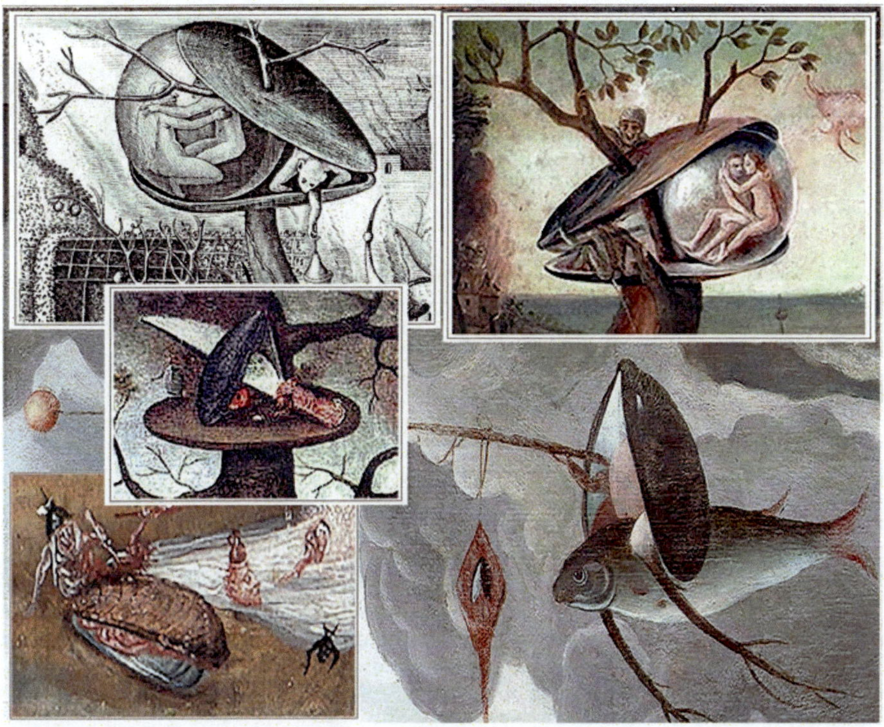

Fig. 13.3 The erotic theme represented by the mussels and the lovers has been reproposed, like countless other more "appealing" and commercial details of Bosch's paintings, by his proselytes and imitators. Among these and those we have selected a series of repropositions, which with the passage of time lose their originality and freshness. From top left, Pieter Bruegel, the Elder, detail from a drawing, later engraved, belonging to the *Seven Deadly Sins* series, in this case the *Lust* [1]. Next to the right, the figure is reproduced by an anonymous follower, second half of the sixteenth century [2]. In the center left and in the background, bottom right, details from paintings attributed to Pieter Huys: The *Temptations of St. Anthony* (1577 and 1546) [3, 4], and at bottom left, detail from The *Temptations of St. Anthony* by an anonymous imitator of Huys, who reproposes, rather clumsily, the mytilus with the two lovers of the woodland version of the *Garden of Earthly Delights* (late sixteenth century) [5]. (Image composition Maurizio Forli)

Chapter 14
Miscellaneous News about Gastropods, from the Earliest Printed Images to the Present Day

Abstract Gastropods are the only mollusks distributed in marine, freshwater, and terrestrial environments. They are mostly benthic marine animals living in close contact with the seabed, even if pelagic, sedentary, or parasitic species are also present. Their presence has been ascertained since the Paleozoic, but those that have stratigraphic importance are relatively few, for example, the group of Nerinea for the Upper Jurassic and Lower Cretaceous. A general indication on their classification and stratigraphic distribution is given. In Encelius (*De re metallica: hoc est, de origine varietate, & natura corporum metallicorum, lapidum, gemmarum, atque aliarum, quae ex fodinis eruuntur, rerum, ad medicinae deservientium*, [...]. Franc. [Frankfurt]: apud Chr. Egenolphum, [15--]. ETH-Bibliothek Zürich, Rar 2259, p 229. https://doi.org/10.3931/e-rara-29357 / Public Domain Mark, 1551), there is what is generally interpreted as the first fossil gastropod represented in printed books, probably an internal model of a Turritella specimen. Another shell, frequently depicted in early printed texts, fossil, or otherwise, is that of the turbinid *Bolma rugosa*. Its importance besides the undoubted beauty, expressed, even more if polished, is due to the fact that it produces a calcareous operculum called Occhio di Santa Lucia, a powerful anti-evil eye. Its use in protective amulets is known from the early Iron Age to the present day. Numerous examples of these apotropaic objects are present in the collection of amulets belonging to Giuseppe Bellucci, in Perugia. They are brought back moreover news about the so-called *Odorous Nails*, horny opercula of gastropods, burnt-like perfumed substances, similar to the incense. Other curiosities are included at the end of the text, about the shells of *Janthina*, pelagic gastropod in commensal relationship with the hydrozoan *Velella velella* Linnæus, 1758, popularly known as St. Peter's boat.

Keywords Gastropoda · Fossils · Encelius · Bellucci collection · Eyes of Saint Lucia

Gastropods are the only mollusks distributed in marine, freshwater, and terrestrial environments. They are mostly benthic marine animals living in close contact with the seabed, endowed with an excellent mobility even if there are also pelagic,

© The Author(s), under exclusive license to Springer Nature Switzerland AG 2022

M. Forli, A. Guerrini, *The History of Fossils Over Centuries*,
https://doi.org/10.1007/978-3-031-04687-2_14

sedentary, or parasitic species. Universally known for their aesthetic beauty and commercial importance, in Italy, especially in the southern regions, they still have a special apotropaic use, against the evil-eye, with the so-called eye of Saint Lucia, the red-colored operculum, of the *Bolma rugosa* (Figs. 14.1 and 14.2).

Fig. 14.1 According to popular tradition, having half a glass of water in the house, from which salt has overflowed after a spoonful has been thrown in, is an indication of negative environmental influences linked to the evil eye. In order to fight it, one of the most effective remedies is the Occhio di Santa Lucia (Eye of St. Lucia), the operculum of the gastropod *Bolma rugosa* (L. 1767). In the background, image modified from Guaccio (1626, ***Compendium Maleficarum***). (Photo and graphic composition Maurizio Forli. Credit from: https://archive.org/details/image4995MiscellaneaOpal/page/n293/mode/2up. under CC/publicdomain/zero/1.0)

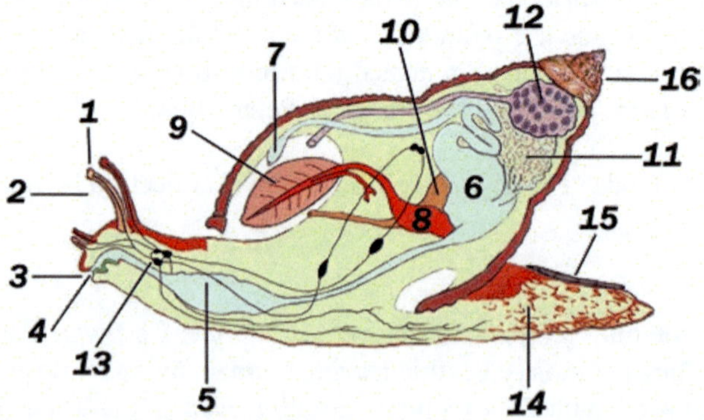

Fig. 14.2 Schematic longitudinal section of a hypothetical gastropod: (1) Eyes. (2) Tentacles. (3) Oral aperture. (4) Radula. (5) Gizzard. (6) Stomach. (7) Anal aperture. (8) Heart. (9) Gill or Lung. (10) Nephridium. (11) Digestive gland. (12) Gonads. (13) Nerve ganglia. (14) Foot. (15) Operculum. (16) Shell. [Drawing Maurizio Forli modified after Vialli (1974)]

Because they are mostly stationary animals closely related to certain environmental conditions and because of the way of life of the present forms, without substantial variations from their fossil representatives, they are of considerable importance in palaeoecological studies. Shells, external impressions, internal models, and opercula are the most common fossil remains, while some indirect traces, such as predation holes produced by mesogastropods and neogastropods, more rarely furrows left by the radula, testify their presence. The gastropods that have stratigraphic importance are relatively few, among these should be mentioned the group of Nerinea for the Upper Jurassic and Lower Cretaceous.

What mainly distinguishes the class of the gastropods from the other mollusks is the phenomenon of the corporal torsion which consists in the rotation of the rear part of the cavity of the mantle and of the anus, laterally and anteriorly, in anticlockwise sense, carrying these last ones over the head. This modification represents the adaptive response to the life in a narrow and elongated shell and grants them an evolutionary advantage, as the repositioning of several organs from a posterior to an anterior arrangement allows the mollusk a greater direct and active control of the environment. For what the soft parts are concerned, we can distinguish those hosted in the shell, visceral mass, and mantle, and those which can come out, head and foot. The head is usually equipped with eyes, at the base or at the top of tentacles, and with other sense organs. The mouth, which hosts the radula, a sort of tongue used for scraping and swallowing the food, formed by an elongated cartilage covered by a membrane on which are inserted series of chitinous denticles, is placed in the lower part of the head or at the extremity of a retractable appendix called proboscis. In the most primitive gastropods, the proboscis consists of a thousand small horny denticles, while in the more evolved forms this number is reduced, so that the most specialized predators have only one row, similar to harpoons and equipped with venom. The prevailing function of the foot is the locomotory one; in the terrestrial gastropods, in order to facilitate the movement, a gland produces a secretion of mucus along which the animal slides, the so-called *snail slime*. Particular modifications of this muscle have allowed some gastropods to adapt to swimming (pteropods) or to acquire the ability to dive. One or two retractile muscles allow the head–foot assembly to retreat inside the shell; an operculum of calcareous or horny composition is able to completely seal the opening. The mantle adheres tightly to the visceral mass near the apex of the shell, while toward the opening it diverges, producing the mantle cavity or pallial cavity. In some gastropods, the mantle is prolonged in an inhaling siphon, sometimes accompanied by a shell siphonal canal, which has the function to irrigate the gills with water. Anteriorly, at the entrance of the cavity of the mantle, there is an organ, called *osphradium*, which, exercising a control on the chemism of the environment and on the sediment, allows the animal to close completely the pallial cavity in dangerous conditions, like, for instance, when the water contains some terrigenous material in suspension. The visceral mass includes the gonads, the digestive system, the heart, and the excretory organs (Fig. 14.2).

The shell or conch of gastropods, almost always elongated trochospiral, is characterized by a pointed end, the apex, and an opening in the opposite position, from which the body portion consisting of the foot–head emerges. The last complete whorl of the shell represents the section inhabited by the mollusk; the part of the shell, above this, is called spire. A central structure called columella, originating in the fusion of the inner part of the spire, is usually distinguished. In some gastropods the last whorls do not meet centrally, so a consequent conical cavity called umbilicus is created, in this case the shell is defined perforated. On the inside of the opening there is often a smooth shell layer (*inductura*) produced by the surface of the mantle which can cover completely or partially the umbilicus, in these cases it is called callus. The opening, whose margin is called peristome, may have a groove, the siphonal canal, in correspondence to the exit of the siphon. Sometimes the shell is characterized by an indentation called sinus slit which consists of an incision at right angle with respect to the axis of the shell, from which the excrements are eliminated. In the various phases of growth, this opening is progressively closed, forming a band with variable morphology, the so-called selenizone (e.g., *Bellerophon*, *Euomphalus,* etc.). Sometimes, in adult forms of some species, the sinus becomes a single hole or a series of orifices called tremata (e.g., *Haliotis*). A wide variety of grooves, spines, tubercles, varices, and ribs, together with the growth lines, constitute the surface ornamentation of the shell. Also inside, the shell can have protuberances or tubercles, often visible on the inner part of the peristome, or folds on the columellar axis.

The shell morphology and the type of coiling can be very variable and usually take their name from the most characteristic genus (e.g., patelliform, turbinate, trochoid, etc.).

The most common mode is the helical coiling around a vertical axis; some gastropods have symmetrical planispiral coiling (i.e., on the same plane). The single whorls are generally in contact with the previous ones, identifying a line of connection called suture. The coiling occurs in a right-handed way, clockwise observing the shell from above, more rarely in a left-handed way. The orientation of the shell must be such as to arrange it with the opening toward the observer and with the apex upwards.

Also the operculum, mainly present in the subclass Prosobranchia, is an element that can fossilize, sometimes with a high taxonomic value, even if it is rarely found preserved associated with its own shell. The shells of the present-day Gastropoda have a horny outer layer, the periostracum, consisting of conchiolin; below is a structure composed of several layers of aragonite, which sometimes form an iridescent nacreous inner layer. The periostracum disappears rapidly after death; therefore, it is rarely conserved in the fossil state; furthermore, following the variation of the range of chemical–physical stability that is established in the fossilization environment, as a result of the diagenetic process, the aragonite generally transforms into calcite. From the point of view of paleontological analysis, pseudoshells and external impressions are of fundamental importance, since the diagnostic characters are generally found on the external surface of the shell; if the gastropod is preserved only as an internal model, a limited amount of information is therefore obtained.

Classification

The classification of current Gastropoda is based mainly on the soft parts, almost never preserved in the fossil state, and on criteria of molecular biology and genetics. It should be remembered, without going into details, that until the second half of the 1990s, their taxonomy was based on the work of Thiele (1929–1935). In 1997 the classification of Ponder and Lindberg was introduced, again following the traditional taxonomy but derived from phylogenetic arguments. Then Bouchet and Rocroi (2010) proposed a new classification that takes its cue from both the traditional classification system and the modern dictates of cladistic taxonomy based on genetic analysis (DNA and RNA). This classification, since 2005, has been further revised by other authors. The classification of fossil gastropods follows a mixed criterion, mostly based on shell morphology, and partly derived from the traditional taxonomy that we adopt in the following of our annotations, as far as possible, dealing with fossils. A valuable character used in taxonomic determinations is the so-called protoconch, that is, the larval shell forming the apical portion of the spire. On the prevailing basis of the respiratory system we distinguish three subclasses: Prosobranchia; Opisthobranchia; and Pulmonata (Figs. 14.3 and 14.4). These subdivisions are outdated but we continue to use them for ease of exposition.

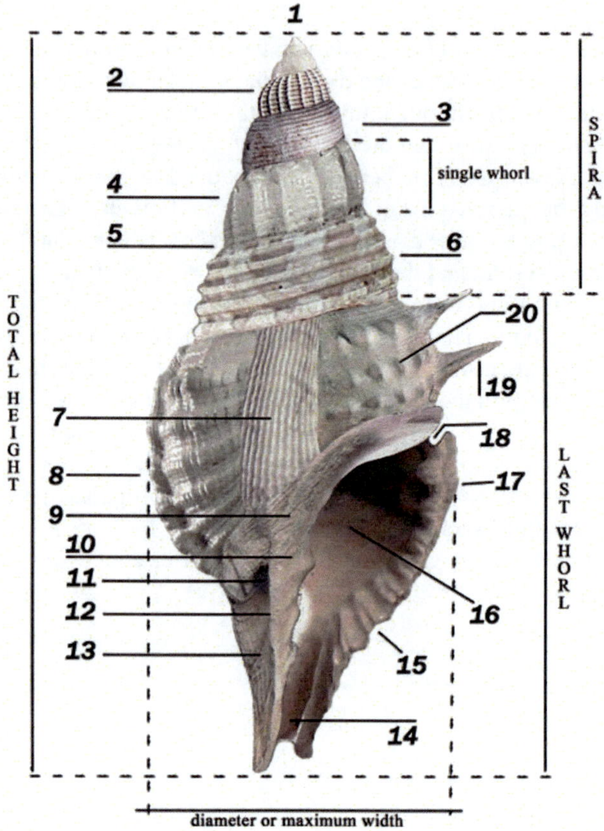

Fig. 14.3 The principal morphological characteristics of a shell, summarized in that of a hypothetical marine gastropod. (1) Apex. (2) Ornamentation with reticulated sculpture. (3) Spiral sculpture constituted by small cords and grooves of equal dimensions. (4) Axial ribs and reticulated spiral sculpture. (5) Suture. (6) Large ribs and spiral grooves (7) Axial striae of growth. (8) Varices. (9) Columellar callus. (10) Columellar teeth. (11) Umbilicus (12) Inner lip. (13) Columella. (14) Anterior siphonal canal. (15) Labial teeth. (16) Aperture. (17) Outer lip. (18) Posterior siphonal sinus. (19) spines. (20) Tubercles. (Drawing Maurizio Forli)

Fig. 14.4 Bottom left, electron microscope image (150×) of the protoconch of a microshell (*Alvania* sp.); right, protoconch and first whorls of the teleoconch of a muricide, *Hexaplex praeduplex* (Landau et al. 2007), showing the first spiral cords that begin to develop after the larval shell (Forli et al. 2013) (courtesy Società Italiana di Malacologia). (1) *Granosolarium millegranum* (Lamarck, 1821), with the large umbilical cavity typical of the genus. (2) *Gibbula verae* Chirli, 2004, showing the apical angle, that is, the angle given by the intersection of two segments tangent to at least two whorls of the spire. (3) Section of *Terebra* sp. with the coiling axis of the columella, equipped with teeth. All the specimens illustrated are of Pliocene age. (Photo and graphic composition Maurizio Forli)

Prosobranchia (Lower Cambrian–Present-day). These are gastropods that have undergone complete torsion and possess the mantle cavity and gill (*ctenidium* or *ctenidia* in the plural) in anterior position. Few possess two gills, in most taxa the right one has disappeared; head with a single pair of tentacles; sexes separate except in a few cases; radula provided with numerous lateral teeth; shell helical or capuliform; operculum usually present. Predominantly marine, although some live in freshwater or terrestrial environments. Most of them are benthic (epifaunal or infaunal); few are pelagic. Half of the known gastropods fall into this Class divided into the following orders:

Archaeogastropoda, Mesogastropoda, and Neogastropoda.

Archaeogastropoda (Lower Cambrian–Present-day). Exclusively marine; lacking inhaling siphon and proboscis; heart provided with two auricles and one ventricle crossed by rectum; they have two kidneys of which the right one is wide while the left one is reduced; shell mostly helical spiral. In the present genera, the spiral is just hinted or completely absent (*Fissurella, Patella*). An example of Archaeogastropoda with planispiral rather than helical spiral shell is *Bellerophon*. They have longitudinal type ornamentation, ribs, and tubercles or radial ribs in conical forms. Particularly widespread during the Paleozoic, we remember some genera besides the above mentioned: *Euomphalus, Worthenia, Haliotis, Pleurotomaria, Gibbula, Turbo, Emarginula, Natiria, Rimula, Diplochilus, Trochus, Naticopsis, Murchisonia, Velates, Laubella, Trachynerita* (Allasinaz 1992).

Mesogastropoda (Ordovician–Present-day). Provided with asymmetrical shell with highly variable morphology, right ctenidium absent, heart characterized by an auricle and not crossed by the rectum, presence of proboscis in many forms, radula with seven rows of teeth (taenioglossa), for example, *Aporrhais, Cypraea, Campanile, Cerithium, Echinophora, Itieria, Naticarius, Nerinea, Rissoa, Strombus, Turritella, Vermetus, Viviparous.*

Neogastropoda (Cretaceous–Present-day). They differ from Mesogastropoda for the radula with three rows of teeth (stenoglossa and rachiglossa) and the inhaling siphon always present, among the most known *Ancilla, Buccinum, Conus, Fusinus, Murex, Mitra, Nassarius, Terebra, Turricula.*

Opisthobranchia (Carboniferous–Present-day). They are marine mollusks, extremely diverse, and adapted to all types of habitats. In some independent evolutionary lines there is a reduction or loss of the shell, and consequently an extensive general reorganization of the body: the pallial cavity moves posteriorly and in some cases disappears; the anus migrates to the right side and reaches a mid-dorsal position; in larger forms, new respiratory structures of various types develop independently. As mentioned above, in addition to the reduction or loss of the shell, other morphological characters of this group are involute spire, convoluted, or covered by a large expansion of the mantle, protoconch generally heterostrophic, frequent columellar folds, absent or spiral ornamentation, operculum frequently absent and gills and auricles posteriorly placed to the ventricle. The group includes the orders Cephalaspidea (Pleurocoela) and Thecosomata (Pteropoda) (e.g., *Actaeonella, Clio).*

Pulmonata or Pulmonates (Carboniferous–Present-day). They represent those gastropods that, thanks to the development of a lung that communicates with the outside through a pore formed by a reduction of the opening of the pallial cavity, have colonized the emerged lands, practically in every kind of environment, from rainforests to deserts. Some groups have then invaded, secondarily, also the freshwaters. Also in the Pulmonata the shell often disappears and when present is without operculum, for example, *Achatina, Arion, Cepaea, Lymnaea*, and *Planorbis.*

Among the many gastropods that have some relevance in folklore events, we have chosen one that, as such, is unknown to most people, despite the fact that its operculum is still appreciated as a good luck charm.

Phylum Mollusca
Class Gastropoda
Superfamily Trochoidea
Family Turbinidae
Genus *Bolma*
Species *Bolma rugosa* (Linnæus, 1767)

In Encelius (*De re metallica*, 1551: p. 229) we find what is generally interpreted as the first fossil gastropod represented in printed books. The description of the author refers, following the Plinian diction, to an unspecified chelonite that in this case takes the form of a shell ... *aquaticam concha tornatilem...* Spiral aquatic shell, this representation is, however, so scarce that it is impossible to attribute it with certainty to a fossil or to a specific entity. The only thing we can deduce from it, if we accept the text, is that it is a stone, similar to a shell with an elongated spiral, found at or near the mouth of the river Elbe. Comparing it with other depictions in later books it could be a terrestrial gastropod similar to a *Cochlodina* or, more likely, a *Turritella*. The same left-handed spiral image, probably due to the fact that the original engraving was made right-handed, that is, without taking into account that it would be reversed in the print, is taken by others, including Gessner and Aldrovandi. One of the first printed images of a gastropod fossil, attributed to Italian authors, is a specimen illustrated in the booklet by Fabio Colonna entitled *Aquatilium et terrestrium aliquot animaliu, aliarumquae naturalium rerum observationes* (1616). The shell is depicted in the upper left corner of the plate on page liii (53), with the denomination of *Buccinum lapideum laeves*. It is probably an internal model of a *Fusinus* whose length should be just under ten centimeters. The author compares it to a similar living one, common in the cliffs around Naples.

[...] *Neapoli frequens in scopulis. Longitudine quatuor pollices aequat.*

We have not found any evidence of the present existence of similar fossils in the Neapolitan area, more likely the origin is to be found in the surroundings of Adria in Apulia, a locality well known to Colonna for his local frequentations. The geological period to which the specimen belongs should be the Pliocene–Pleistocene and the species, if any, could be attributed to *Fusinus longiroster* (Brocchi, 1814) (Figs. 14.5, 14.6, and 14.7).

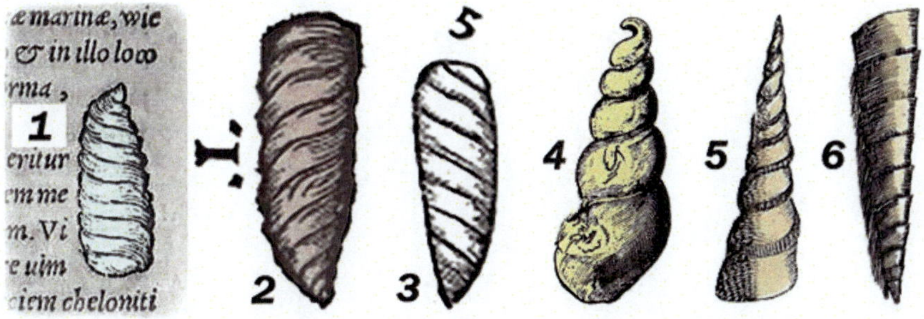

Fig. 14.5 (1) The Chelonite which is generally interpreted as the first fossil gastropod represented in printed books (Encelio, *De re metallica*, 1551: p. 229). (2) The same by Gessner (1565: p. 1050, Fig. I). (3) by Aldrovandi (1648: p. 845, fig. 5). (4) Internal cast of a turreted gastropod, similar in shape to the one illustrated by Encelio (De Laet 1647: p. 191). (5–6) Two other gastropods with elongated spiral shell, taken among others, from Moscardo: "*Turbine, e Bucine di varie spetie, delle quali pongo queste poche in disegno, acciò si veda parte della gran varietà d'impietriti, ch'io conservo*" (1656: pp. 179, 181). [Credit from: (1) https://doi.org/10.3931/e-rara-29357 / Public Domain Mark (2) https://doi.org/10.3931/e-rara-4176 / Public Domain Mark (3) https://doi.org/10.3931/e-rara-13017 / Public Domain Mark (4) https://doi.org/10.3931/e-rara-35487 / Public Domain Mark (5–6) https://www.biodiversitylibrary.org/bibliography/157058 (not in copyright)]

Fig. 14.6 Among the earliest printed woodcuts dedicated to mollusks in general, and to gastropods in particular, are those present in the various versions of the *Hortus Sanitatis* dedicated to the *Blatta Bisanzia* (Blacte byzantine). In the background a detail from a manuscript copy (1489), on the left from the 1497 edition and on the right from the 1491 edition (all modified). [Credit from: (1489) Strasbourg. B.N.U.S.Ms.2.152288 www.calames.abes.fr/pub/#details?id=D47A14797 (Wonnecke von Kaub 1491) https://sammlungen.ub.uni-frankfurt.de/inc/content/titleinfo/7714347 urn:nbn:de:hebis:30:2-257502 (Public Domain Mark 1.0) (1497) https://www.biodiversitylibrary.org/bibliography/61747 (not in copyright)]

Fig. 14.7 Fabio Colonna, *Aquatilium et terrestrium aliquot animaliu, aliarumquae naturalium rerum observationes* (1616). This is probably the first printed representation of a fossil gastropod by Italian authors, is placed in the upper left corner of the plate on page iii, named *Buccinum lapideum laeves*. In our opinion it is a representative of the family Fasciolaridae, belonging to the genus *Fusinus*. The other shells, not fossil, are as follows: top right *Turritella* sp.; middle, the large buccinid *Charonia variegata* (Lamarck, 1816); left, the terebra *Oxymeris maculata* (L., 1758); middle, *Cerithium nodulosum* Bruguière, 1792; bottom left *Cymatium* sp.; bottom right a Fasciolaridae, perhaps *Fasciolaria filamentosa* (Röding, 1798) (https://www.biodiversitylibrary.org/bibliography/723 **Public domain**)

In the *De omni rerum fossilium genere* (Gessner 1565: p. 164) another naturalia is described and depicted, of those that somehow become distinctive icons perpetuated in several subsequent texts. In this case it is a *Bolma rugosa*, a Mediterranean representative of the family Turbinidae, seen from above, with the operculum in recto-verso position. It is not a fossil specimen nor a shell in its natural appearance as it has been polished by hand. In his other works, (*Historiae animalium*, 1558 and *Nomenclator Aquatilium*, 1560) this shell is repeated several times by Gessner in the version borrowed from Rondelet (*Universae aquatilium Historiae pars altera*, 1555: pp. 98–99). An earlier representation is due to Bellon, who shows it with the operculum in place (*De Aquatilibus*, 1553: p. 430). Gessner's polished specimen is instead taken up by Boetius in his *Gemmarum et lapidum historiae* (1609) and in the slightly later *Le parfaict Joaillier ou histoire des pierres* (1644: p. 451). From our point of view, the interesting thing does not concern the shell but its calcareous operculum, the navel of Venus:

> *Le nombril marin ressemble un nombril d'homme, & en a la grosseur, d'où quelques-uns l'ont appellé nombril de Venus* [...] *Ceux de Crete n'appellent pas cette pierre nombril, mais oeil marin, & croyent qu'estant porté il profite aux yeux.*

The opercula of *B. rugosa* have played various parts in the theatre of nature: sea beans, sea bellies or navels, Venus' navel, sea eye, St. Margaret's stone, and it is under a denomination that seems to overlap with these last definitions that even today they are known as *St. Lucy's eyes*. In addition to the protection for the eyes, Boetius also reports other uses and beliefs related to them: the girls who adorned themselves with the navel of Venus were more graceful; therefore, more sought after by men, they were not subject to disease or they were cured. Worn as an amulet preserved from erysipelas, an inflammatory skin disease that in the past was the cause of many victims, also if placed wet with saliva on a wound, blocked the escape of blood. Its medicinal properties were also manifested as a curative in women with problems of abundant menstrual flow, which was administered after having been finely pulverized with appropriate water or preserves. Powdered with red coral and deer horn, taken in an egg, it was an adjuvant against vomiting and intermittent menstruation (de Boodt 1644: pp. 452–453).

Lucia, the patron saint of sight and eyes in Christian iconography is depicted holding a saucer in her hand on which are placed the two eyeballs symbolizing martyrdom. The opercula, identified as *beneficent eyes*, have therefore become talismans capable of blocking any other "malevolent eyes," the *evil eye* (Fig. 14.8).

Fig. 14.8 The sacred image of Saint Lucia in a holy card of 1898. (Photo Maurizio Forli, courtesy private collection)

The beliefs, related to the ability to negatively influence and make a curse transmitted through a simple glance, are very old and even date back to mythology, such as the legend of Medusa who petrified anyone who crossed her gaze.

This devilish power possessed by the look was attributed especially to women suspected of witchcraft. Of course, the anti-evil eye rituals are not limited to the possession of the amulet but are part of more complex actions that usually merge religion with popular magic in which, among the many talismans considered effective, the eyes of Saint Lucia were counted among the most powerful and their protective function, still somehow considered valid and more or less manifest, makes the tradition of adorning oneself with opercula mounted in gold or silver still continues to perpetuate. In Bellucci (1907: p. 37, fig. 17) two amulets made of the opercula of the *Bolma rugosa are* represented photographically, one coming from a necropolis of the early Iron Age, from the surroundings of Ascoli, and another, of age contemporary to the printing of the volume, coming from Papiano in the province of Perugia.

In Aldrovandi (*Musaeum metallicum*, 1648: p. 800) seven different types of marine umbilicus are shown in an image illustrating both sides, under the denomination of *Fabae marinae*, sea beans. Also in the *Metallotheca Vaticana*, in valuable and realistic images, the opercula of the *Bolma rugosa* are depicted under the usual denomination of *marine navel* or as *ocularis lapis*, the stones of the eyes, to which Mercati recommends giving them the name of *Sanctae Luciae lapides*. In chapter XXV: p. 138 some fifteen present ones are shown, in chapter LXXVII: p. 343, four fossils, recto and verso, from the Pliocene soils in the neighborhood of Siena, near the river Arbia, in the locality Coroncina, now in the immediate southern suburbs of Siena:

> *...ad Arbiae flumen prope Coroncinam locum ita dictum in agro Senenſi...*

Of fossil opercoletti from Capo Milazzo, in the province of Messina, writes instead Agostino Scilla citing them as *stones of Santa Margherita*. In addition to giving a detailed description, in support and confirmation of his idea that the marine remains on land were once living in the sea, he states:

> *Ella non si curi delle tre conche* [...] *ma consideri sopra tutto, e raffiguri nella stessa Tavola un'opercoletto di lumaca marina* A. *detto pietra di S. Margherita* [...] *che per me sarà impossibile, che la natura generante scherzi di sasso, scherzi con tanta puntualità in tutte le cose, e con insoffribile bizzaria formi infiniti opercoletti, per applicarli alle lumache impietrate, che ne sono prive* (Scilla 1670: pp. 132–133, pl. XVII).

Numerous other fossil gastropods, many of them of Pliocene age, are represented in the illustrated volumes of our major authors of the sixteenth and seventeenth centuries. From our point of view of experts in Pliocene fossil malacology, and of muricids in particular, a plate in the work of Mercati seemed to be extremely interesting. In Accordi (1980: 11), there is already a proposal of classification with which, however, we are not totally in agreement. It is about the figures enclosed under the denomination of *Porphiroides, et Muricites* at page 299. There we can easily recognize, at the top *Strombus coronatus* Defrance, 1827, at its left *Bolinus brandaris* (Linnæus, 1758), at the top right *Hexaplex trunculus* (Linnæus, 1758), and at the bottom left, what we believe to be the first image of *Chicoreus aitus* (de Gregorio, 1885) current name, but not definitively accepted, of the one universally known until a few years ago as *Murex brevicanthos* Sismonda, 1847, one of the rarest and most beautiful neogenic *muricidae*. The locality of origin of these specimens is *in Hetruria trans Arnum fluvium ad Castrum Cerretum*, the present municipality of Cerreto Guidi not far from Empoli and San Miniato, where even today along the river Arno, clays and Pliocene sands are exposed in which these types of fossils are still present. It is practically a few kilometers from the birthplace of Leonardo da Vinci, who knew these places well and described the niches in the cut of Colle Gonzoli (Collegonzi), more or less in correspondence with the present disused clay quarry at Sovigliana, on the immediate periphery of Empoli (Figs. 14.9, 14.10, 14.11, and 14.12).

Fig. 14.9 Living and fossil (F) gastropods, with today's specific names, distributed according to the old subdivision of orders. ARCHAEOGASTROPODA: (1) Bellon, *Altera Patella maior*, 1553: p. 395. (2) Rondelet, *De Aure marina*, 1555: p. 3 > *Haliotis tuberculata* L., 1758. (3) Lister, 1678: p. 214 > *Pleurotomaria* sp. (F). (4) Sowerby, 1821: p. 113 > *Emarginula* sp. (F). (5) Colonna, 1616: p. 16 > *Fissurella* sp. (6) Sowerby, 1821: p. 145 > *Euomphalus* sp. (F). (7) Colonna, 1616: p. 16 > *Trochus* sp. (8) Scilla, 1670: pl. XIII > *Gibbula* sp. (F). MESOGASTROPODA: (9) Rondelet, 1555: 64 > *Bolinus brandaris* (L., 1758) with operculum. (10) Scilla, 1670: pl. XII > *Thylacodes arenarius* (L., 1758) sometimes believed to be petrified snakes, were recognized as marine organisms by Scilla (F). (11) Aldrovandi, *Strombites*, 1648: pl. 845 > *Neverita olla* de Serres, 1829 (F). (12) Gessner, 1565: p. 166 > *Cypraea* sp. (F). (13) Mercati, 1717: p. 301 > *Turritella* sp. (F). (14)

(continued)

(Fig continued) Aldrovandi, ***Murex albus é monte effossus***, 1648: **p. 472 >** ***Thetystrombus coronatus*** **(Defrance, (Fig continued) 1827) (F). (15) Scilla,** ***Pendedactylus***, 1670: **pl. XVI >** ***Aporrhais pespelecani*** **(L., 1758) (F).** **Neogastropoda: (16) Rondelet,** ***De Turbine angulato***, 1555: **pl. 89 >** ***Tarantinaea lignaria*** **(L., 1758). (17) Bellon,** ***Cochlea marina***, 1553: **p. 429 >** ***Buccinum undatum*** **L., 1758. (18) Aldrovandi,** ***Turbinites***, 1648: **p. 472 >** ***Conus*** **sp. (F). (19) Scilla,** ***Turbine***, 1670: **pl. XVI >** ***Clavatula*** **cfr** ***C. romana*** **(Defrance, 1826) (F). (20) Bellon,** ***tertia Nautili species***, 1553: **pl. 383 >** ***Tonna galea*** **(L., 1758). (21) Rondelet,** ***Buccino***, 1555: **p. 81** ***Charonia lampas*** **(L., 1758). (22) Bellon,** ***Coclea marina altera***, 1553: **p. 429 >** ***Conus*** **sp. (23) Mercati,** ***Strombite***, 1717: **p. 303 >** ***Conus mercati*** **Brocchi, 1814 (F). (Color and graphic composition Maurizio Forli. Credit from: (1, 17, 22)** https://www.biodiversitylibrary.org/bibliography/5765 **(Public domain). (2, 9, 16, 21)** https://archive.org/details/ARes071032/page/n13/mode/2up **(Public Domain 1.0). (3)** https://www.biodiversitylibrary.org/bibliography/65746 **(Public domain). (4, 6)** https://www.biodiversitylibrary.org/bibliography/14408 **(not in copyright). (5, 7)** https://www.biodiversitylibrary.org/bibliography/723 **(Public domain) (8, 10, 15, 19) personal copy Maurizio Forli (11, 14, 18)** https://doi.org/10.3931/e-rara-13017 **/ Public Domain Mark (12)** https://doi.org/10.3931/e-rara-4176 **/ Public Domain Mark (13)** https://archive.org/details/bub_gb_DC_E6UVWg7EC **(Public Domain Mark 1.0) (23) personal copy Maurizio Forli)**

(Fig continued) mode/2up **(Public Domain 1.0) (3)** https://doi.org/10.3931/e-rara-16853 **(Public Domain Mark) (4)** https://www.biodiversitylibrary.org/bibliography/157058 **(not in copyright) (5)** https://doi.org/10.3931/e-rara-4176 **/ Public Domain Mark (6)** https://babel.hathitrust.org/cgi/pt?id=gri.ark:/13960/t8w96041q&view=1up&seq=194&skin=2021 **(Public domain) (7)** https://doi.org/10.3931/e-rara-12490 **/ Public Domain Mark (8)** https://www.biodiversitylibrary.org/bibliography/119999 **(Public Domain) (9–10)** https://babel.hathitrust.org/cgi/pt?id=ucm.5323847476&view=1up&seq=217 **(Public domain)]**

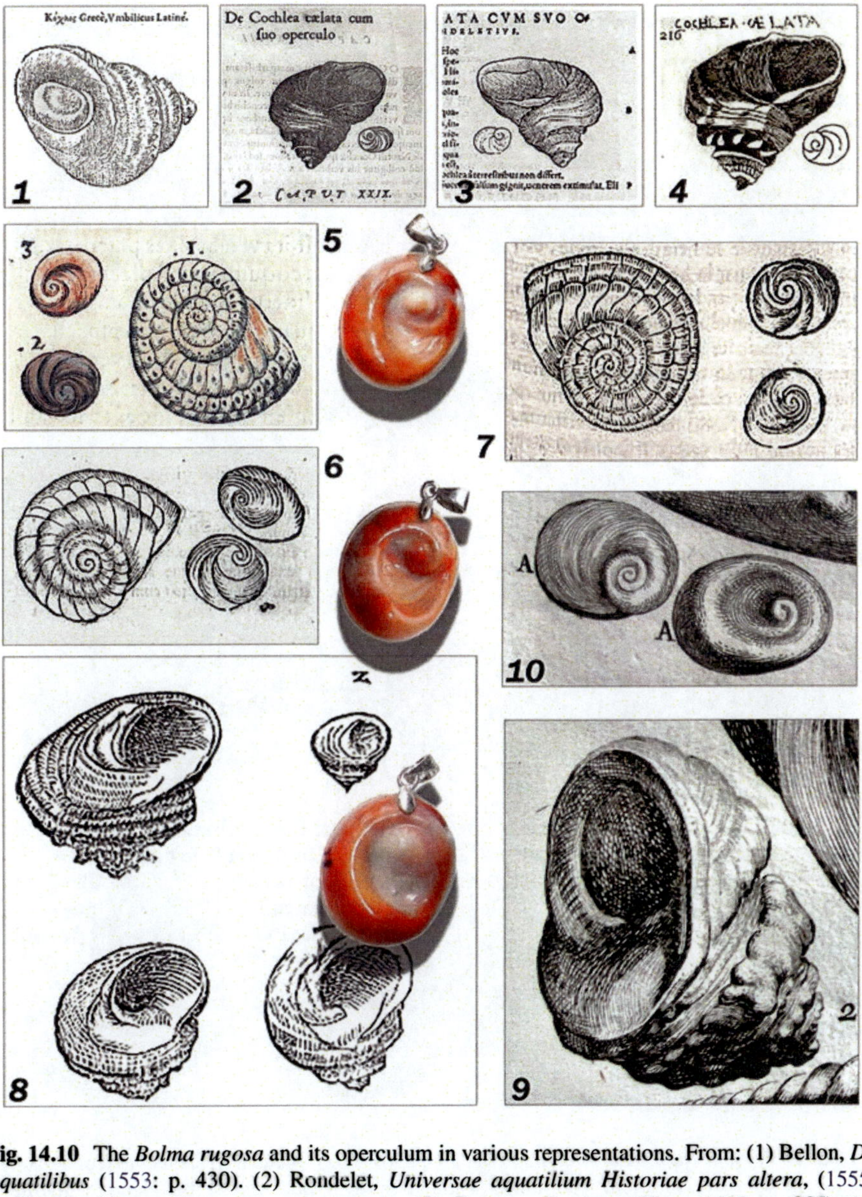

Fig. 14.10 The *Bolma rugosa* and its operculum in various representations. From: (1) Bellon, *De Aquatilibus* (1553: p. 430). (2) Rondelet, *Universae aquatilium Historiae pars altera*, (1555: p. 98), *Cochlea caelata* cum *suo operculo*. (3) Gessner, *De Aquatilibus* (1558: p. 285). (4) Moscardo, *Memorie del museo di Lodovico Moscardo* (1656: p. 216). (5) Gessner, *Omni rerum fossilium genere* (1556: p. 164). (6) Boëtius, *Gemmarum et lapidum historia* (1609: p. 176). (7) Boëtius, *Le parfaict Joaillier ou histoire des pierreries* (1644: p. 451). (8) Aldrovandi, *De Reliquis Animalibus exanguibus* (1606: p. 393). (9–10) Scilla, specimen and opercula definitely fossils, *De corporibus marinis lapidescentibus quæ defossa reperiuntur* (1759: pp. 132–133, pls XVI, XVII). [Photo and graphic composition Maurizio Forli. Credit from: (1) https://www.biodiversitylibrary.org/bibliography/5765 (Public domain) (2) https://archive.org/details/ARes071032/page/n13/

(continued)

Fig. 14.11 In the *Metallotheca Vaticana* (1717: pp. 301–302), Mercati gives us the iconography of Pliocene opercula, different from those of the classical *Bolma rugosa*, belonging to *Astraea fimbriata* (Borson, 1821), a rare and elegant turbinid characteristic of the deep clays of the lower-middle Pliocene, whose opercula have a characteristic roughness, well recognizable in the print. Here are some specimens of different species compared: (1) *B. rugosa* (L., 1767), surroundings of San Gimignano (SI, Tuscany), Pliocene. (2) *B. rugosa*, present-day, Castiglioncello (LI, Tuscany). (3) *B. tuberculata* (de Serres, 1829), with operculum, Monte Antico (GR, Tuscany), Pliocene. (4) Above *Astraea fimbriata* with its operculum, front and back, and in the center four others, from the same locality mentioned by the author from San Gimignano, Coroncina (SI, Tuscany) natural size. [Photo and graphic composition Maurizio Forli. Plate from Mercati: https://archive.org/details/bub_gb_DC_E6UVWg7EC (Public Domain Mark 1.0)]

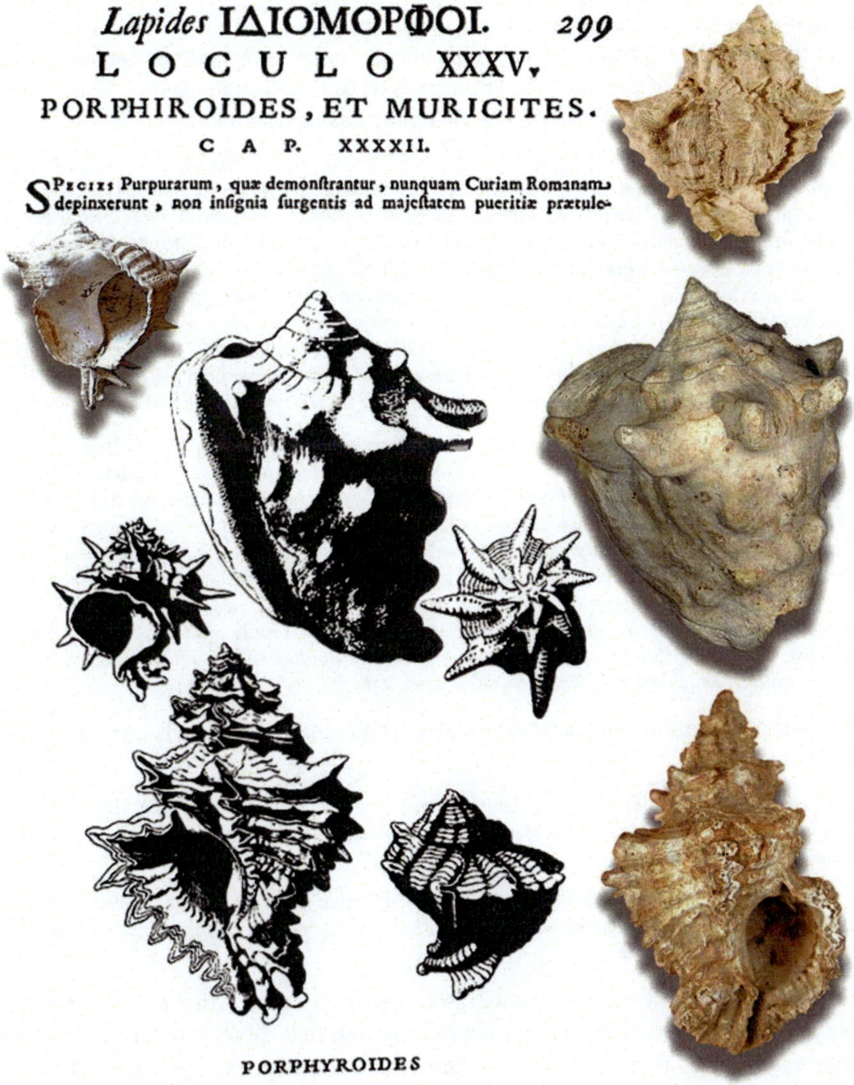

Fig. 14.12 The Pliocene gastropods illustrated by Mercati and their counterparts, in the center at the top *Tethystrombus coronatus* (Defrance, 1827), on its left *Bolinus brandaris* (Linnæus, 1758), on the right at the top and *Hexaplex trunculus* (Linnæus, 1758) on the left at the bottom, the first image of *Chicoreus aitus* (de Gregorio, 1885) current name, but not definitively accepted, of the one universally known until a few years ago as *Murex brevicanthos* Sismonda, 1847, one of the rarest and most beautiful neogenic murice. The locality of origin is in *Hetruria trans Arnum fluvium ad Castrum Cerretum*, the present commune of Cerreto Guidi, not far from Empoli and San Miniato, where still today, along the Arno river, there are clays and sands which preserve malacofauna with such types of fossils. [Photo and graphic composition Maurizio Forli. Plate from Mercati: https://archive.org/details/bub_gb_DC_E6UVWg7EC (Public Domain Mark 1.0)]

In the ancient pharmacopoeia the use of gastropods was quite extensive, but almost exclusively concerned the use of living specimens. Mollusks and their shells were administered for the most varied needs as, for example, can be seen from the numerous methods of preparation and use set forth in Passera's *Nuovo Tesoro degli Arcani Farmacologici* (1688: pp. 838–841), for example, with snails.

> *Le Lumache cavate fuora dal gusso, & cotte nell'orzata, levano i dolori del costato* [...] *Lumache levate fuora del guscio, & nette da ogni viscosità & cotte perfettamente nel brodo di Pollo senza Sale, & tritturatovi dentro un poco di Farfara si diano à mangiare, che è ottimo alli Tisici. Lumache cotte nella Decottion d'Orzo, e cosa utile al Petto, & al Polmone. Lumache di Bosco purgate dalla viscosità cotte nel latte Vaccino, sanano l'Asma.* [...] *Lumache crude, piste con la loro scorza, levito, e radice di Canna, e poste sopra le cose insite, le tira fuori.* [...] *Lumache trite con rannoforte, e poste sopra le scrofole, le sanano. Trite, & applicate, provocano il Mestruo.* [...] *Olio, ò liquor di Lumache* [...] *Fa crescere i capelli caduti, e li fa divenir biondi.* [...] *Lumache, & Buttiro fresco pisti insieme, & applicata alle Morene infiammate, & addolorate, che sono fuori, le sana sperimentato.*

> *Snails taken out of their shells and cooked in barleywater, relieve the pains of the ribs* [...] *Snails taken out of their shells and free of all viscosity, perfectly cooked in chicken broth without salt, with a little bit of Farfara chopped in it, are to be eaten, which is excellent for Tisicians. Snails cooked in Barley Decoction, which is useful for the Chest and Lung. Wood snails purged of their viscosity, cooked in cow's milk, heal Asthma.* [...] *Raw snails, pounded with their rind, yeast, and cane root, and placed on top of the inherent things, pull them out.* [...] Snails *crushed with strong lye, and placed on the scrofula, heal them. Chopped, & applied, they cause menstruation.* [*Oil or liquor of snails* [...] *makes fallen hair grow and become blond.* [...] Snails, *& fresh Butter together, & applied to the inflamed, & pained hemorrhoids, which are outside, healthy experienced.*

and with the actual snails, particularly with the Red Snails [*Arion rufus* (Linnæus, 1758)]

> *Questi lumaconi rossi senza guscio, stesi per schena, cioè per mezo, e di quel liquore che n'esce, si componi con sale, e si sfregano, e stropicciano li Porri, e le Veruche, che sene anderanno* via.

> *These red snails without their shells, placed belly up", that is, by the middle, and the liquor that comes out of them is mixed with salt, and they are rubbed and crumpled and the Leeks and the Warts, will go away.*

The revival of snail slime as the main ingredient in regenerating and anti-wrinkle beauty creams is a recent phenomenon; having your face directly sprinkled with the slimy secretions of the cute mollusks, free to crawl on the skin, is one of the most desired (and expensive) and, so they say, effective treatments.

Moreover, among the fossil things were included what were believed to be stones with multiple virtues, actually internal shells of some snails, such as the already mentioned genus *Arion*, considered by Gessner as Chelidonie, that is, stones innate in swallows, with all the doubts of the case (1565: p. 107). Three pairs of *lapis ex nudo Limace prone & supine pictae* are correctly described by Aldrovandi (1648: p. 799) and others by Mercati in the paragraph *De Lapide limacis* (1717: p. 186).

Chelidonites, on the whole, were mysterious, in the sense that we do not really understand what the ancient writers were referring to, understood as small stones more or less spherical, shiny, and variously colored, for example, Aldrovandi himself proposes them of two different qualities under the names of *Lapis Chelidonius fossilis* and *Lapis Chelidonius animalis*, probably a pair of hemispherical teeth of sparids and two gastrolites found in the crop of a swallow.

Plinio (II.6.37) *In ventre Hirundinum pullis lapílli, candida, aut rubenti colore, qui Chelidonii appellantur, reperiuntur.*

In the belly of young swallows, pebbles have been found, either white or reddish in color, which are called Chelidonie.

Chelidonio è pietra trovata nel ventre delle Rondini. Laquale è di due sorti. Percioche quello, ch'è vermiglio, portato in panno di lino mondo giova a lunatici e pazzi, e guarisce le lunghe infirmità. Quei, che lo portano fa eloquenti e grati. Trito nell'acqua, a guisa de' coliri a cio deputati, sana i mali de gli occhi. Il nero conduce le facende al disiderato fine. Accheta l'ire. Fa altresì chi lo porta piacevole e gratioso. Mitiga l'ira de' Signori. Legato al collo in pāno di lino giallo, scaccia le febbri: e frena e spenge ogni nocevole humore. S'è veduto, ch'appeso al collo, sana gli Epilentici, e gl'Ilerici. Alcuni dicono, che si dee involgere in pelle di vitello, o di cervo non nato, e legare al sinistro braccio. Debbono cotali pietre esser cavate, quando le Rōdini sono picciole nel nido, e del mese d'Agosto, accio che elle piu perfette siano: e i rōdinelli non tocchino terra, ne vi sia la madre, mentre che queste pietre si cavano. (Dolce 1565, II: p. 36).

Chelidonium is a stone found in the belly of the Swallows. Which is of two kinds. Therefore that which is vermilion, worn in linen cloth, is good for lunatics and madmen, and cures long infirmities. Those who wear it are eloquent and grateful. Mashed in water, like the eye drops appointed for it, it heals the ills of the eyes. Blackness leads to the desired end. It reduces the wrath. It also makes him who wears it pleasant and gracious. It mitigates the anger of the Lords. Tied around the neck in a yellow linen garment, it drives away fevers: and it curbs and extinguishes every harmful humour. It has been seen that hanging it around the neck heals the Epileptics and the Ilerics. Some say that it should be wrapped in the skin of a calf or an unborn deer, and tied to the left arm. These stones must be extracted when the swallows are small in the nest, and in the month of August, so that they are more perfect: and the rhodlings do not touch the ground, nor is the mother there, while these stones are extracted.

From the classic definition of swallow stone, the use of the term has been extended to include micro pebbles of variegated agate, gastropod opercula, or the inner shells of large snails. Despite the fact that the following brief notes do not concern fossil specimens, it seemed interesting to us, almost as a completion of the treatment of gastropods, to include some curious and unusual things, at least according to our judgment.

De Lepore marino is the subject of chapter XI in *Libri de piscibus marinis, in quibus verae piscium effigies expressae sunt* (Rondelet 1554: p. 520) where an image of *Aplysia depilans* Gmelin, 1791, a nudibranch gastropod, placed immediately below the title, introduces us to its description with annotations and quotations from Dioscorides and Pliny.

There is no lack of atrocious poisons either, one of which is the sea hare: in the Indies Sea it is so pernicious, that by simple contact it causes vomiting and dissolution of the stomach (a). In our sea this hare is a shapeless mass which has nothing in common with the animal of that name except the colour [...] Pliny, LXXII, 48. *It is called a sea hare, because when it is huddled up, holding its two hind parts out, it looks like a hatching hare, with its ears upright. It is so smelly, that the ancients, because of their disgust, did not have the courage to make attempts to collect the coloured juice which spreads from the edge of the lid; on the contrary, they attributed to it some poisonous properties, which it does not have at all* (Berini. *Essay of the translation of the Natural History of Cajo Pliny the Second*, 1824: p. 83).

Aplysias can reach a maximum size of thirty centimeters, despite their appearance they can swim by doing so with a characteristic oscillatory wave-like movement of their expanded foot. They are herbivorous animals, typical of algal environments at small depths, widespread throughout the Mediterranean. The common name of sea hares is due to the presence of two elongated sensory appendages located at the back of the head, which resemble the ears of hares. Their color varies from brown to black, more or less marbled. As a defense, they emit a viscous, purplish liquid that inhibits the sensory capabilities of natural attackers. Their contact was believed to cause the loss of any body hair.

For an in-depth study of the period, see Aldrovandi (*De reliquis animalibus exanguibus libri quator*, 1606: pp. 78–90), who discusses the subject at length and also describes species other than those of Rondelet (Fig. 14.13).

Fig. 14.13 De Lepore marino, from Rondelet, *Libri de piscibus marinis, in quibus verae piscium effigies expressae sunt* (1554: p. 520) https://www.biodiversitylibrary.org/bibliography/64229 (not in copyright)

Odorous nails and *Byzantine Blatta Byzantia* ... explaining how nails and "cockroaches" can be associated with gastropods seems a difficult task but, with the help of ancient writers, it will not be so difficult, despite a few tricks.

> *Also called* Onyx indicus *in the medical matter of the good ancients is a term used by the Greek writers to express what is usually called* odorous nail, unguis odoratus. *By some it is called* onycus indicus [...] *the same writer speaks of the* Blata byzantia, *and tells us that it is not the same thing as the Indian onyx, but the Italians distinguished with such a name the nose bone of the purple fish* [...] *translates the expression* Blatta byzantia *of the Writers of the most ancient times by means of the phrase* os nasi purpurae*; and by the Interpreters of the Arabic Writers the same name is assigned to what they called* unguis odoratus, *or* onyx indicus, *and this just because the Arabic voice is so important.* [...] *The expression* Blatta Byzantia *means nothing more than the* Byzantine Purple*; and when the purple colour was removed, the tongues, or other parts of the niches of these fishes, were used in medicine* [...] *Dioscorides says, as it is a part of a niche fish, or a testaceous fish resembling in many respects the Purple fish. Some people suppose that Dioscorides expressed the whole niche, or shell of that fish by the word* Pama [...] *The voice* Pama [...] *means* operculum, *lid; and as all testaceous fishes of the spice of whelks, to which spice, or class this purple fish belongs, have their lids* [...] *to close the orifices, so it is full, and clear enough, that the ancients for the voice* Pama *meant this same lid* [...] *addimandarolo* Onyx, *or* Unguis, *Onyx or nail, from its being thin, and flat, and not disconcertingly resembling* [...] *in the figure to a human nail.* (Scott *George Lewis, George Lewis's Supplement to Efraimo Chambers's Universal Dictionary of Arts and Sciences...* 1764: p. 80).

Some aromatic nails can be observed in Rondelet, associated with their shells (*Universae aquatilium Historiae pars altera*, 1555: pp. 83, 86), also proposed by Aldrovandi in a single graphic composition (1606: 346). Summarizing and updating the definitions, the famous odorous or aromatic nails and the Blatta Byzantia of the ancient Greeks, are nothing but opercula of shells of the muricids *Hexaplex trunculus* and *Bolinus brandaris*, widely used in the production of purple, and of some strombids, mostly representatives of the genus *Lambis*, very common in the Indian Ocean. That more than two thousand years ago these objects could appear mysterious and with desirable pharmacological powers we can understand, a little less, or perhaps only with a little more wonder, we can conceive how still today it is possible to buy through web services, some kilograms of *odorous nails* or Blatta Byzantia, sometimes at a high price. But why odorous? The shape of almost all horny opercula reminds us of human nails, but the adjective odorous needs to be clarified. It does not refer to the natural "scent" of this part of the mollusk because, even if at first glance we would be led to consider them as nauseating objects, in reality, once dried up, they are almost without any smell. The fragrant definition is justified by the fact that they were and are burned as a total or partial substitute for incense (Fig. 14.14).

Fig. 14.14 *Unguis odoratus*, from Rondelet (1555: p. 86). Above an operculum, recto and verso, of *Lambis* sp. and, below, one of *Hexaplex trunculus* (L., 1758), in external view. [Photo and graphic composition Maurizio Forli. Credit from: https://archive.org/details/ARes071032/page/n13/mode/2up (Public Domain 1.0)]

Even if it has nothing to do with fossils, in a self-respecting book, a gastronomic recipe cannot be missing, and therefore, we point out one taken from Imperato that has as main ingredient a coelenterate, that is a colony of Hydrozoa of intense blue color, with size oscillating on average around five centimeters and discoidal body surmounted by a vertical appendage, called velella or St. Peter's boat that moves passively transported on the sea surface (*Velella velella* Linnæus, 1758). These organisms undergo massive beaching, coloring the beaches blue, sometimes for hundreds of meters. It is on these occasions that mixed with their bodies can be found the shells of an otherwise rare pelagic gastropod, the *Janthina*, which adhering to the lower part of the disc of the velella, as well as being transported, establishes with them a relationship of commensalism parasitism.

> *The sea-sail is among the number of marine animals and plants, of cartilaginous consistency, two ounces in length, covered with a tender membrane, of a lively cerulean colour, while it is alive, it has the name of sail, because, being triangular, wherever it is seen it shows some of its faces similar to an outstretched sail. It is rarely seen, and when it is seen, it is seen in great numbers. There are some who use it as food, fried, and seasoned with oil and salt. It has not been shown, as far as we know, by other writers* (Imperato 1599: pp. 773, 783).

It is Colonna who gives a more complete description of the velella and the gastropods associated with it, the *Janthina pallida*. In the area of the Strait of Messina, the phenomenon of the massive beaching of these organisms, which, when they putrefy, release a characteristic odor and leave behind whitening congealed remains, sometimes associated with pebbles of pumice, is said to have given rise to a phenomenon cited in the Odyssey [...] *it is a great pile of putrid men's bones, with shriveling skin* (for more information on this topic, see Vazzana 2016: pp. 54–56) (Figs. 14.15, 14.16, 14.17, 14.18, 14.19, 14.20, and 14.21).

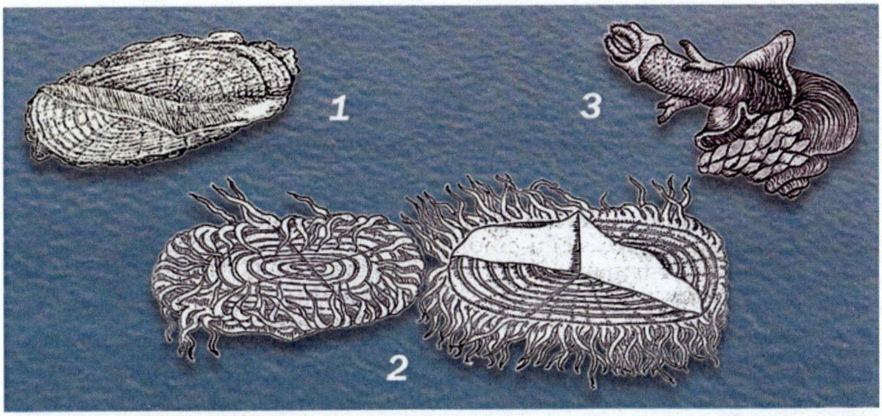

Fig. 14.15 (1) *Velella* seen from above (Imperato 1599). (2–3) Specimen of *V. velella* with soft parts highlighted; individual of *Janthina pallida*, associated with *Velella*, with the ovigerous capsules and the body of the mollusk emerging from the shell (Colonna 1616: xxii). [Graphic composition Maurizio Forli. Credit from (1) https://www.biodiversitylibrary.org/item/183177#page/5/mode/1up (not in copyright); (2–3) https://www.biodiversitylibrary.org/bibliography/723 (Public domain)]

Fig. 14.16 Massive beaching from the Beach of the Abyss, between Punta Pezzo and Cannitello (RC) (detail). (Photo credit Angelo Vazzana)

Fig. 14.17 Interactions between velella and the gastropod *Janthina pallida*. The latter lives in flotation, without ever touching the seabed, hanging from a mass of bubbles it emits or adhering, especially in the growth phases, to the lower surface of the velella. (Photo credit Angelo Vazzana)

Fig. 14.18 Detail of a *Janthina pallida* adhering to the lower part of *Velella velella*. (Photo credit Angelo Vazzana)

Fig. 14.19 Two opercula from Bellucci (1907): on the left from a Neolithic burial and on the right an amulet from the early twentieth century. Superimposed on a silver ring from the second half of the same century. (Photo and graphic composition Maurizio Forli, courtesy private collection)

Fig. 14.20 Recto-verso images of three amulets from the Bellucci collection: on the sides, an amulet composed of a fossil shark tooth and an operculum of Turbinidae (probably *Turbo petholatus* Linnaeus, 1758) held together by a satin ribbon passing through the hangers, which combines the power attributed to the *Tongue of St. Paul* and the *Eye of St. Lucy*. The bezel of the tooth is rather unusual, for the subject, as it is a heart-shaped sheet of silver open at the back, with the tooth embedded by means of a worked serration, and closed at the front, decorated with floral motifs, except for a circular hole with a serrated edge, from which one can see the central part of the fossil. The operculum is instead enclosed in an elliptical silver ring. Measures 35 × 56 × 6 mm and 21 × 32 × 9 mm, respectively. Provenance Penna San Giovanni, Macerata, late nineteenth century [Coll. G. Bellucci. M.A.N.U. (PG) Inv. 3447 B]. In the center there are two *Bolma rugosa* opercula, the classic *eye of Saint Lucia* used against eye diseases and by extension also against the evil eye, set in two elliptical rings with a hanger [Coll. G. Bellucci. M.A.N.U. (PG) Inv. 6889, 6934]. (Reproduced by kind permission of the Ministry for Cultural Heritage and Activities and for Tourism, Regional Directorate Museums of Umbria—National Archeological Museum of Umbria)

Fig. 14.21 Tuscany, handcrafted apotropaic necklace, in wrought silver, set with three opercula of *Turbo petholatus* Linnaeus, 1758, from the Philippines. Second half of the '900. (Photo Maurizio Forli, courtesy private collection)

References

Accordi B (1980) Michele Mercati (1541–1593) e la Metallotheca. Geol Romana 19:1–50

Aldrovandi U (1606) De Reliquis Animalibus exanguibus libri quatuor, post mortem eius editi: nemp' De Mollibus, Crustaceis, Testaceis, et Zoophytis. Apud Io: Baptistam Bellagambam, Bononie. https://www.biodiversitylibrary.org/bibliography/119999 (Public Domain)

Aldrovandi U (1648) Ulyssi Aldrovandi patricii Bononiensis Musaeum metallicum in libros IIII: distributum Bartholomaeus Ambrosinus ... labore, et studio composuit cum indice copisosissimo Arcus Antonius Bernia. Bononiae: typis Io. Baptistae Ferronii, 1648. ETH-Bibliothek Zürich, Rar 1027. https://doi.org/10.3931/e-rara-13017 / Public Domain Mark

Allasinaz A (1992) General palaeontology and systematics of invertebrates. Edizioni Culturali Internazionali Genova

Bellon P (1553) De Aquatilibus, Libri duo. C. Stephanum, Parisiis. https://www.biodiversitylibrary.org/bibliography/5765 (Public domain)

Bellucci G (1907) Il feticismo primitivo in Italia e le sue forme di adattamento. Unione Tipografica Coop. Editrice, Perugia

Berini G (1824) Saggio della traduzione della Storia Naturale di Cajo Plinio Secondo. F.lli Mattiuzzi, Udine

Bouchet P, Rocroi JP (2010) Nomenclator of bivalve families. Malacologia 52(2):1–184

Colonna F (1616) Fabii Columnae Lyncei Minus cognitarum rariorumque nostro coelo orientium stirpium ekphrasis ... https://www.biodiversitylibrary.org/bibliography/723 Public domain

Boodt A Boëtius de (1644) Le parfaict Ioaillier ou histoire des pierreries: ou sont amplement descrites leur naissance, iuste prix, moyen de les cognoistre, et se garder des contrefaictes, Facultez medicinales, et proprietez curieuses. Iean-Antoine Huguetan, Lyon. ETH-Bibliothek Zürich, Rar 1010. https://doi.org/10.3931/e-rara-12490 / Public Domain Mark

Boodt A Boëtius de (1609) Gemmarum et lapidum Historia. Claudiam Marnium, Hanoviae https://hdl.handle.net/2027/gri.ark:/13960/t8w96041q (Public domain)

De Laet J (1647) De Gemmis et Lapidibus, libri Duo. Quibus praemittur Theophrasti, liber de Lapidìbus Graece & latine cum Brevibus Annotationibus. Ioannis Maire, Lugduni Batavorum. https://doi.org/10.3931/e-rara-35487 / Public Domain Mark

Dolce L (1565) Libri tre ne i quali si tratta delle diverse sorti delle Gemme, che produce la Natura, della qualità, grandezza, bellezza, et virtù loro. Gio Battista, Marchio Sessa, et Fratelli, Venetia

Encelio C (1551) De re metallica: hoc est, de origine varietate, & natura corporum metallicorum, lapidum, gemmarum, atque aliarum, quae ex fodinis eruuntur, rerum, ad medicinae deservientium, [...]. Franc. [Frankfurt]: apud Chr. Egenolphum, [15--]. ETH-Bibliothek Zürich, Rar 2259. https://doi.org/10.3931/e-rara-29357 / Public Domain Mark

Forli M, Brunetti MM, Reitano A (2013) Due specie del Pliocene italiano a confronto:Hexaplex dumosus (de Stefani, 1875) ed Hexaplex praeduplex Landau, Houart & da Silva, 2007 (Gastropoda: Muricidae). Bollettino Malacologico, 49(1): 49-57.

Gessner C (1558) Historiae Animalium Liber IIII. qui est de Piscium & quatilium animantium natura. Christoph. Froschoverum, Tiguri. https://doi.org/10.3931/e-rara-16853 / Public Domain Mark

Gessner C (1560) Nomenclator aquatilium animantium: icones animalium aquatilium in mari et dulcibus aquis degentium, plus quam DCC. cum nomenclaturis singulorum Latinis, Grecis, Italicis, Hispanicis, Gallicis, [...]. Tiguri: excudebat Christoph. Froschoverus ..., 1560. ETH-Bibliothek Zürich, Rar 9734: 1. https://doi.org/10.3931/e-rara-48090 / Public Domain Mark

Gessner C (1565) De omni rerum fossilium genere, gemmis, lapidibus, metallis, et huiusmodi, libri aliquot, plerique nunc primum editi. Tiguri.. https://doi.org/10.3931/e-rara-4176 / Public Domain Mark

Guaccio FM (1626) Compendium Maleficarum. Ex Collegii Ambrosianii Typographia. Mediolani. https://archive.org/details/image4995MiscellaneaOpal/page/n293/mode/2up (under CC/publicdomain/zero/1.0)

Hortus Sanitatis (1489) Manuscrit. Strasbourg. B.N.U.S.Ms.2.152288 feuillets, en partie à 2 colonnes. www.calames.abes.fr/pub/#details?id=D47A14797

Imperato F (1599) Dell'Historia Naturale di Ferrante Imperato napolitano. Libri XXVIII. Nella quale ordinatamente di tratta della diversa condition di miniere, e pietre. Con alcune historie di Piante, et Animali fin'ora non date in luce. Costantino Vitale, Napoli. https://www.biodiversitylibrary.org/item/183177#page/5/mode/1up (not in copyright)

Lewis SG (1764) George Lewi's supplement to Efraimo Chambers's Universal Dictionary of Arts and Sciences, containing the figures, spices, properties, productions, preparations, and uses of things both natural and artificial. Exact translation from English. Tome One. Giambattista Pasquali, Venice

Lister M (1678) Angliae Historiæ animalium Angliæ, tres tractatus. Unus de araneis. Alter de cochleis tum terrestribus, tum fluviatilibus. Tertius de cochleis marinis. Quibus adjectus est quartus de lapidibus ejusdem insulæ ad cochlearum quandam imaginem figuratis. Joh. Martyn, Londini. https://www.biodiversitylibrary.org/bibliography/65746 (Public domain)

Mercati M (1717) Metallotheca Vaticana. Opus posthumum, Autoritate et Munificentia Clementis Undecimi Pontificis Maximi. E tenebris in lucem eductum: Opera autem, & studio Joannis Mariae Lancisii, Archiatri Pontificii illustratum. Ex Officina Jo: Mariae Salvioni Romani, Romae. https://archive.org/details/bub_gb_DC_E6UVWg7EC (Public Domain Mark 1.0)

Meydenbach J (1491) Hortus sanitatis. Mainz. Digitalisiert durch die Universitätsbibliothek J.C. Senckenberg Frankfurt am Main [2011]: urn:nbn:de:hebis:30:2-13126. https://sammlungen.ub.uni-frankfurt.de/inc/content/titleinfo/7714347 (Public Domain Mark 1.0)

Moscardo L (1656) Note overo memorie del museo di Lodovico Moscardo nobile veronese. Academico Filarmonico, dal medesimo descritte, Et in Tre Libri distinte. Per Paolo Frambotto, Padoa. https://www.biodiversitylibrary.org/bibliography/157058 (not in copyright)

Passera F (1688) Il Nuovo Tesoro degli Arcani Farmacologici, Galenici. & Chimici, ò Spargirici. Giovanni Parè, Venetia

Prüss J (1497) Ortus sanitatis. Goff, Strasbourg. [Johann Prüss, not after 21 Oct. 1497]. https://www.biodiversitylibrary.org/bibliography/61747 (not in copyright). https://doi.org/10.5962/bhl.title.61747

Rondelet G (1554) Libri de Piscibus Marinis, in quibus verae Piscium effigies expressae sunt. Matthiam Bonhomme, Lugduni. https://www.biodiversitylibrary.org/bibliography/64229 (not in copyright)

Rondelet G (1555) Universae aquatilium Historiae pars altera, cum veris ipsorum Imaginibus. Matthiam Bonhomme, Lugduni. https://archive.org/details/ARes071032/page/n13/mode/2up (Public Domain 1.0)

Scilla A (1670) La vana speculazione disincantata dal senso. Lettera responsiva Circa i corpi marini, che Petrificati si trovano in varij luoghi terrestri. Andrea Colicchia, Napoli

Scilla A (1759) De corporibus marinis lapidescentibus quæ defossa reperiuntur. Addita Dissertatione Fabii Columnæ de Glossopetris. Joannis Zempel, Romae

Sowerby G (1821) The Mineral Conchology of Great Britain; or colored figures and descriptions of those remains of testaceous animals or shells, which have been preserved at various times and depths in the Earth. Vol. III, W. Harding, London. https://www.biodiversitylibrary.org/bibliography/14408 (not in copyright)

Thiele J (1929–1935) Handbuch der Systematischen Weichtierkunde. Voll. 1-4. Gustav Fischer, Jena

Vazzana A (2016) Fenomeni Naturali e Miti nell'area dello Stretto. Gangemi Editore, Roma

Vialli V (1974) Appunti di Paleontologia. Pitagora Editrice, Bologna

von Wonnecke KJ (1491) Hortus Sanitatis, de Herbis et Plantis, de Animalibus, de Avibus, de Piscibus, de Lapidibus & de Urinis, cum figuris. Jacobus Meydembach, Moguntiae

von Wonnecke KJ (1499) Hortus Sanitatis. Liber de Herbis, Animalibus, Avibus, Piscibus, Lapidibus & c. cum eorum Fig. https://www.biodiversitylibrary.org/bibliography/61747 (not in copyright)

Chapter 15
Scaphopoda: from Cannelli Simpatici, Shrieking Bones, to Tusk Shells

Abstract Scaphopods are exclusively marine conch mollusks that live embedded in incoherent sediments, protruding from the bottom. The shell is tubular, cylindrical-conical, more or less curved, reminiscent of elephantine tusks. They vary in size from a few millimeters in length and diameter, up to over 100, with a larger diameter around 1 cm. They are known in the fossil state since the Paleozoic Era (Silurian). Among the first images referred to the scaphopods we believe are those found in the oldest illustrated editions of Dioscorides, translated by Jean Ruel (1543). The French author shows two incomplete ones, paired together, under the name of Buccini, reporting more than anything else their medicinal virtues. Their most frequent use as an ingredient in recipes for medical use was linked to the production of an ointment, recommended to soften or remove freckles, sunburns, boils, pustules, and furuncles from the face. What we consider the first certain representation of fossil specimens coming from the Italian Plio-Pleistocene soils is found in Scilla (*La vana speculazione disingannata dal senso. Lettera responsiva Circa i corpi marini, che Petrificati si trovano in varij luoghi terrestri*. Andrea Colicchia, Napoli, 1670). The peculiar structure of scaphopoda attracted attention not only as an ornamental object but also as a magical object endowed with beneficial and healthy powers and virtues. In particular, they were believed to be particularly effective against inflammation of the oral cavity and respiratory tract. Numerous examples of amulets made with these shells can be found in the Bellucci Collection, in Perugia.

Keywords Mollusca · Scaphopoda · Tusk Shells · Citrine ointment · Amulets · Bellucci collection

Scaphopods (from Greek ΣΚΑΦΟΣ boat and ΠΟΔΙ foot) are exclusively marine conch mollusks that live sunken in the incoherent sediments, sandy or muddy, protruding from the bottom with their upper part; the shell is tubular, cylindrical, more or less curved, reminiscent of elephantine tusks. The ornamentation is inconspicuous, being, when present, more or less pronounced longitudinal striations and reliefs, up to the formation of ribs. The background coloration may be uniformly pink, pale

© The Author(s), under exclusive license to Springer Nature Switzerland AG 2022

M. Forli, A. Guerrini, *The History of Fossils Over Centuries*,
https://doi.org/10.1007/978-3-031-04687-2_15

green, whitish, at times there are ringed bands shaded of green gray or brick red, more or less intense. They vary in size from a few millimeters in length and diameter to over a hundred, with the largest diameter around 1 cm (Fig. 15.1).

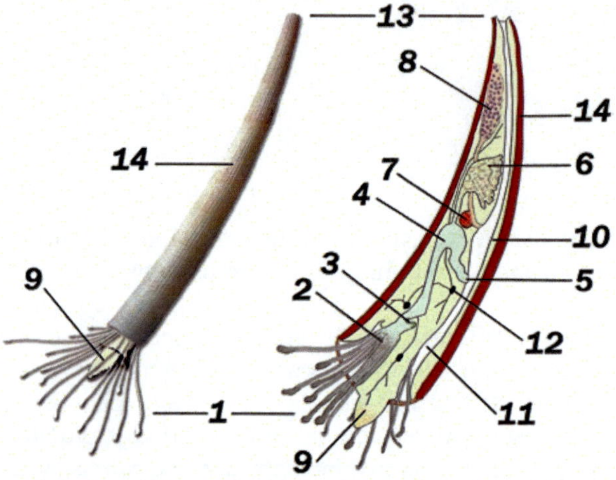

Fig. 15.1 Simplified morphological scheme of a hypothetical scaphopod: (1) Captacles. (2) Oral aperture. (3) Radula. (4) Stomach. (5) Anal aperture. (6) Digestive gland. (7) Nephridia. (8) Gonads. (9) Foot. (10) Mantle. (11) Mantle cavity. (12) Nerve ganglia. (13) Posterior aperture. (14) Shell. [Drawing Maurizio Forli modified after Vialli (1974)]

The body, developed anteroposteriorly, distinguishes them from all other mollusks, even if it has bilateral symmetry and anatomical similarities with that of gastropods; they lack operculum and gills and the mantle, which wraps around the whole body, is used for gas exchange.

The characteristic trilobed foot with the median projection vaguely reminiscent of the prow of a ship, the cause of the name given to them by ancient authors, serves to move and sink. The head and the foot overlook the lower anterior opening where, in the same position, there are thin adhesive prehensile tentacles, called *captacles*, exclusive and typical of scaphopods, which are used to introduce food, mainly foraminifera, toward the oral opening.

They are known in the fossil state since the Paleozoic era, with the genus *Laevidentalium* Cossmann 1888, starting from the Silurian.

From the systematic point of view, they are currently divided in two Orders Dentaliida da Costa, 1776 and Gadilida Starobogatov, 1974. In the Dentaliida are included the commonly called dentalium, while in the Gadilida are grouped those of smaller size, with the shell sometimes smooth and swollen centrally, and with some anatomical differences on which we do not dwell, being insignificant in the optics of our analysis (Fig. 15.2). What we want to highlight is that the first representations in the printed texts are to be attributed to representatives of tusk shells.

Below is the nomenclature of a common Plio-Pleistocene fossil species, often used as an amulet.

Phylum Mollusca
Class Scaphopoda
Order Dentaliida
Family Dentaliidae
Genus *Fissidentalium*
Species *Fissidentalium rectum* Gmelin, 1791 †

Their peculiar tubular structure, which lends itself particularly well to being pierced for threading, has certainly facilitated their use as ornaments; known and used, albeit sporadically, since the Upper Paleolithic, they became more frequently used in the Mesolithic–Neolithic period, as pendants or necklace beads.

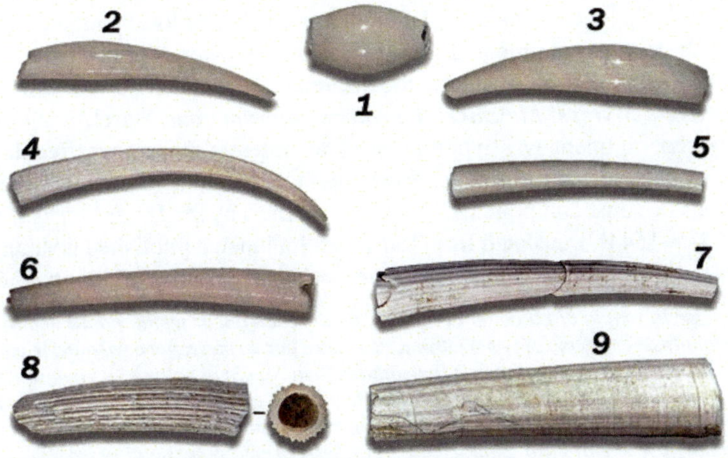

Fig. 15.2 Some of the most common Pliocene Scaphopoda from the surroundings of Orciano Pisano (PI). Order Gadilida: (1) *Cadulus tumidosus parvulinus* Sacco, 1897. (2) *Gadila ventricosa* (Bronn 1827). (3) *G. gadula Sacco*, 1897. Order Dentaliida: (4) *Entalina tetragona* (Brocchi 1814). (5) *Pseudoantalis rubescens* (Deshayes, 1825). (6) *Omniglypta jani* (Hoernes 1856). (7) *Antalis inaequicostata* (Dautzenberg 1891). (8) *Antalis* sp. (9) *Fissidentalium rectum* (Gmelin 1791). (Graphic composition Maurizio Forli, photo credit Massimo Cresti)

With the advent of the availability of metals there is a strong reduction of this use, even if this tradition continues, albeit marginally, until today. Almost always these are fossil specimens, sometimes collected directly on the site of settlements or more often, coming into the possession of users from distant places. Their use is widely documented in the European prehistory and protohistory and, as for the other finds, their presence in sites different from the place of supply, allows us to investigate and know the socio-economic relations, connected to the distribution and circulation of goods, among the primitive populations.

Their ancient medicinal use is extremely limited; they are remembered in particular for entering the composition of an *ointment*, the so-called *Citrine Ointment*. Of the late medieval evidence of this use is credited, although the fact is not universally acknowledged, Nicolaus or Nicolaos Myrepsus, a pharmacologist and physician of the thirteenth century, who bringing together the news of the ancient Greek and Oriental pharmacopoeia, and other sources contemporary with him, composed an antidotary in Greek in which he reported the recipe for the composition of this particular product, in the ingredients of which are mentioned the Antali *αυταλι* and the Dentali *τευταλι*. The interpretation and attributions to the real dentalium species, as we consider them nowadays, in the ancient texts is controversial and confused, the descriptions do not correspond between those of an author and another, and often they are referred to different shells, in particular to gastropods of the family Muricidae, those which were then called buccini and porpore.

Leonhart Fuchs, a German botanist and physician, who lived in the first half of the sixteenth century, commented and transcribed the work from Greek to Latin, adding his own annotations, realizing the discrepancies handed down by the ancient sources (*Nicolai Myrepsi Alexandrini medicamentorum opus*, Basel, 1549: p. 167). This particular ointment was recommended to mitigate or remove freckles, sunburns, boils, pustules, and furuncles from the face. Below is the recipe, and some clarifications on some components, handed down to us by the *Dispensario of* Valerio Cordus (1515–1544), translated from Latin into Italian and published posthumously (*Il Dispensario cioe il modo di comporre le medicine...* Venice, 1558: pp. 256-258):

> *Far liquefar la songia di porco, et di gallina, quali sendo ancor tepidi, ponili dentro una bona quantita de corteccie di pomo citrone. et macera per un mese, poi cava fuori le corteccie et R. lb. i.s. della songia di porco liquefatta et 3. di songia di gallina. et giongili il sevo di capra, poi ponili il restante, sottilissimamente pisto. et ultimamente li giongierai la canfora sottilissimamente trita. Nota in questo onguento che non si concordano i testi nei pesi.* [...] Borace] *è spetie di nitro, qual usano gli orifici.* Amento dolce] *è di gesso, il quale si divide in vene capillari.* Ombilico marino] *sono pietricelle trovate su la riva del mare, hanno la forma de fave grosse.* Entalo] *è un animale crustoso marino, longo et concavo come una fistola, di fuori stretto, et è longo come il detto picolo.* Dentalio] *è una conchetta marina picola, et ha la sfessura con denti.* Nitro] *è il sale di pietra.* Gersa di serpentaria] *è la radice di serpentaria vera preparata, come diremo di seguito. Scaccia le macchie della faccia, e la scabia, ovunque si ritrovi. Sottiglia le cicatrici, cava* via *il rosso de gli occhi, et polisce le aspredini, et cose asperse.*
>
> *Liquefy pork and chicken fat, which, being still warm, put them in a good quantity of citron apple bark and macerate for a month, then remove the bark and R. lb. i.s. of the liquefied pork meat and R. lb. i.s. of the goat meat, then put the rest, very thinly, into a pot. 3. of liquified pork fat and 3. of chicken fat. and add the goat's tallow, then put them the rest, pounded more than thin. and finally you will add them the camphor minced more than thin. Note in this ointment that the texts do not agree in the weights.* [Borax] *is a nitrous spice, which is used by the goldsmiths.* Sweet ointment] *is made of chalk, which is divided into capillary veins.* Ombilico marino] *are small stones found on the seashore, in the form of large beans.* Entalus] is a *crustose marine animal, long and concave like a fistula, narrow on the outside, and is as long as the aforesaid little one.* Dental] *is a small sea conch, and it has a fissure with teeth.* Nitro] *is stone salt.* Gersa di serpentaria] *is the root of true serpentaria prepared,*

as we will say below. It drives away the stains of the face, and scabia, wherever it is found. It thins the scars, takes away the red of the eyes, and polishes the roughness.

Following the Cordian version, the description of the *entalium* is more or less consistent with what we now call scaphopods while that of the *dentalium*, at least in this representation, seems to correspond more to a small cypraea shell or a generic gastropod with the lip internally denticulate:

[Dentalio] *è una conchetta marina picola, et ha la sfessura con denti.*
[Dental] *is a small seashell, and has a slit with teeth.*

The first image referring to scaphopods we believe is the one found in the oldest illustrated edition of Dioscorides, translated by Joannis Ruellio (1543). The French author shows two incomplete, paired, under the name of *Buccini*, broad grouping with which in antiquity were indicated many shellfish, reporting more than anything else the medicinal virtues in the treatment of eyes and dog bites, after calcination, purification, and administration with honey (*Pedanii Dioscoridis Anazarbei, de medicinali materia, libri sex;* 1543, II: pp. 95–96). About 10 years later Guillaume Rondelet gives a different iconography (1555: p. 110); under the heading *de vermibus in tubuli delitescentibus*, he describes generic red worms, living in tubules of various shapes, both on rocks and on old sea shells. Since, as is clear from the illustration accompanying the text, these are polychaete worms, only the "small horns" in the upper part of the woodcut can be referred to real scaphopods. As was usually done in the texts of the time, also in this version of Rondelet are given indications on the pharmaceutical use of these tubules and their therapeutic virtues. Since these shells are shaped like curved canine teeth, their use was justified in the production of citrine ointment pro dentali, that is, for the treatment of dental infections.

... in compositione unguenti citrini pro dentali. (1555: p. 111).

In Gessner (*Nomenclator acquatilum animantium*, 1560: p. 256) instead, the figure clearly shows a dentalium, even if, as for the previous ones mentioned, it is not a fossil specimen.

Also in Aldrovandi (*De Reliquis Animalibus exanguibus*, 1606: pp. 282–283) are mentioned and figured, although not in an excellent way, some real Scaphopoda. In the text the news and opinions of the preceding authors are reported with the various names given to the shells: Antali, Dentali, and Enthali. Like Rondelet and Gessner, the Bolognese scientist also mentions their use in the preparation of citrine ointment and reproposes the figure of the *horns* associated with sea worms (1606: p. 562).

After more than a century of printed publications, in the texts of pharmacopoeia is still unclear what is meant by the term *Dental*: in another antidotary subsequent to that of Cordus (1558), in the same entry is given the description of what is, evidently, a muricide that today we call *Bolinus brandaris*:

Dentale, i.e. *Buccino is an animal similar to a Snail, with many spurs around, with a long beak like a channel, with some spurs.* – (Ceccarelli Ippolito, *Antidotario Romano Latino, e volgare*, Rome, 1639: p. 325).

As for the other word, *antale,* we continue to give a description all in all consistent with the morphology of scaphopods even if sometimes continue to be assimilated to porpores, understood here as shells:

> *Entali,* i.e. *Porpore, is a long animal with a mouth from head to foot of it, of the form of a smooth mouth, penticulated with different colors, called by the people a sea slug* [ibidem].

In the Moscardo (*Notes overo memorie del museo di Lodovico Moscardo*, 1656: p. 211) the Scaphopoda are described, as we have seen in the previous texts, under the names of Antali and Dentali, always reiterating the same notions:

> *The Antals are placed in the number of Testacei:* [...] *they are born in the depths of the sea in some caves: they do not pass the length of a finger: they are concave, upright, bent, like white horns, striped, and of a rather hard material.* [...] *The Dentals have almost the same shape as the Antals: they are a little more curved, they are striped, but they are vowels, and these too are born, like the Antals in the depths of the sea in some stone caves: if some people want these to be the teeth of the Toothfish.*

The association with the canine teeth of the *Dentex* (*Dentex dentex* L., 1758), the Toothfish, is also suggested by other authors, for example, in Bellon (*De Aquatilibus*, 1553: pp. 180-181). Finally, in Scilla we find what we consider the first certain representation of fossil specimens coming from the Italian Plio-Pleistocene soils. (1670: pp. 136-138, pl. XVIII, figs VI–VIII).

> *Tra le cose* [...] *cavate in una valle, detta dello Sperone, vicino la Terra Varapodi di Calabria, dieci miglia lungi dal mare, ho riconosciuto, oltre d'infiniti altri curiosissimi gusci, tutte le spezie de' dentali, o antali consevatissimi.* [...] *non occorre, ch'io li descriva; perciocchè l'Aldrovandi nel Terzo de Testacei, li mostra puntualmente espressi da varii Autori* [...] *Io credo di non essermi ingannato, equivocando nel nome; perciocchè questi sono istessissimi a' descritti, e portati dall'Aldrovandi.*
>
> *Among the things* [...] *excavated in a valley, called Sperone, near the Land of Varapodi of Calabria, ten miles from the sea, I have recognized, in addition to infinite other curious shells, all the sort of the most common dentals, or antals.* [...] *it is not necessary for me to describe them; because Aldrovandi in the Third of Testacei, shows them punctually expressed by various Authors* [...] *I believe that I was not deceived, misunderstanding in the name; because these are the same as those described, and brought by Aldrovandi.*

Mercati compares them to the canine teeth of beasts or to small horns, *cornets*. In the figure representing them in the *Metallotheca Vaticana* (1717: pp. 301–302), several Pliocene fossil specimens from the surroundings of San Quirico d'Orcia, in the Siena area, are reproduced.

A different conception of their mode of formation is present in Lang (1708: p. 30) who assimilates them to hollow tubular stalactites, thus defining the two partial remains figured in plate 40, under the names of *Stalactites tubulatus niveus & striatus* and *Stalactites tubulatus cinereus & striatus*.

Still beyond the half of the eighteenth century, in the scientific publications, these organisms were associated to vermetids under the name of *Tubulites*. For example, Carlo Allioni describes them as fossils from the Pliocene soils of Piedmont (*Oryctographiae pedemontanae*, 1757: pp. 47–50), referring to the figures of Plate X of the *Index Testarum Conchyliorum* of Niccolò Gualtieri:

> *Tubulus marinus regulariter intortus, arcuatim incurvatus, & versus unam extremitatem acuminatus, Dentali dictus, laevis, candidus*. Gualt. Ind. T. 10 C

Soon afterwards Linnaeus gave us his version in the *Systema Naturae* (1758: p. 785, no. 297), *Dentalium*—Vermes, Testacea. Four species are *elephantinum*, *dentalis*, *entalis*, and *minutus* (Figs. 15.3 and 15.4).

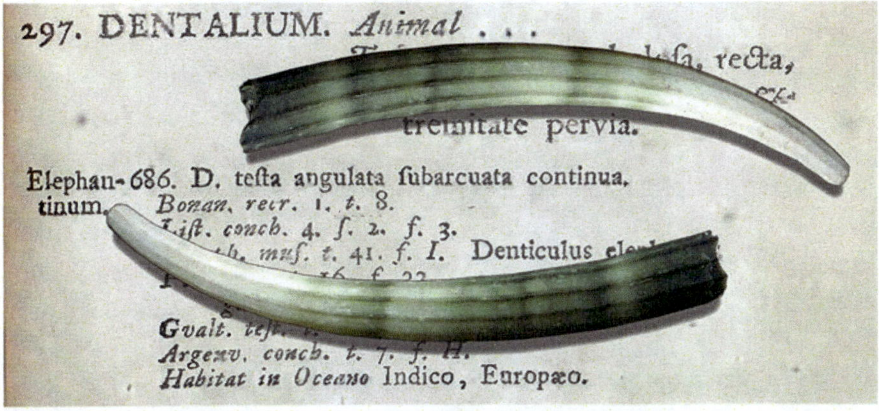

Fig. 15.3 A specimen of *Dentalium elephantinum* Linnaeus, 1758, within the background the original description of the great Swedish author, from page 785, to number 297, of the tenth edition of his *Systema Naturae*. Length ~ 90 mm, Indo-Pacific Ocean, Philippines. [Photo and graphic composition Maurizio Forli. Credit from: https://www.biodiversitylibrary.org/bibliography/542 (Public domain)]

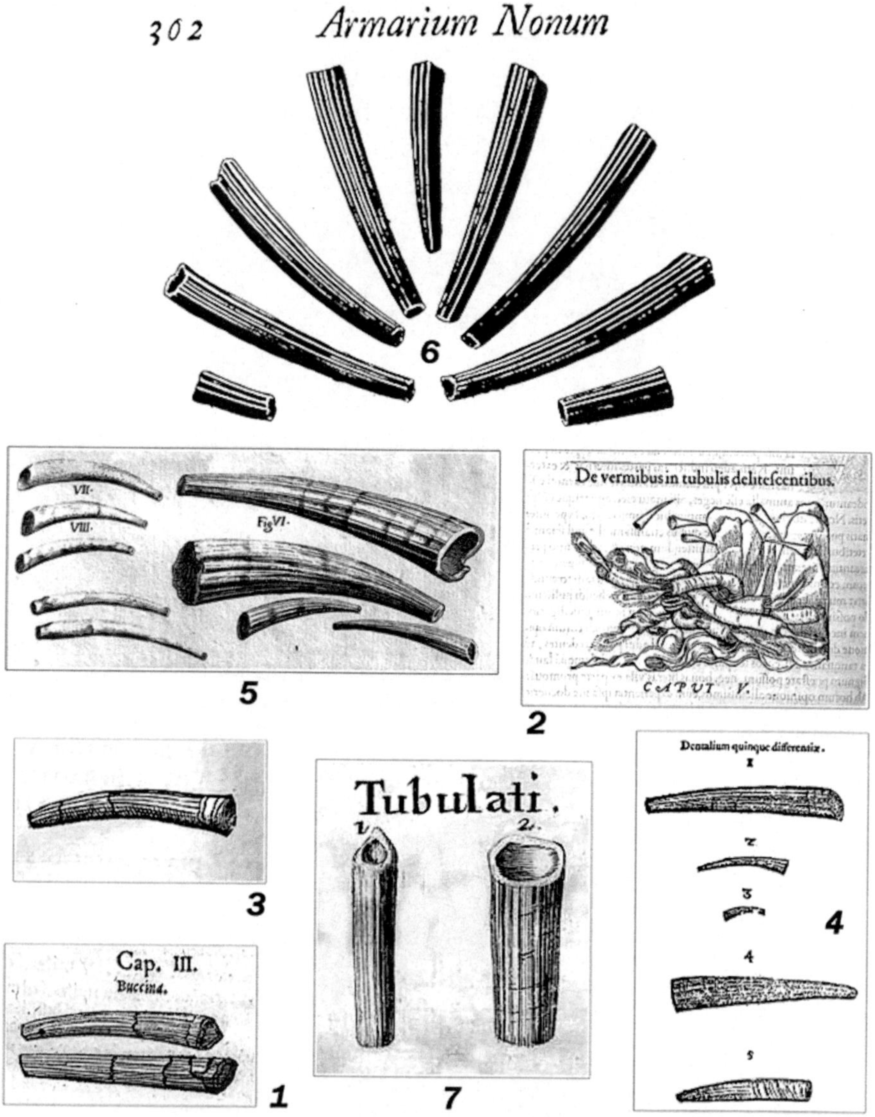

Fig. 15.4 From the earliest prints in which they appear, we have taken these illustrations which we consider to be among the oldest and most significant. Current specimens: (1) Buccini, from Ruellio (1552: 145). (2) *Cornetti*, at the top in the representation of sea worms (Rondelet 1555: p. 110). (3) from Gessner (1560: p. 256). (4) from Aldrovandi (1606: p. 283). Fossil specimens: (5) from Scilla (1670: pl. XVIII, figs VI–VIII). (6) from Mercati (1717: p. 302). (7) Tubulates, from Lang (1708: pl. 4, figs 1–2) (all modified). [Graphic composition Maurizio Forli. Credit from: (1) https://www.biodiversitylibrary.org/bibliography/9317. Public domain (2) https://archive.org/details/ARes071032/page/n13/mode/2up (Public Domain 1.0). (3) https://doi.org/10.3931/e-rara-48090 / Public Domain Mark. (4) https://www.biodiversitylibrary.org/bibliography/119999 (Public Domain). (5) https://babel.hathitrust.org/cgi/pt?id=ucm.5323847476&view=1up&seq=221&skin=2021 (Public Domain). (6) https://archive.org/details/bub_gb_DC_E6UVWg7EC (Public Domain Mark 1.0). (7) https://doi.org/10.3931/e-rara-10398 / Public Domain Mark)]

The peculiar structure of the scaphopods has attracted attention not only as an object of ornamentation but also as a *magical* object and, as such, endowed with beneficial and healthy powers and virtues. In particular, they were believed to be particularly effective against inflammations of the oral and respiratory cavity that manifested themselves with difficulty in swallowing and breathing. The popular imagination has coined for them different names that today are certainly curious, among these we think rather unique the epithet of "bones big scream" (osso strillone) *...designated with the above name, perhaps because of the cries of pain, which are usually emitted by patients with arthritis and tooth, being valid amulets to prevent and heal such diseases* (Bellucci, 1898: p. 60 and 1907: p. 46, fig. 26), and *Cacazzi di Luna*, lunar defecations, certainly because of their shape and color, a term attested in the central and southern part of Sicily. In the Agrigento area they are also known as *cannaruzzeddu di San Brasi,* that is *cannarozzu*, throat, *trachea of San Biagio*, a saint venerated as protector of the throat (Pitrè 1940) (Figs. 15.5 and 15.6).

Fig. 15.5 Moon dejections from various versions of the *Hortus Sanitatis* (1491). Such phenomena were believed to give rise to selenite, or Moonstone, from the Greek name of the lunar goddess Selene, a milky crystalline chalk. As witnessed by Pitrè (1940), the dentals were also believed to have been similarly produced, but estimated to be "less noble parts" (all modified). (Credit from: https://sammlungen.ub.uni-frankfurt.de/inc/content/titleinfo/7714347 urn:nbn:de:hebis:30:2-257502 Public Domain Mark 1.0 https://wellcomecollection.org/works/zuph7pum/items?canvas=748 https://www.biodiversitylibrary.org/bibliography/61747 not in copyright)

Fig. 15.6 Lithographic prints representing classic devotional images of Saint Blaise in the act of performing the miracle for which he is famous: the rescue from certain death of a child, now half-dead from a fishbone. (Graphic composition Maurizio Forli, courtesy private collection)

They are also known as Cannelli Simpatici:

> *Portano il nome di Cannelli Simpatici per la similitudine, che hanno co' i piccioli cannelli, e perché sono creduti operare per simpatia: e per la similitudine, che hanno coll'aspra arteria della gola, gli attribuiscono virtù speziale contro i mali della gola stessa* (Gimma 1730: p. 268)

> *They bear the name of Cannelli simpatici* (Sympathetic small tubes) [1]*because of their resemblance to small tubes, and because they are believed to work by affinity (sympathy): and because of the resemblance which they have to the sour artery of the throat, they are attributed with the special virtue against the illness of the throat itself* (Gimma 1730: p. 268).

Although the author is not at all convinced of the therapeutic virtues attributed to them, he reports them, among the many news about them, in the chapter *De' Corpi Impietriti.*

> *Delle virtù, che a' Cannelli Simpatici attribuiscono, molti veramente raccontano maraviglie, ed affermano ricavarne giovamento; anzi vi sono di quei, che hanno la pazienza di portarne molti legati intorno al collo. A tutte queste cose, che sospette si portano, e sono dette Amuleti, niuna fede vien data da' Medici, e dagli Uomini dotti* […] *Tre ricette abbiamo avute sotto'occhio* […] *ma non uguali nella relazione delle virtù; poiché una gli dice solo giovevoli al male della gola, altre a varj mali* [...] *Dicono, che vale questo Cannello per quelle persone, che patiscono il male della gola: portandolo al collo che tocchi la carne, non patirà mali di gola o dolori: leva l'enfiagione di essa, non patirà male di Scaranzia, o Candioni causati dal Riscaldarsi o raffreddarsi. Vale per lo granco, portan-*

[1] For the Italian meaning of Cannelli Simpatici, referring to scaphopods in their protective function against the evils of the throat, does not find exact correspondence in the English language. We propose with many doubts this version: Sympathetic small tubes

dolo in luogo, che tocchi la carne. È buono per istagnare il sangue dal naso: rischiara la vista, fa buona voce, e buona memoria. Che alcuni l'hanno sperimentato per la Sciatica, doglia di testa, vertigine; con far toccare la testa, come sopra: e tutte le Ricette conchiudono che si potrà chiamare fortunato chi lo conosce.

Of the virtues attributed to the Cannelli simpatici, many truly tell marvels, and claim to derive benefit from them; indeed there are those who have the patience to wear many of them tied around their necks. To all these things, which are suspected to be worn, and are called Amulets, no faith is given by physicians, and by learned Men [...] *Three recipes we have had under our eye* [...] *but not equal in the relation of virtues; since one says they are only beneficial to the illness l of the throat, others to various ills* [...] *They say that this blowpipe is good for those people who suffer from a sore throat: if you wear it on your neck, so that it touches the flesh, you will not suffer from sore throats or pains: if you lift the swelling of the throat, you will not suffer from Scaranzia, chilblains caused by heating or cooling. It is valid for the cancer, bringing it to a place where it touches the flesh. It is good for stagnating the blood from the nose: it illuminates the sight, makes a good voice, and good memory. That some have experimented with it for Sciatica, headache, vertigo; by touching the head, as above: and all the Recipes conclude that he who knows it may be called fortunate.*

Usually the amulets consist of a *Dentalium* crossed and tied by a red ribbon. The tradition of using whole specimens or variously cut and reduced, as pendants or elements of necklaces, has been handed down to the present day. These products, for example, can easily be found in research activities in the field of experimental archaeology or in trade, traditional or by the web, as ethnic products (Figs. 15.7, 15.8, and 15.9).

Fig. 15.7 In the middle there is the so-called *cacazzo di luna*; therefore, an object considered sacred decorated with a red ribbon. Also known as the *cannaruzzeddu of San Brasi*, it was worn as a condom for sore throats (Pitrè 1978: p. 206, anastatic reprint (modified)). On the sides, two copies specially made for comparison, using Pliocene specimens of *Fissidentalium rectum* (Gmelin 1791). (Photo and graphic composition Maurizio Forli)

Fig. 15.8 *Shrieking bones*. Specimens from the Bellucci collection (M.A.N.U. (PG) Inv. 4742, 4880, 5013, 5975). (Reproductions by kind permission of the Ministry for Cultural Heritage and Activities and for Tourism, Regional Directorate Museums of Umbria – National Archeological Museum of Umbria)

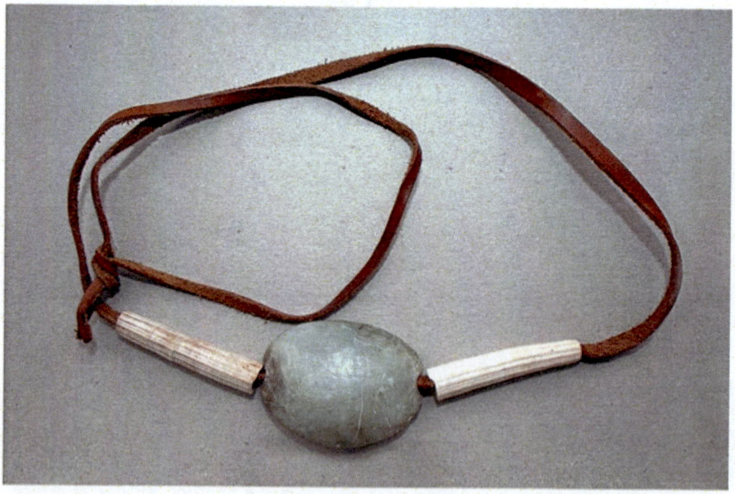

Fig. 15.9 Necklace with central pendant in steatite and two lateral teeth. Reproduction of a prehistoric necklace made as part of studies on experimental archaeology. (Photo Maurizio Forli, courtesy private collection)

References

Aldrovandi U (1606) De Reliquis Animalibus exanguibus libri quatuor, post mortem eius editi: nemp' De Mollibus, Crustaceis, Testaceis, et Zoophytis. Apud Io: Baptistam Bellagambam, Bononie. https://www.biodiversitylibrary.org/bibliography/119999 (Public Domain)

Allioni C (1757) Oryctographiae pedemontanae specimen, Exhibens corpora Fossilia Terrae adventitia. Cl. J. B. Bauche, Parisiis, ad ripam Augustinorum

Bellon P (1553) De Aquatilibus, Libri duo. C. Stephanum, Parisiis. https://www.biodiversitylibrary.org/bibliography/5765 (Public domain)

Bellucci G (1898) Amuleti italiani contemporanei. Catalogo descrittivo della collezione inviata all'Esposizione Nazionale di Torino. Unione Tipografica Cooperativa, Perugia

Bellucci G (1907) Il feticismo primitivo in Italia e le sue forme di adattamento. Unione Tipografica Coop. Editrice, Perugia.

Ceccarelli I (1639) Antidotario Romano Latino, e Volgare. Tradotto da Ippolito Ceccarelli. Li ragionamenti, e le aggiunte dell'elettione de' Semplici, e Pratica delle Compositioni. Con le Annotationi del Sig. Pietro Castelli Romano. E Trattati della Teriaca Romana, e della Teriaca Egittia. E nuova aggiunta di molte ricette ultimamente publicate dal Collegio de' Medici di Roma. Pietro Antonio Facciotti, Roma

Cordus V (1558) Il Dispensario, cioe il modo di comporre le medicine et conoscer quelle insieme con li semplici, con una breve dichiaratione, che gli accommoda alle infmità. Pietro Bosello, Venetia

Fuchs L (1549) Nicolai Myrepsi Alexandrini medicamentorum opus in sectiones quadragintaocto digestum, hactenus in Germania non visum, omnibus tum medicis, tum seplasiarijs mirum in modum utile, Basileae

Gessner C (1560) Nomenclator aquatilium animantium : icones animalium aquatilium in mari et dulcibus aquis degentium, plus quam DCC. cum nomenclaturis singulorum Latinis, Grecis, Italicis, Hispanicis, Gallicis, [...]. excudebat Christoph. Froschoverus, Tiguri. ETH-Bibliothek Zürich, Rar 9734: 1, https://doi.org/10.3931/e-rara-48090 / Public Domain Mark

Gimma G (1730b) Della Storia Naturale delle gemme, delle Pietre, e di tutti i Minerali, ovvero della Fisica Sotterranea. Tomo II, Felice Mosca, Napoli

Gualtieri N (1742) Index Testarum Conchyliorum quae adservantur in Museo Nicolai Gualtieri. Ex Typographia Caietani Albizzini, Florentiae https://gdz.sub.uni-goettingen.de/id/PPN367583143

Lang KN (1708) Historia lapidum figuratorum Helvetiae, eiusque viciniae : in qua non solum enarrantur omnia eorum genera, species et vires aeneisque tabulis repraesentantur, sed insuper adducuntur eorum loca nativa, in quibus [...]. sumptibus authoris, typis Jacobi Tomasini, Venetiis. ETH-Bibliothek Zürich, Rar 5680, https://doi.org/10.3931/e-rara-10398 / Public Domain Mark

Linnæus C (1758) Systema Naturae per Regna Tria Naturae, secumdum Classes, Ordines, Genera, species, cum characteribus differentiis, synonimis, locis. Tomus I. Editio decima, Reformata. Laurentii Salvii, Holmiae. https://www.biodiversitylibrary.org/bibliography/542 (Public domain)

Mercati M (1717) Metallotheca Vaticana. Opus posthumum, Autoritate et Munificentia Clementis Undecimi Pontificis Maximi. E tenebris in lucem eductum: Opera autem, & studio Joannis Mariae Lancisii, Archiatri Pontificii illustratum. Ex Officina Jo: Mariae Salvioni Romani, Romae. https://archive.org/details/bub_gb_DC_E6UVWg7EC (Public Domain Mark 1.0)

Moscardo L (1656) Note overo memorie del museo di Lodovico Moscardo nobile veronese. Academico Filarmonico, dal medesimo descritte, Et in Tre Libri distinte. Per Paolo Frambotto, Padoa. https://www.biodiversitylibrary.org/bibliography/157058 (not in copyright)

Pitrè G (1940) Medicina Popolare Siciliana, vol unico. G. Barbèra Editore, Firenze

Pitrè G (1978) La famiglia, la casa, la vita del popolo siciliano. Ristampa anastatica dell'edizione di Palermo, 1870–1913

Rondelet G (1555) Universae aquatilium Historiae pars altera, cum veris ipsorum Imaginibus. Matthiam Bonhomme, Lugduni. https://archive.org/details/ARes071032/page/n13/mode/2up (Public Domain 1.0)
Ruel J (1543) Dioscorides Pedanius, Anazarbeus. De medica materia libri sex, Ioanne Ruellio Suessionensi interprete. Chr. Egenolphum, Frankfurt
Ruellio I (1552) Pedanii Dioscoridis Anazarbei, De medicinali materia libri sex. Apud Balthazarem Arnolletum, Lugduni. https://www.biodiversitylibrary.org/bibliography/9317 Public domain
Scilla A (1670) La vana speculazione disingannata dal senso. Lettera responsiva Circa i corpi marini, che Petrificati si trovano in varij luoghi terrestri. Andrea Colicchia, Napoli
Vialli V (1974) Notes of paleontology. Pitagora Editrice, Bologna

Chapter 16
Belemnoidea: from Lyncurium, Lynx Stone, to Cephalopoda

Abstract The Belemnoidea are extinct marine mollusks included in the class Cephalopoda. Very important from the paleontological point of view, they had their greatest diffusion and diversification in the Mesozoic; in particular, their fossil remains are numerous in the lands that extend from the Jurassic to the Cretaceous. The shell consists of three parts called rostrum, phragmocone, and pro-ostracum. The rostrum is the most frequently preserved part in the sediments, and the specific classification has been mainly based on its characteristics. Since rostrum is mostly cylindrical and pointed, and there is no evidence of them in living organisms, they have been interpreted in the most imaginative ways, both in a folkloristic and scientific sense. Generally regarded as thunder or lightning stones, they are also known as gnome candles, stone swords, bat stones, St. Peter's fingers, or devil's claws. They were first drawn in a printed book on fossils by Gessner (*De omni rerum fossilium genere, gemmis, lapidibus, metallis, et huiusmodi, libri aliquot, plerique nunc primum editi.* Tiguri, 1565. https://doi.org/10.3931/e-rara-4176 / Public Domain Mark) who, in doing so, established the exact correspondence between the name and the shape of the object "belemnite."

Keywords Fossilia · Fossils · History · Belemnite · Lyncurium · Ceraunia

The Belemnoidea are extinct marine mollusks included in the class Cephalopoda. Their remains in folklore, in particularly the rostrum, are generally regarded as thunder or lightning stones but they are also known with others fancy name (Fig. 16.1). The Belemnoidea, from the Greek βέλεμνον, *bélemnon* javelin, dart, are extinct molluscs presently inserted in the subclass Coleoidea, included in the class of the Cephalopoda Cuvier, 1795. Very important from the paleontological point of view, even if they appeared in the Carboniferous, they had their major diffusion and diversification in the Mesozoic, in particular their fossil remains are numerous in the soils extending from the Jurassic to the Cretaceous, disappearing at the end of this geological era, during the mass extinction at the end of the period. They were a sort of squid with an inner shell, for this reason also called endocochleate, completely covered with skin and soft tissues and, like the other cephalopods similar to them, they must have led a planktonic or

© The Author(s), under exclusive license to Springer Nature Switzerland AG 2022

M. Forli, A. Guerrini, *The History of Fossils Over Centuries*,
https://doi.org/10.1007/978-3-031-04687-2_16

nectobenthic life in a pelagic marine environment. The shell consists of three parts, rarely preserved together in fossil finds, which are called rostrum, phragmocone, and pro-ostracum. The rostrum is the most massive element of the skeleton and is calcitic in nature, a structure that seems to have no counterpart in Ammonoidea and Nautiloidea shells. It represents the most frequently preserved part of the sediments, and in this regard the classification has been mainly based on its characteristics. Besides having cylindrical form, the rostrum can also be lanceolate, mucronate or flattened. The rostrum is composed of calcite fibers oriented at right angles to the surface and radiating the apical line, that is, the axis joining the base of the alveolus with the apex of the rostrum. In addition to the radial structure, we notice in section also a structure with dark concentric rings, formed by calcite with organic substance, alternated with light rings, formed by pure calcite, which testify the various stages of growth. The surface of the upper part of the rostrum is characterized by a variable number of longitudinal grooves. Jurassic belemnites possessed an additional long and slender structure, the epirostrum, which enveloped the tip of the rostrum and differed in shape and internal structure. The epirostrum lacks the distinction between light and dark layers; moreover, it is poorly calcified inside, but has a strong calcareous crust (Figs. 16.1 and 16.2).

Fig. 16.1 Belemnite rostrums were generally regarded as thunder or lightning stones, petrified pieces of lightning, according to some versions referring to Greek mythology, remnants of darts thrown by Jupiter that fell to the ground during thunderstorms [Graphic composition Maurizio Forli. The Jupiter image is from Jacopo Zucchi, Jupiter Fighting the Giants (1540–96). Credit Metropolitan Museum of Art Collection API, https://www.metmuseum.org/art/collection/search/337896 Public Domain]

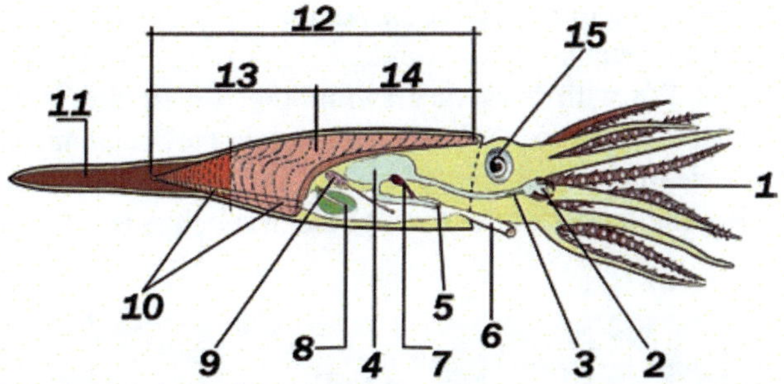

Fig. 16.2 Simplified morphological scheme of a hypothetical belemnite. (1) Anterior part of body, hooked tentacles. (2) Beak-like jaws. (3) Radula. (4) Stomach. (5) Anal aperture. (6) Siphon. (7) Ink gland. (8) Compound heart and gills. (9) Gonad. (10) Siphuncle. (11) Rostrum. (12) Conotheca. (13) Phragmocone. (14) Pro-ostracum. (15) Eye. [Drawing Maurizio Forli modified after Vialli (1974)]

The phragmocone, of aragonitic nature, is the concamerated part of the shell, which, due to its structure with successive chambers similar to that of the nautiloids, is the most familiar to us. It is represented by a cone more or less elongated or arched, subdivided into chambers by simple superimposed septa, more or less distant from each other, convex and perimetrically edged like that of a watch glass, perforated by the siphuncle in the ventral area, differing for this from similar septa of orthoceratids in which the hole is placed centrally. The apical angle of the cone varies from 5° to 30°. The dorsal part of the phragmocone projects beyond the last septum, with a laminar extension called, considered as a vestigial dwelling chamber that had to receive and protect dorsally the visceral mass of the organism. The proostracum is of horny nature encrusted by aragonite and shows growth lines, with such a course, that we can distinguish four different regions: one dorsal, one ventral, and two lateral. On the ventral side, there are convex transversal lines toward the cephalic part of the mollusc, each of which marks the successive positions reached by the extreme margin of the pro-ostracum. This type of ornamentation allows to reconstruct the shape starting also from small fragments of the phragmocone. The part of the shell composed of the phragmocone and the pro-ostracum constitutes a very delicate structure and is easily subject to crushing during the accumulation of sediments and is therefore exceptionally preserved in its fossil state.

It is evident that the various parts of the shell had different functions. The pro-ostracum, which protected the underlying visceral mass, was also the base of attachment of the various types of muscles, while the phragmocone certainly had an important hydrostatic function, regulating with the action of the siphuncle, the density of the gas, and the quantity of liquid in the chambers, similarly to the analogous parts of the nautilus and ammonoid shells. The rostrum, on the other hand, had to serve as a counterweight, to balance that of the body and the arms, so that the animal could maintain the most suitable position for swimming, and also served as a protection for the apex of the phragmocone. The function of the epirostrum, found only in the Jurassic species, can only be object of conjectures. According to some authors, it might have been related to reproduction. As for the general type of organization, that of the belemnites recalls that of squids, both for the constitution of the body, and for the way of life with backward swimming. Belemnites are useful stratigraphic fossils from the Jurassic to the Cretaceous, in particular, the standard zonal biostratigraphic scale of these geological periods is based on their associations with ammonites, the leading fossils of the Ammonoidea subclass (Fig. 16.3).

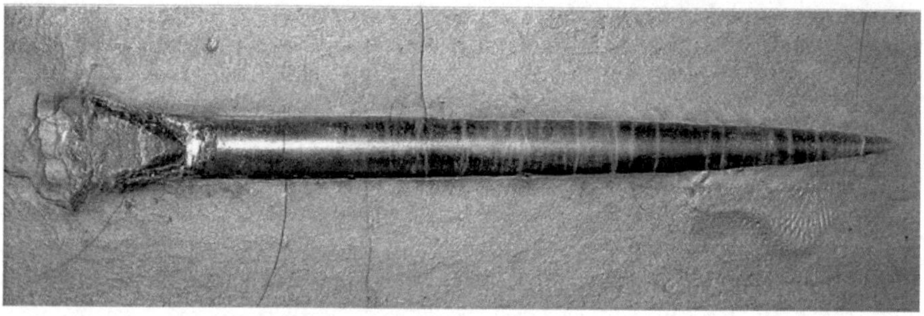

Fig. 16.3 From left to right: final part of the phragmocone in connection with the rostrum, in a belemnite of the genus *Youngibelus*. Lower Jurassic, Toarcian (174.1–182.7 Ma), Germany. Length ~ 140 mm. (Photo credit Geocollection)

We can mention stratigraphically important forms, such as those belonging to the genus *Duvalia* Bayle, 1878 with shells characterized by a wide and flattened rostrum shape, for example, *D. lata* (Blainville 1827) and *D. emerici* (Raspail 1829) from the Lower Cretaceous (Barremian–Ativan) and *Belemnitella mucronata* (Schlotheim 1813) from the Upper Cretaceous (Campanian–Maastrichtian). Typical belemnites became extinct at the end of the Cretaceous. Few genera, and species, had been described for the Cenozoic, limited to the Eocene period, but the most recent studies have shown that the species involved by these findings are to be considered belonging to other families, such as Sepiidae and Spirulidae (Fuchs and Lukeneder 2014). In Italy they are found in the same lithotypes where the ammonites or their aptychi are found, in particular, in the nodular limestones of the Ammonite Red or in the flinty ones, generally white, of the Majolica.

Phylum Mollusca
Class Cephalopoda
Superfamily Belemnoidea †
Family Belemnitellidae †
Genus *Belemnitella* †
Species *Belemnitella mucronata* (Schlotheim 1813) †

Since the rostrums are mostly of pointed cylindrical shape, and there is no evidence of them in living beings, they have been interpreted in the most imaginative ways, both in a folkloric and scientific sense. An ancient evidence of the particular attention they aroused in humans is attested by their allochthonous presence already in a prehistoric site of chipping of lithic tools, referable to a population of Neanderthals, dated to about 60,000 years ago (1993). Generally considered as thunder or lightning stones, petrified pieces of lightning, according to some versions referring to Greek mythology, remains of the darts thrown by Jupiter during the fight between the Gods and the Giants fallen to the ground during thunderstorms.

Called in a variety of other ways, depending on the countries in which they are found, they are also known as gnome candles, stone swords, bat stones, St. Peter's fingers, or devil's claws.

> *First of all, there is a quantity of "petrified sticks" of every size from one and two centimetres up to fifteen and twenty in length. These black, pinned "sticks" are called "stones of thunder" by the vintners. They were also considered as freaks of nature, stony concretions, stalactites, fish teeth and sometimes they were also called "devil's claws", etc.; they are so numerous that there is nothing to do but stoop down to pick them up.* – (from Flammarion, The World before the Creation of Man. ed. 1934)

Powdered, they have been used till the eighteenth century, for curing distemper and worms in horses, while in man they were used for avoiding nightmares, neutralizing spells, mitigating acute pains, rheumatisms, and for healing sores, in particular those of the eyes. In addition, due to their phallic shape they were considered aphrodisiac. Still today they are used in jewellery as lucky charms.

> *Molti stimano le sue facoltà esser così valevoli per romper le Pietre, come quelle della Pietra Giudaica. Vien commendata nel mal della Ponta, per esser pontata. Guarisce le piaghe, abbrucciata è buona per nettar li denti. Alcuni la soffiano nelli occhi de Cavalli per levare le cicatrici. Bevuta credasi giovar contro le strigherie, e per impedir li travagli dell'Incubo, e vanegiamenti della notte* (Passera 1688: p. 275).

> *Many esteem its faculties to be as valid for breaking stones as those of the Judaic Stone. It is commended in the malady "of the Ponta", for being "pontata". It heals sores, and when burnt it is good for cleaning teeth. Some blow it into the eyes of horses to remove scars. Drunk, it is believed to help against the "strigherie", and to prevent the travails of the Nightmare, and the delirium of the night.*

In particular from the paragraph "De la Pietra del Lenzo" that is the stone of the Lynx, where among the medical virtues it indicates, among other things, excellent for the mal de la Ponta, understood in Venetian dialect, that is for pleurisy, flank pain, and pain of the ribs and kidneys

We owe to the French zoologist and naturalist de Blainville, a detailed chronological account on the interpretation of the fossil remains of belemnites which, during the centuries, have been attributed to the most various objects.

> *Parmi les corps organisés que l'on rencontre fossiles dans les sein de la terre à des profondeurs variables, il n'en est certainement aucuns qui aient autant occupé les naturalistes que ceux auxquels on a donné le nom de Bélemnites* [...].

Among the organized bodies which are found fossilized in the bosom of the earth at different depths, there are certainly no others which have so much occupied the naturalists, as those to which they have given the name of Belemnites [...] (1827: p. 1).

So says the beginning of the second edition of his comprehensive work, in seven chapters, dedicated to the knowledge of the belemnites, of course, until the time of his writing, *Mémoire sur les Bélemnites considérées zoologiquement et gèologiquement*, in which he reports all the previous citations, from the first Greek author identified in Theophrastus, to Pliny and so on, up to his contemporaries, citing a total of more than 90 works in which, in a more or less thorough and dedicated, it deals with these fossil remains. Following Blainville, in Theophrastus [*De Lapidibus*, translation from Greek into Latin by Fridericus Wimmer (1866) *in Theophrasti Eresii Opera quae supesunt, omnia graeca recensuit, latine interpretatus est*] the description of the belemnites does not appear at all, and the only quotation which gave rise to this interpretation is that of a phantom substance called *lyngurium* or *lyncurium*, which would be obtained by means of the solidified and buried urine of the lynx, which is also interpreted, perhaps more correctly by other authors after the Greeks, to include amber. The one who in our opinion appears instead more important in the view of the writing of the present text, which is obviously found in De Blainville, is the one who first understood, correctly, what the *thunderstones* really were:

Ainsi il me semble à peu près certain que les bélemnites n'étaient pas connues des anciens, et que par conséquent l'origine et toutes les vertus qu'on leur a attribuées sont de date moderne. C'est en effet dans les auteurs de la renaissance des lettres, et surtout chez les Allemands, qu'il faut en chercher la source, probablement dans la dernière moitié du quinzième et dans la première moitié su seizième siècle (1827: 5).

It is therefore in the second half of 1400 and the beginning of 1500, particularly among Germanic authors, that we must find the origin of the virtues attributed to them, as it seems almost certain that the ancient Greek authors did not know the belemnites.

It is in Agricola (1546) that we find for the first time the name *Belemnite* associated with the petrified bodies that today we understand under this name. The author does not agree with the ancient attribution to *lyncurium* and describes them by observing the differences in size, shape, color, transparency, and opacity but, like many other natural fossil things, he considers them simply stone objects, at most useful from the medicinal point of view. He also gives us an account of the reasoning behind the confusion between lyncurium identified as amber, succinus, or belemnite.

Tramezzino Michele (1550: p. 267), vernacular translation of the Agricola text:

Il Belemnite mostra la effigie di una saetta: onde i Sassoni il chiamano d'un nome composto da la saetta, e da l'ephialte, che è l'incubo, ò spirito maligno, che diciamo: Onde bevendosi questa pietra dicono che giovi contra queste cosi fatte fantasme notturne, che pare che opprimano e travaglino il corpo humano; e contra le fascinationi, e malie: Alcuni medici tengono hoggi questa pietra, e lo usano per lo lyncurio, che giamai de la urina del lynce si congelò : per che non si ritrovò mai un tale lyncurio. È ben vero che gli antichi scrittori chiamarono di questo nome il succino, come s'è già detto di sopra; [...] *Se alcuno Greco però vidde mai il lyncurio, che non fosse il succino; egli, come io penso, questi Belemniti vidde: per che alcuni di loro ne hanno il colore del succino falerno, e tralucono: et alcuni, à la guisa che fa il succino; attraheno à se la paglia e le cose minute. Egli si ritrova il Belemnite in molti luoghi de la Germania* [...].

The Belemnite shows the effigy of a thunderbolt: from which the Saxons call it with a name composed of the thunderbolt and the ephialtes, which is the nightmare, or evil spirit, that we say: Therefore, drinking this stone, they say that it helps against these nocturnal phan-

tasms, which seem to oppress and torment the human body; and against fascination, and evils: Some phisicians nowadays keep this stone, and use it for the lyncurium, which has never been frozen in the urine of the lynce: because such a lyncurium has never been found. It is true that the ancient writers called the succinus by this name, as has already been said above; [...] *If, however, any Greek ever saw the lyncurium, which was not the succinus, he, as I think, saw these Belemnites: because some of them have the color of the "falerno succinus" (amber of intense red color), and they shine: and some, in the manner of the succinus, attract straw and small things. Belemnite is found in many places in Germany* [...].

The first representation in a printed book is, as for many other fossils, thanks to Gessner (1565) who, in doing so, establishes exactly the correspondence between the name and the shape of the object "belemnite," to which he refers the four illustrations on page 91, although the one indicated with the number three is obviously different from the others and could refer to a different organism. Nevertheless, Ferrante Imperato a few years later still uses the name Lyncurio to describe the rostrums (Fig. 16.4):

Fig. 16.4 The first printed image of belemnites is believed to be this one proposed by Gessner in *De omni rerum fossilium genere, gemmis, lapidibus, metallis* (1565: p. 91) (https://doi.org/10.3931/e-rara-4176 / Public Domain Mark)

Tra le figurate di natura manifestatamente è la pietra detta Lyncurio, che imita nella figura l'obelisco, ò cono pinea aguzza, Diciamo che il lyncurio e pietra di figura naturalmente lunga aguzza, di esquisita rotundezza, e liscezza, e di sustanza più e men trasparente; perciò che vi si veggono alcuni di essi nel biondo chiari [...] *Adoprasi il lyncurio da medici per cacciar l'arenella dal corpo, nelle difficoltà di urina, altri le adoperano nelle punture delle costole* (1599: 254–255).

Among the figurations of nature there is the stone called Lyncurio, which imitates the obelisk in the figure, or the pointed pine cone, We say that the lyncurio is a stone of naturally long pointed figure, of exquisite roundness, and smoothness, and of more and less transparent substance; therefore we see some of them in the clear blond [...]. *The lyncurium is used by physicians to expel the renal gravel from the body, in the difficulties of urine, others use them in the puncture of the ribs* (1599: 254–255).

Boëtius de Boodt Anselmus begins their description, on page 614 of his book *Le parfaict joaillier ou histoire de pierreries* (1644), thus:

> *La pierre belemnite, ou dactilus Ideus ressemble tellement à une flesche* [...] *qu'elle paroit etre une pointe de flesche d'or, d'argent, ou de bois changée en pierre. Car elle est quelquefois de couleur de fer, d'or, ou d'argent, & a au dedans quelque chose de semblable au bois,*

which translated and updated to today's language reads *The stone belemnite, or dactilus ideus looks so much like an arrow* [...], *that it seems to be an arrowhead of gold, silver, or wood changed into stone. For it is sometimes of the color of iron, gold, or silver, and has in it something resembling wood.*

In the continuation of the description, in addition to reporting the various names by which they are known, among the curious things attributed to them, he says that, although they rarely exceed the size of a finger and the thickness of a common arrow, they have nevertheless been found to be the size of *a man's thigh, de la grosseur d'une cuisse d'homme,* so that it can be thought by the way that they are large arrows, parts of war machines of the most ancient peoples, changed into stone. It is curious to note that in the 1609 and 1636 Latin versions of his *Gemmarum & lapidum Historiae*, written in Latin, the same point of the description devoted to the belemnites in chapter CCLVIII, respectively, in the two editions on page 235 and 477, states

> *Inventae tamen sunt tibiae, humanae tibiae crassitudine, ita ut quis balistarum & machinarum bellicarum sagittas veterum in lapides commutatas non ineptè exstimare possit.*

However, they found some as tibiae, about the size of a human shin, so that they can be mistaken not improperly with any old arrows, changed into stone, of crossbows and military machines.

The difference in interpretation due to the different translation of the word *tibiae*, understood as leg bone or leg as a whole, leads to give disproportionate dimensions for the rostrums. To date, the largest known rostrums measure just over 45 cm in length with species belonging to the genus *Megateuthis*, in particular, the species *Megateuthis gigantea* (Schlotheim 1820) is considered the largest known belemnite with a total length of the animal estimated at around three meters. In *Le parfaict joaillier*— writing about the organoleptic properties, so to speak, of the rostrums, he effectively informs us, like other previous authors after Agricola, about the odors they emit when burnt *When burning it has a heavy and bad smell, like horns, or burnt bone, or cat urine*, which still allows us to perceive the smell just by reading the description. The figures used to illustrate the above texts, in order of printing on pages 236, 477, and 616, are always the same and are a recomposition of the illustration taken from Gessner. Aldrovandi dedicates chapter XIII of the *Musaeum Metallicum* to the description of the belemnites and illustrates some of them at its end, on page 621, and an isolated rostrum at the end of chapter XXII, on page 650, among the pyrite objects. As for the other stone bodies collected and studied by Aldrovandi, the belemnites, too, are only natural curiosities in imitation of living things, and his text is limited to reporting what was then known. A beautiful original illustration of 12 rostrums composed in three rows of four specimens each, inclined and staggered, is reported by Mercati (*Metallotheca Vaticana*, 1717: p. 281). In the subdivision according to the cabinets in which the objects of the collection described by San Miniato were contained, the belemnites were in the IX dedicated to the idiomorphic stones.

> *The description of cupboard IX constitutes the most important part of the Metalloteca, both for the richness and variety of the specimens, and for the passion and enthusiasm with which the Author described, in no less than 132 pp., the countless pieces distributed in 68 "species", listed according to the order of the drawers at the end of the chapter* (Accordi 1980: p. 27).

Unfortunately, no particular contribution is proposed by Mercati, who mostly deals with distinguishing or characterizing them into the two types by which they were generally known, *Belemnites*, that is, dart shaped, or *Dactyli*, that is, finger shaped. He naturally reports the beliefs connected with them, including some as to their use as a medicine, and a large number of localities scattered over Europe, in which they are most easily and frequently found.

The definition of *Dactyli* or more strictly *Dactyli idaei* to indicate the rostrums lends itself to at least two interpretations, which although used indiscriminately to describe them seem to have in common only part of the etymology and not their meaning.

In Greek antiquity, the *Curets* or *Coribanti were* called *Dactyli idaei.*

> *... who (it is said) were the first to introduce music into the festivals of the Greek Deities. Their music consisted of loud cries accompanied by drums and other romantic instruments. According to the fable, it was precisely these Coreti who, keeping the child Jupiter hidden on Mount Ida, so that he would not be discovered for his cries, made a great noise with their instruments, beating at the same time with proportionate steps, called Dactyli. From these steps and from Mount Ida they got the name of Dactyli idaei* (Lichtenthal 1826: p. 221).

The generic name of *Dactyli*, which was usually used for elongated finger-shaped shells, seems appropriate to indicate belemnites. The similarity is correct because, like many rostrums, the date shell, the bivalve *Lithophaga lithophaga* (L., 1758), is elongated like a finger and golden brown like the palm fruit.

It is obvious that belemnites are better known in countries where their fossils are more abundant and it is for this reason that quite numerous appear illustrated in books published by French-Anglo-Saxon authors as, for example, in Worm (*Museum wormianum*, 1655: p. 70) the same illustration derived from Gessner is given, also used Boëtius de Boodt; Scheuchzer (*Specimen litographiae Helveticae curiosae*, 1702) some original figures in the first plate at numbers 10 and 14 and in the third at number 14; Valentini (*Museum Museorum*, 1704: p. 53), a rostrum indicated as *Lyncurium vulgaris*; Lang (*Historia lapidum figuratorum Helvetiae eiusque viciniae*, 1708) a plate on page 131, where in addition to loose rostrums others can be seen on a rocky matrix; Lhwyd (*Lithophylacii Britannici Ichnographia*, 1760) at plates 21 and 25 some rostrums and remains of phragmocones are illustrated. The first description of belemnites as fossil remains of marine organisms similar to nautiluses and ammonites was made by Balthasar Ehrhart in 1724:

> *Belemnites, qui tam crebò inter alias marina fossilia occurrit, est nil aliud quam domicilium seu tegumen, solidum, concentricis striis insignitum, fovendo Alveolo, testaceo univalvi, conico, in concamerationes distincto, et siphunculo instructo, destinatum. Ipse autem*

> *Alveolus, à congeneris suis, Nautilo et Cornu Ammonis, non nisi figurà erectà, fastigiatà, non spirali, differt*. – (Ehrhart, Belemnitis suevicis dissertatio, 1727: p. 41)

> *Belemnite, which is so frequently met with among other marine fossils, is nothing else than the habitation destined for a univalve testaceous, with a solid outer layer, marked by concentric striae, conical, distinct in concamerations, and provided with a siphuncle. The same alveolus as the Nautilus and the Horn of Ammon, its congeners, from which it differs only by the straight, conical, unspiralled figure*.

In a publication contemporary to that of De Blainville, of a markedly scientific character, where are described also several new species, published a couple of years after his first presentation, in the *Transactions of the Geological Society of London*, the English J.S. Miller devised a singular experiment. Rightly attributing to the chambers of the phragmocone, the possibility to fill up and empty at will according to the requirements of life of the mollusc, as well as in the nautilus, in order to verify how big should be the portion of the chambers to support the total weight, he inserted in the cavity of a rostrum the most complete he could get, a cone of oiled paper, fastened with rubber, filled with cotton, which counterbalanced the weight of the rostrum by floating in the water, thus showing that the size required was reasonably similar to that of the most complete fossils found, proving that their dimensions were reliable:

> *To ascertain the size of the chamber which would be requisite to compensate for the weight of the guard, I made the following experiment. I fastened with gum a piece of oiled paper of a conical form in the cavity from which the chambered part had fallen out of a large Belemnite; and putting into this cone some cotton, to prevent the paper from collapsing by the pressure of the water*.
>
> *I found that the apparatus floated with a paper of no larger size than from some known fine specimens we may fairly conclude the outer chamber to have attained* (Miller 1826, *Observations on Belemnites*).

However, following also to these considerations, they imagined that these cephalopods could stay in vertical position. This contrasts with the modalities of life and movement in the waters, considered already then, equal or very similar to those of the present squids and cuttlefishes (Figs. 16.5, 16.6, 16.7, 16.8, 16.9, 16.10, and 16.11).

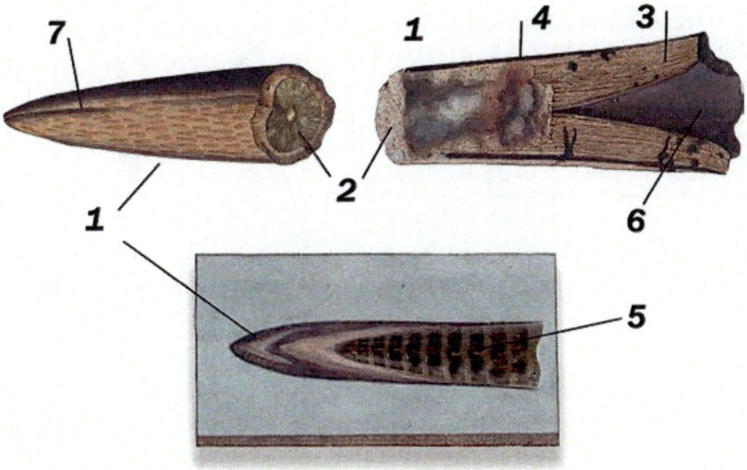

Fig. 16.5 Main parts of the rostrum [1] and phragmocone [5]: (2) Radiated fibrous structure, in transverse section. (3) The same in longitudinal section. (4) Outer cortical part. (5) Chambers and septa of the phragmocone. (6) Conotheca. (7) Longitudinal sulcus. (Graphic composition Maurizio Forli. Phragmocone section [5], credit from: en.wikipedia.org/wiki/File:Belemnoidea_fossil_characters.jpg; other images from Walch: 1768-1773, https://doi.org/10.3931/e-rara-10246 / Public Domain)

Fig. 16.6 Belemnite from the Jurassic of Evreux, Normandy, France. Length of central rostrum 140 mm. (Photo Maurizio Forli)

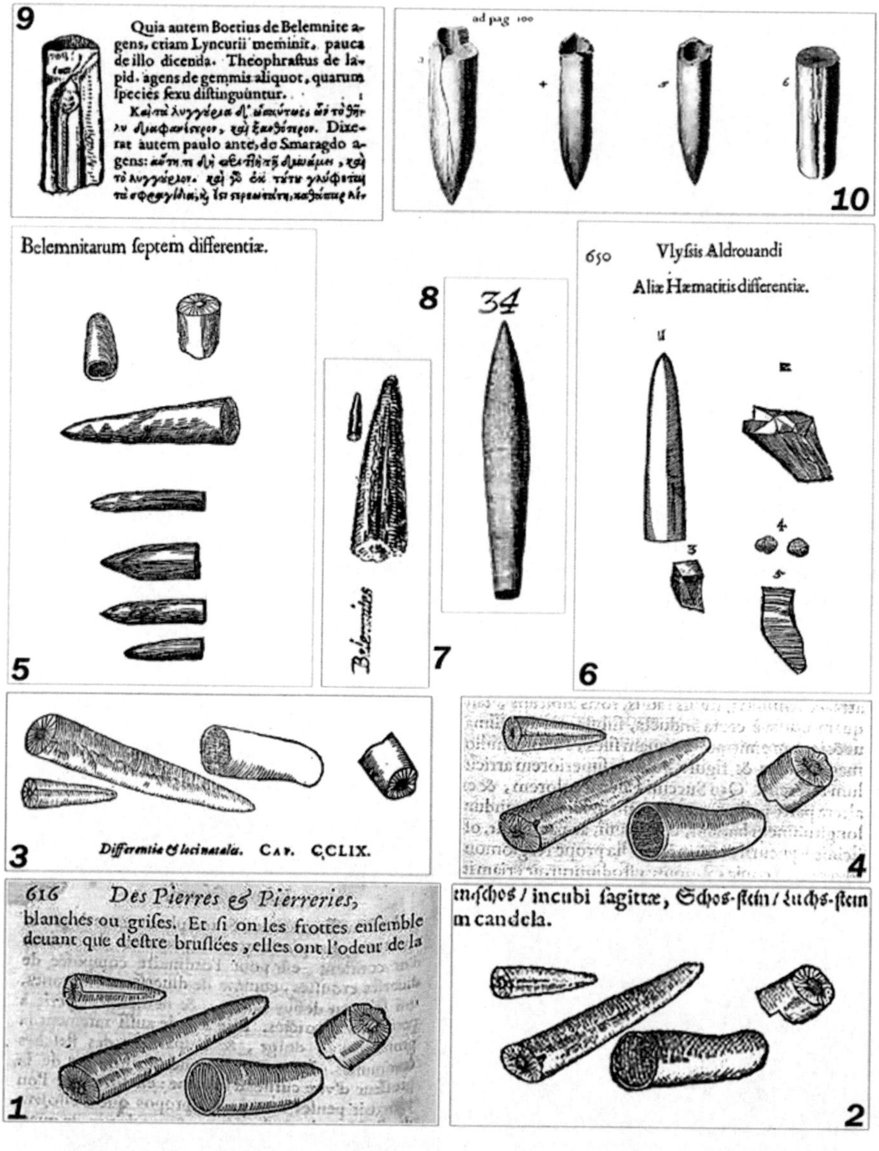

Fig. 16.7 Illustrations referred to Belemnite rostrums, from the earliest printed images: (1) De Boodt (1644: CCLVIII: p. 616), *Du Belemnites des boutiques, du Lyncurium, et du Dactilus Ideus*. (2) Worm (1655: p. 70) *Belemnites à sagitta* sic *dictus*. (3) De Boodt (1609: CCLVIII: p. 236) *De Belemnite officinarum, Lyncurio et dactiloideo*: (4) De Boodt (1636: p. 477). (5) Aldrovandi (1648: p. 621) *Belemnitarum septem differentiae*. (6) Aldrovandi (1648: p. 650) *Haematites forma Belemnitis ab arte accepta*. (7) Boccone (1674: p. 307) *Belemnites, ou Lapys Lincis*. (8) Scheuchzer (1702: p. 25) *Belemnites minor cinereus*. (9) De Laet (1647: p. 154) *In Belemnite interdum reperiri lapidem in conum acutum, turnibatu, etiam concavum, nec minus terra vel arena repletum*. (10) Plot (1677: pp. 93–95) Belemnites. [Graphic composition Maurizio Forli. Credit from: (1) https://doi.org/10.3931/e-rara-12490 / Public Domain Mark (2) https://doi.org/10.5962/bhl.title.63844

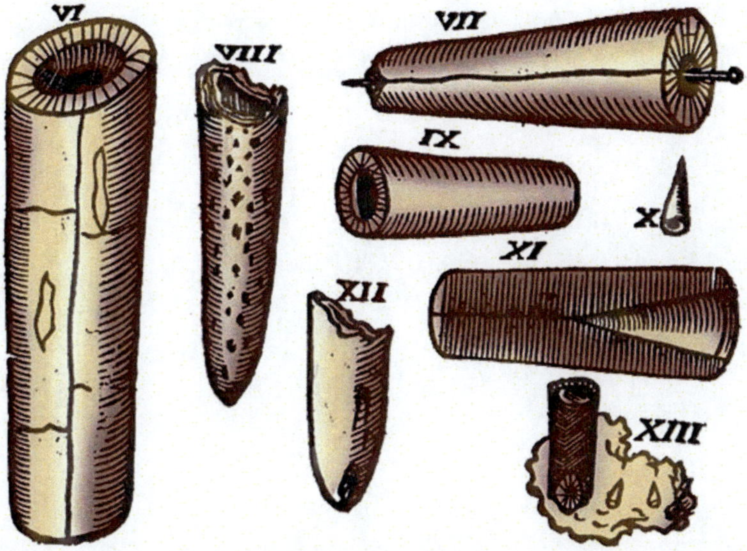

Fig. 16.8 Various portions of rostra proposed by Friedrich Lackmund, a scholar born in Hildesheim, Germany (1653–1676), in his *Oryktographia Hildesheimensis; sive, admirandorum fossilium quae in tractu Hildesheimensi reperiuntur, descriptio iconibus illustrata, cui addita sunt alia de calculis, de fontibus, etc* (*De Belemnite*, 1669, X: pp. 24–28). (Colour Maurizio Forli modified after: https://wellcomecollection.org/works/rs34c8db/items?canvas=55 Public domain)

Fig. 16.7 (continued) (not in copyright). (3) https://babel.hathitrust.org/cgi/pt?id=gri.ark:/13960/t8w96041q&view=1up&seq=254&skin=2021 (Public domain). (4) https://doi.org/10.3931/e-rara-15217 / Public Domain Mark. (5–6) https://doi.org/10.3931/e-rara-13017 / Public Domain Mark. (7) https://doi.org/10.3931/e-rara-21877 / Public Domain Mark. (8) https://doi.org/10.3931/e-rara-12116 / Public Domain Mark. (9) https://doi.org/10.3931/e-rara-35487 / Public Domain Mark. (10) https://www.biodiversitylibrary.org/bibliography/23488 (not in copyright)]

Fig. 16.9 Partial preservation of a belemnite phragmocone, in the Jurassic rocks of the surroundings of Sasso Rosso (Lucca). Length of the find ~80 mm. (Photo Maurizio Forli, courtesy Carlo Chirli collection)

Fig. 16.10 Rostra of *Belemnitella mucronata* (Schlotheim 1813), Upper Cretaceous, Ukraine. Length ca. 80 mm, Ø 10 mm. (Photo Maurizio Forli)

Fig. 16.11 Rostrum of belemnite, contained among other objects, in the bag of a sanpaolaro (quack in the name of St. Paul), preserved in the Bellucci collection. This fossil is the only certain and concrete example of a belemnite used in the apotropaic-medicinal field, in the Italian context, of which we are aware. We recall that the bag and its contents became part of the collection around 1884 (Coll. Bellucci, M.A.N.U (PG). Inv. 338 G. Reproductions by kind permission of the Ministry for Cultural Heritage and Activities and for Tourism, Regional Directorate Museums of Umbria – National Archeological Museum of Umbria) (Enlarged rostrum, not to scale)

References

Agricola G (1546) De ortu & causis subterraneorum Lib. V. De natura eorum quae effluent ex terra Lib. IIII. De natura fossilium Lib. X. De veteribus & novis metallis Lib. II. Bermannus, sive de re metallica dialogus. Interpretatio Germanica vocum rei metallicae, addito Indice foecundissimo. Basel: Hieronymus Froben.

Accordi B (1980) Michele Mercati (1541–1593) e la Metallotheca. Geol Romana 19:1–50

Aldrovandi U (1648) Ulyssi Aldrovandi patricii Bononiensis Musaeum metallicum in libros IIII: distributum Bartholomaeus Ambrosinus ... labore, et studio composuit cum indice copisosissimo Arcus Antonius Bernia. typis Io. Baptistae Ferronii, Bononiae. ETH-Bibliothek Zürich, Rar 1027, https://doi.org/10.3931/e-rara-13017 / Public Domain Mark

Boccone PS (1674) Recherches et observations naturelles touchant le corail, la pierre étoilée, les pierres de figure de coquilles, le corne d'ammon, l'astroïte Undulatus, [...]: examinées à diverses fois dans l'Assemblée de Messieurs de Societé Royale de Londres, & dans les conferences de Monsieur l'Abbé Bourdelot à Paris. chez Jean Jansson à Waesberge, A Amsterdam. ETH-Bibliothek Zürich, Rar 5705, https://doi.org/10.3931/e-rara-21877 / Public Domain Mark

Boodt A Boëtius de (1609) Gemmarum et lapidum Historia. Claudiam Marnium, Hanoviae. https://hdl.handle.net/2027/gri.ark:/13960/t8w96041q (Public domain)

Boodt A Boëtius de (1636) Gemmarum et lapidum historia. ex officina Ioannis Maire, Lugduni Batavorum [Leiden]. ETH-Bibliothek Zürich, Rar 1881, https://doi.org/10.3931/e-rara-15217 / Public Domain Mark

Boodt A Boëtius de (1644) Le parfaict Ioaillier ou histoire des pierreries: ou sont amplement descrites leur naissance, iuste prix, moyen de les cognoistre, et se garder des contrefaictes, Facultez medicinales, et proprietez curieuses. Iean-Antoine Huguetan, Lyon. ETH-Bibliothek Zürich, Rar 1010, https://doi.org/10.3931/e-rara-12490 / Public Domain Mark
De Laet J (1647) De Gemmis et Lapidibus, libri Duo. Quibus praemittur Theophrasti, liber de LapidìbusGraece & amp;amp; latine cum Brevibus Annotationibus. Ioannis Maire, Lugduni Batavorum. https://doi.org/10.3931/e-rara-35487 / Public Domain Mark
Ducrotay de Blainville MH (1827) Mémoire sur les Bélemnites, considérées zoologiquement et géologiquement. Levrault, Strasbourg
Ehrhart B (1727) Belemnitis Suevicis Dissertatio, qua imprimis in obscure hactenus fossilis naturam inquiritur; dein et haud paucae observationes universum marino terrestrium censum. David Raymund Merz & Johann Jacob Mayer, Augustae Vindelicorum
Flammarion C (1934) Il Mondo prima della Creazione dell'Uomo. Casa Editrice Sonzogno, Milano
Fuchs D, Lukeneder A (2014) A Cenozoic coleoids (Cephalopoda) from Austria – a review of Schultz's Catalogus Fossilium Austriae. Denisia 157:23–32
Gessner C (1565) De omni rerum fossilium genere, gemmis, lapidibus, metallis, et huiusmodi, libri aliquot, plerique nunc primum editi. Tiguri. https://doi.org/10.3931/e-rara-4176 / Public Domain Mark
Imperato F (1599) Dell'Historia Naturale di Ferrante Imperato napolitano. Libri XXVIII. Nella quale ordinatamente di tratta della diversa condition di miniere, e pietre. Con alcune historie di Piante, et Animali fin'ora non date in luce. Costantino Vitale, Napoli. https://www.biodiversitylibrary.org/item/183177#page/5/mode/1up (not in copyright)
Lachmund F (1669) Oryktographia Hildesheimensis sive admirandorum fossilium, quae In tractu Hildesheimensi reperiuntur, Descriptio Iconibus Illustrata, Cui addita sunt alia de calculis, de fontibus, & c. Jacobi Mülleri, Hildesheimii. https://wellcomecollection.org/works/rs34c8db (Public Domain Mark)
Lang KN (1708) Historia lapidum figuratorum Helvetiae, eiusque viciniae. Inqua non solum enarrantur omnia eorum genera, species et vires aeneisque tabulis repraesentantur, sed insuper adducuntur eorum loca nativa, in quibus reperiri solent, ut cuilibet facile sit eos colligere, modo adducta loca adire libireeat. Jacobi Tomasini, Venetiis
Lichtenthal P (1826) Dictionary and bibliography of music. Antonio Fontana, Milano
Luidus E (1760) Lithophylacii Britannici ichnographia sive Lapidum aliorumq; Fossilium Britannicorum singulari figura insignum, quotquot hactenus vel ipse invenit vel ab amicis accepit, Distributio Classica, Scrinii sui lapidarii repertorium cum locis singulorum natalibus exbidens. Additis rariorum aliquot figuris aere incisisis. Editio Altera. Typographeo Clarendoniano, Oxonii
Meignen L (ed) (1993) L'Abri des Canalettes: un habitat moustérien sur les Grands Causses, Nant (Aveyron). Éditions du CNRS, Paris
Mercati M (1717) Metallotheca Vaticana. Opus posthumum, Autoritate et Munificentia Clementis Undecimi Pontificis Maximi. E tenebris in lucem eductum: Opera autem, & amp; studio Joannis Mariae Lancisii, Archiatri Pontificii illustratum. Ex Officina Jo: Mariae Salvioni Romani, Romae. https://archive.org/details/bub_gb_DC_E6UVWg7EC (Public Domain Mark 1.0)
Miller JS (1826) Observations on belemnites. Trans Geol Soc Lond Second Ser 2:45–62
Passera F (1688) Il Nuovo Tesoro degli Arcani Farmacologici, Galenici. & Chimici, ò Spargirici. Giovanni Parè, Venetia
Plot R (1677) The natural history of Oxfordshire, being an essay toward the natural history of England. Printed at the Theater, Oxford. https://www.biodiversitylibrary.org/bibliography/23488 (not in copyright)
Scheuchzer JJ (1702) Specimen Litographiae Helveticae Curiosae, Quo Lapides ex Figuratis Helveticis Selectissimi Aeri incisi. Typis Davidis Gessneri, MDCCII, Tiguri. Zentralbibliothek Zürich, NNN 1883, https://doi.org/10.3931/e-rara-12116 / Public Domain Mark
Tramezzino M (1550) Di Giorgio Agricola De la generatione de le cose che sotto la terra sono e de le cause de' loro effetti e nature, lib. V; De la natura di quelle cose che dalla terra scorrono, lib.

IIII; De la natura de le cose fossili e che sotto la terra si cauano, lib. X; De le minere antiche e moderne, lib. II; Il Bermanno, o De le cose metallice, dialogo / recato tutto hora dal latino in buona lingua volgare. In Vinegia: per Michele Tramezzino. https://archive.org/details/hin-wel-all-00002012-001 (not copyright found)
Valentini MB (1704) Museum Museorum. Johann David Zunners, Frankfurt am Mayn
Vialli V (1974) Appunti di Paleontologia. Pitagora Editrice, Bologna
Walch JEI (1768–1773) Walch Johann Ernst Immanuel: Die Naturgeschichte der Versteinerungen: zur Erläuterung der Knorrischen Sammlung von Merkwürdigkeiten der Natur. gedruckt mit Felsseckerischen Schrifften, Nürnberg. ETH-Bibliothek Zürich, Rar 399, https://doi.org/10.3931/e-rara-10246 / Public Domain
Wimmer MF (1866) Theophrasti Eresii Opera quae supesunt, omnia graeca recensuit, latine interpretatus est. Ambrosio Firmin Didot, Parisiis
Worm O (1655) Museum Wormianum, seu Historia rerum rariorum, tum Naturalium, quam Artificialium, tam Domesticarum, quam Exoticarum; quae Hasniae, Danorum in aedibus Authoris Servantur. Iohannem Elsevirium, Lugduni Batavorum. https://doi.org/10.5962/bhl.title.63844. (not in copyright)

Chapter 17
Ammonoidea: From the Ram's Horns of the God Jupiter Ammon to the Ammonites

Abstract The Ammonoidea constitute a subclass of the Cephalopoda, included in the Phylum Mollusca. They appear in the Devonian and spread in all the seas with a wealth of species of the most varied forms, the most common of which is a flat spiral with a degree of coiling more or less covering the successive whorls. Because of their cosmopolitan spread and great specific diversification in a relatively short time are used as guide fossils. They became extinct at the end of the Cretaceous. Anciently called Horns of Ammon, curled ram's horns attribute of the Egyptian god Amon-Ra, from which will derive then the actual name of ammonites, they are mentioned by Pliny as Ammonis corni, a gem of golden color, rather obscure definition that will be cause of many doubts about their exact nature until the beginning of the nineteenth century. During the Middle Ages and beyond, until the end of the seventeenth century, they were believed to be remains of petrified snakes. It is under this guise that they gave rise to numerous superstitions and their use as thaumaturgical objects. Their first image in a printed book on fossilia is in Gessner (*De omni rerum fossilium genere, gemmis, lapidibus, metallis, et huiusmodi, libri aliquot, plerique nunc primum editi. Tiguri.* https://doi.org/10.3931/e-rara-4176 / Public Domain Mark, 1565).

Keywords Fossilia · Fossils · Ammon's Horn · History · Ammonites

Ammonoidea constitute a subclass of Cephalopods, included in the Phylum Mollusca. They appear in the Devonian and spread throughout the seas with a wealth of species with the most varied shell shapes, with composition aragonitic. Some, typically cretaceous, called heteromorphic, have atypical, developed or more complex coiling (baculicona, cryocona, scaficona, torticona, eg.), but the most common morphology is a flat spiral with a degree of coiling covering more or less or successive whorls. It is precisely in this guise that during the Middle Ages and beyond for much longer, that they gave rise to numerous superstitions and their use as thaumaturgical objects (Figs. 17.1, 17.2, and 17.3).

© The Author(s), under exclusive license to Springer Nature Switzerland AG 2022

M. Forli, A. Guerrini, *The History of Fossils Over Centuries*,
https://doi.org/10.1007/978-3-031-04687-2_17

Fig. 17.1 *Stones of a form similar* [...] *to the serpent in itself wrapped up; but for the great resemblance still, which has with the horn of Aries, are called horn of Ammon, in that way formed by Nature* (Stelluti 1637: p. 11). Jurassic Ammonite (1) from Sassorosso (Lucca, Italy) with a viperid (2). [Photo credit (for the fossil) Massimo Cresti, (for the snake) Matteo Riccardo Di Nicola. Graphic composition Maurizio Forli]

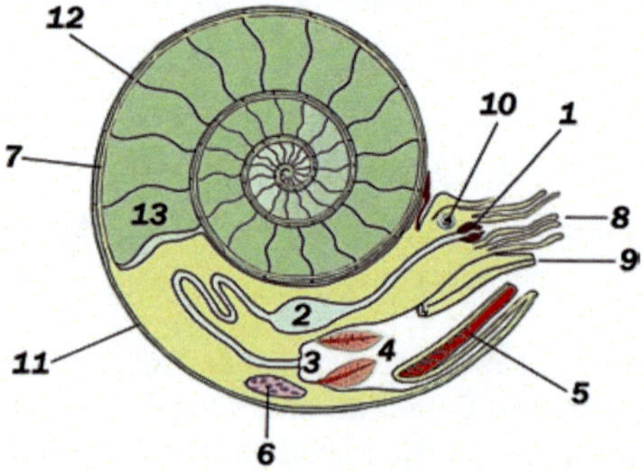

Fig. 17.2 Simplified morphological scheme of a hypothetical ammonite: (1) Oral aperture, beak-like jaws. (2) Stomach. (3) Anal aperture. (4) Gills. (5) Aptychus. (6) Gonad. (7) Siphuncle. (8) Anterior part of mollusk, tentacles. (9) Hyponome. (10) Eye. (11) Concamerated shell. (12) Septum. (13) Dwelling chamber. [Drawing Maurizio Forli modified after Vialli (1974)]

Fig. 17.3 Simplified external morphological scheme of the shell: (A) Ventral side in adapical view, that is, toward the center of the umbilicus. (B) Ventral side in abapical view, that is, toward the opening. (C) Lateral or umbilical view; (1) Oral aperture. (2) Whorl height. (3) Umbilical diameter. (4) Maximum diameter. (5) Apex. (6) Part of the spiral whorl partially covered by the next one. (7) Part of free whorl. (8) Ribs. (Central photo Maurizio Forli, lateral side modified after: https://commons.wikimedia.org/wiki/File:Ammonite_morfologia_3.JPG Public domain)

Their dimensions are very variable, reaching the maximum around three meters in diameter. Like all cephalopods, the shell is always formed by three main parts with morphological changes characteristic for its subclasses: an initial chamber called protoconch, which represents the shell in the larval state, a portion divided by septa, the phragmocone, and a terminal chamber where the mollusk resided, the *body chamber*.

All the chambers were crossed by a siphuncle, usually marginal and ventral, whose external portion, called ectosiphuncle, is formed by a folding of the septum called prochoanitic septal neck, that is turned toward the opening even if rarely it could be turned toward the apex, and several small tubes, connection rings, with porous wall which originated from the protoconch and connected the septal neck between them. Its internal part, called endosiphuncle, allowed the mollusk to be physiologically connected to all the chambers of the shell, thus being able to fill or empty the chambers of the liquid contained therein and to regulate, through an osmotic process, the internal pressure for maintaining the hydrodynamic balance and the relevant way of floating.

The Ammonites had developed one or two opercular structures, calcareous or horny and with different morphology and sculptures, which allowed them to close their own opening. These opercula, sometimes abundant as isolated fossils, are instead rather rare preserved in place together with the shell. The single operculum is defined anaptic (*anaptycus*), if they are paired, haptic (*aptychi*), paired and soldered, and synaptic (*synaptychi*) (Fig. 17.4).

Fig. 17.4 Singles and one pair of aptychi. The singles from Furlo surroundings (Pesaro and Urbino, Umbria, Italy) Jurassic, and the pair from generic "Kimmeridgian" of Umbria. (Photo credit Stefano Dominici, graphic composition Maurizio Forli)

Sexual dimorphism is manifested with large shells for females and small ones for males, while the ornamentation is constituted by relief elements of a very varied type. Especially in the Paleozoic, the shell is generally smooth and shows only growth lines. Starting from the Triassic and throughout the Mesozoic, more and more complex structures develop, especially with a radial course, such as ribs, constellations, nodes, tubercles, bullae, clubs, spines, and horns. A character well observable in fossil specimens preserved as internal models is the suture lines. They are derived from the intersection of the septa with the inner surface of the shell along a line, also called lobal line. In Ammonites, unlike nautiloids, the suture lines are usually complex. The course of this character reflects the introflections and extroflections of the mantle during the construction of the septa, with different courses according to the groups. The suture lines are a very important diagnostic character for the taxonomy of these mollusks and in their course we can distinguish two main elements, the saddles and the lobes: the saddles are inflections turned toward the opening, that is forward, with the convexity toward this last and the concavity in the opposite direction called adapical, while the lobes are inflections turned toward the apex of the shell, concave in aboral direction, that is toward the opening.

The suture lines represent the *fingerprints* of each species of ammonite and, for this characteristic, they have been much studied and subdivided differentiating them according to morphology, from the simplest and most primitive to the most complex and evolved. The so-called *Goniatitic* suture, with linear, non-jagged lobes and saddles, that is, without secondary subdivisions, even if very rarely the ventral lobe may present denticulate structures, is dominant in the Paleozoic, and disappears in the Triassic, reappearing in the Middle Cretaceous. The *Ceratitic* suture, with linear saddles and indented lobes, appears in the Lower Carboniferous (Mississippian) and during the Upper Carboniferous transitional forms develop between Ceratitic and Ammonitic suture characterized by saddles and indented or jagged lobes, often subdivided into accessory elements, in structures sometimes very complex. The first appearance of this particular typology dates back to the Permian, even if in this period the goniatitic suture still dominates, while starting from the Upper Trias there is a wide diffusion of the *Ammonitic* suture until the Jurassic, when only forms with ammonitic suture are known until the Middle Cretaceous, when the ceratitic and goniatitic sutures reappear together with forms with a new type of very complex suture called *Phylloid*, characterized by saddles subdivided into more or less leaf-shaped elements, with a high number of umbilical lobes, up to a maximum of six, and a geometry similar to the fractal one (Figs. 17.5 and 17.6).

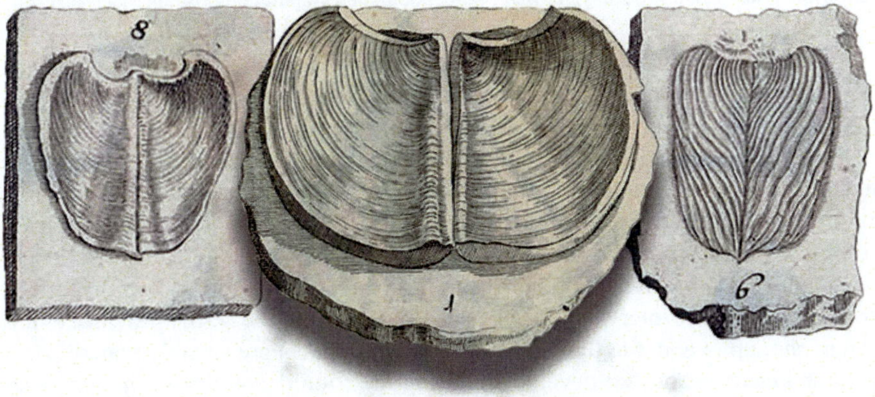

Fig. 17.5 Aptychi of ammonites interpreted as fossil bivalves (Baier Ferdinand Jakob, *Oryctographiae noricae supplementi*, 1757: pl. XIV). (Graphic composition Maurizio Forli modified after: https://doi.org/10.3931/e-rara-14953 / Public Domain Mark)

Fig. 17.6 Main morphologies of suture lines: (1) Goniatitic suture, Goniatite indeterminate, Erfoud, Morocco, Devonian. (2) Ceratitic suture, (detail), *Ceratites spinosus* (Philippi 1901) Lindach, Germany, Triassic, [Photo credit Geocollection) (3) Ammonitic suture (detail), *Calliphylloceras* sp., Umbria, Italy, Jurassic (Other photos and graphic composition Maurizio Forli]

Their attitude in life was with the shell more or less vertical and the opening at the bottom like the present Nautili. They moved thanks to the hyponome, a very mobile funnel-shaped muscle, located in the ventral region, through which the water was forcefully expelled from the cavity of the mantle, the violent expulsion determined the retrograde movement of the mollusk. The hyponome leaves on the peristome a mark called hyponomic sinus which, usually, in the post-Paleozoic forms, is replaced by a rostrum. Depending also on their morphostructural characteristics, shell thickness, ornamentation, and greater or lesser spiral coiling, they lived in a variable bathymetric interval of the oceanic domain, from the neritic province to the pelagic one, up to about 850 m of depth. The Ammonites presumably nourished of fishes and crustaceans, but developed, during their long evolutionary history, also other alimentary adaptations, such as the microphagy, the necrophagy, and even the cannibalism. In their turn, these mollusks were preyed by marine reptilians or by bigger cephalopods.

In Italy and in the Alpine chain in general, there is a formation deposited from the Middle Triassic to the Upper Jurassic, composed of limestone and marly limestone, sometimes badly stratified, with a nodular texture, generally characterized by a considerable frequency of fossil ammonites, red or pinkish in color, sometimes with violet and green tones, called "Rosso Ammonitico."

Phylum Mollusca
Class Cephalopoda
Superfamily Hildoceratoidea †
Family Hildoceratidae †
Genus *Hildoceras* †
Species *Hildoceras bifrons* Bruguière, 1789 †

The question of the exact origin of the name of these fossil cephalopods, even if substantially concordant among the various authors who have given information about them, is, however, rather debated because of the many different versions that are reported about them. We thought it appropriate to stick to the Plinian text and to the Servian one quoted by Aldrovandi, who was the first author to use the term ammonite for *Ammonis cornu.*

Servius (*Servius Marius Honoratus*), Roman grammarian and commentator, who lived at the end of the fourth century, in his apparatus of notes of the book IV of Virgil's Aeneid, gives us his version of the reasons at the origin of the etymon of those stones which will take later the name of *Ammon's horns* and later of *ammonites*, quoting the following legend:

> *Filius Iovis Ammonis. Liber quum Indos peteret, et per Lybiam exercitu duceret fatigatu siti, Iovis sui patris imploravit auxilio, et statim viso ariete fons secutus est ab arenis: unde factum est Iovi Ammoni ab arenas dicto simulachru* cum *capite arietino. ἀμμος enim Graecè arena dicitur, quòd ideo fingitur, quia satis eius sunt involuta responsa* (Servius, Edition H. Curion 1544: p. 305).

The following is a quote from the passage reported by Aldrovandi:

> *Ammonis cornu dicitur, propterea quom* cum *Bacchus, ex sententia Servij in Virgilium, unà* cum *exercitu Indiam peteret, in arenis Lybiaea siti stimulatus, Iovis auxilium imploravit & illico visus Aries illum ad fontem duxit, ubi in monumentum tanti beneficij, simulacrum Iovis* cum *capite arietino fabricatum est, & Iuppiter Ammon ab arena cognominatus,* cum *ἀμμος arenam significet. Qua mobrem rerum naturalium observatores lapides ex arena a natura consictos in modum arietini cornu, cornu Ammonis indigitarunt.* (*Musaeum metallicum*, 1648: p. 53) and its translation
>
> *usually called Ammon's Horn for the reason that Bacchus, according to what Servio says in Virgil, heading with his army to India and stimulated by thirst in the desert of Libia, implored the help of Jupiter. Immediately a ram appeared and led him to a spring, where a statue of Jupiter with the head of a ram was erected in memory of such a benefit, called Jupiter Ammon, since Ammos means sand. Therefore the observers of natural things indicate the* stones composed by the work of nature *in the manner of a ram's horn as* Ammon's Horn (Sarti 2003: p. 162).

The origin of the coiled ram's horn that adorned the temples of the Roman god *Juppiter Ammon* or *Jupiter Ammon*, is derived from the Egyptian Theban god *Amon* then joined by the twelfth dynasty with the Sun God of Heliopolis, *Ra,* thus becoming *Amon-Ra*, one of the most important deities of the Egyptian pantheon. The story reported by Servius draws a historical basis from an episode in the life of Alexander the Great following his conquest of ancient Egypt and its proclamation as the new Pharaoh. In 332 B.C., Alexander went with his army to the oasis of Siwa, (currently in the Egyptian desert on the border with Libya, where there was, and still partially preserved, a temple dedicated to the god Amon, built about 200 years before by Pharaoh Amasis), to listen to the divination of the local oracle, which declared him the son of the god Amon, a title due to the pharaoh, and consequently, in Greek mythology, the son of Jupiter Ammon. His image with ram's horns is quite widespread and is also shown on some ancient coins. The reference to *Bacchus*, comes from one of the many names of Alexander, who called himself Diòniso, the Bacchus of Latin mythology (Figs. 17.7 and 17.8).

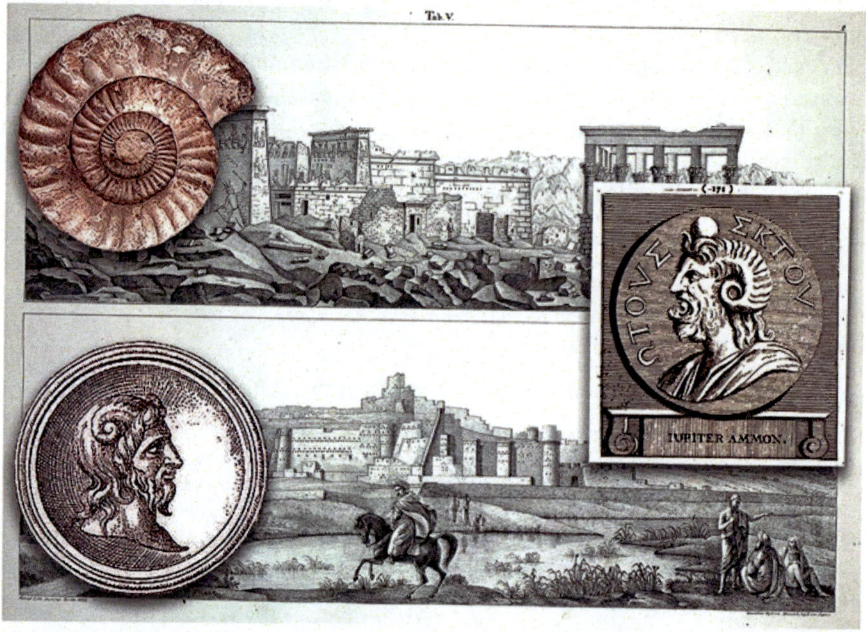

Fig. 17.7 Nineteenth-century plate with the representation of the temple of the oracle of Amun (or sanctuary of Zeus Amun) and the oasis of Siwa. Jupiter Ammon represented in the classic iconography with ram's horns, these last, compared with an ammonite from the French Jurassic. (On the background (1) the temple of the Oracle and the oasis of Siwa, from von Minutoli, *Reise zum Tempel des Jupiter Ammon, in der Libyschen Wüste* 1824: pl. 5. Juppiter Ammon: on the left (2) from *L'Art pour Tous*, 1862: 37, p. 147, to the right (3) from *Favissae utriusque Antiquitatis* 1707: p. 171). [Graphic composition Maurizio Forli. Credit from: (1) https://digitalcollections.nypl.org/items/510d47e2-6f41-a3d9-e040-e00a18064a99 (Public domain) (2) https://digi.ub.uni-heidelberg.de/diglit/art_pour_tous1862/0059 (3) https://libraryimage.nga.gov/uv/?manifest=https://libraryimage.nga.gov/manifest/mms/991836413504896.json#?c=0&m=0&s=0&cv=183&xywh=-1%2C-1455%2C7730%2C7709 (Public domain)]

Fig. 17.8 ***Alexander, not satisfied with the human condition, wanted us to believe he was descended from Jupiter;* [...] *therefore beyond to the royal diadem sees itself with the horns of ram in conformity the simulacrum of Jupiter Ammone represented from those of Egypt*. (Canini, *Iconografia*,** 1669: **p. 386). Portraits of Alexander: (1) Drawing of coin, recto and verso (Lachmund (*Orictographia Hildesheimensis*,** 1669: **p. 33). (2) ancient silver coin (3) Alexander with Ram horns (Canini, *Iconografia*,** 1669: **p. 36). [Graphic composition Maurizio Forli. Credit from: (1)** https://wellcomecollection.org/works/rs34c8db/items?canvas=61 **(Public domain)**
(2) https://commons.wikimedia.org/wiki/File:Alexander-Coin.jpg **(Public domain)**
(3) https://babel.hathitrust.org/cgi/pt?id=dul1.ark:/13960/t6350084m&view=1up&seq=91&skin=2021 **(Public domain)]**

Pliny the Elder writes about it in volume XXXVII of his *Naturalis Historia*. In this book dedicated to gems and precious stones, without, however, specifying anything else that can make us understand the correspondence with the objects we know, he says that the *horn of Ammon* makes us dream of future things:

> *Ammonis cornu inter sacratissimas Aethiopae gemmas, aureo colore, arietini cornu effigiem reddens, promittitur praedivina somnia repraesentare.*
>
> Translated by Domenichi (1844: p. 1486) in *Il corno d'Ammone, che va tra le sacratissime gioie d'Etiopia, è di colore d'oro, e ha forma di corno di montone. Dicono ch'ella eccita sogni che presagiscono l'avvenire.*
>
> *The horn of Ammon, which goes among the most sacred jewels of Ethiopia, is golden in colour, and has the shape of a ram's horn. They say that it excites dreams that presage the future.*

In reality, so far it has been impossible to ascertain that Pliny's gems actually correspond to ammonites, which, among other things, on the basis of what is written, *a golden colour*, should be pyrite. Many authors have analyzed the story, for example, Johann Reiske, in his book *Exercitatio historico-physica de cornu ammonis* (1689), retraces this story with the texts of Pliny and the ancient Greeks, comparing it with the opinions of his contemporary scholars, coming to the conclusion that Pliny's definition is vague, the story has no foundation and there is no certainty of what he meant when speaking of *Cornu Ammonis*. In a recent article by Vavra (2004) it is argued that rather than being an ammonite, Pliny's *Cornu Ammonis* is almost certainly an Eocene gastropod. Indeed, all those petrefacts in the shape of arietine or bovine horns were considered Ammonian horns. Of course, from the point of view of the current definition, when by now the concept of ammonite is consolidated and universally accepted, all this is just a historical exercise, a semantic paleontological curiosity.

The passage from a totally positive interpretation to a negative one is due to a cultural process in which the ammonites, especially during the medieval period, for their resemblance to the horns of the rams and the coiled snakes, become symbols of the struggle between good and evil, between God and the Devil. The ram, or better the bearded goat, whose horns are in the image of the ammonites, is considered one of the devilish animals under whose form the lord of darkness manifests himself; furthermore, some of the most common species of ammonites, and in particular the planispiral ones, can actually resemble the shape of a coiled snake and this is how they were often represented and considered as work or concrete presence of the Devil, symbol and transmutation of the snake itself.

Obviously, the *stone serpents* did not have the head and for confirming such a rumor, in some cases the final part of the coil has been carved as resemblance, in order to induce to believe in the capacity of the saints to transform the serpents in stone and therefore to be capable to defeat the Evil. One of these particular modified ammonites is depicted in *The Mineral conchology of Great Britain* by the English naturalist Sowerby, who in 1821 (pl. 106, fig. 2), also intrigued by the legends about ammonites, had a specimen with a modeled head brought from the surroundings of Whitby, now kept in the British Museum in London. This characteristic representation was particularly widespread in the Elizabethan period, that is, between the middle of 1500 and the beginning of 1600.

In the Anglo-Saxon folklore, between the numerous saints capable of this miracle, is mainly remembered St. Hilda, who lived in mid-600 AD, but her legend seems to have originated around the mid-1500. This Saint, in Whitby, England, disinfested the place where later on an abbey was to be built, transforming countless evil serpents in stone. The scientific name of a genus of these molluscs, *Hildoceras,* is honored to it with the meaning of *Hilda's horn* from its name and from *ceras*, Greek word (κέρας), whose meaning is horn. For those interested in having more information about St. Hilda we recommend the text by Alfred Kracher (2012).

Another saint, about whom, however, less information is available, is St Keyna, who at Keynsham near Bristol performed the same miracle as Hilda by petrifying the snakes of a wood with her prayers. This information and the other concerning St Hilda are mentioned by William Camden in his work *Britannia* (1590), on page 169, respectively:

> *Secundum eundem Avonam (qui nunc limes est inter hunc et Glocestrensem agrum) in occidentali ripa Cainshan cernitur,* id est *Cangorum, ni fallor, habitatio. Sunt tamen qui existimant à Keina Britannica Virgine Sanctissima denominatam, quam serpentes in lapides mutasse superiori aeui credulitas multis persuasit, eo quod eiusmodi ludentis naturae miracula ibi in latomijs nonnunquam inveniantur. Vidimus enim lapidem hinc delatum serpentis in spiram revoluti effigie; cuius caput in circumferentia prominuit, estrema cauda centro occupante*
>
> and on page 585 *Lapides hic inveniuntur, serpentium in spiram revolutorum effigie, naturae ludentis miracula; quae, ut inquitille, Natura* cum *veris et feriis negotiis quasi fatigata, ludendo efformat. Serpentes olim suisse crederes, quos lapideus cortex interxisset. Hildae autem precibus adscribit credulitas, tanquam illa commutasset,*

where describing the *snakes,* he says that they are found coiled with the head at the outer extremity and the tail at the center of the coil, not sparing some irony about the credulity of the people. Always in the Anglo-Saxon area, for the same miraculous capacities, were very much considered also St. Cutbert, who freed the island of Lindisfarne (Northumberland, North East of England), and St. Patrick, who disinfested Ireland. Regarding the formation of these beliefs, we cannot help connecting them to the Celtic culture, dominant for a long time in the area of Central-North Europe, which has, among others, as distinctive sign, just the meanders and the snake-like circles.

Petrified snakes are many, so it is not surprising that where ammonites abound, they are found in as much quantity. On the French side of the Channel, for example, on the lintel of the door of the chapel dedicated to St. Peter in the cathedral of Bayeux, in the fifteenth century, a large ammonite was affixed with a Latin inscription underneath, explaining that it was a snake transformed into stone by a miracle (Fig. 17.9):

Fig. 17.9 The ammonite hanging on the lintel of the door of the chapel dedicated to Saint Peter, in the cathedral of Notre-Dame of Bayeux (France). (Photo credit Arnaud Brignon. For other information about see Brignon 2020)

> *Credite mira Dei, serpes fuit hic lapis extans.* Sic *trasformatum Bartholus attulit hùc*

which translated means

> *Believe the miracle of God, I was a snake now stone protruding (hanging). So transformed Bartolo adduced (offered) in this place.*

The custom of placing ammonites on the walls of buildings, especially above doors, for protection against misfortune and evil spells is documented in other parts of France, and perpetuated in recent times, naturally favored by the availability of fossil specimens, induced by the legacy of ancient beliefs and undoubted aesthetic beauty. Ammonites were often placed on the doors of barns and stables. Similar rites of protection, especially for cattle, performed with the help of ammonites, are reported for other places, for example, on the Isle of Skye in Scotland, where during the sixteenth and seventeenth centuries cows' cramps were cured by bathing the aching parts of the animals with water in which ammonites, called crampstones, had been immersed for a few hours (Duffin 2008). In some regions of Germany, called Drachenstein, Dragon Stones, were considered as a remedy against evil and were used to cure cows when they had no more milk, by putting an ammonite in the collection bucket. In this way the witches could no longer drink it and perform their malevolent spells. The association between *Ammonis cornu* and *Draconites* is due to Albertus Magnus (1200 ~) who, transposing the term from Pliny, describes the stone that was believed to be extracted from the head of the dragon at the same time of its killing *Draconites autem lapis est, à capite draconis extractus*, as a stone in the effigy of a snake. Pliny's *Draconites* was probably a dark-colored pyramidal glassy gemstone, while Albertus Magnus's, although also dark-colored and pyramidal, is a serpentiform stone, perhaps a turreted ammonite from the Upper Cretaceous (Accordi 1981: 143). One of these precious stones, later given as a gift to Albertus Magnus himself, according to the account of its discovery, came from a meadow in the Swabian mountains, under the head of a large snake, where there was a rock (almost certainly a rocky slab with copious fossilized ammonite remains), on which more than 500 of them stood, cut into pieces by the soldiers of the local lord. (1569: pp. 135–136). It is not quite clear what was meant by *Draconites*, except that it was a gemstone with fantastic powers, including that of nullifying any kind of poisoning, as it has been described as a stone of the most disparate forms and characteristics, for example, in Spada (1744: p. 61) six different qualities are reported for the Veronese alone.

The local limestones, known universally as *Red Marble of Verona*, are fossiliferous and it is not uncommon to see in them, elegant and evident spiral sections, of the shells of cephalopods ammonoidea in good evidence. Being used in the past, mainly for constructions of a certain value, it has been very much used for palaces and churches for ornamental purposes and, especially in places of worship, one wonders if the serpentiform spirals of the ammonites have been in some way related to the construction and the internal environmental expressions, strictly linked to religiosity or have been placed by chance. The placement of some marble tiles containing ammonites in strategic positions related to the arrangement of statues of saints and of the Madonna, near baptismal fonts or altars, is documented, for example, in some churches in the Brescia area, in particular, in the parish church of Rovato, where a tile containing an ammonite was significantly placed just in front of an altar where there is a statue of the Madonna crushing a snake (Dimitriadis et al. 2006). Ammonites, in some churches, were also used directly as containers for holy water and this particular use refers to the same symbolism of some stoups in which a snake is carved, often with remarkable naturalistic movements, as a snake submerged in water or unable to get out, warning and reassurance of the superiority of the Faith over the Evil one for believers who dip their fingers in it at the entrance of the house of God, *...the Devil and the Holy Water...*.incompatible with each other but invariably united (Fig. 17.10).

Fig. 17.10 A naturalistic and suggestive effigy of snake from the Acquasantiera of the Cathedral of Santa Maria Assunta of Guardialfiera (Campobasso, Italy). (Photo credit Franco Valente)

Of course, like many other naturalia and mirabilia, marvellous and strange natural objects, ammonites, also believed to be bearers of beneficial and magical effects, were used for pharmaceutical purposes. Some quotations concerning the medicinal use of *Ammonite horns*, for example, as an antacid in stomach ailments, are, however, to be attributed to other fossils with which they have been confused, sometimes mistaking them for the remains of fossilized tusks of ancient elephants. On this subject, we propose what was written in his time by Paolo Silvio Boccone (1697: p. 62):

> *Dell'Unicorno Fossile. Perchè questo Fossile è noto al Volgo sotto nome di Leocorno Minerale, e agli intelligenti, e Fisici sotto quello di Corno di Amone, di Litomarga, e di Lapis Arabicus non mi dilatarò con lungo ragionamento in questo luogo.* Circa *la sua Origine si può ridurre ugualmente a finissima marga, legata insieme, e disposta a figura di Corno, ò di Dente recurvo, che a Parte distinta, e speciale d'Animale vivente, e poi alterata, e calcinata nelle viscere della Terra, atteso che la Natura più volte ha prodotto per mezzo delle sue sinuosità, e cavità coniche, che s'incontrano nella Terra, quello stesso, che risulta dalla figura; e Calcinazione del Dente degli Animali, ed in specie dell'Elefante, sepolto nella medesimaTerra; E per i suoi salutari effetti è stata ascritta l'una, e l'altra Produzione, ò l'una, e l'altra alterazione fra i Medicamenti Antacidi. Trovasi in Palermo, (ed in altre Contrada della Sicilia) in un luogo detto li Pirreri.*

> *Of the Fossil Unicorn. Because this Fossil is known to the People under the name of Mineral Leocorn, and to the intelligent, and Physicists under that of Horn of Amon, of Lithomarga, and of Lapis Arabicus I will not dilate with long reasoning in this place.*

Regarding its origin, it can be reduced to a very fine marl, bound together and arranged into the shape of a horn or curved tooth, which is a distinct and special part of a living animal, and then altered and calcined in the bowels of the Earth, since Nature has often produced through its sinuosity and conical cavities, which meet in the Earth, the same thing that results from the figure; and Calcination of the Tooth of Animals, and especially of the Elephant, buried in the same Earth; and for its salutary effects has been ascribed the one, and the other Production, or the one, and the other alteration among the Antacid Medicaments. It is found in Palermo, (and in other parts of Sicily) in a place called Pirreri.

In this locality of the surroundings of Palermo, from a geological point of view, there are no Mesozoic rocks, so it is to be excluded that in this case, the Horn of Ammon is to be referred to one of these cephalopods, while it is certain that this name can be attributed to the elephant's tooth, present in the fossil state in the Pliocene-Pleistocene soils of the area. This possible identification of Ammon's Horn with the elephant remains, and not with the real ammonites, could have been made also by other authors, starting from Ferrante Imperato, as we will see shortly below.

As far as we know, the first printed representation of an ammonite in a book is that of Conrad Gessner, who illustrates an unidentifiable shell resembling a *Lythoceras*, on page 159 of his *De omni rerum fossilium genere, gemmis, lapidibus, metallis* published in 1565, while a first description is given on page 31 of the same volume under the heading *Ammonis Cornu*. On the following page 164 a pyritized shell is illustrated, of an iron-like color, corresponding to an ammonite of the Genus *Amaltheus*, in all probability an *A. margaritatus* Montfort, 1808, a guide fossil of the Pleinsbachian (~190.8 and ~182.7 Ma) in the Lower Jurassic. The same illustration is also given in Boëtius de Boodt Anselmus (*Gemmarum et lapidum historia*, 1636: p. 437) and *Le parfaict ioailllier ou histoire de pierreries* (1644: p. 562). In the pages immediately following Gessner's work, other ammonites are presented, one coming from Veneto, pinkish in color, sent to him by the Veronese Domenico Montesauro and Francesco Calzolari, sometimes interpreted by recent authors as an internal model or gastropod *tout court* and, a few lines later, in figure A of the following page, a block of rock with some shells partially embedded is shown, very similar to some *Dactylioceras*, like the ones in our figure; moreover, on page 167, another ammonite is shown as a *snake-shaped stone*.

This is the ammonite depicted in Fabio Colonna (*Aquatilium et terrestrium aliquot animaliu, aliarumquae naturalium rerum observationes*, 1616, p. XXXVI, lower left), which is supposed to be the oldest representation in Italian printed books. This is a specimen described on page XLIX, also belonging to the genus *Amaltheus*, which he saw among the natural and mineral things in the collection of Alessandro Ciombolo, from Chieti. This genus is characterized by the peculiar shape of the siphonal canal, externally raised at the periphery of the shell, resembling a *twisted* cord*: some of them have a crest accompanying their length similar to a twisted cord* (Ferrante Imperato, *Dell'Historia Naturale*, 1599: p. 667) (Figs. 17.11, 17.12, 17.13, 17.14, 17.15, 17.16, 17.17, 17.18, and 17.19).

In 1622 Benedetto Ceruti and Andrea Chiocci, in their catalog *Musaeum Franc. Calceolari jun veronensis*, reproduce on page 424 the figure of an ammonite

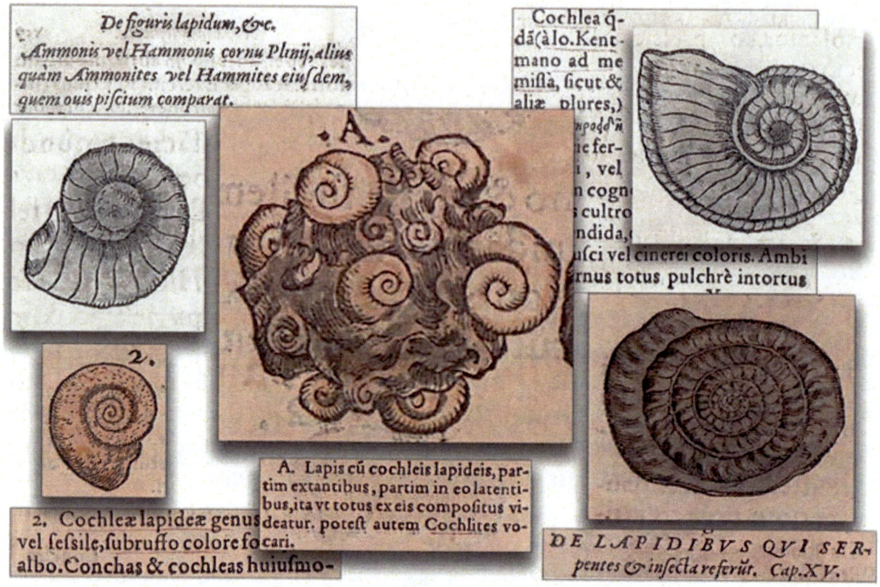

Fig. 17.11 From *De omni rerum fossilium genus ...* (1565): top left the first ammonite in the book (p. 159), top right the specimen of *Amaltheus* (p. 164); bottom left the specimen in Rosso Ammonitico from Veneto (p. 165); bottom right the ammonite representing a petrified ophid (p. 167); in the center the block containing several small specimens (p. 165). (Graphic composition Maurizio Forli credit from: https://doi.org/10.3931/e-rara-4176 / Public Domain Mark)

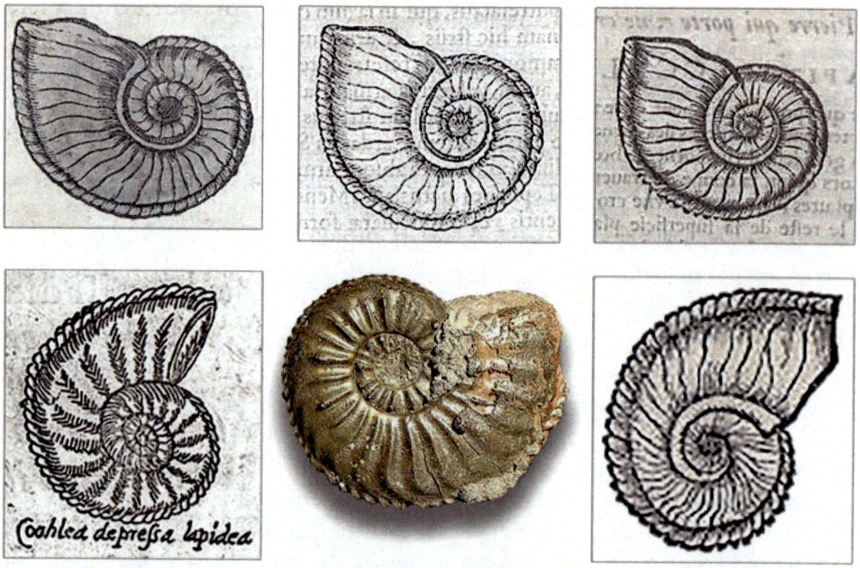

Fig. 17.12 Depictions of *Amaltheus* sp.; top, from left to right (1) Gessner (1565: p. 164). (2) De Boodt (1636: p. 457). (3) De Boodt (1644: p. 562); bottom, from left to right (4) Colonna (1616: p. 36), in all likelihood the oldest representation in Italian printed books. (5) a pyrite specimen from the Lias of Eislingen, Germany, diameter ~1 cm. (6) Worm (1655: p. 91). [Graphic composition Maurizio Forli credit from: (1) https://doi.org/10.3931/e-rara-4176 / Public Domain Mark (2) https://doi.org/10.3931/e-rara-15217 / Public Domain Mark (3) https://doi.org/10.3931/e-rara-12490 / Public Domain Mark (4) https://www.biodiversitylibrary.org/bibliography/723 (Public domain) (5) (Photo credit Geocollection) (6) https://doi.org/10.5962/bhl.title.63844 (not in copyright)]

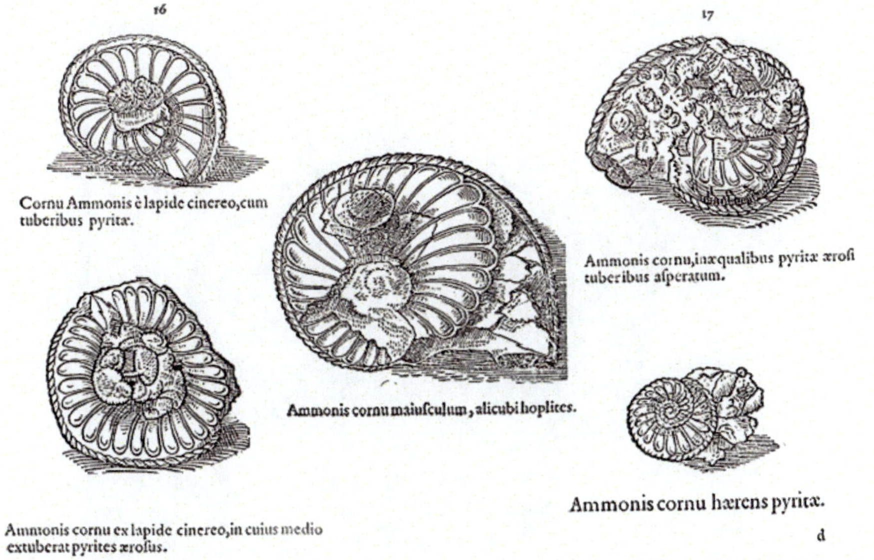

Fig. 17.13 Other pyritized ammonites of the same type as the above are found in the fourth book of Historia novi et admirabilis fontis balneique Bollensis in Ducatu Vvirtembergico ad acidulas Goepingenses by Johann Bahuin (1598). (Modified after: https://archive.org/details/historianoviet-ad00bauh/page/n4/mode/2up)

Fig. 17.14 *Ammonis cornu hoplites multiplex, obscurè lucens, in lapide fissili*. In the foreground, pyritized ammonites, from Bahuin (1598, IV: p. 7). In the background, a similar find with ammonites of the genus *Dactylioceras* from Holzmaden, Germany. (Photo credit Massimo Cresti, graphic composition Maurizio Forli, credit from: https://archive.org/details/historianoviet-ad00bauh/page/n4/mode/2up)

Fig. 17.15 Two arenaceous little blocks with some Jurassic ammonites, probably of the species *Dactylioceras commune* Sowerby, 1815, from Germany. Above the illustration by Gessner (1565: p. 165, verso) described as rock composed of shells partly covered and partly exposed, therefore called Cochlites, "*Lapis cù cochleis lapideis, partim extantibus, partim in eo latentibus, ita ut totus ex eis compositus videtur. potest autem Cochlites vocari.*" Below is a similar find (width ca. 170 mm) from Schlaifhausen, Germany, Lower Jurassic, Toarcian, age ~ 180 Ma. (Photo and graphic composition Maurizio Forli, credit from: https://doi.org/10.3931/e-rara-4176 / Public Domain)

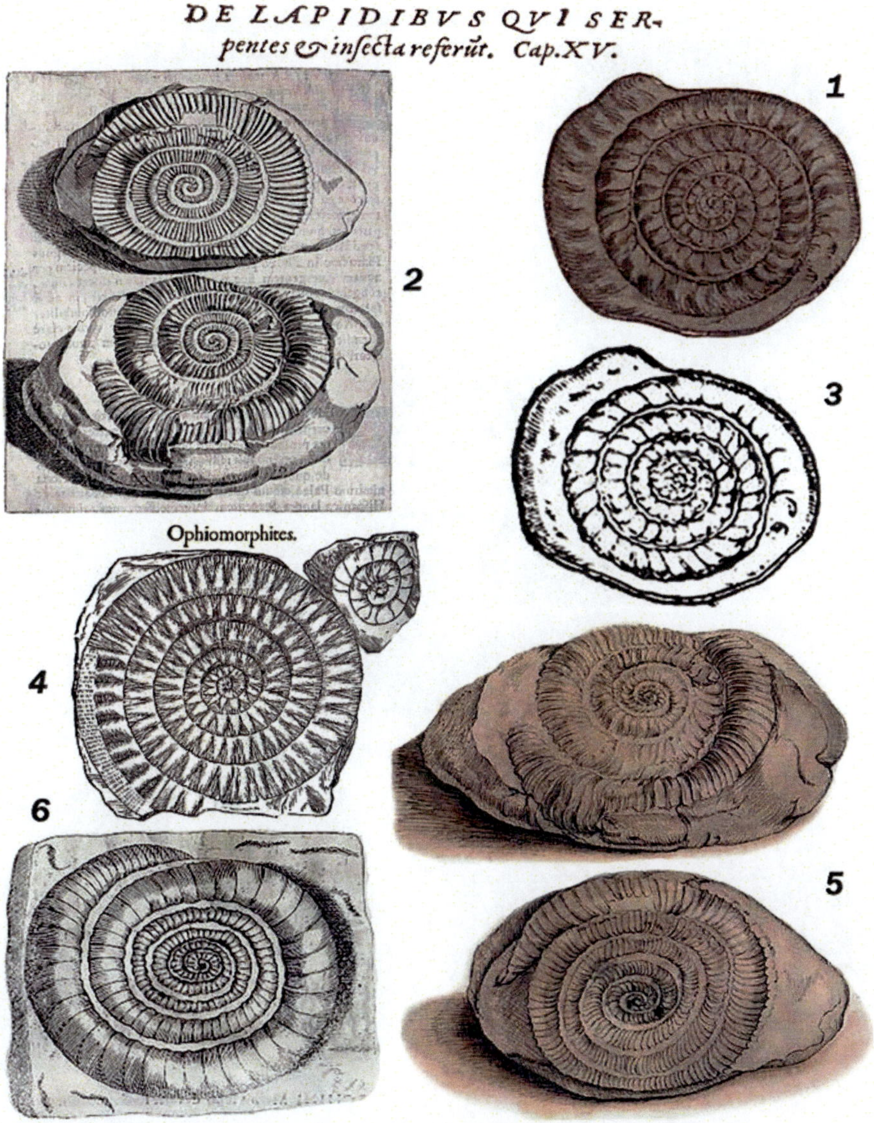

Fig. 17.16 Petrified snakes or stone referred to snakes: (1) this image from Gessner (1565: p. 167), copied and variously reproduced, inspired the subsequent repropositions of similar finds. (2) Cerutus and Chioccus *Museum Calceolari* (1622: p. 424). (3) De Laet (1647: p. 107). (4) *Ophiomorphites* (Aldrovandi 1648: p. 450). (5) From the illustrations of the objects of the Settala Museum (1660 circa). (6) *Adorned likewise my Museum are snakes of various species, converted into very hard stone, which preserve their natural horridness*; Moscardo (1656: p. 178). [Graphic composition Maurizio Forli, credit from (1) https://doi.org/10.3931/e-rara-4176 / Public Domain. (2) https://doi.org/10.3931/e-rara-50568 / Public Domain Mark. (3) https://doi.org/10.3931/e-rara-35487 / Public Domain Mark. (4) https://doi.org/10.3931/e-rara-13017 / Public Domain Mark. (5) https://archive.org/details/museoogaleriaadu00terz_0 (no copyright page found). (6) https://www.biodiversitylibrary.org/bibliography/157058 (not in copyright)]

Fig. 17.17 Ketteleness, Whitby, Yorkshire, England, Lower Jurassic, Toarcian, 180 Ma.: bottom, *Dactylioceras commune* (Sowerby 1815); top, *D. tenuicostatum* (Young and Bird 1822): middle, *Hildoceras lusitanicum* (Meister 1913). According to English folk tradition, these are the classic Snake-Stones. (Graphic composition Maurizio Forli, photo credit Geocollection)

Fig. 17.18 (1) A depiction of the Saint, carved into a partition of the monument known as Caedmon's Cross, in the Yorkshire town of Whitby. (2) In the background, a plate by James Sowerby (*Mineral Conchology of Great Britain*, 1818: pl 107, p. 10) with the illustration of an ammonite with the terminal part carved in imitation of a snake. The English scholar, particularly impressed by these legends about ammonites, had one sent to him specially carved, which later became the type species of the Genus *Dactylioceras* Hyatt, 1867. In the center (3) an Umbrian Jurassic ammonite with the same natural serpentiform effigy, probably an *Eodactylites*. Top right (4) ammonite from Ketteleness, Whitby, Yorkshire, England. [Graphic composition Maurizio Forli, credit from: (1) Photo by Wilson44691/Wikimedia (Public domain) (2) https://www.biodiversitylibrary.org/bibliography/14408 (not in copyright) (3) photo Misa Forli (4) photo Geocollection]

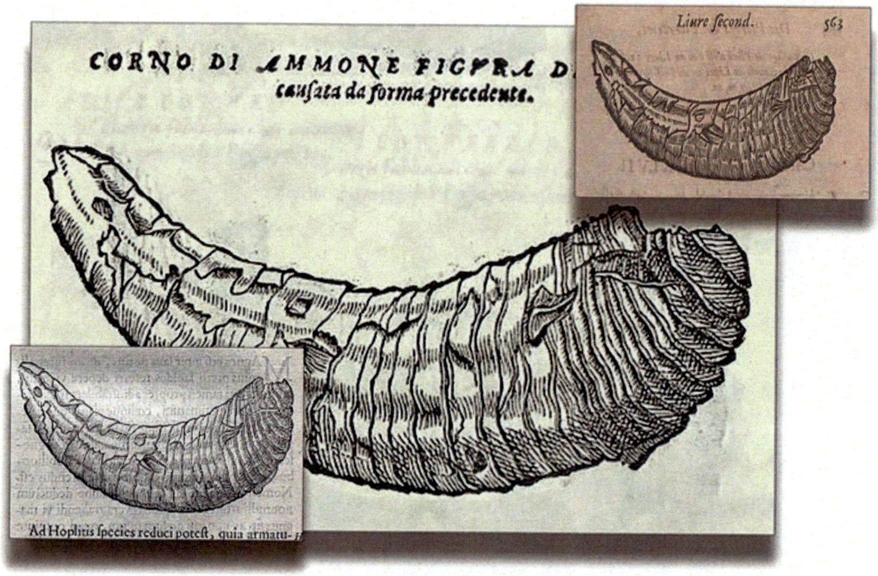

Fig. 17.19 (1) In the center, *Horn of Ammon* from Ferrante Imperato (1599: p. 667). (2–3) Bottom and top, the same reproposed by De Boodt (1636: p. 437; 1644: p. 563). *It is not of dissimilar generation from the stones known as the Horn of Ammon, formed in the manner of the Horn of Ram, or of Dragon, also crossed by wrinkles not otherwise than the said horn, some of which have a ridge that accompanies their length similar to a twisted rope. Those, therefore, of which in a short time the whorl ends, and in a short time they become thicker, are called horns by their resemblance, and those that have many whorls are more conveniently resembled snakes. They take their consistency from viscous earth, and are of the consistency of lime stone, others of a reddish color, others of an ashy color: and one sees some of the said forms attached to marchasita, and dressed in metallic color, which they call armor* (Imperato 1599: p. 667). [Graphic composition Maurizio Forli, credit from: (1) https://www.biodiversitylibrary.org/item/183177#page/5/mode/1up (not in copyright) (2) https://doi.org/10.3931/e-rara-15217 / Public Domain Mark (3) https://doi.org/10.3931/e-rara-12490 / Public Domain Mark]

believed to be a stone snake. This is a fossil on matrix consisting of both impressions, the inner model and relative outer impression, which in type and form approaches that figured by Gessner as a *snake-shaped stone* and by our authors presented as of the same category as a *petrified serpent*. It is curious to note the similarity of the two drawings in that that of Gessner seems to be the same ammonite, from the raised, contoured part of the rocky matrix, figured for the Calzolari Museum. Gessner and Calzolari exchanged correspondence and materials, and therefore, it is probable that they had news about their respective collections. Moreover, this same specimen seems to be reproduced, or used as a reference model, to show an ammonite belonging to the collection of Manfredo Settala from Milan, whose family is mentioned in Alessandro Manzoni's Promessi Sposi (The Betrothed), in one of the catalogs composed of drawings and paintings that he had made to document the contents of his collection, dating back to around 1660. A representation of ammonites coming from the area of the Umbrian-Marche Apennines is unexpectedly given in a work by Francesco Stelluti di Fabriano, dated 1637, entitled *Trattato del Legno Fossile Minerale Nuovamente Scoperto* [...] in which he reports:

> *Nel seguente foglio xiij vi sono intagliate alcune pietre di forma simili alla Chiocciola, & alla Serpe in se stessa ravvolta; ma per la gran somiglianza ancora, c'ha col corno dell'Ariete, son chiamate corno d'Ammone, in quella guisa formate dalla Natura: e son fatte con tanta proporzione quelle circolazioni spirali, ch'io l'haverei sempre stimate artificiose, e non altrimenti naturali, se non l'havessi così ritrovate, non solo nel territorio di Acquasparta, ma ancora nel territorio di Fabriano mia patria; oltre che ne scrivono anche molti Autori naturalisti; si come si può vedere nel Gesnero, in Plinio, nell'Agricola, Bellonio, Imperato, & altri.*

> *In the following folio xiij there are carved some stones similar in shape to the Snail, & to the Serpent wrapped up in itself; but because of the great resemblance they have to the horn of Aries, they are called the Horn of Ammon, in that way formed by Nature: and those spiral circulations are made with such proportion, that I would always have considered them artificial, and not otherwise natural, if I had not found them in this way, not only in the territory of Acquasparta, but also in the territory of Fabriano, my birthplace; moreover many naturalist Authors write about them; as can be seen in Gesnero, Plinio, Agricola, Bellonio, Imperato, & others.*

In the *Musaeum Metallicum* three different species of ammonites, of yellow colored, are illustrated, perhaps pyritized, called *Chrysammonites duplex*, *Chrysammonites foliaceus* and *Chrysammonites, et a figura ammonis cornum* that, following what is reported in Foresti (1887: p. 87), would correspond, respectively, to *Harpoceras radians* (Reinecke 1818), *Phylloceras doderleinianum* Catullo, 1853, and to an unidentifiable species. Page 144 shows another ammonite called *Siderammonites duplex*, which according to Foresti (1887: p. 90) would partially correspond to *Harpoceras serpentinum* Schlotheim, 1813, and an indeterminable *Lytoceras* from the limestone rocks of the Rosso Ammonitico Veronese formation. On page 450, however, two ammonites are drawn joined together, one on the back of the other, called *Ophiomorphites*, that is, snake shaped, and described as similar to that illustrated by Gessner. The best preserved part is an external model of the species *Skirroceras bayleanum* (Oppel 1857), a species characteristic of the Lower Bajocian (this Plane extends from ~170 to ~168 Ma) coming from the surroundings of Gubbio as specified by Aldrovandi and reported by Sarti:

> *"Here, under number 2, we have the drawing of a yellowish coloured stone in the shape of snake we can call Spirite. In Umbria in the Eugubini Mountains that are part of the Apennines, this coil stone was found covered with fine carving, almost as if done with a chisel. The coils near the centre are similar to the figure of the Ionic Epistilio volute, and are all marked by rays. On the other side the stone looks like an imperfect, coiled snake, as can be seen in the figure. Therefore we called this stone ofiomorfite"* (Musaeum metallicum, p. 450). *In fact the specimen of smaller ammonite that is represented beside the Ophiomorphites on page 450, is a deeply eroded and indeterminable ammonoidea that is found on the back of the above-mentioned piece* (Sarti 2003: p. 162).

Of the ammonites figured by Aldrovandi, only the *Siderammonites duplex* and the *Ophiomorphites are* still preserved and currently exhibited in the Museum of Palazzo Poggi in Bologna. Worm in 1655, on page 86 of his *Museum wormianum seu Historia rerum rariora*, reproduces the drawing of an ammonite with saddles and lobes of the septa clearly evident and therefore considered a skeleton of a petrified snake; another one on page 91 is the same as the one depicted in Gessner (1565: p. 164) and described as similar to that of Ferrante Imperato (1599: p. 667) which, however, figures a remnant other than an ammonite, in all probability a rudist or a fragment of a fossil elephantine defense. The arched banana shape and the clearly visible concentric growth plates suggest this identification, a belief also confirmed by the description, among other objects in his father's museum, of a similar find made by his son Francesco (1628: p. 6).

> [...] *& anco un pezzo di dente, ò per dir' meglio d'un corno di elefante,* [...] *qual' di dentro si fende in pezzi circolari, il superiore è di colore negro, e l'inferiore bianco, & con questo istesso modo si giunge infino al meditulio, e nella superficie ancor tiene il suo proprio colore, quando è invecchiato*

> [...] *and also a piece of tooth, or better to say of an elephant's horn,* [...] *which from the inside is split into circular pieces, the upper is of a black colour, and the lower white, and in this same way it reaches down to the intermediate part, and in the surface still keeps its own colour, when it is aged.*

The same specimen is reported again among the Ammon's Horn by de Boodt (1644: p. 563). Among other things, the iconography of this fossil remnant of the Imperato Museum was considered the first representation of a specimen of Rudista (Mesozoic bivalve mollusk) by Prof. Carlo Parona (1918: 7). A year after Worm, Moscardo illustrates two true ammonites (*Note overo memorie del Museo di Lodovico Moscardo nobile veronese*, 1656: p. 178) commenting thus

> *My museum is also adorned with snakes of various sizes, converted into very hard stone, which preserve their natural horror.*

Already aware of the facts, we can affirm a posteriori that Paolo Silvio Boccone from Palermo, mainly a botanist but also a naturalist interested in physical and medical matters, has the merit of proposing a plausible reconstruction of the body and shell of ammonites. In 1674, on page 307 of his *Recherches et Observations Naturelles* [...], he published images of parts and whole specimens of *Ammon's Horns* and *corn de bélier*, ram's horns, a French term here referring to Nautiloids. In the text in the form of a letter, disquisition on the mode of their formation and conservation compares them with the internal models of other fossil shells and with the structure of nautiluses. Boccone, at the end of the short text, asserts that the two types of *horns* resemble the shells called by the Neapolitan fishermen of the time *Muscardino* or *Muscarolo*, inhabited by a species of cuttlefish, the current *Argonaut Argus*, and others called *Nauplium* or *Nautilus*, showing that he had in mind what kind of animal could be the one that inhabited and produced the shells of these petrefacts, not straying too far from modern reconstructions.

In Mercati's work (*Metallotheca Vaticana*, 1717), published 124 years after his death, a number of ammonites are illustrated, one pyritized specimen described as *Ammonis Cornu Aureum* and four others, drawn in different perspectives each, as *Ammonis Cornu Lapideum et Ophioides* (1717: p. 307–311). In particular, the latter are to be referred, respectively, to *Hildoceras* gr. *bifrons* (Bruguière 1789*)* on the left and *Hildaites praecursor* on the right (1), *Mercaticeras umbilicatum* Buckman 1913 (2), *Mesodactylites sapphicus* Renz, 1912 (3), *Phylloceras doderleinianum* Catullo, 1853 (on the left), and *Calliphylloceras capitanei* (Catullo 1853) in the middle and lower right of the plate representing them, coming from Monte Petrano, in the municipality of Cagli, in the Umbria-Marchigiano Apennines, where there are extensive outcrops of the typical Rosso Ammonitico Formation, chronologically referable to the Lower Jurassic. A very important genus of these fossil cephalopods is dedicated to *Mercaticeras* Buckman, 1913, known from many European fossil localities of the Paratetide, used as a guide fossil of the Lower Jurassic, and the species *Mercaticeras mercati* (Hauer 1856).

Part of the work of the Englishman Robert Hooke is also printed afterward, edited by Richard Waller in 1705 *The Posthumous Works of Robert Hooke* [...], in which ammonites and other fossil remains are reported in the chapter devoted to earthquakes and their causes, discussing the possible reasons for their formation, mentioning among others, the Italians Aldrovandi and Imperato.

Two images of ammonites with the spiral wound in a conical shape, as probably was the *Draconites* of Albertus Magnus, are found at plate 32, figures 6 and 7 of the *Historia lapidum figuratorum Helvetiae eiusque viciniae* [...] (Lang 1708). They are described by the Swiss author in the section dedicated to the shells with the turreted coil, that is, wound in a more or less elongated spiral, according to the size of the single whorls in contact with each other, under the name of *Turbinites*. This particular type of shells has the characteristic of having the sense of winding left, that is, counterclockwise. The number 6 could be a *Turrilites costatus* Lamarck, 1801, and the number 7 a *Hoplites* sp.

Lang treats separately, like others before and after him, the fragments of the internal models of the coil of some ammonites, in which the ornamentation and subdivisions of the jagged lines produced by the saddles and lobes of the sutures between the laps are clearly visible, considering them almost obviously, vertebrae of serpents (see, e.g., Worm, *Museum wormianum*, 1655: p. 86, *Lapis sceleton serpentis*).

In these same years (1726) was published the *Lithographiæ Wirceburgensis* edited by Johann Bartholomäus Adam Beringer, a book written with the scientific method of the time but made on carved fossil remains, believed to be true by the author, more or less unconscious victim of two of his colleagues, in which at plate V is represented an ammonite, recto and verso, portrayed as a strange snake, regularly described on the previous page 57. The proposition of ammonites in the guise of petrified snakes in such a context of mockery and science at the same time is symptomatic of how much it was still unclear what they actually were, not finding in nature animals equal or similar to which to compare them. It is true that by the middle of the eighteenth century, hypotheses had been put forward, which later proved to be correct, regarding their similarity and close correlation with nautiluses and their being petrified marine bodies, but despite this, there was still some uncertainty about their actual nature. The news of the discovery of some actual marine microfossils that resembled the shape of ammonites, made by Bartolomeo Beccari around the first decade of 1700, spread after the publication of Giovanni Bianchi, also known as Jano Planco or Simone Cosmopolita, of *Corpus minùs notis* [...] (1760), where the author confirmed the existence of small Ammon's Horn, among the microscopic remains of marine debris collected on the beaches of the Adriatic Sea. Recent discoveries of the materials examined by Beccari and some of his manuscripts regarding the methods of study and related observations on foraminifera allow us to attribute him the definition of "*founder of micropaleontology*" (Vaiani et al. 2019b). The lectotype of *Ammonia beccarii* (Linnaeus 1758) was selected, chosen, and described, from Beccari's original samples to clarify the taxonomic status of this species (Vaiani et al. 2019a). Ambrogio Soldani himself, considered together with Beccari a progenitor of Italian micropaleontology scholars, in his work *Saggio orittografico ovvero osservazioni sopra le terre nautilitiche ed ammonitiche della Toscana,* published in 1780 wrote about some microfossils (nummulite foraminifera) calling them *ammonites*, distinguishing himself with this from the "vulgo" who continued to use the ancient term *corno di Ammone*.

As far as we know, Aldrovandi was the first to introduce, in an appropriate way, the word *ammonites* (Sarti 2003), for example, on page 53 of *Musaeum Metallicum* he uses the expression *Lapis Ammonites* for these extinct cephalopods. An example of the different interpretations related to the nature of ammonites, also present at the same time in the same text, can be found among others, in a paper by Carlo Allioni from Piedmont, *Oryctographiae Pedemontanae specimen* [...] (1757), in which in a scientific way under the name AMMONITES of the relevant chapter, are described real ammonites and foraminifers, using also the term *cornu ammonis*. From the systematic point of view, it is the French Bruguière who, in 1789, instituted the genus *Ammonites*. The Sienese scholar Giuseppe Baldassarri, in his *Saggio di Produzioni Naturali dello Stato Sanese che si ritrovano nel Museo del Nobile Sig. Cav.re Giovanni Venturi Gallerani* (1750), commenting on the presence of Ammon's Horn in the surroundings of San Casciano dei Bagni, takes stock of the scientific knowledge then available on the subject:

> *About the Horns of Ammon we read at Pliny the following words Ammonis Cornu* [...]. *From this it seems to arise reason to suspect that the stones called in this name, are different from the Horns of Ammon described by Pliny, not having any brand, for which they deserve to be counted among the gems. In order not to enter into such a thorny issue, it will be sufficient to note that there are many species of Horns of Ammon, different from each other in shape,*

size, color, and material, as can be seen in Aldrovandi, Wormio, Scheuzero, Langio, and Listero. Their figure is spiral, and in the guise of a compressed snail, which decreases in size in proportion as it approaches the Center, and is divided into many cells by an intermediate division. One can observe their surface either smooth, or striated, or marked by curls, or tubercles. They are reddish or white, or ashy, or iron, or copper, and sometimes are covered by a golden crust, and are composed of petrous matter, or iron, or copper, or Marchesita. It is certain that these must be placed in the class of marine bodies that have been petrified or converted into metal, but the difficulty is reduced to determining the precise quality of the animal from which they originated. There are two opinions concerning this article which are the most plausible. The first is that of those who maintain that the Horns of Ammon are nothing more than petrified Indian nautiluses. Boccone in his researches, and Natural Observations lett. 20. asserts to have observed a shell of Nautilus attached to a Horn of Ammon. The same is also believed by M. Jusseu in the Memoirs of the Accad. Real. of Paris of the year 1722. Who comparing the Horns of Ammon with the Nautilus, shows that the first ones are nothing but petrified Nautilus; and since there are several species of Horns of Ammon, and comparing three of them with three of the others, he shows their exact correspondence [...]. Others believe that there is a great difference between the Nautilus and the Horns of Ammon. 1. Because the surface of the former is smooth, that of the latter divided into many parts; 2. Because the spines of the Nautilus are fewer in number than those of the Horns of Ammon; 3. Because the spines do not extend equally from the centre to the extremity of the mouth, so that at equal distances from the centre the Nautilus dilates more, and terminates in a wider orifice than the Ammon's Horns. We omit the opinions of those who have believed these to be either fish spines twisted into the shape of coils, or the skeletons of snakes, or worms, or those who have believed them to be a production of the Earth, ensuring that they have observed the first rudiments, and taken measurements of them with a compass, having seen them grow from one day to the next. The very learned Mr. Beccari observed a large copy of the very small fossil Horns of Ammon in a certain yellow sand of Bologna, as it costs from the first volume of the Commentaries of the Institute of that City, and the famous Mr. Giano Planco observed many different species of these very small native horns in the sand of the Lido of Rimini, from which he gave news to the public in his book de Conchis minùs notis.

Practically, it is only after the first half of the 1800s that the name *ammonite* fully assumes the meaning it is currently recognized with, the use of the archaic term *Cornu Ammonis* remaining until the end of the eighteenth century. An examination of the subject is reported by Romano (*From petrified snakes, through giant foraminifers*, 2014) (Figs. 17.20, 17.21, 17.22, 17.23, 17.24, 17.25, 17.26, 17.27, 17.28, 17.29, 17.30, 17.31, 17.32, 17.33, 17.34, 17.35, 17.36, and 17.37).

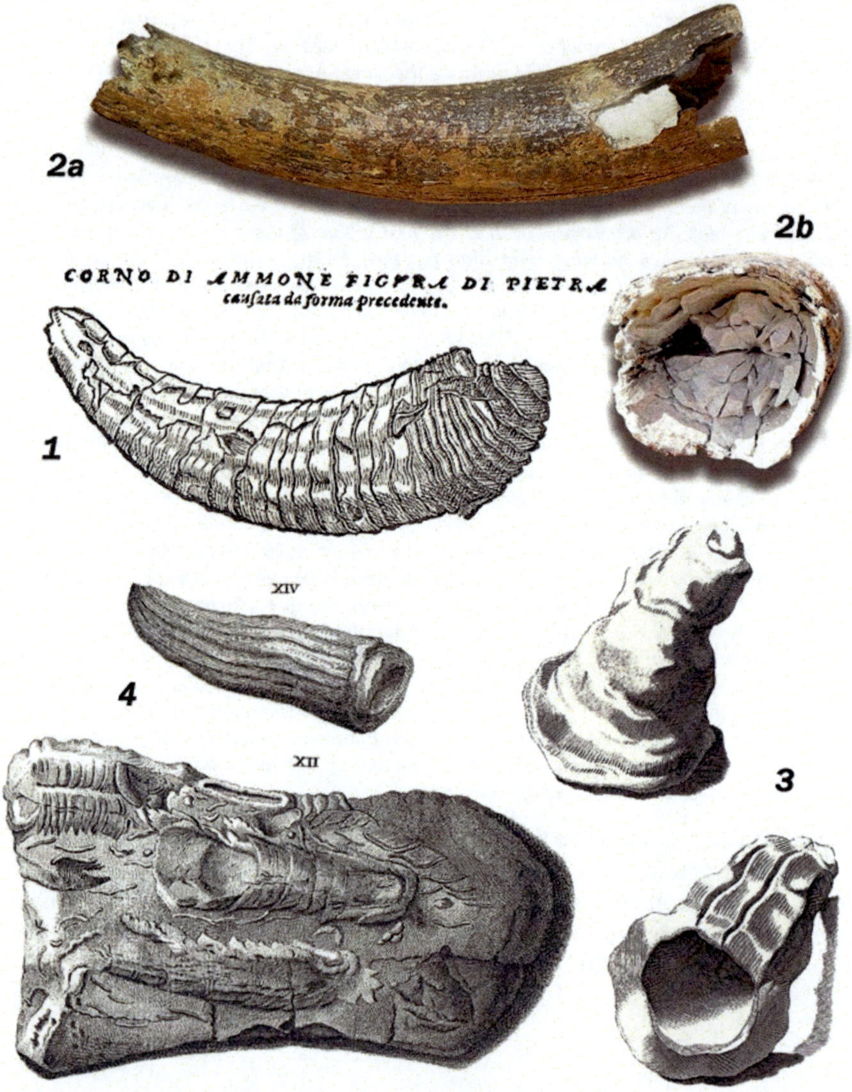

Fig. 17.20 The image from Imperato (1) is considered the first referable to a Rudista, probable internal cast of a radiolithid, Parona (1918) and Cestari (2008: p. 42). (2a–b) Fossil fragment of elephant tusk, in lateral and frontal view (Pleistocene, North Sea). (3) Radiolitidae, *Lapides monstruosi seu Tubuliteae maximi, cinerei, pedis caballini faciem gerentes* (Spada 1744: p. 43, pl. III). (4) Rudiste described as particular Orthoceras, fig. XII and fig. XIV (Fortis 1774: pp. 174–175, pl. VII). Although less likely, a similar horn of ammon, could be a portion of elephantine fossil tusk, a find that we know to have been present in the museum, *…& also a piece of tooth, or to say better of an elephant's horn,* [...] *which inside is divided into circular pieces* (Imperato 1628: p.6). (Graphic composition Maurizio Forli. Credit from: (1) https://www.biodiversitylibrary.org/item/183177#page/5/mode/1up (not in copyright) (2) photo credit Misa Forli (3) https://doi.org/10.3931/e-rara-50868 / Public Domain Mark (4) https://doi.org/10.3931/e-rara-40493 / Public Domain Mark)

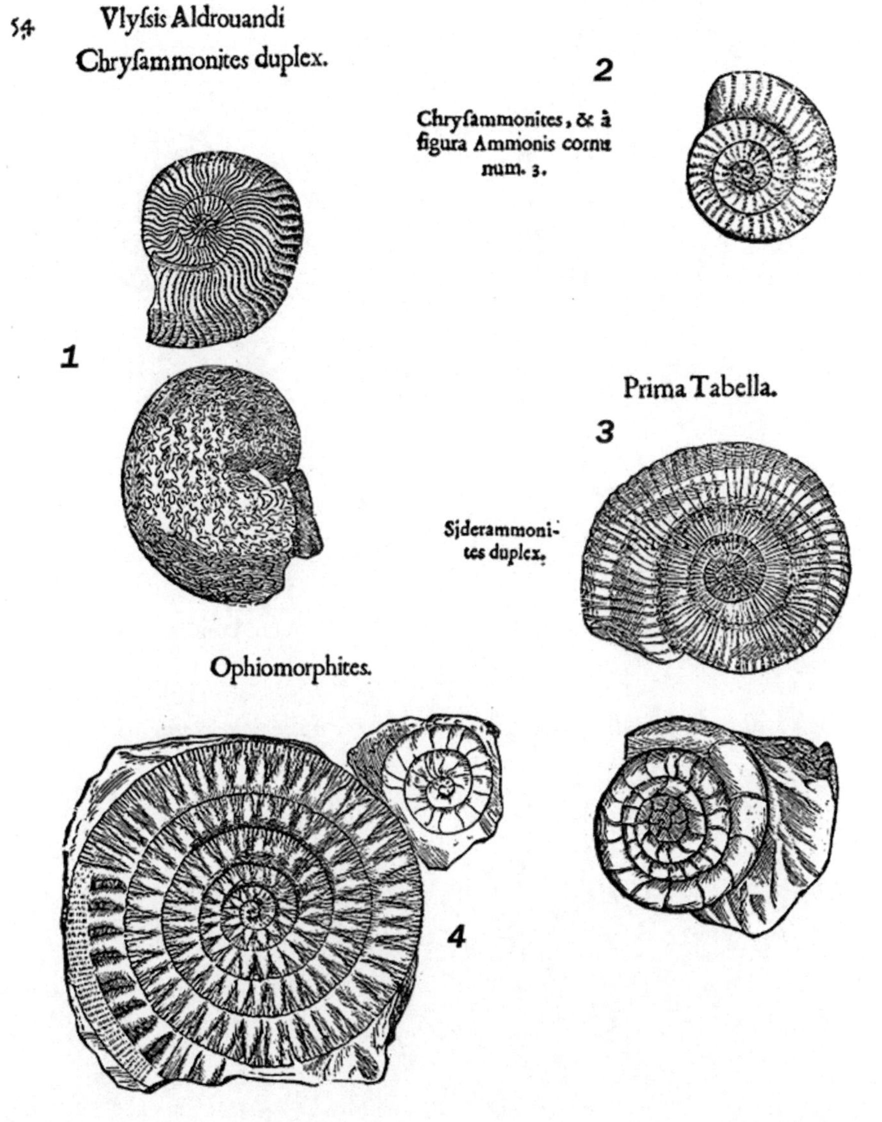

Fig. 17.21 Specimens of ammonites figured in the Aldrovandian *Musaeum Metallicum* (1648): (1) *Chrysammonites duplex*; (2) *Chrysammonites, et a figura ammonis cornum*; (3) *Siderammonites duplex*; (4) *Ophiomorphites*. Only the *Siderammonites duplex* and the *Ophiomorphites* are still preserved, and currently exhibited, in the Museum of Palazzo Poggi in Bologna (https://doi.org/10.3931/e-rara-13017 / Public Domain Mark)

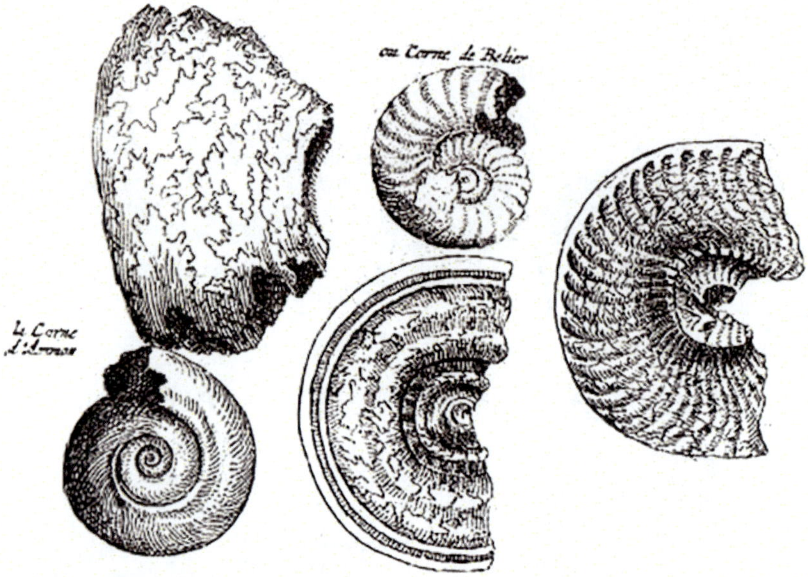

Fig. 17.22 Boccone Paolo Silvio (*Recherches et observations naturelles*... 1674: p. 307) ammonite remains (modified) (https://doi.org/10.3931/e-rara-21877 / Public Domain Mark)

Fig. 17.23 Boccone Paolo Silvio (*Recherches et observations naturelles*… 1674: p. 313) *Les Coquille aux quelles les dites* Pierre *sont semblales sont appllées par les Neapolitains* Muscardino, *ou* Muscarolo, *& par plusieurs autheurs* Pompiles, & Nauplium *ou* Nautilus *ou est logé un poisson de la nature de la* Sepia. Image from Bellon (1551: p. 52 v) [https://www.biodiversitylibrary.org/bibliography/5751 (Public domain)]

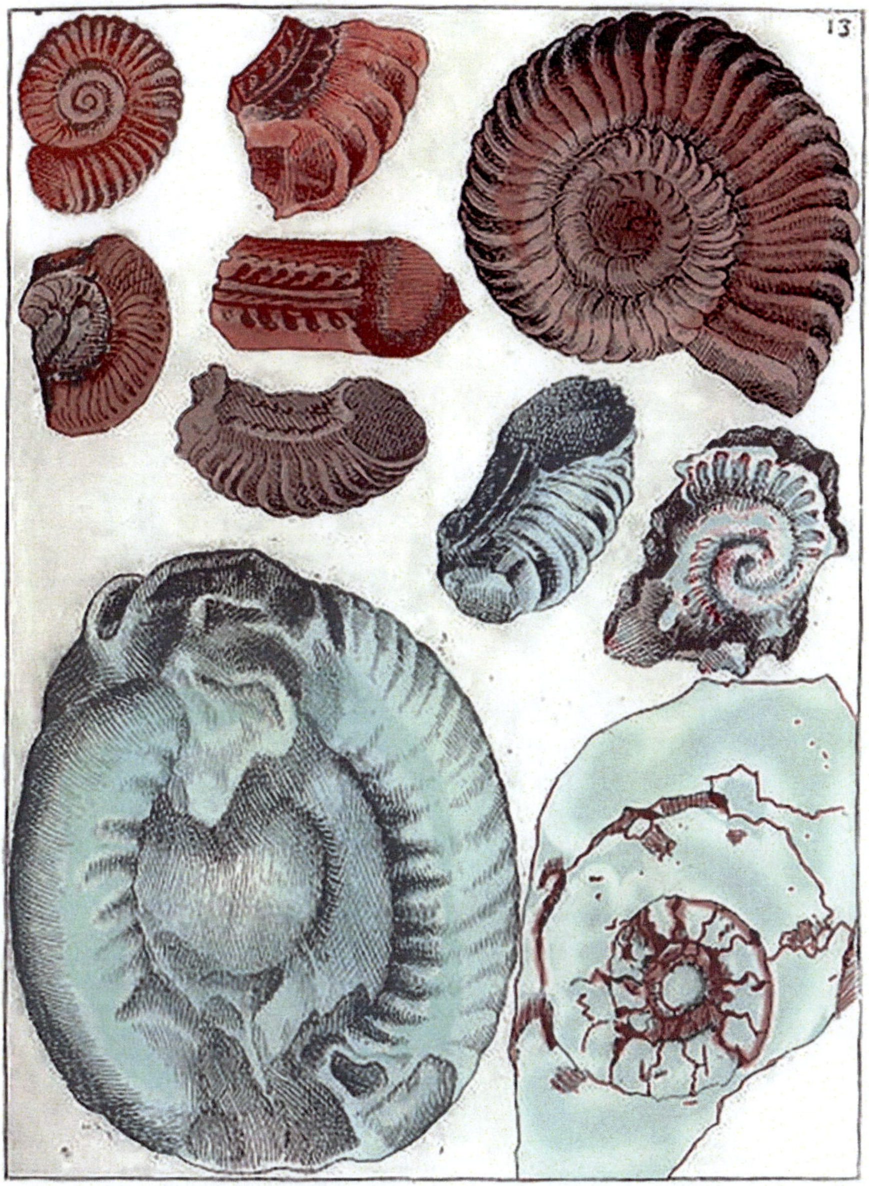

Fig. 17.24 Francesco Stelluti of Fabriano, folio XIII, 1637, *Trattato del Legno Fossile Minerale Nuovamente Scoperto*. [...] *those spiral circulations are made with such proportion, that I would have always considered them artificial, and not otherwise natural, if I had not found them in this way, not only in the territory of Acquasparta, but also in the territory of Fabriano, my country.* [Colour Maurizio Forli, modified by original plate in black and white. Credit from: https://digital.slub-dresden.de/werkansicht/dlf/16093/4 (Public Domain Mark 1.0)]

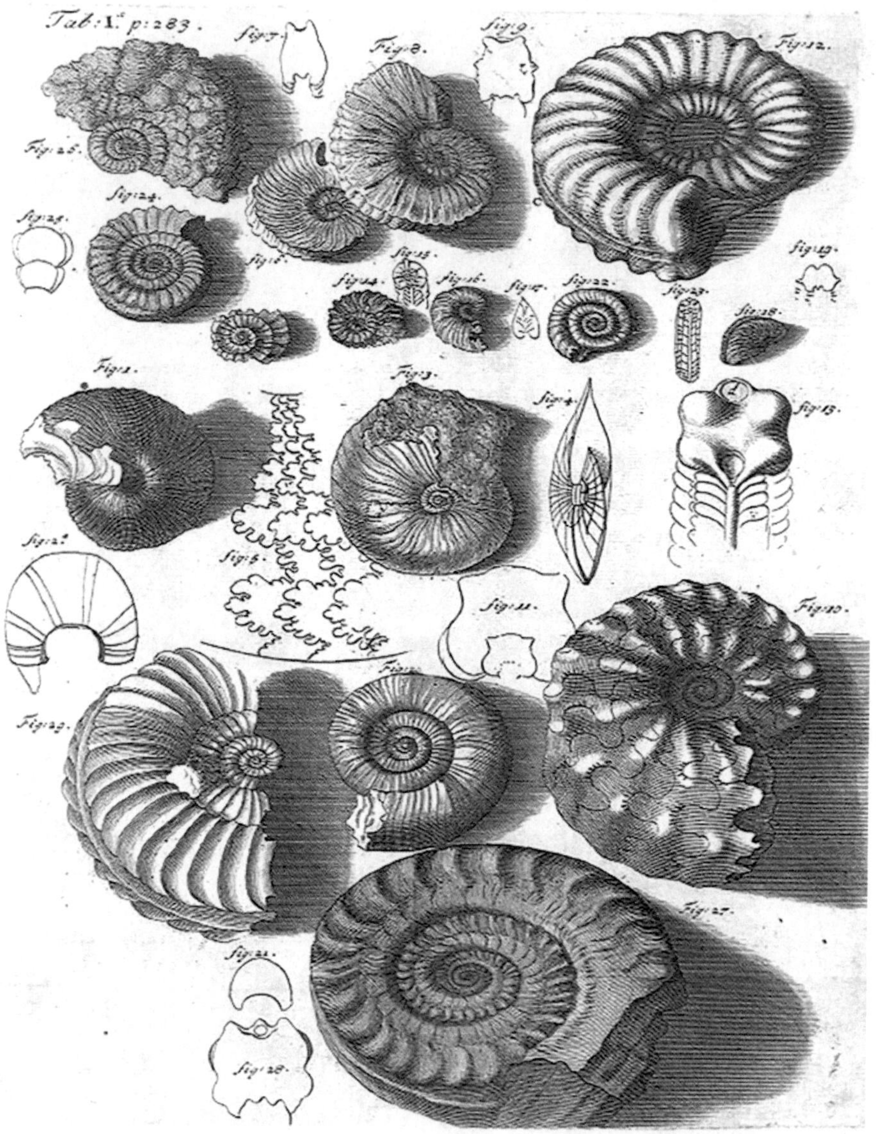

Fig. 17.25 *I have designed 15 several sorts of snail rather than Snake-stones, call'd by some Authors Cornua Ammonis, or Sceleta Serpentum* (Robert Hooke). Plate I, inserted between pages 282 and 283, from Hooke's drawings, printed upside down from right to left from the originals, in which ammonites are excellently represented. [Modified after Richard Waller, *The Posthumous Works of Dr. Robert Hooke*, 1705. https://wellcomecollection.org/works/xxggvyj3 (Public Domain Mark)]. This topic is also discussed in Kusukawa (2013)

Fig. 17.26 Mercati's *Ophioides* referred, respectively, to (1) *Hildoceras* gr. *bifrons* (Bruguière 1789) on the left and *Hildaites praecursor* Venturi & Ferri, 2001 on the right; (2) *Mercaticeras umbilicatum* Buckman, 1913; (3) *Mesodactylites sapphicus* Renz, 1912; *Phylloceras doderleinianum* Catullo, 1853 (left), and *Calliphylloceras capitanei* (Catullo 1853) in the center and lower right of the plate representing them, from the Jurassic of Monte Petrano, in the municipality of Cagli, in the Umbrian-Marche Apennines. The ammonites compared, while similar, may not correspond exactly to those engraved in the *Metallotheca Vaticana* (1717: p. 310). [Photo and graphic composition Maurizio Forli. Credit from: https://archive.org/details/bub_gb_DC_E6UVWg7EC (Public Domain Mark 1.0)]

Fig. 17.27 *Hoploscaphites nebrascensis* (Owen 1852), Fox Hills Formation, South Dakota, USA. Upper Cretaceous, Maastrichtian, dimensions 75 × 27 mm. (Photo credit Geocollection)

Fig. 17.28 *Hoploscaphites nicolletii* (Morton 1842), Fox Hills Formation, South Dakota, USA. Upper Cretaceous, Maastrichtian, dimensions 65 × 30 mm. (Photo credit Geocollection)

Fig. 17.29 *Hoploscaphites nicolletii* (Morton 1842), Fox Hills Formation, South Dakota, USA. Upper Cretaceous, Maastrichtian, dimensions 62 × 44 mm. (Photo credit Geocollection)

Fig. 17.30 *Hoploscaphite*s sp., Fox Hills Formation, South Dakota, USA. Upper Cretaceous, Maastrichtian, dimensions 95 × 65 mm. (Photo Maurizio Forli)

Fig. 17.31 An imitation of ammonite, recto and verso, a carved object like the others depicted in plate V, from Beringer's *Lithographia Wirceburgensis* (1767). The proposition of the ammonites as petrified snakes in such a context of mockery and science at the same time is symptomatic of how much it was still unclear what they actually were, not finding in nature, equal or similar animals to which to compare them [https://www.biodiversitylibrary.org/bibliography/150131 (Public domain)]

Fig. 17.32 Above left, pyritized ammonite still partially embedded in its matrix, taken from Besler Michaelis Rupert, *Gazophylacium Rerum Naturalium* (1642); in the center, a specimen similar to the one illustrated with the internal imprint of the central rocky rind removed and, on the right, the shell freed from encrustation: *Pleuroceras spinatum* Bruguière 1789, surroundings of Hannover, Germany, Jurassic. (Photo and graphic composition Maurizio Forli. Credit from: https://doi.org/10.3931/e-rara-56503 / Public Domain Mark)

Fig. 17.33 *Pliny's Draconites was probably a dark-colored pyramidal glassy gemstone while Albertus Magnus's, while also dark-colored and pyramidal, is a serpentiform stone, perhaps a turritellata ammonite of the Upper Cretaceous* (Accordi et al. 1981: p. 143). Left: (1) portion of *Turrilites costatus* Lamarck, 1801 (Upper Cretaceous, South-Eastern France). (6) the same, *Turbinites striatus striis transversis densioribus...*; (7) *Hoplites* sp., *Turbinites striatus striis transveris & in medio in duplice papillarum seriem divisis* [Lang 1708: pp. 111–112 (modified)]. Bottom right (2) an obsidian pyramid, base side 50 mm, still sold today as a powerful talisman or for crystal therapy. (Graphic composition Maurizio Forli. Credit from: (1) photo Massimo Cresti (2) photo Misa Forli (6–7) https://doi.org/10.3931/e-rara-10398 / Public Domain Mark)

Fig. 17.34 *Crioceratites nolani* (Kilian 1919), ammonites criocone, that is, with open spiral, from the French Cretaceous (Castellane surroundings, Alpes de Haute-Provence). Upper specimen × 0.5, lower specimen at full size. (Photo Maurizio Forli)

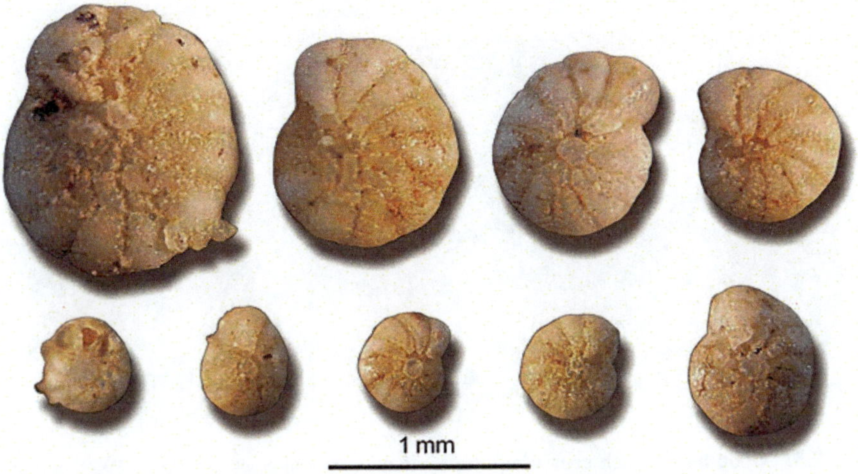

Fig. 17.35 *Ammonia beccarii* (Linnaeus 1758). Type material from what remains in Beccari's original samples. The third specimen of the upper row has been selected as lectotype of *A. beccarii*. (Graphic composition Maurizio Forli modified after Vaiani et al. 2019a, b. Courtesy GB Vai)

Fig. 17.36 Foraminiferal assemblages, handpicked from the Beccari's original sample, catalog number 20, including selected specimens of *Ammonia beccarii*, with original label: *Arena crivelata dove apariscono distintamente i corni d'Amone – Sieved sand where horns of Hammon are distinctly present*. (Graphic composition Maurizio Forli modified after Vaiani et al. 2019a, b. Courtesy GB Vai)

Fig. 17.37 Good luck brooch consisting of a pyritized ammonite set in gold (~ 1960, courtesy private collection). Currently the ammonite has altered and the original beauty has disappeared. Below at the sides, the serpentine ring, in brass, part of the material included in the bag of sanpaolaro (quack in the name of Saint Paul. (Coll. Bellucci, M.A.N.U. (PG). Inv. 338 G. Reproductions by kind permission of the Ministry for Cultural Heritage and Activities and for Tourism, Regional Directorate Museums of Umbria—National Archeological Museum of Umbria). (Photo and graphic composition Maurizio Forli)

References

Accordi B, Stacchiotti L, Tagliaferro C (1981) Traduzione e commento del De Lapidibus di Gerolamo Cardano (dal libro VII del De Subtilitate, edizione del 1560). Geol Romana 20:125–169

Albertus Magnus (1569) De mineralibus et rebus metallicis libri quinque. Auctore Alberto Magno Summo Philosopho. Solerti cura repurgati, & rerum natural. studiosis publicati. Apud Ioannem Birckmannum & Theodorum Baumium, Coloniae.

Aldrovandi U (1648) Ulyssi Aldrovandi patricii Bononiensis Musaeum metallicum in libros IIII: distributum Bartholomaeus Ambrosinus ... labore, et studio composuit cum indice copisosissimo Arcus Antonius Bernia. typis Io. Baptistae Ferronii, Bononiae. ETH-Bibliothek Zürich, Rar 1027, https://doi.org/10.3931/e-rara-13017 / Public Domain Mark

Bahuin J (1598) Historia novi et admirabilis fontis balneique Bollensis in Ducatu Vvirtembergico ad acidulas Goepingenses. Apud Iacobum Foilletum, Montisbeligardi. https://archive.org/details/historianovietad00bauh/page/n4/mode/2up

Baier JJ (1757) Monumenta rerum petrificatarum praecipua oryctographiae noricae supplementi loco iungenda interprete filio Ferdinando Iacobo Bai [...]. in commissione Georgii Lichtenstegeri, Norimbergae [Nürnberg]. ETH-Bibliothek Zürich, Rar 2333, https://doi.org/10.3931/e-rara-14953 / Public Domain Mark

Baldassarri G (1750) Saggio di Produzioni Naturali dello Stato Sanese che si ritrovano nel Museo del Nobile Sig. Cav.re Giovanni Venturi Gallerani. Stamperia del Pubblico, Siena

Bellon P (1551) L'histoire naturelle des estranges poissons marins, avec la vraie peincture & description du daulphin, & de plusieurs autres de son espece. De l'imprimerie de Regnaud Chaudiere, A Paris. https://www.biodiversitylibrary.org/bibliography/5751 Public domain

Beringer JBA (1767) Lithographiæ Wirceburgensis, ducentis lapidum figuratorum, a potiori insectiformium, prodigiosis, imaginibus exornata. Tobiam Goebhardt, Francofurti et Leipzig. https://www.biodiversitylibrary.org/bibliography/150131 (Public Domain)

Besler MR (1642) Gazophylacium Rerum Naturalium E Regno Vegetabili, Animali & Minerali depromptarum, nunquam hactenus in Lucem editarum, Cum Figuris Æneis Ad Vivum Inc [...]. Lipsiae & Francofurti: Apud Ioh. Herebordum Klosium, Bibliopol., Anno M.DCC. XVI. Zentralbibliothek Zürich, NNN 61, https://doi.org/10.3931/e-rara-56503 / Public Domain Mark

Boccone PS (1674) Recherches et observations naturelles touchant le corail, la pierre étoilée, les pierres de figure de coquilles, le corne d'ammon, l'astroïte Undulatus, [...]: examinées à diverses fois dans l'Assemblée de Messieurs de Societé Royale de Londres, & dans les conferences de Monsieur l'Abbé Bourdelot à Paris. chez Jean Jansson à Waesberge, A Amsterdam. ETH-Bibliothek Zürich, Rar 5705, https://doi.org/10.3931/e-rara-21877 / Public Domain Mark

Boccone PS (1697) Museo di fisica e di esperienze variato, e decorato di osservazioni naturali, note medicinali, e ragionamenti secondo i principij de' moderni. Giacomo Combi, Venetia.

Boodt A Boëtius de (1636) Gemmarum et lapidum historia. ex officina Ioannis Maire, Lugduni Batavorum [Leiden]. ETH-Bibliothek Zürich, Rar 1881, https://doi.org/10.3931/e-rara-15217 / Public Domain Mark

Boodt A Boëtius de (1644) Le parfaict ioaillier ou histoire des pierreries ou sont amplement descrites leur naissance, iuste prix, moyen de les cognoistre, et se garder des contrefaites, [...]. chez Iean-Antoine Huguetan, A Lyon. ETH-Bibliothek Zürich, Rar 1010, https://doi.org/10.3931/e--rara-12490 / Public Domain Mark

Brignon A (2020) Les premières découvertes de vertébrés jurassiques aux Vaches Noires (Calvados, France). Actes du premier colloque de l'APVSM, "Paléontologie et Archéologie en Normandie", 5–6 octobre 2019. Bulletin de l'Association paléontologique de Villers-sur-Mer, 2020, p 7–40

Camden W (1590) Britannia, sive, Florentissimorum regnorum Angliae, Scotiae, Hiberniae et insularum adiacentium ex intima antiquitate chorographica descriptio. Francofurdi: apud Ioannem Wechelum, impensis P. Fischeri & haeredum Henrici Tackii. Zentralbibliothek Zürich, Rn 269, https://doi.org/10.3931/e-rara-28418 / Public Domain Mark

Canini GA (1669) Iconografia cioè disegni d'Imagini de' famosissimi Monarchi, Regi, Filosofi, Poeti e Oratori dell'Antichità. Stamperia Ignazio de' Lazari, Roma. https://babel.hathitrust.org/cgi/pt?id=dul1.ark:/13960/t6350084m&view=1up&seq=91&skin=2021 (Public Domain)

Cerutus B, Chioccus A (1622) Musaeum Franc. Calceolari iun Veronensis. Apud Angelum Tamum, Veronae. https://doi.org/10.3931/e-rara-50568 (Public Domain Mark)

Cestari R (2008) Le Rudiste come vincolo stratigrafico per il rilevamento geologico in alcune località dell'Appennino Centrale. Memorie Descrittive della Carta Geologica d'Italia 77:41–60

Colonna F (1616) Aquatilium et terrestrium aliquot animalium, aliaruq. naturalium rerum observationes. In Ekphrasis altera. Apud Jacobum Mascardum, Romae

De Laet J (1647) De Gemmis et Lapidibus, libri Duo. Quibus praemittur Theophrasti, liber de Lapidìbus Graece & latine cum Brevibus Annotationibus. Ioannis Maire, Lugduni Batavorum. https://doi.org/10.3931/e-rara-35487 / Public Domain Mark

Dimitriadis G, Marini V, Massetti G (2006) Considerazioni preliminari sulla presenza di ammoniti negli edifici di culto del 1500 in Franciacorta (BS). Storiadelmondo, 42.

Domenichi L (1844) Della Storia Naturale di C. Plinio Secondo, Libri XXXVII. Traduzione di M. Lodovico Domenichi emendata per la prima volta secondo il testo latino con l'aggiunta di un nuovo indice generale, vol Secondo. Venezia, Giuseppe Antonelli Ed.

Duffin C (2008) Fossils and Folklore. Ethical Record, 113(3):17–21

Foresti L (1887) Sopra alcuni fossili illustrati e descritti nel Musaeum Metallicum di Ulisse Aldrovandi. Boll Soc Geol Ital 6:81–121

Fortis A (1774) Viaggio in Dalmazia. in Venezia : presso Alvise Milocco ..., 1774. Zentralbibliothek Zürich, NR 842, https://doi.org/10.3931/e-rara-40493 / Public Domain Mark

Minutoli H v (1824) Reise zum Tempel des Jupiter Ammon, in der Libyschen WüsteGeneral Research Division, The New York Public Library 1. Ansicht des Hauptortes der Oase des Ammon, Siwah-Kebir, von der Morgenseite, ... ; 2. Ansicht des Tempels der Isis und eines kleinen peripterischen Templels auf der Insel Philä [Philae]. Retrieved from https://digitalcollections.nypl.org/items/510d47e2-6f41-a3d9-e040-e00a18064a99

Gessner C (1565) De omni rerum fossilium genere, gemmis, lapidibus, metallis, et huiusmodi, libri aliquot, plerique nunc primum editi. Tiguri. https://doi.org/10.3931/e-rara-4176 / Public Domain Mark

Imperato F (1599) Dell'Historia Naturale di Ferrante Imperato napolitano. Libri XXVIII. Nella quale ordinatamente di tratta della diversa condition di miniere, e pietre. Con alcune historie di Piante, et Animali fin'ora non date in luce. Costantino Vitale, Napoli. https://www.biodiversitylibrary.org/item/183177#page/5/mode/1up (not in copyright)

Imperato Fr (1628) Discorsi intorno a diverse cose naturali. Opera non meno curiosa, che utile, e necessaria à Professori della natural Filosofia. Egidio Longo, Napoli

Kracher A (2012) Ammonites, legends and politics. The snakestones of Hilda of Whitby. Eur J Sci Technol 8:51–66

Kusukawa S (2013) Drawings of fossils by Robert Hooke and Richard Waller. Notes Records 67:123–138

Lachmund F (1669) Oryktographia Hildesheimensis sive admirandorum fossilium, quae In tractu Hildesheimensi reperiuntur, Descriptio Iconibus Illustrata, Cui addita sunt alia de calculis, de fontibus, &c. Jacobi Mülleri, Hildesheimii. https://wellcomecollection.org/works/rs34c8db (Public Domain Mark)

Lang KN (1708) Historia lapidum figuratorum Helvetiae, eiusque viciniae : in qua non solum enarrantur omnia eorum genera, species et vires aeneisque tabulis repraesentantur, sed insuper adducuntur eorum loca nativa, in quibus [...]. sumptibus authoris, typis Jacobi Tomasini, Venetiis. ETH-Bibliothek Zürich, Rar 5680, https://doi.org/10.3931/e-rara-10398 / Public Domain Mark

Mercati M (1717) Metallotheca Vaticana. Opus posthumum, Autoritate et Munificentia Clementis Undecimi Pontificis Maximi. E tenebris in lucem eductum: Opera autem, & studio Joannis Mariae Lancisii, Archiatri Pontificii illustratum. Ex Officina Jo: Mariae Salvioni Romani, Romae. https://archive.org/details/bub_gb_DC_E6UVWg7EC (Public Domain Mark 1.0)

Moscardo L (1656) Note overo memorie del museo di Lodovico Moscardo nobile veronese. Academico Filarmonico, dal medesimo descritte, Et in Tre Libri distinte. Per Paolo Frambotto, Padoa. https://www.biodiversitylibrary.org/bibliography/157058 (not in copyright)

Parona CF (1918) Saggio bibliografico sulle rudiste, con indici dei nomi di autore, di genere e di specie. Boll R Com Geol Ital 46(1–4):1–78

Planco J (1760) Corpus minùs notis. Liber cui accessit specimen aestus reciproci maris superi ad littus Portumque Arimini. Editio Altera. In Aedibus Palladis, Romae

Reiber E (eds) (1862) L'Art pour Tous. 37(10). https://digi.ub.uni-heidelberg.de/diglit/art_pour_tous1862/0059

Reiske JJ (1689) Exercitatio historico-physica de Cornu Hammonis, quod Veteres vocârunt, Agri Brunshusani et Ganderheimensis lapide, quem vulgo Drakenstein nominant

Romano M (2014) From petrified snakes, through giant 'foraminifers', to extinct cephalopods: the early history of ammonite studies in the Italian peninsula. Hist Biol 27(2):1–22

Sarti C (2003) Le Collezioni di geologia nel Museo di Ulisse Aldrovandi. In: Vai GB, Cavazza W (eds) Quadricentenario della parola Geologia. Ulisse Aldrovandi 1603 Bologna, Minerva, p 152–167

Servius MH (1544) P. Virgilii Maronis latinorum poetarum, principis, bucolica, georgica, et aeneis. Per Hieronymus Curionem, Basileae

Soldani A (1780) Saggio Orittografico ovvero osservazioni sopra le terre nautilitiche ed ammonitiche della Toscana. Con Appendice o Indice Latino Ragionato de' piccoli Testacei, e d'altri fossili d'origin marina per schiarimento dell'Opera. Stamperia di Vincenzo Pazzini Carli e Figli, Siena. https://doi.org/10.3931/e-rara-14788 / Public Domain Mark

Sowerby G (1818) The mineral conchology of Great Britain; or coloured figures and descriptions of those remains of testaceous animals or shells, which have been preserved at various times and depths in the earth, vol II. W. Harding, London. https://www.biodiversitylibrary.org/bibliography/14408 (not in copyright)

Sowerby G (1821) The mineral conchology of Great Britain; or coloured figures and descriptions of those remains of testaceous animals or shells, which have been preserved at various times and depths in the earth, vol III. W. Harding, London. https://www.biodiversitylibrary.org/bibliography/14408 (not in copyright)

Spada G (1744) Corporum lapidefactorum agri Veronensis catalogus. Typis Dionysii Ramanzinii: Bibliopolæ apud S. Thomam, Veronæ. Zentralbibliothek Zürich, NG 58 I G, https://doi.org/10.3931/e-rara-50868 / Public Domain Mark

Spoor H (1707) Favissae utriusque Antiquitatis tam Romanae tam Graecae, in quibus reperiuntur simulacra deorum, icones magnorum ducum, poetarum, ... Typis Gerardi Muntendam. https://libraryimage.nga.gov/uv/?manifest= https://libraryimage.nga.gov/manifest/mms/991836413504896.json#?c=0&m=0&s=0&cv=5&xywh=3275%2C475%2C3727%2C3717 (Public domain)

Stelluti F (1637) Trattato del Legno Fossile Minerale, Nuovamente Scoperto, nel quale brevemente si accenna la varia, & mutabil natura di detto Legno, rappresentvi con alcune figure, che mostrano il luogo dove nasce, la diversità dell'onde, che in esso si vedono, e le sue così varie e maravigliose forme. Vitale Mascardi, Roma. https://digital.slub-dresden.de/werkansicht/dlf/16093/4 Public Domain Mark 1.0

Vaiani SC, Vai GB, Borsetti AM (2019a) The lectotype of Ammonia beccarii (Linnaeus 1758) from Jacopo Bartolomeo Beccari's original samples. Micropaleontology 65(6):561–566

Vaiani SC, Vai GB, Borsetti AM, Sarti C (2019b) From ammonites to ammonia, early history of micropaleontology by Jacopo Bartolomeo Beccari (1682–1766). Micropaleontology 65(6):551–560, text-figures 1–3, appendix 1.

Vavra N (2004) Die "Ammonshörner" des PLINIUS – Gastropodensteinkerne zu Ehren von Baal Hammon? Jahrb Geol Bundesanst 144(1):137–140

Vialli V (1974) Appunti di Paleontologia. Pitagora Editrice, Bologna

von Minutoli HF (1824) Reise zum Tempel des Jupiter Ammon, in der Libyschen Wüste, und nach Ober-Aegypten in den Jahren 1820 und 1821/von Heinrich Freiherrn von Minutoli ... Bei August Rucher, Berlin

Waller R (1705) The posthumous works of Robert Hooke ... containing his Cutlerian lectures, and other discourses, read at the meetings of the illustrious Royal Society ... Illustrated with sculptures. To these discourses is prefixt the author's life, giving an account of his studies and employments, with an enumeration of the many experiments, instruments, contrivances and inventions, by him made and produc'd as Curator of Experiments to the Royal Society. Printed by Sam. Smith and Benj. Walford (Printers to the Royal Society) at the Prices Arms in St. Paul's Church-yard, London. https://wellcomecollection.org/works/xxggvyj3 (Public Domain Mark)

Worm O (1655) Museum Wormianum, seu Historia rerum rariorum, tum Naturalium, quam Artificialium, tam Domesticarum, quam Exoticarum; quae Hasniae, Danorum in aedibus Authoris Servantur. Iohannem Elsevirium, Lugduni Batavorum. https://doi.org/10.5962/bhl.title.63844 (not in copyright)

Chapter 18
Histerolithus Lapis, from Strange Formed Stone to Brachiopods

Abstract The phylum Brachiopoda is composed of exclusively marine invertebrates. They have bilateral symmetrical shells, formed by a ventral or peduncular valve, and a dorsal or brachial valve, called so because the brachidium is positioned there, a peculiar structure that supports the lophophore, the organ responsible for respiration and nutrition. They are known from the Paleozoic to the Present. The first images that can be interpreted as fossil shells of brachiopods are found in Gessner (*De omni rerum fossilium genere, gemmis, lapidibus, metallis, et huiusmodi, libri aliquot, plerique nunc primum editi*. Tiguri. https://doi.org/10.3931/e-rara-4176 / Public Domain Mark, 1565), but Fabio Colonna (*Fabii Columnae Lyncei Minus cognitarum rariorumque nostro coelo orientium stirpium ekphrasis …* https://www.biodiversitylibrary.org/bibliography/723 Public domain, 1616) is responsible for their representation as fossils distinct from other testacei. Their systematics has been elaborated only since the nineteenth century and is still subject to further study. Due to the resemblance of some of their fossils to the female genital organ, they were called hysterical stones, but their contribution to popular folklore is very scarce.

Keywords Fossilia · Fossils · Brachiopoda · History · Hysterolites · Lamp shells

© The Author(s), under exclusive license to Springer Nature Switzerland AG 2022

M. Forli, A. Guerrini, *The History of Fossils Over Centuries*,
https://doi.org/10.1007/978-3-031-04687-2_18

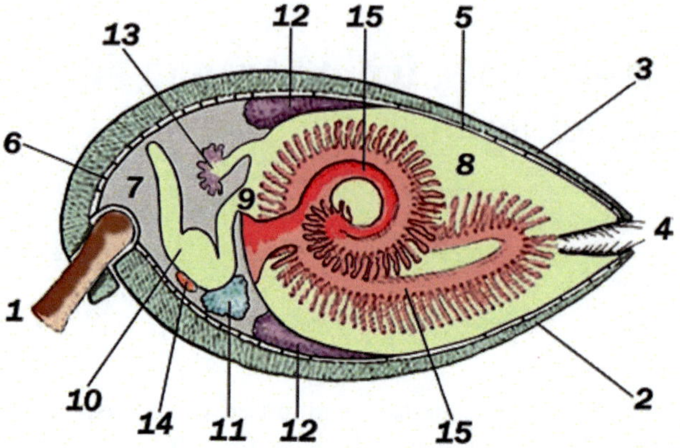

Fig. 18.1 Simplified anatomical structure of a brachiopod: (1) Peduncle. (2) Brachial valve. (3) Peduncular valve. (4) Ciliate opening. (5) Mantle. (6) Mantle expansions. (7) Visceral cavity. (8) Mantle cavity. (9) Oral aperture. (10) Stomach and intestine. (11) Digestive gland. (12) Gonads. (13) Nephridium. (14) Heart. (15) Lophophore. [Drawing Maurizio Forli modified after Vialli (1974)]

The *phylum* Brachiopoda is composed of exclusively marine invertebrates with inequivalved and equilateral shells, that is, with bilateral symmetry. The shell is formed by a ventral or peduncular valve, and a dorsal or brachial one, so called because it supports the brachidium, which is generally smaller. The shell is secreted by the epithelial cells of the mantle and can be formed either by calcite or by apatite, a complex form of calcium phosphate. In the phosphatic shells it is often present also an organic substance, the chitin. In this case, it is more appropriate to speak of chitinophosphatic shells. The plane of symmetry is perpendicular to the junction plane of the two shells, that is, to the union of the ventral valve with the dorsal one, unlike the bivalve mollusks that are symmetrical with respect to the right and left valves. The size of the living species measurements range from about 1–8 centimeters. The representatives of two Paleozoic genera, *Gigantoproductus* and *Titanaria*, widespread in the Carboniferous, are considered to date the giants of the category, being greater than 30 centimeters. They appeared for the first time in the Cambrian, becoming widespread and important in the Paleozoic, and to a lesser extent in the Mesozoic, leaving us with abundant fossilized remains. Few hundreds of genera and about 360 current species contrast sharply with more than 3000 genera and about 30,000 fossil species, showing that the phylum is facing a sharp decline even if they are still quite common in today's seas, distributed at all depths, with a preference for cold and deep waters, up to about 6000 meters. They are benthic filter-feeding organisms, that do not tolerate large variations in salinity, which live generally attached to hard substrates on the seabed by means of the peduncle. Some forms without the peduncle (e.g., *Crania*) attach themselves to their substrate by means of one of the valves; the anchorage is almost always permanent, being mobile only during the larval stage. Exceptions to this way of life are the Lingulids, considered "living fossils" because they seem to have remained unchanged from the Paleozoic to the Present. These species thrive in muddy–sandy environments, with wide range of salinity. In the Paleozoic, many brachiopods were able to live on various soft substrates developing particular adaptive strategies, such as spines or wing expansions, which prevented the sinking of the shell in the more or less incoherent sediments of soft bottoms.

In these invertebrates, the internal cavity delimited by the two valves, encloses the soft parts divided into the visceral cavity (posterior) and the mantle cavity (anterior). In the visceral cavity, are located the main organs, digestive tract, excretory, nervous and circulatory systems, genital glands, etc., and the muscles responsible for the opening and closing of the valves. All the soft parts are enveloped by the mantle, formed by two epithelial membranes adhering to the inner surface of the shell. The visceral cavity prolongs in the mantle by means of long and thin tubular extensions which can leave inside the valves some impressions with ramified course called, in analogy to those present in bivalve mollusks, pallial impressions. The composition, the morphology of the shell, and the anatomy of the brachiopods are rather complex; therefore, we shall not go into the description except for two particulars characterizing them: the brachidium and the lophophore. The brachidium is a sort of rigid skeletal structure, internal expansion of the shell, formed by two elongated elements joined together. In some primitive brachiopods there were only two short calcareous processes, which probably did not reach the lophophore, called *brachiophores*. Longer, blade-shaped apophyses, which provide support to the lophophore, are called *crura*. In some forms only the base of the lophophore was supported by the crura, while in others there were more complex ribbon-like structures, collectively referred to as brachidia. There are two types of brachidia, a simple ribbon-shaped one, variously developed, called *loop*, and a complex one, *spiralia*, formed by a pair of spiral cones oriented in different ways. In many Spiriferidae there is a bridge structure, called *jugum*, which connects the first turns of the spiralium. These particular structures protrude into the cavity delimited by the mantle and serve as a support for the lophophore, a soft organ of variable shape, responsible for respiration and nutrition, consisting of an elongated structure carrying numerous vibratile cilia that create a continuous flow of water inside the shell. The cilia thus trap particles of nourishment that are entangled in a mucous fluid that carries them toward the digestive system. The shape of the brachidium, which in some way mirrors that of the corresponding lophophore, is one of the distinctive characters, often preserved even in the fossil state, used in the classification of the various species (Fig. 18.1, Fig. 18.3).

Classification

Fossil brachiopods possess a great diversity but few discriminatory features while the present genera, although presenting a less pronounced diversity, have many in both soft and hard parts: these two factors are limitations that make a complete classification of the phylum difficult. Some taxonomists think it is still premature to define the highest taxonomic levels as the order and recommend a basic approach that identifies groups of genera and these in intermediate level groups (families), while others argue that some types of characters are sufficiently stable and that there are prerequisites to adopt a classification approach at the highest hierarchical levels. The taxonomic group Brachiopoda was established by Duméril in 1806, while the so-called "traditional" classification, based mostly on the morphology of the shell, was the work of Huxley (1869) who distinguished two groups, the Articulata, which have hinges with teeth between the valves, and the Inarticulata, whose valves are bound only by muscles. More recent classifications, nowadays more used, are mainly based on the composition of the shell and on the anatomy.

- Articulated Class (Cambrian–Current).

Calcareous shells, valves articulated along a cardinal axis by two ventral teeth and two dorsal dimples. Valves open by abductors and closed by adductors. Lophophore always supported by a mineralized structure, usually crura, sometimes prolonging into a brachidium. The peduncle develops directly from an embryonic structure. The digestive tract terminates blind, without anus.

- Class Inarticulate (Middle Ordovician–Current).

Brachiopods with calcareous shell, lacking teeth and dimples. Reduced or absent peduncle.

Some scholars consider the Lingulata as a subclass of the Inarticulata, others as a systematic category equivalent to the other two:

- Class Lingulata (Cambrian–Current).

Brachiopods with chitinophosphate shell, without teeth and dimples. The valves are held together only by the muscles and the body wall. The peduncle is formed by a protrusion of the ventral mantle. The lophophore is without a mineralized support structure. The alimentary canal has an anal opening.

Hereafter, the classification of the species whose representation is considered the first explicit printed figure of a fossil brachiopod, taken from *Aquatilium et terrestrium aliquot animalium, aliarumquae naturalium rerum observationes* Colonna (1616), even if some authors (e.g., Davidson, 1884) propose a first edition of the *Aquatilium* for the year 1606 (Fig. 18.2, Fig. 18.5).

Phylum Brachiopoda
Class Rhynchonellata
Family Pygopidae
Genus *Pygope*
Species *Pygope diphya* de Buch, 1834 †

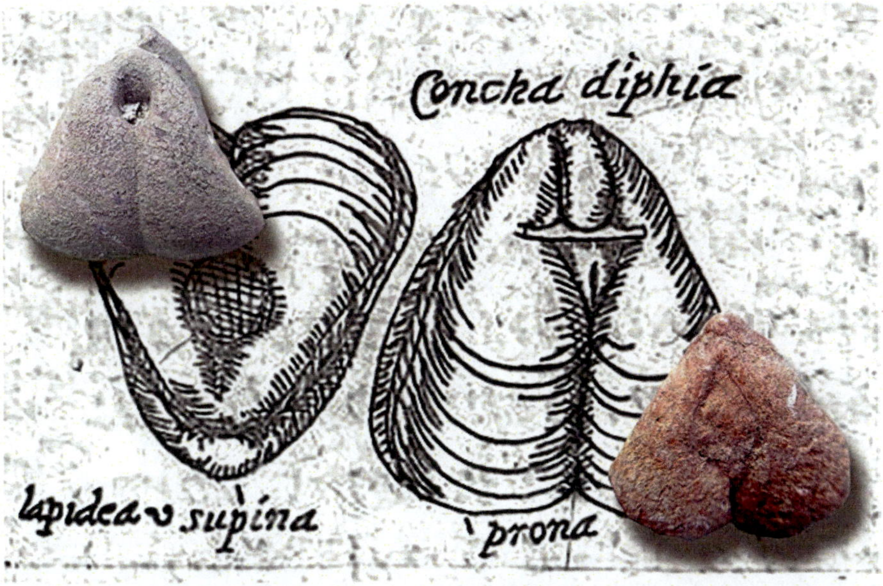

Fig. 18.2 *Concha diphia* by Colonna (1616). Aquatilium et terrestrium aliquot animalium, aliaruq. Naturalium rerum observationes. In Ekphrasis altera pl. XXXVI. On the sides, two similar Jurassic specimens from Sierra Subetica (Cabra), Spain. [Graphic composition Maurizio Forli, photo credit Mauro M. Brunetti; https://www.biodiversitylibrary.org/bibliography/723 (Public domain)]

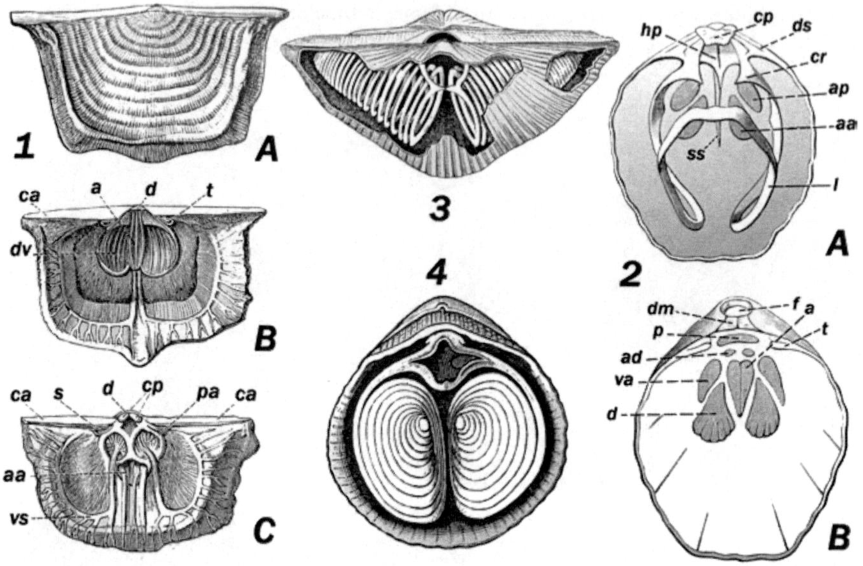

Fig. 18.3 Shells and internal structures of some brachiopods: (1) *Leptaena rhomboidalis* (Paleozoic, Silurian). (A) Exterior of ventral valva: (B) Interior of ventral valva. (a) footprint adductor (occluder) muscle. (d) pseudo-deltidium. (dv) retractor muscle. (ca) hinge area. (t) teeth. (C) Interior of dorsal valva. (aa) anterior adductor muscles. (pa) posterior adductor muscles. (ca) hinge area. (cp) cardinal process. (d) chilidium (s) dental dimple. (vs) vascular sinuses. (2) *Waldheimia flavescens*. (A) Interior of ventral valva. (a) impressions of adductor muscles. (va) impressions of ventral regulators muscles. (d) impressions of retractor muscles, (ad) accessory retractors muscles. (p) peduncular muscle. (dm) deltidium. (f) foramen. (t) teeth. (B) Interior of dorsal valva. (aa) scars of anterior adductor muscles. (ap) scars of posterior adductor muscles. (cp) cardinal process (cr) crura. (ds) dental dimple. (hp) plane of hinge. (l) brachidium rings. (ss) septum. (3) *Spirifera striata* (Paleozoic, Carboniferous) double-spiral shell and brachidium. (4) *Atrypa reticularis* (Paleozoic, Silurian) inside of valves from dorsal side with brachidium coils in evidence. [Graphic composition Maurizio Forli, modified after Reed FRC, Brachiopods (Fossil), (1895). https://www.biodiversitylibrary.org/item/214013#page/11/mode/1up (copyright not provided)]

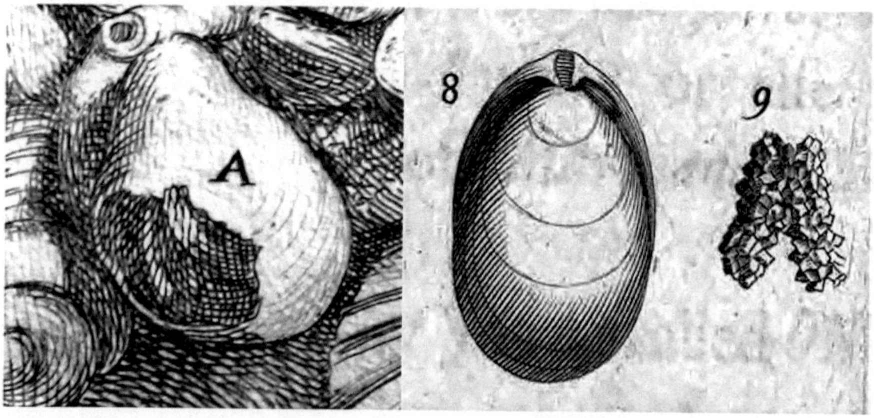

Fig. 18.4 (1) Illustrations of brachidium in two seventeenth-century works, left from Scilla, *what can be seen from a small break*. A. [...] *with the aid of the transparency of the shell* (1759: p. XIX, fig. 2), and on the right (2) from Plot, (1677, pl. IV, figs 8–9). [(1) personal copy Maurizio Forli (2) https://www.biodiversitylibrary.org/bibliography/23488 (not in copyright)]

Sporadically some manually drilled brachiopods, similar to other fossil types, have been recovered from Paleolithic contexts; given their apparent practical uselessness, their use must have been ornamental (Moncel et al. 2012).

These particular invertebrates generally do not seem to be recognized as different from the fossil remains of acephalic mollusks and therefore are not listed separately in ancient lapidaries. Some hints in Pliny to stones called *Diphyes*, with a double valence, have convinced the Renaissance and later authors, to associate them with the *Histerolithos*, petrous bodies similar to the Bucardie, in imitation of the female genital organ, commonly called *Diphyen*, some of which today are actually included among the brachiopods. As we will see shortly, in retrospect, the succinct Plinian description fits perfectly the specimen illustrated by Colonna with a similar name, corresponding precisely to a characteristic Mesozoic brachiopod.

> *Diphyes duplex, candida ac nigra, mas ac femina, genitale utriusque sexus distinguente linea.* (Pliny, XXXVII: LVII)

> *The difie is of two sorts, black and white, male and female; and has a line to distinguish its sex.* (Domenichi 1844, 2: p. 1432)

The earliest images that can be interpreted as fossil shells of brachiopods are found in Gessner, (1565: p. 166) although they cannot be attributed with certainty because of the inaccuracy of the illustrations and the corresponding descriptions. These are the specimens numbered 2 and 3, named *Pectunculus ferreolus* and *Alius lapideus totus,* respectively. What leads to think that they may be brachiopods is the profile of the shell, rounded with two narrow expansions in correspondence and coincidence of what, compared to the symmetry and the morphology of the shell of brachiopods, should be the umbonal area, with the deltidium and the foramen (visible in the specimen in Figure 3), and the frontal area. Moreover, the arrangement of the rayed ribs, starting from the central axis, even if improper, recalls more closely the ornamentation of brachiopods than that of bivalve molluscs. The same figure is reproposed in Aldrovandi (*Musaeum metallicum*, 1648: p. 145) without adding anything to what Gessner had written, even if in an earlier work of the Bolognese author brachiopods are well recognizable, evidently not considered different from other marine shells, among which they are included, and without comparing them to other previous representations. In fact, in *De Reliquis Animalibus exanguibus libri quator* (Aldrovandi 1606: p. 477), even if the engravings render minuscule prints, actual Mediterranean specimens are evident, probably *Megerlia truncata* (Linnæus 1767) (18.6). Still in the *Musaeum Metallicum* at page 445 you can see three engravings of fossils called Hieracite, actually a Permian brachiopod of the genus *Productus*, *Productus geinitzianus*, currently kept in the Capellini Museum:

> *Hieracites tres differentiae* [...]: *these are three specimens of the Permian brachiopod* Productus geinitzianus *Kon., one of which bears the bryozoan* Stenopora columnaris *Schloth on its valva. The description of the specimens is given on page 444 of the MM: Observantur alius lapis, qui colore pennas, veloculos Accipitris, sive Hieracis imitans, appellatur Hieracites, & ut plurimùm ad colorem cinereum tendit...* (Sarti 2003: p. 162)

in the continuation of the book, other figures could represent brachiopods, in particular, that numbered 5 on page 727 and those numbered 3 and 4 on page 838.

As anticipated, it is in the works of Colonna (1616) that these fossils are represented with certainty distinct from the other testacei, illustrated together with them in the respective plates. The first and most characteristic, depicted in two positions, prone and supine, named *Concha diphia*, was later included in the synonymy of what was first classified as *Terebratula diphya* and then *Pygope diphya* (von Buch 1834). Representatives of the genus *Pygope* Link, 1830 belong to a particular family (Pygopidae) of brachiopods, which lived and became extinct in the Mesozoic, in a period between the Jurassic and Cretaceous periods (~160 and ~95 Ma). They are characterized by a biconvex shell with a rounded trigonal profile, formed by two lateral expansions which, during growth, join at the front, leaving a central depression or opening. Fossils are found in the calcareous sediments of ancient marine environments.

From the original text we extrapolate this short description that, in its immediacy, gives a good idea of the morphology of this particular polymorphic brachiopod:

> *... ex genitaliium maris et foemine effigie; quam in summo vertice exprimi putatur, sed diphya quia duplex sive bifida aut bipertita sive gemina concha videatur, veluti si binos Mytulos latere coniunctuos natura produxisset* [Colonna 1616: pl. 1 (50)].

effigy of the genitals of males and females; always showing a shell composed of two parts or bifid divided into two or bipartite double shell, if broken and divided into two by the upper part, as if nature had produced the lateral union of two mussels.

The other brachiopods described by Colonna are part of *De Purpura, Alisq; Testaceis rarioribus*, printed and appended together with the *Aquatilium* ..., and are included and illustrated at the top of the plate on page 22. Named *Concha anomia apex overtice?*, the specimen on the upper left was referred to by Linnaeus (1758: p. 703, no. 201) as the first reference of his *Anomia terebratula* and also later used by Brocchi for the synonymy of the species he proposed under the name *Anomia sinuosa* (1814: p. 469, pl. 10, fig. 7), thus causing a taxonomic problem. The specimen on the right would correspond to the *Anomia ampulla* established by the same author a few pages earlier (Brocchi 1814: p. 466, pl. 10, fig. 5). However, the systematic position of these two brachiopods, from the surroundings of Andria in Apulia and of Pliocene age, is rather complicated and not yet definitively clarified. Specialists do not agree on the validity of all the species, these and others related to them, established later. The third and largest shell depicted centrally is instead unanimously attributed to a rhynchonellid.

According to Worm, the correct original name to be given to the corresponding Plinian *Diphyes* is *Diphris* (double), as would also be reported in Agricola and Gessner, representing included in this category, coincidentally, some internal models actually remains of brachiopods, calling them *Histerolithos lapis* (Worm, *Museum Wormianum*, 1655: pp. 83–84).

In reality, under these names, or other similar ones, are included different stones, not always attributable to brachiopods, and also slags resulting from metallurgical processing. Often with these names are identified internal models of bivalves that also recall to the imagination of the observer, feminine and masculine attributes: androgynous fossils of undoubted charm, both for shape and meaning, often remembered as *Hysterolithi, Hysterolithes*, and *Hysterapetra*, deriving their name from the similarity with the "shameful" feminine part.

It is so, in fact, that for a good part of the 1600s and 1700s several fossilized remains are still indicated, which can be referred to brachiopods but also, and above all, to internal casts of shells, which we cannot distinguish on the basis of written explanations if lacking the relevant iconographies.

An example of this way of proceeding is given by a short passage of the Agricola, reported among others by Gessner and de Boodt, where the discovery of these particular impudic stones is mentioned, but without any explanatory images, thus making it impossible for our purposes to understand what it was really about:

> *Agricola a raconte que lors qu'on demolissoit les fondmens de la Citadelle Erebreistemensis dans le Diocese de Trevir, furent trouvées des pierre noires & dures, qui representoient & exprimoient les parties hontuses d'une femme. Cardan les appelle à cause de cette similitude hysterapetra. Plus à propos on les pourroit appeller hysterolites. Agricola adiouste que l'on les trouve aussi a Mariebourg. Elles n'ont aucun usage que je sçache* (de Boodt 1644: p. 525).

Agricola said that when the foundations of the Citadel of Erebreistemen in the Diocese of Trier were demolished, hard black stones were found, representing and expressing the dishonorable parts of a woman. Cardano, because of this resemblance, calls them *hysterapetra*. More to the point, they might be called *hysteroliths*. Agricola adds that they are also found at Mariebourg. To my knowledge they have no use.

The opposite case is the one we find in Moscardo (1656: pp. 185–186), where brachiopods are included in the plate together with mollusk shells but are left undetermined. An iconic representation without correlative description.

> [...] *the Mitulo, & others, of which I do not find any particular mention by Writers, all made of stone.* [...] *Other Conches* [...] *Petunculi with many others, which being unknown by name, are also unknown to the eye, I place them.*

About 200 years later, in a much more explicit way, the German scholar von Schlotheim (1820: p. 247) established the genus *Hysterolites* on a fossil brachiopod with a distinctly vaginal shape, giving the species the name of *H. vulvarius*, taxon still considered valid under the present denomination of *Schizophoria vulvaria* (von Schlotheim 1820).

In the second half of the seventeenth century, morphological differences with other similar fossil remains began to be recorded and noted with increasing frequency, and this was reflected in the subsequent cataloguing and depictions. For example, in Frederic Lachmund (1635–1676) a series of brachiopods, although still called *Conchites* and roughly distinguished according to physiognomy, are grouped together in a single plate (1669: pp. 41–43). The next print shows a single rock specimen containing remains of ammonites and brachiopods, apparently of the Rhynchonella; these too are described by a vague term of broad meaning, *Lapis Megaricus*, referring the reader to earlier authors who had used this appellation: the Johnston:

> *Lapis Megaricus, in cujus toto corpore conchae marinae sparsae sunt* (1661: p. 65),

and especially Agricola: *Questa maniera di sassi massimamente i molli, e i mediocri sogliono essere a molti vitij* (vizi) *soggetti: perché à le volte in se contengono marga; la quale, quando si toccano i sassi col ferro per scolpirli, ò impianarli; si scuote, e ne va via fuori, onde ne restano pertugiati i sassi* [...] *a volte hanno in se picciole pietre à gemme somigliantissime* [...] *Anzi in Megara in un sasso bianchissimo e più di ogni altra pietra, molle si veggono d'ogni intorno conche marine attaccate dentro; onde il chiamano perciò pietra Conchite, de la quale* (*come scrisse Pausania*) *fu fatto il monumento di Phoroneo; e molte altre opere ne la citta di Megara* [traduzione da Agricola (Tramezzino 1550: pp. 327–328)].

This type of stone, especially soft stones and mediocre ones, are subject to many vices: *because sometimes in themselves they contain marl, which, when you touch the stones with iron to carve them, or level them, shakes, and it goes away, so that the stones remain riddled* [...] *sometimes they have in themselves small stones with very similar gems* [...].*Indeed in Megara in a very white stone and more than any other stone, are seen all around sea shells attached inside; so they call it Conchite stone, of which* (*as Pausanias wrote*) *was made the monument of Phoroneo; and many other works in the city of Megara.*

A corresponding find, a rock block containing Rhynchonella, is described and depicted by Robert Plot (1640–1696), an English naturalist and author of *The Natural History of Oxfordshire* (1677: pp. 99, pl. 3, fig. 13), still with the exact same definition and designation as Lachmund. The specimens shown are most likely to be attributed to *Tetrarhynchia tetrahedra*, a Jurassic brachiopod from the vicinity of Adderbury, a town in Oxfordshire, a county in southeastern England. Curiously, but not too curiously, in keeping with the point of view and method of study of the time, three other brachiopods are shown in the following plate, all different from each other, and of the one shown at number 8 [now classified as *Lobothyris punctata* (Sowerby, 1812)] a partial brachidium is also shown, but Plot does not suggest anything about this unusual internal structure of the shells, merely describing it and being surprised by what he had never seen before (Fig. 18.4, Fig. 18.7).

And first, the above-mention'd Conchites *found in Hornton quarry, and represented in Tab. 4. Fig. 8. which is not a solid stone within (as all the Cochle-Stones hitherto described have been) but hollow, and filled with spar; somtimes shot into irregular figures, but for the most part forked, as in fig. 9. the basis, or place where the branches of the fork are conjoyned, being rooted (in all that I have yet seen) at the commissure or hinge of the valves, and the branches extending themselves in the broader parts of the Conchites; of which operation of Nature I can give no other account, but that it was first observed, by the Reverend and Ingenious Mr. Clark, Rector of Dreyton near Banbury, from whom, befide other favors, I received many of them* (1677: p. 103, pl. IIII, figs 8–9).

Actually, at least as far as we know, a similar structure had been observed, described and partially drawn a few years earlier by Agostino Scilla, a keen observer, attributing it to the remains of the animal contained inside the shells. In this case there are brachiopods of the Pleistocene age, incorporated together with other remains, in a "*firm stone.*"

[...] *confondendo in un groppo molte ossa simili a stinchi d'animale, conchiglie semplici, e striate, turbini con casuale posizione, e molte conchiglie petrificate, e non petrificate* [...] *mantiene alcune conchigliette, che per esser rimaste vacue di loto, conservano ben custodito l'animal petrificato dentro, rimirandosi manifestamente le membranuzze necessarie, e*

> *proprie di quello. Mi dispiace però, che non posso portare in disegno una veduta totalmente graziosa* [...] *quel che si scorge da una piccola rottura.* A. *della detta conchiglia, con l'aiuto della trasparenza del guscio* (1670: p. 139, pl. XIX, fig. 2a)

> [...] *mixing in a tangle many bones similar to animal shanks, simple and striped shells, whirlwinds with a random position, and many petrified and non-petrified shells* [...] *it keeps some small shells, which, because they have remained empty of* dirt, *keep the petrified animal well preserved inside, clearly showing the necessary and proper membranes of the animal. I am sorry, however, that I cannot draw a totally gracious view* [...] of *what can be seen from a small aperture.* A. *of the said shell, with the help of the transparency of the shell.*

Another specimen, apparently conspecific to the one containing the remains of brachidium, is depicted adhering to a matrix that also includes echinid remnants, in plate XXIII. Two other brachiopods, superbly reproduced, are depicted separately in the print on page XIV, called *Anomie*, as *said by Colonna*. These same illustrations, referring, however, to the corresponding plates of *De Corporis marinis lapidescentibus* ... (1759: pl. XIV, fig. 6) were used as a reference to establish the species *Terebratula scillae*, in 1871, by Giuseppe Seguenza, geologist and naturalist, like Scilla also a native of Messina (1833–1889).

The denomination *Terebratula* appears to us to be used for the first time by Eduard Llwyd in his book *Lithophylacii Britannici ichnographia* (1699: p. 40), with reference to fossil specimens represented, to tell the truth rather clumsily, in plates VI and VII. At present, the valid designation of the Genus *Terebratula* is attributed to Müller (*Zoologiae Danicae prodromus*, 1766: p. 249, no. 3006).

Linnaeus himself distributed the shells of brachiopods in various genera included among the bivalve molluscs, among which, for example, that called *Anomia craniolaris* (1758: p. 700), a fossil brachiopod, belonging to the genus *Crania*, peculiar for its shape and for some beliefs derived from it, known as *Nummulus brattensburgensis*, Brattinsborg coin. The ventral valva of this particular species, flattened and rounded like a coin, bears engraved on the inside a *dead man's head*, an effigy in the shape of a skull. Known since the middle of the first millennium, this singularity has given rise to many medieval legends along the lines of those linked to nummulites, also believed to be petrified coins due to witchcraft, apparently developed from the island of Ivö in Scania (Sweden), where these particular fossils are abundant. From this locality come the specimens used by Retzius (1781) to describe the genus *Crania which is* still scientifically valid.

The formalization of the systematics of Brachiopoda was rather complicated and it is only since the early 1800s that their differential characteristics are more explicitly understood. We owe to the French zoologist André Marie Constant Duméril (1774–1860) the definition of the group *Brachiopodes* distinguished according to the following characteristics: *Mollusques Testacés sans tête, à tentacules ciliées, rentrant dans l'intérieur de la coquille qui est fixée* (1806: p. 171). Still, however, in the same systematic grouping, are included the Cirripedia, completely different organisms, belonging presently to the Crustacea in the phylum Arthropoda.

Correctly, instead, are included the representatives of the Lingulata, at least apparently more similar to the Lepads, belonging instead to the Cirripedia. Certainly this consideration was influenced by the opinions and studies of his illustrious contemporary colleagues, Bruguière, Lamarck and Cuvier who, basing themselves on living specimens, created the name *Lingula* (1791), described the first species, *Lingula anatina* (1801), and studied its anatomy in detail (1802) respectively.

The first report of fossil *Lingula*, on the contrary, is due to the English naturalist James Sowerby, who in 1812 described and represented some specimens, coming from the Jurassic of Great Britain.

The brachiopods, due to their various forms, especially in the English folklore, have assumed various names among which the most known are lamp shell, shells shaped like an ancient oil lamp, and *Delabole butterfly*, from the locality where the remains of a spiriferid brachiopod were found, considered by the popular imagination as a petrified butterfly. Their utilization for medicinal and curative purposes, at least in the European area, is practically nil. There are not diffused popular traditions and also their employment as amulet or talisman, we suppose may have been possibly occasional (Fig. 18.8, Fig. 18.9, Fig. 18.10, Fig. 18.11, Fig. 18.12, Fig. 18.13, Fig. 18.14, Fig. 18.15, Fig. 18.16, Fig. 18.17, Fig. 18.18, Fig. 18.19, Fig. 18.20).

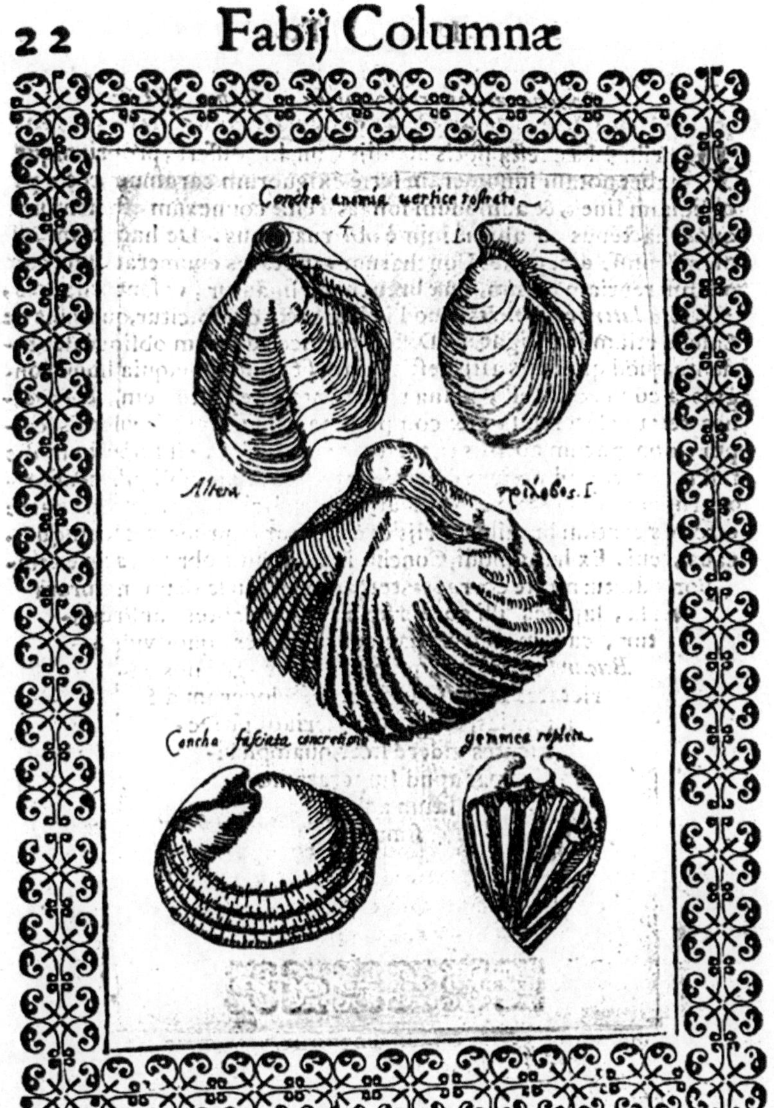

Fig. 18.5 Fabio Colonna (*De Purpura, aliisq. Testaceis rarioribus*, 1616: p. 22), top and middle illustrations from which one can easily identify terebratulae named "*Concha anomia vertice rostrato,*" and a rhynconella. The other illustrations at the bottom of the plate are instead to be attributed to casts of bivalves; in the one on the right is highlighted the crystalline filling, in all probability, consisting of calcite. [https://www.biodiversitylibrary.org/bibliography/723 **(Public domain)**]

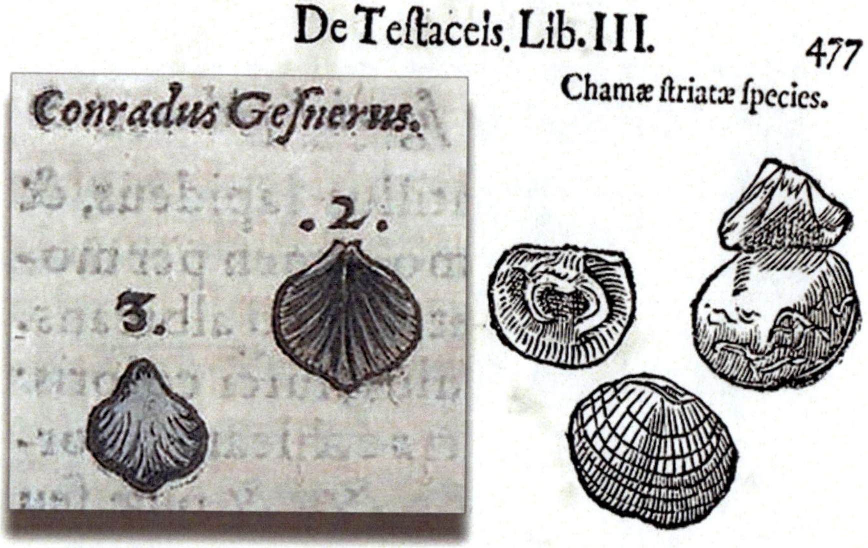

Fig. 18.6 Illustrations of fossil brachiopods from Gessner (1565: p. 166) (2) *Pectunculus ferreolus* (3) *Alius lapideus totus*. (1) Current Mediterranean specimens, probably *Megerlia truncata* (Linnæus, 1767) from Aldrovandi (1606: p. 477). [Graphic composition Maurizio Forli. Credit from: (1) https://www.biodiversitylibrary.org/bibliography/119999 Public Domain (2–3) https://doi.org/10.3931/e-rara-4176 / Public Domain Mark]

Fig. 18.7 On the sides two fossil brachiopods from the Tuscan Pliocene, in the middle two similar specimens opened to highlight the crystallized structure of the brachidium. Largest height 48 mm. (Photo and graphic composition Maurizio Forli)

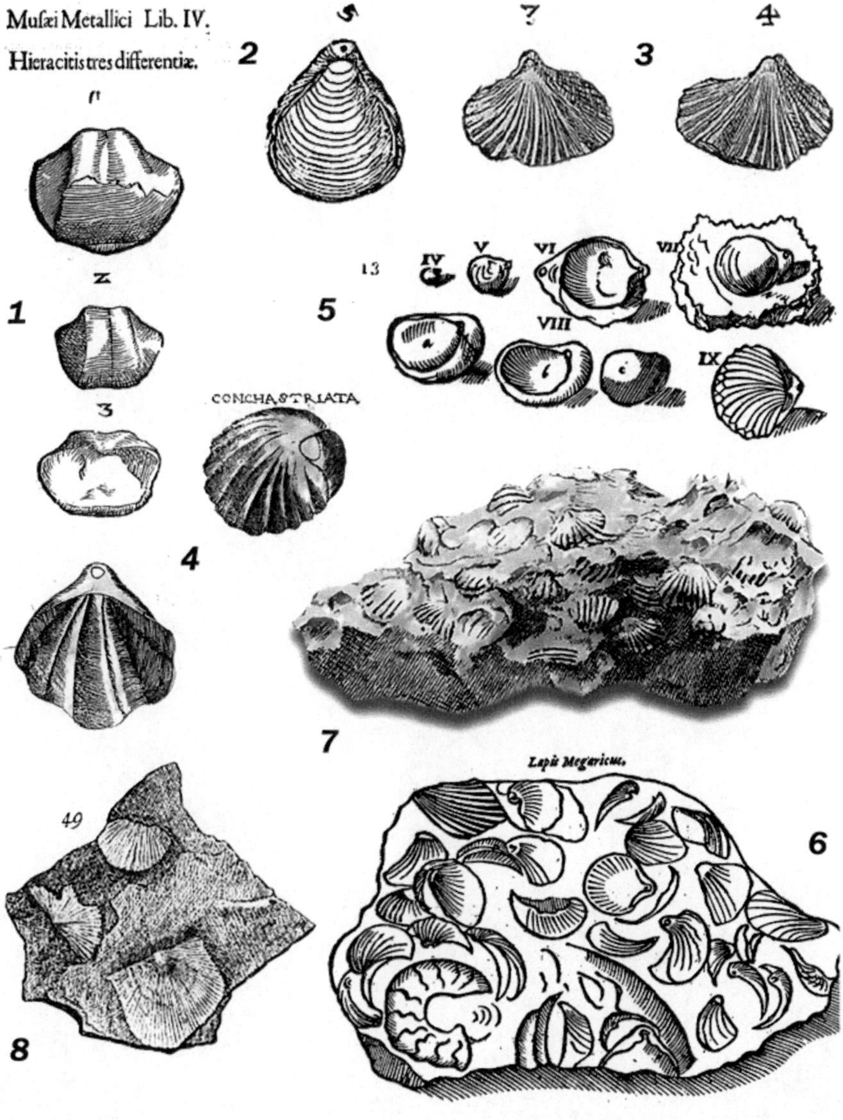

Fig. 18.8 Some of the earliest printed illustrations of brachiopods: (1) *Hieracitis, Productus geinitzianus*, a Permian brachiopod still preserved in the Capellini Museum, Bologna. (2) *Mirabilis icones silicum*. (3) *Differentiis Chamarum petrificatarum* (Aldrovandi 1648: pp. 445, 727, 838). (4) Moscardo, 1656: pp. 185–186; *striated conches*, probable brachiopods. (5) *Myites and conchites*. (6) *Lapis cinereus, qui totus ex conchis rotundis striatis et laevibus constat*; *Lapis Megaricus* (Lachmund 1669: pp. 42–46). (7) *Cockles in their full proportion*, probable *Tetrarhynchia tetrahaedra*, lower Jurassic of Adderbury, England (Plot 1677: p. 100, pl. III, Fig. 13). (8) Brachiopods from Lister (1678: p. 42). [Graphic composition Maurizio Forli. Credit from: (1–3) https://doi.org/10.3931/e-rara-13,017 / Public Domain Mark. (4) https://www.biodiversitylibrary.org/bibliography/157058 (not in copyright). (5–6) https://wellcomecollection.org/works/rs34c8db (Public Domain Mark). (7) https://www.biodiversitylibrary.org/bibliography/23488 (not in copyright) (8) https://www.biodiversitylibrary.org/bibliography/65746 (Public domain)]

Fig. 18.9 *Conchites lunatus*, a Paleozoic brachiopod, similar to the one illustrated by Aldrovandi, included by Mercati among the moon-shaped shells, together with Jurassic bivalves of the species *Gryphaea arcuata* (*Methalloteca Vaticana*, 1717: p. 292). Above the terebratulae, named *Myites* by Mercati, and by Lancisi referring to the corresponding specimens illustrated by Colonna and Lhwyd (Lancisi, *Appendix ad Metallotheca Vaticana*, 1719: pp. 17–18). [Color and graphic composition Maurizio Forli. Credit from: [https://archive.org/details/bub_gb_DC_E6UVWg7EC (Public Domain Mark 1.0)]

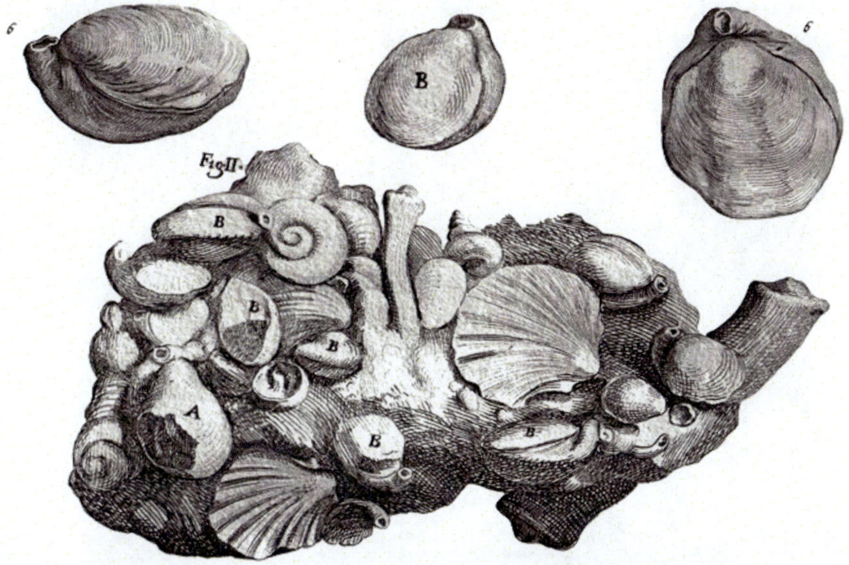

Fig. 18.10 Agostino Scilla, (1759) representations of Plio-Pleistocene brachiopods: top (6) from Plate XIIII, the others from Plate XIX. (Personal copy Maurizio Forli, modified)

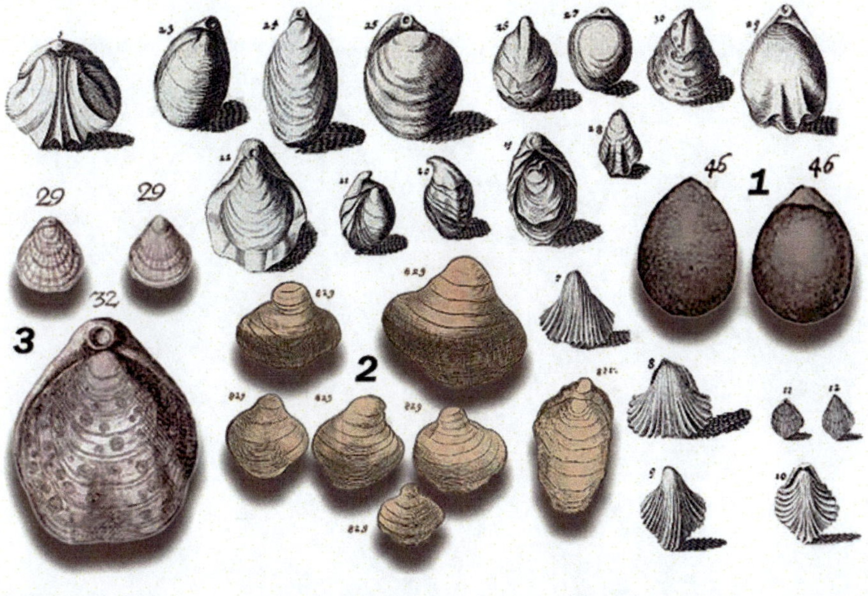

Fig. 18.11 Brachiopods of different ages and species in six-eighteenth-century illustrations: (1) Lister (1678: 230, fig. 46). (2) Lhwyd (1699, pl. 6). (3) Scheuchzer (1702, pl. 3, figs 29, 32) (4) All other images from Baier, *Oryktographia Norica*, (1708): pl. V (modified). [Graphic composition Maurizio Forli. Credit from: (1) https://www.biodiversitylibrary.org/bibliography/65746 Public domain (2) https://doi.org/10.3931/e-rara-30,541 / Public Domain Mark (3) https://doi.org/10.3931/e-rara-12,116 / Public Domain Mark (4) https://doi.org/10.3931/e-rara-18,618 / Public Domain Mark]

Fig. 18.12 Arenaceous block containing Jurassic brachiopods of the species *Isjuminella decorata* Schlotheim 1820 (ex *Terebratulites decoratus*); Mézières France, age ca. 165 Ma., mean size of specimens 15–20 mm. (Photo Maurizio Forli)

Fig. 18.13 Internal and external impressions of Paleozoic brachiopods of the genus *Orthis*. Ordovician (age 478 to 438 million years ago), length about 25 mm. (Photo Maurizio Forli)

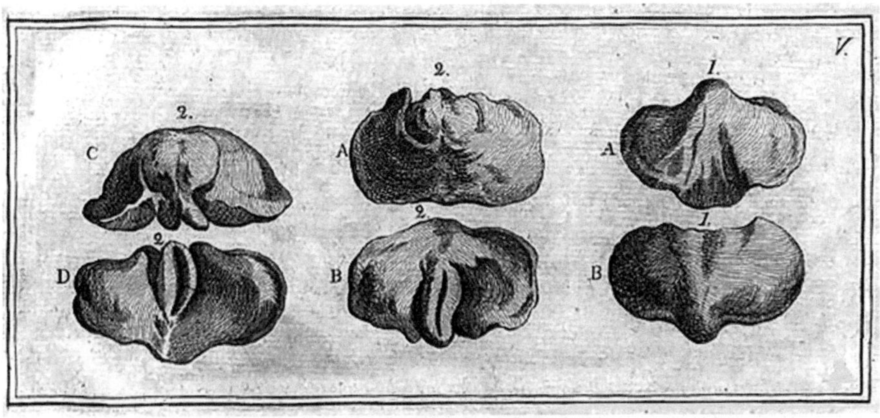

Fig. 18.14 Partial reproduction of figure V from *Museum Tessinianum*: (1) *Anomia caputserpentis*. (2) *Histerolithus.* Herr Gr. Carl Gust. Tessins Naturalie-Samling. Holmia [Stockholm]: apud Laurentium Salvium, (1753). [Modified after ETH-Bibliothek Zürich, Rar 3295, https://doi.org/10.3931/e-rara-41248 /Public Domain Mark]

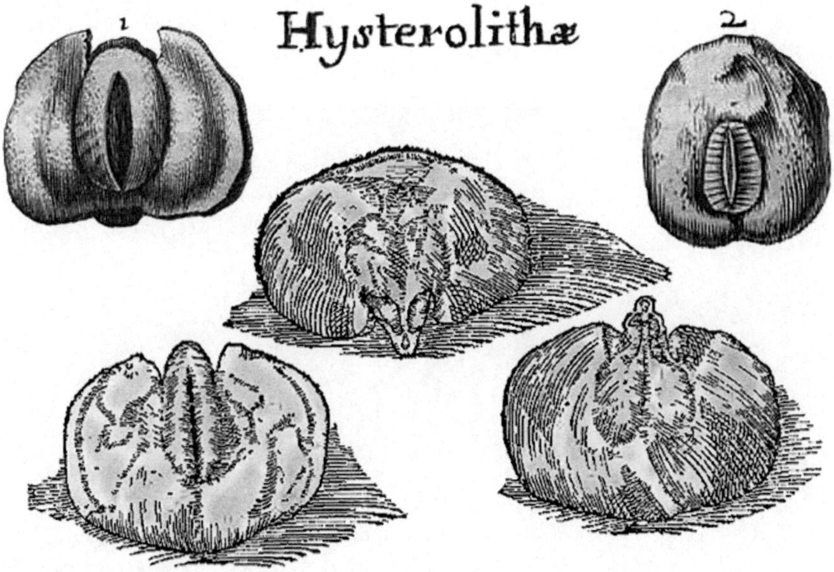

Fig. 18.15 (1) Bottom and center: *Histerolithos lapis* (Worm 1655: p. 83). (2) Top: *Hysterolithae* (Lang 1708: p. 48). Petrous imitations of female sex organs corresponding to casts of brachiopods. [Graphic composition Maurizio Forli. Credit from (1) https://doi.org/10.5962/bhl.title.63844 (not in copyright) (2) https://doi.org/10.3931/e-rara-10,398 / Public Domain Mark]

Fig. 18.16 Hysterical stones, Paleozoic brachiopods from Port-Bail (Normandy, France). Greater length about 30 mm. (Photo Maurizio Forli)

Fig. 18.17 (*Kienella* sp.) Close-up of a Paleozoic brachiopod imprint, also resembling the exterior of human female genitalia. (Photo credit Massimo Cresti)

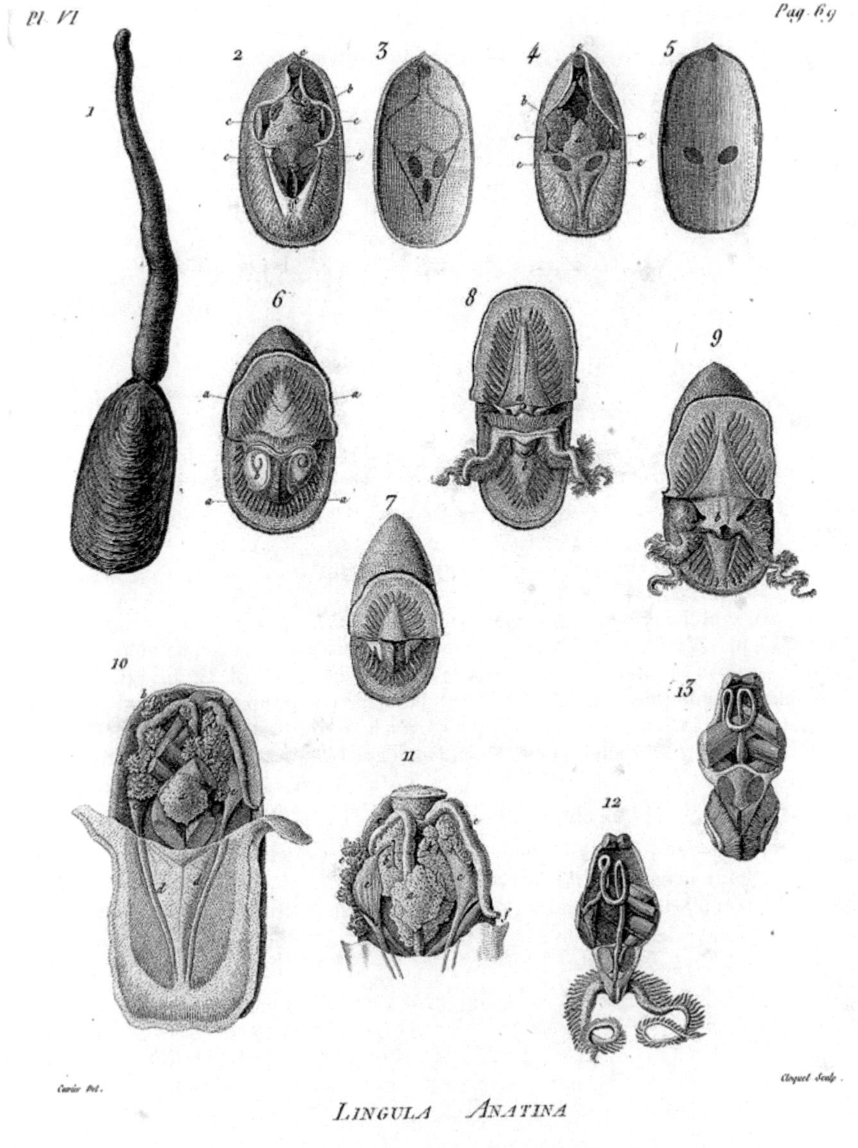

Fig. 18.18 Anatomy of the brachiopod *Lingula anatina* by Georges Cuvier: En effet, les coquilles des Lingule, quoique d'une forme assez particulière, ne pouvoient faire soupconner les grandes différences qui séparent leur animal des autres genres de sa classe; et tant qu'on n'a connu qu'elles, on les ballottées arbitrairement de genre en genre (Mémoire sur l'animal de la Lingula... 1802: p. 69, pl. VI). [http://hdl.handle.net/2268.1/2205 **(Public domain)**]

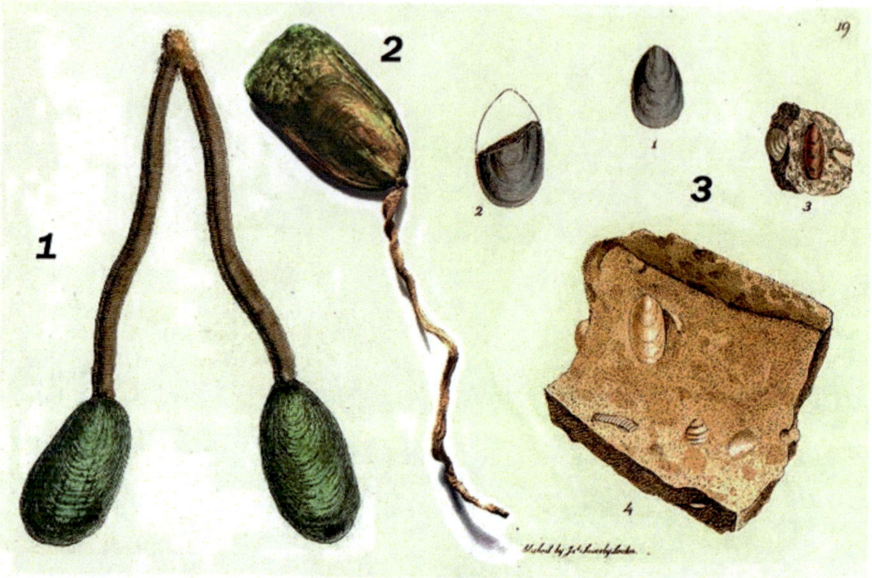

Fig. 18.19 Images of living and fossil Lingula: (1) Albert Seba (*Locupletissimi Rerum Naturalium Thesauri*, 1758, pl. XVI, fig. 4). (2) A present specimen (Burma, shell height 45 mm). (3) Fossils from James Sowerby [*The Mineral Conchology of Great Britain*, 1812, pl. 19, fig. 4 (Clay stratum near Pakefield in Suffolk, by Mr. John Thurtell)]. [Graphic composition Maurizio Forli. Credit from: (1) https://www.biodiversitylibrary.org/bibliography/62760 (not in copyright) (2) photo Misa Forli (3) https://www.biodiversitylibrary.org/bibliography/14408 (not in copyright)]

70 ANNALES DU MUSÉUM

pourquoi, qu'elle devoit passer dans le genre des jambonneaux, et la nomma *pinna unguis*. *Bruguière* est le premier auteur systématique qui ait su que ces deux valves sont naturellement attachées à un pédi… …neux, comme celles des *térébratules* et des … …ait fait en conséquence, dans les … …die, un genre particulier, dont il … …ption, parce que son voyage et … …duire jusque-là son dictionnaire … …citoyen *Lamark* a adopté et car… …ra d'autant plus sûrement, que … …oir par ma description, diffère … …eux des bivalves ordinaires.

Il est assez s… …systématiques aient été si long-temps dans l'erreur au sujet de la *Lingule*, tandis que cette coquille étoit déja parfaitement représentée avec ses deux valves et son pédicule, dans Séba, *tome III*, *pl.* 16, n°. 4; mais l'indication que cet auteur en donne en peu de

Fig. 18.20 In the background a passage taken from Georges Cuvier, (*Mémoire sur l'animal de la Lingula*... 1802: p. 70.) with in the foreground, a fossil specimen from the Tuscan Pliocene (height about 40 mm). [Photo and graphic composition Maurizio Forli. Credit from: http://hdl.handle.net/2268.1/2205 (Public domain)]

References

Aldrovandi U (1606) De Reliquis Animalibus exanguibus libri quatuor, post mortem eius editi: nemp' De Mollibus, Crustaceis, Testaceis, et Zoophytis. Apud Io: Baptistam Bellagambam, Bononie. https://www.biodiversitylibrary.org/bibliography/119999 (Public Domain)

Aldrovandi U (1648) Ulyssi Aldrovandi patricii Bononiensis Musaeum metallicum in libros IIII : distributum Bartholomaeus Ambrosinus ... labore, et studio composuit cum indice copisosissimo Arcus Antonius Bernia. Bononiae : typis Io. Baptistae Ferronii, 1648. ETH-Bibliothek Zürich, Rar 1027, https://doi.org/10.3931/e-rara-13017 / Public Domain Mark

Baier JJ (1708) Oryktographia Norica sive rerum fossilium et ad minerale regnum pertinentium, in territorio Norimbergensi eiusque vicinia observatarum [...]. Norimbergae : impensis Wolfgangi Michahellis, 1708. ETH-Bibliothek Zürich, Rar 3707, https://doi.org/10.3931/e-rara-18618 / Public Domain Mark

Boodt A Boëtius de (1644) Le parfaict ioaillier ou histoire des pierreries ou sont amplement descrites leur naissance, iuste prix, moyen de les cognoistre, et se garder des contrefaites, [...]. A Lyon : chez Iean-Antoine Huguetan, 1644. ETH-Bibliothek Zürich, Rar 1010, https://doi.org/10.3931/e-rara-12490 / Public Domain Mark

Brocchi GB (1814) Conchiologia fossile subapennina con osservazioni geologiche sugli Apennini e sul suolo adiacente. Stamperia Reale, Milano

Colonna F (1616) Fabii Columnae Lyncei Minus cognitarum rariorumque nostro coelo orientium stirpium ekphrasis … https://www.biodiversitylibrary.org/bibliography/723 Public domain

Cuvier G (1802) Mémoire sur l'animal de la Lingule (Lingula anatina lam.). Annales du Muséum National d'Histoire Naturelle 1:69–80. http://hdl.handle.net/2268.1/2205 (Public domain)

Davidson T (1884) A Monograph of the British Fossil Brachiopoda. Appendix to the Supplements. General Summary. Catalogue of British Fossil Brachiopoda. Palaeontographical Society, London, 5 (3): 243-476.

Domenichi ML (1844) Della Storia Naturale di C. Plinio Secondo, Libri XXXVII. Traduzione di M. Lodovico Domenichi emendata per la prima volta secondo il testo latino con l'aggiunta di un nuovo indice generale. Volume Secondo. Giuseppe Antonelli Ed. Venezia

Duméril AMC (1806) Zoologie Analytique ou Méthode Naturelle de Classification des Animaux, rendue plus facile a l'aide de tableaux synoptiques. Allais, Paris

Gessner C (1565) De omni rerum fossilium genere, gemmis, lapidibus, metallis, et huiusmodi, libri aliquot, plerique nunc primum editi. Tiguri. https://doi.org/10.3931/e-rara-4176 / Public Domain Mark

Huxley TH (1869) An introduction to the classification of animals. John Churcill and Sons. London 147 pp.

Lachmund F (1669) Oryktographia Hildesheimensis sive admirandorum fossilium, quae In tractu Hildesheimensi reperiuntur, Descriptio Iconibus Illustrata, Cui addita sunt alia de calculis, de fontibus, &c. Jacobi Mülleri, Hildesheimii. https://wellcomecollection.org/works/rs34c8db (Public Domain Mark)

Lancisi MG (1719) Appendix ad Metallothecam Vaticanam Michaelis Mercati […]. Apud Mariam Salvioni […] Romae

Lang KN (1708) Historia lapidum figuratorum Helvetiae, eiusque viciniae : in qua non solum enarrantur omnia eorum genera, species et vires aeneisque tabulis repraesentantur, sed insuper adducuntur eorum loca nativa, in quibus [...]. Venetiis : sumptibus authoris, typis Jacobi Tomasini, [1708]. ETH-Bibliothek Zürich, Rar 5680, https://doi.org/10.3931/e-rara-10398 / Public Domain Mark

Linnæus C (1758) Systema Naturae per Regna Tria Naturae, secumdum Classes, Ordines, Genera, species, cum characteribus differentiis, synonimis, locis. Tomus I. Editio decima, Reformata. Laurentii Salvii, Holmiae. https://www.biodiversitylibrary.org/bibliography/542 (Public domain)

Lister M (1678) Angliae Historiæ animalium Angliæ, tres tractatus. Unus de araneis. Alter de cochleis tum terrestribus, tum fluviatilibus. Tertius de cochleis marinis. Quibus adjectus est quartus de lapidibus ejusdem insulæ ad cochlearum quandam imaginem figuratis. Joh. Martyn, Londini. https://www.biodiversitylibrary.org/bibliography/65746 (Public domain)

Luidus E (1699) Eduardi Luidii Lithophylacii Britannici ichnographia : sive lapidum aliorumque fossilium Britannicorum singulari figura insignium quotquot hactenus vel ipse invenit vel ab amicis accepit, distributio classica, [...]. Londini : ex officina M.C ; Lipsiae : sumt. Joh. Ludw. Gleditsch & Weidmann, 1699. Zentralbibliothek Zürich, NG 1883, https://doi.org/10.3931/e--rara-30541 / Public Domain Mark

Mercati M (1717) Metallotheca Vaticana. Opus posthumum, Autoritate et Munificentia Clementis UndecimiPontificis Maximi. E tenebris in lucem eductum: Opera autem, & studio Joannis Mariae Lancisii, Archiatri Pontificii illustratum. Ex Officina Jo: Mariae Salvioni Romani, Romae. https://archive.org/details/bub_gb_DC_E6UVWg7EC (Public Domain Mark 1.0)

Moncel M-H, Chiotti L, Gaillard C, Onoratini G, Pleurdeau D (2012) Non-utilitarian lithic objects from the European paleolithic. Archaeol Ethnol Anthropol Eurasia 40(1):24–40

Moscardo L (1656) Note overo memorie del museo di Lodovico Moscardo nobile veronese. Academico Filarmonico, dal medesimo descritte, Et in Tre Libri distinte. Per Paolo Frambotto, Padoa. https://www.biodiversitylibrary.org/bibliography/157058 (not in copyright)

Müller OF (1766) Zoologiae Danicae Prodromus, seu Animalium Daniae et Norvegiae indigenarum, characteres, nomina, et synonyma imprimis popularium. Typis Hallageriis, Hauniae

Plot R (1677) The natural history of Oxfordshire, being an essay toward the natural history of England. Printed at the Theater, Oxford. https://www.biodiversitylibrary.org/bibliography/23488 (not in copyright)

Reed FRC (1895) The Cambridge natural history. Brachiopods (Fossil). Macmillan and Co., London. 533 pp. https://www.biodiversitylibrary.org/item/214013#page/11/mode/1up (copyright not provided)

Retzius AJ (1781) Crania oder Todtenkopfs-Muschel. Schriften der Berlinischen Gesellschaft Naturforschender Freunde, vol. 2, p. 66-76.

Sarti C (2003) Le Collezioni di geologia nel Museo di Ulisse Aldrovandi. In: Vai GB, Cavazza W (eds) Quadricentenario della parola Geologia. Ulisse Aldrovandi 1603 Bologna, Minerva, pp 152–167

Seba A (1758) Locupletissimi rerum naturalium thesauri accurata descriptio, et iconibus artificiosissimis expressio, per universam physices historiam : opus, cui, in hoc rerum genere, nullum par exstitit. III https://www.biodiversitylibrary.org/bibliography/62760 (not in copyright)

Scheuchzer JJ (1702) Specimen Litographiae Helveticae Curiosae, Quo Lapides ex Figuratis Helveticis Selectissimi Aeri incisi. Tiguri : Typis Davidis Gessneri, MDCCII. Zentralbibliothek Zürich, NNN 1883, https://doi.org/10.3931/e-rara-12116 / Public Domain Mark

von Schlotheim EF (1820) Die Petrefactenkunde, auf ihrem jetzigen Standpunkte durch die Beschreibung seiner Sammlung versteinerter und fossiler überreste des Thierund. Gotha, in der Becker'schen Buchhandlung

Scilla A (1670) La vana speculazione disincantata dal senso. Lettera responsiva Circa i corpi marini, che Petrificati si trovano in varij luoghi terrestri. Andrea Colicchia, Napoli

Scilla A (1759) De corporibus marinis lapidescentibus quæ defossa reperiuntur. Addita Dissertatione Fabii Columnæ de Glossopetris. Joannis Zempel, Romae

Tessin CG (1753) Museum Tessinianum : Herr Gr. Carl Gust. Tessins Naturalie-Samling. Holmia [Stockholm] : apud Laurentium Salvium, 1753. ETH-Bibliothek Zürich, Rar 3295, https://doi.org/10.3931/e-rara-41248 / Public Domain Mark

Tramezzino M (1550) Di Giorgio Agricola De la generatione de le cose che sotto la terra sono e de le cause de' loro effetti e nature, lib. V ; De la natura di quelle cose che dalla terra scorrono, lib. IIII; De la natura de le cose fossili e che sotto la terra si cauano, lib. X ; De le minere antiche e moderne, lib. II ; Il Bermanno, o De le cose metallice, dialogo / recato tutto hora dal latino in buona lingua volgare. In Vinegia : per Michele Tramezzino. https://archive.org/details/hin-wel-all-00002012-001 (not copyright found)

Vialli V (1974) Appunti di Paleontologia. Pitagora Editrice, Bologna

Worm O (1655) Museum Wormianum, seu Historia rerum rariorum, tum Naturalium, quam Artificialium, tam Domesticarum, quam Exoticarum; quae Hasniae, Danorum in aedibus Authoris Servantur. Iohannem Elsevirium, Lugduni Batavorum. https://doi.org/10.5962/bhl.title.63844 (not in copyright)

Chapter 19
Echinodermata

Abstract In the next chapters we will discuss echinodermata. After each introduction, concerning information about the anatomy and the geological distribution of the members of the single classes, the same are discussed from the historical and scientific point of view, with notes on some folkloristic aspects related to them.

Keywords Fossils · History · Folklore · Echinodermata

© The Author(s), under exclusive license to Springer Nature Switzerland AG 2022

M. Forli, A. Guerrini, *The History of Fossils Over Centuries*,
https://doi.org/10.1007/978-3-031-04687-2_19

Fig. 19.1 Summary plate of the classes belonging to the Phylum Echinodermata, most common in the fossil state. From top left, bottom right: Echinoidea, (1) *Clypeaster* from Scilla (1759). Asteroidea (2) impressions from Walch (1771). Ophiuroidea (3) an incomplete specimen from Bourguet (1742). Crinoidea (4) a partially pedunculated calyx from Rosinius (1719). In the background detail from the antiporta of the famous work by Agostino Scilla, in which among other fossils, an echinid is recognized (1759). Another class to be included in the Phylum Echinodermata is the one named Holothuroidea. In our paper, almost exclusively devoted to fossils found in ancient publications, we have omitted their treatment because their remains, almost always consisting of spicules and microscopic plates, were not known. (Graphic composition Maurizio Forli. For credit see references below)

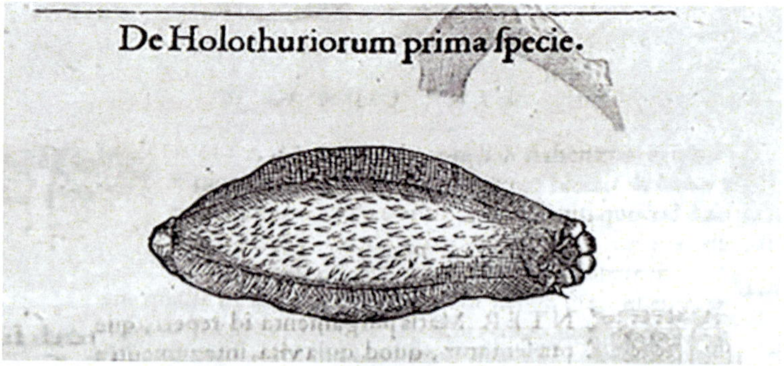

Fig. 19.2 The first printed image we have found of one of these marine organisms is that relating to a living specimen, depicted by Rondelet, in his ***Libri de Piscibus Marinis, in quibus verae Piscium effigies expressae sunt*** (1554: p. 125). [https://www.biodiversitylibrary.org/bibliography/64229 (not in copyright)]

References

Bourguet L (1742) Mémoires pour servir à l'histoire naturelle des petrifications dans les quatre parties du monde : avec figures, et divers indices aussi méthodiques que nécessaires. A la Haye : chez Jean Neaulme, 1742. ETH-Bibliothek Zürich, Rar 2642, https://doi.org/10.3931/e--rara-16053 / Public Domain Mark

Rondelet G (1554) Libri de Piscibus Marinis, in quibus verae Piscium effigies expressae sunt. Matthiam Bonhomme, Lugduni. https://www.biodiversitylibrary.org/bibliography/64229 (not in copyright)

Rosinius MR (1719) Tentaminis de lithozois ac lithophytis, olim marinis jam vero subterraneis, prodromus : sive, De stellis, marinis quondam, nunc fossilibus. Typis Nicolai Sauer, Hamburgi. https://doi.org/10.5962/bhl.title.14427 (Public domain)

Scilla A (1759) De corporibus marinis lapidescentibus quæ defossa reperiuntur. Addita Dissertatione Fabii Columnæ de Glossopetris. Joannis Zempel, Romae

Walch JEI (1768–1773) Walch Johann Ernst Immanuel: Die Naturgeschichte der Versteinerungen : zur Erläuterung der Knorrischen Sammlung von Merkwürdigkeiten der Natur. Nuremberg : printed with Felsseckerischen Schrifften, 1768–1773. ETH-Bibliothek Zurich, Rar 399, https://doi.org/10.3931/e-rara-10246 / Public Domain

Chapter 20
Snake Eggs, Judaic Stones, Ombriae: Echinoidea from Magic, to Folk Beliefs and Science

Abstract Echinoids, or echinids, are marine invertebrates which, at the adult stage, have a regular or irregular shape, usually globular, heart shaped, or flattened. Their body is supported by a shell, called theca or crown, formed by a rather complicated set of calcareous plates firmly united, externally provided with spines, the radioli, which in the fossil state are found more easily isolated (*Lapis judaicus*). They are known from the Paleozoic to the present day. Because of their particular morphological diversifications, they have been described in various ways and the individual parts believed to be things in their own right. According to the currently available documentation, these particular fossils attracted human attention from around 40,000 years ago. Found almost everywhere in the fossil state, they are particularly common in the Cretaceous soils of the Franco-English area. The specimens still found there are generally spheroidal in shape and whitish in color due to their preservation in the local sediments. It was in this area that the Celtic civilization developed the cult of the egg attributed to the Druids as reported by Pliny, the ovum anguinum, the petrified egg of the snake. Gessner (1565) illustrates two fossils of this, ab Echini marini simile. Very beautiful pictures of fossil echinids can be found in the work of Agostino Scilla (1670 and subsequent editions). Because of their magical aura they have given rise to many popular beliefs and have been used as amulets.

Keywords Fossilia · Fossils · History · Echinoidea · Ovum anguinum · Jew's Stone

The echinoids, echinids (from the Greek ἐχῖνος *ekhinos*), or sea urchins, at the adult stage may have a regular or irregular shape, usually globular, heart shaped, or flattened; their skeleton is formed by a cuirass, called theca or crown, formed by a fairly complicated set of firmly joint calcareous plates, on which the spines, called radioles, are grafted. These are more easily found isolated, when in the fossil state. On account of these peculiar morphological diversifications they have been described in various ways, and the individual parts believed to be independent organisms themselves (Fig. 20.1).

© The Author(s), under exclusive license to Springer Nature Switzerland AG 2022

M. Forli, A. Guerrini, *The History of Fossils Over Centuries*,
https://doi.org/10.1007/978-3-031-04687-2_20

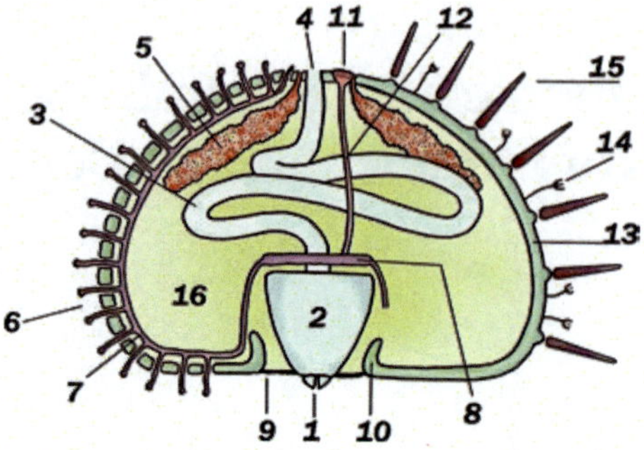

Fig. 20.1 Simplified anatomical structure of an echinid: (1) Teeth. (2) Aristotle's Lantern. (3) Digestive tube. (4) Anal aperture. (5) Gonads. (6) Pedicels. (7) Radial canal. (8) Aquiferous ring. (9) Peristomal plates. (10) Auricle. (11) Madreporite. (12) Hydrophore canal. (13) Dermaskeleton. (14) Pedicellaria. (15) Aculei or Radioli. (16) Coeloma. [Drawing Maurizio Forli modified after Vialli (1974)]

According to the currently available fossil evidence, the role of the oldest echinoids belongs to the regulars, which, appeared in the Upper Ordovician, become quite common from the Lower Carboniferous. Equipped with a flexible theca, difficult to fossilize, which remains until the forms of the Upper Devonian, they are poorly represented among the fossils of marine benthic epifauna, that is, among the organisms living on the surface of the seabed. During their evolution, they underwent structural modifications in the size of the radioli, in the number of columns of the ambulacra, of the interambulacra, etc. Their biodiversity is greatly reduced in the Upper Carboniferous, with only a few species exceeding the Permian-Triassic limit. From the Upper Triassic onwards, they spread again with species predominantly equipped with rigid thecae. The irregular echinoids appeared in the Mesozoic, in the lower Jurassic, developing more and more starting from the Cretaceous, and in the end dominated over the regular ones. The process that determines the "transformation" from regular to irregular echinoids consists of a series of morphostructural changes, such as the acquisition of bilateral symmetry, migration of the anus toward the posterior area and consequent exit from the apical system, changes in the thecal morphology, increase in the number of radioles, the petaloid shape of the ambulacra, etc. These processes allowed irregular echinoids to adapt to life within the substrate, an infaunal environment denied to regular echinoids.

In the regular echinoids, the body is supported by the calcium carbonate skeleton that forms a spherical theca, more or less flattened at the opposite ends. Externally, the theca is divided into ten meridian sectors, arranged as segments that join the upper discoidal part, where the anal opening is located, with the lower oral one, where the mouth and the related apparatuses are located. The segments or sectors are alternated between them, five, called ambulacral because perforated in various ways in order to allow the exit of the organs of movement, the pedicels, and as many inter-ambulacral sectors, not perforated and containing on the inner side the gonads. Furthermore, the endoskeleton carries externally attached several spines, of various shapes and sizes, mainly with protective and locomotory function. The shape and arrangement of the plates and their pores are used in systematics and are particularly useful for the description and distinction of fossil species. In the inner part of the theca of the echinoderms, especially of the regular ones, is clearly visible and recognizable also to the non-experts, the pentaradial symmetry, or pentamera, common to the organisms of the class Echinodermata. In fact, if we cut horizontally a shell in half and observe it internally, the sectors with the ambulacral areas resemble the five arms of a starfish.

The two main openings of the theca or crown are technically called peristome and periproct.

The peristome is circular and central on the lower surface and it is there that the mouth opens. At the waist, much of this opening is covered by a thick skin that incorporates imbricated plates.

At the apex of the theca, at the point where the ambulacral areas originate, is placed the apical disc. This ring of ten plates contours the anal opening, and is always placed on the aboral surface. The apical disc consists of five genital plates and five ocular plates, which surround a flexible membrane with plates containing the anus. This is the structure properly called periproct. The arrangement of the plates of the apical disc also forms an important taxonomic character, especially for distinguishing living species.

All the plates bear radioli, the so-called sea urchin spines, which attach and articulate on the tubercles. The internal anatomy, in the apparent simplicity of the functions of the animal, is very complex; we will limit ourselves to mention a characteristic structure called the *lantern of Aristotle*. The lantern is an apparatus used for biting and grinding and is composed of 50 skeletal elements moved by 60 muscles. It is almost completely internal, although the tips of the teeth extend outside the mouth. There are two main groups of muscles that connect the lantern to the shell by anchoring to the peristomal flange, the ejectors, which drive the lantern out, and the retractors, which lift it up. This works like the jaws of animals, opening or closing during feeding and connecting the outside with the inside. The most specialized and modified lanterns of all the echinids are found in the so-called *Sand Dollars* (Figs. 20.2, 20.3 and 20.4).

Fig. 20.2 Exemplified morphology of regular echinid of the genus *Cidaris*. Top and bottom left: aboral view and detail of apical disc or periproct; top and bottom right: oral view and detail of teeth and peristomal plates. In the center: lateral view with ambulacral areas and mamelons in evidence (graphic composition and photos Maurizio Forli)

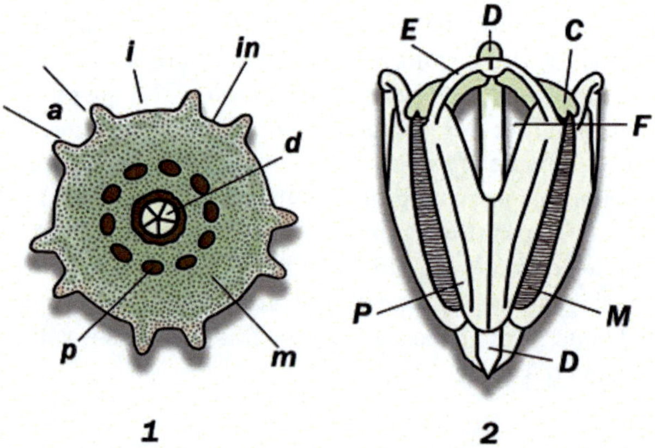

Fig. 20.3 (1) Scheme of peristome of a regular echinoid: (a) ambulacral area. (d) teeth, outer part. (m) peristomal membrane. (p) oral plate. (i) interspace between ambulacra. (in) peristomal incisura. (2) Aristotle's Lantern: (C) Compass. (D) Tooth. (E) Epiphysis. (F) Window. (M) Interpyramidal muscles. (P) Pyramid. [Drawing Maurizio Forli modified after Vialli (1974)]

Fig. 20.4 Details of the structure of a Cidarid tubercles. (1) Mamelon. (2) Button. (3) Central ligament pit. (4) Areole. (5) Parapet. (6) Scrobiculus (graphic composition and photo Maurizio Forli)

Echinids are exclusively marine and benthic, closely related to the bottom on which they rest or sink. Especially some irregular echinids, for example, those belonging to the genus *Schizaster* L. Agassiz, 1835, live sunken in the sand, where they move leaving various traces, sometimes found also as fossils (Fig. 20.5).

Fig. 20.5 Irregular fossorial echinid, recovered stranded along the Tirrenia coast (Pisa); length 55 mm (graphic composition and photos Maurizio Forli)

Some echinoderms are sometimes preserved with a spheroidal or heart shape, without spines that generally disperse during fossilization, giving rise to round objects similar to eggs. This phenomenon is particularly evident in the fossil echinids of the white cliffs on both sides of the English Channel, in France and in Great Britain: it is just in these places that the beliefs and superstitions about their meaning and use as magical snake eggs have mostly developed. Important echinological fossil faunas are present almost everywhere in Europe and also in Italy, especially in Veneto, Sardinia, and Calabria.

According to the currently available documentation, these particular fossils attracted human attention starting about 40,000 years ago. The oldest finds in anthropized contexts date back to the French Aurignacian: it is in fact in France that we find the largest number of archaeological sites that show how echinid fossils were variously used; Demnard and Néraudeau (2001) have surveyed more than one hundred sites involved in this practice, mainly concentrated in what was once considered Gaul, but with evidence also from other European, African, and Middle Eastern countries, with ages ranging from the Upper Paleolithic to the Gallo-Roman age.

Phylum Echinodermata
Class Echinoidea
Ordo Holasteroida
Family Echinocorythidae †
Genus *Echinocorys*
Species *Echinocorys ovatus* (Leske, 1778) †

The above classification refers to one of the most representative species, at archaeological level, of the cult connected to the sacred egg in the Celtic area. More than 150 specimens of this species together with another echinoid, *Micraster coranguinum*, have been placed in a circle around the Bronze Age burial of a young woman and a child in the Dunstable Downs mound at Berkhamsted, in the English county of Hertfordshire, just north of London. Regarding this specific type of burial, as laymen, we immediately think that the arrangement of the bodies in the center of the circle made with the "eggs," has a meaning manifestly linked to rebirth. (Fig. 20.6)

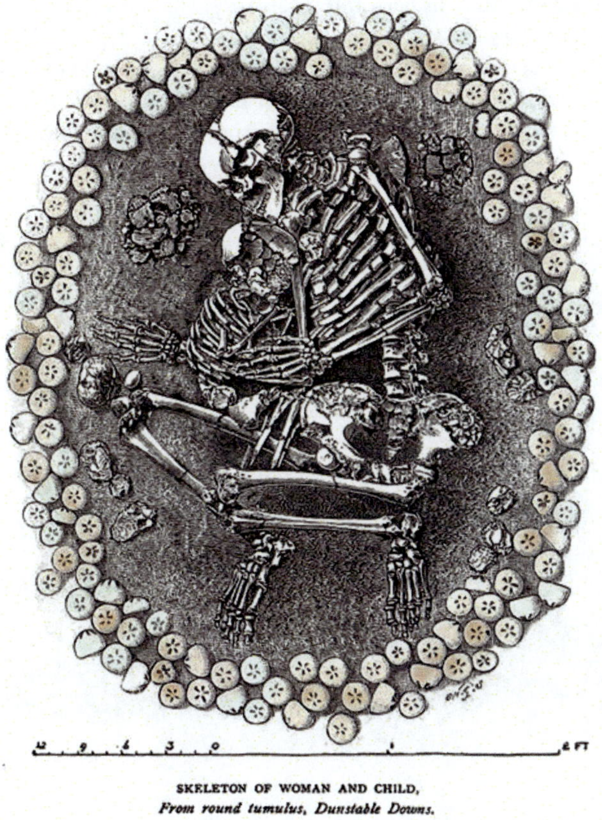

Fig. 20.6 Bronze Age burial arrangement of a young woman and child in the Dunstable Downs mound at Berkhamsted, in the English county of Hertfordshire, located just north of London [Frontispiece modified after Smith, 1894 (original black-and-white engraving)]. Echinids were represented by two species, *Echinocorys ovatus* (Leske, 1778) and *Micraster coranguinum* (Leske, 1778) (https://wellcomecollection.org/works/sc2hhxdm/items Public Domain Mark)

For prehistoric man, besides having a psychological value as a strange and different object, their shape and consistency had a direct relationship with their functional use. In particular in the Paleolithic, some silicified echinids were exploited as raw material for the construction of instruments, such as bifaces, scrapers, percussion instruments, or cores. Among the best known and most significant examples is a biface from Swanscombe, England, dated to about 400,000 years ago, attributed to *Homo heidelbergensis*, a circular scraper, made directly on an echinid, from Saint-Just-des-Marais, France, also attributed to the Lower Paleolithic, and a flint core on which a specimen has been preserved, so that it was prominent, from Tercis (France), attributed to the Mousterian (Demnard and Néraudeau 2001).

The ornamental use of perforated echinids consists instead of elements of *parures* or pendants. Their greater symbolic importance is expressed in protohistoric and historical periods with offers of perforated echinids, either alone or accompanied by other traditional funerary objects, in burials or set aside in deposits in sites of magical significance. A curious, and to our eyes, fascinating example of this use is given by an *Echinolampas africanus*, a find from ancient Egypt now kept at the Egyptian Museum of Turin, whose lower part bears a hieroglyphic inscription (Fig. 20.7).

Fig. 20.7 The echinid of Tjanefer: an *Echinolampas africanus* de Loriol, 1880, a fossil species from the Eocene of North Africa, offered to the Sun God by the Egyptian priest Tjanefer, as specified by the hieroglyphic inscription engraved in the ventral part and indicated by the place of discovery, the Temple of Heliopolis (Karlshausen and De Putter 2017). [Photo credit Nicola Dell'Aquila/Egiptian Museum of Turin. Licensed byCC 2.0 (Modified after Karlshausen and De Putter)]

The cult of the egg has found fertile ground in Celtic civilization, Pliny himself reports that *ovum anguinum* was particularly appreciated by the Druids, as evidenced by archaeological findings:

> *Praeterea est ovum genus in magna Galliarum fama, omissum Graecis. Angues innumeri aestate convoluti, salivis faucium corporumque spumis artifici complexu glomerantur, anguinum appellatur* [...] *Vidi equidem id ovum mali orbiculati modici magnitutine, crusta cartilaginis, velut acetabulis brachiorum polypi crebris, insigne Druidis* (Pliny XXIX, 12).
>
> *Oltre ai detti, c'è una sorte d'uova, la quale ha gran riputazione in Francia, di cui i Greci non hanno fatto menzione. La state infinite serpi avviluppate insieme, con la saliva della gola e con la schiuma del corpo si aggruppano con certo artificioso abbracciamento: l'uovo che ne fanno si chiama anguino. Dicono i Druidi, che le serpi con fischi le gettano in alto, e che bisogna riceverle nel grembo della veste che si chiama sago, innanzi che toc-*

chino terra; e che chi le toglie a questo modo dee fuggire a cavallo con ogni prestezza, perché le serpi lo seguitano, fin che al valicar di qualche fiume elle non possono ir più innanzi [...] *Io ho veduto questo uovo grande quanto una piccola mela tonda, con crosta di cartilagine, con ispesse cavità quali son nelle braccia del polpo, cosa notabile ai Druidi* (Domenichi 1844: pp. 800-802).

In addition to the already described, there is a kind of egg, which has great reputation in France, never mentioned by the Greeks. In the midsummer infinite snakes wrapped together, with the saliva of the throat and with the foam of the body are grouped with certain artificial embrace: the egg that they make is called anguinus (the Snake's egg). The Druids say that the snakes throw them upwards with whistles, and that they must be received in the lap of the robe, which is called "sago", before they touch the ground; and that whoever removes them in this way must flee on horseback with all haste, because the snakes follow him, until, at the crossing of some river, they cannot get further [...] *I have seen this egg as big as a small round apple, with a cartilage crust, with thick cavities as they are in the arms of the octopus, something noteworthy to the Druids.*

This famous Plinian narration of the snake egg, the so-called *ovum anguinum*, is the basis of many subsequent descriptions of fossil specimens of echinids, and of their magical and medicinal properties. In retrospect, the representation that the Roman author makes is perfectly suited to that of a member of the family Cidaridae, characterized by the large circular shape of the insertion area of the large central tubercle, which is reminiscent of the suckers of the octopus.

Phylum Echinodermata
Class Echinoidea
Family Hemicidaridae †
Genus *Hemicidaris* †
Species *Hemicidaris quenstedti* Mérian in Desor, 1856 † (Radwanska, 2014)

Since the classification of these fossil echinids is rather complex and controversial, this above should be the actual denomination of the specimen that Gessner imagines to correspond to the one seen and described by Pliny (1565: p. 168). Another cidarid of different species, partially aculeate, is reproduced in the *Dell'Historia Naturale* of Ferrante Imperato under the denomination of sea urchin converted into cement stone (1599: p. 671), except for our oversights, it is the first fossil figure for this genus, in the Italian printed texts (Figs. 20.8 and 20.9).

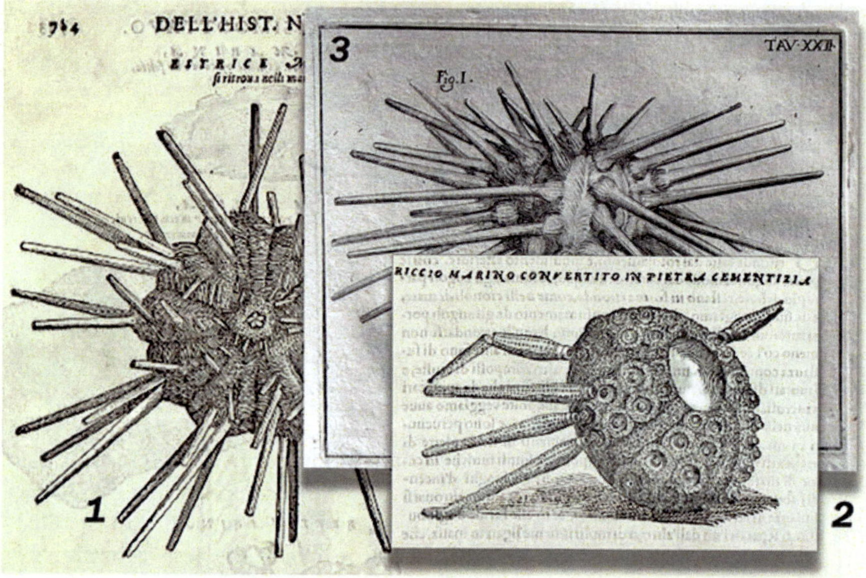

Fig. 20.8 From the *Dell'Historia Naturale* of Ferrante Imperato: (1) Present specimen, probably *Stylocidaris affinis* (Philippi, 1845), characterized by having spines as long as the diameter of the body. (2) Fossil specimen of "*Cidaris*" sp., figured and described by Imperato as a *sea urchin converted into a cement stone* (1599: p. 671). (3) Actual specimen, probably *Cidaris cidaris* ((Linnaeus, 1758), distinguishable by radioli twice as long as the diameter of the theca, compared in the text by Scilla with the Maltese Miocene fossil *Stylocidaris melitensis* (Forbes in Wright 1855) [*Sea urchin, the like of which is fished in the seas, which bathe Sicily* (De corporibus marinis lapidescentibus, 1759: pp. 145–146, p. XXII, Fig. I)] [graphic composition Maurizio Forli. (Credit from: (1–2) https://www.biodiversitylibrary.org/item/183177#page/5/mode/1up (not in copyright) (3) personal copy Maurizio Forli)]

Fig. 20.9 The recovery of the *ovum anguinum* among the entangled snakes, as reported by Pliny, in an engraving dated 1497 (?), according to Basset (1982: p. 16), of which we found no direct reference to the original. Curiously, the shape of the echinid (1) is extremely similar to that of the exemplar illustrated and referred by Gessner, to this particular "gem," sent to him by his correspondent in Antwerp (Belgium) Petrus Coldebergus (1565: p. 198 verso). (2) *De lapide Ovum anguinum appellato*, by Anselmus Boëtius de Boodt (1609: p. 174). (3) Another version reproposed by de Boodt (1636: p. 347) [graphic composition Maurizio Forli. (Credit from: (1) https://doi.org/10.3931/e-rara-4176 / Public Domain Mark (2) https://hdl.handle.net/2027/gri.ark:/13960/t8w96041q Public domain (3) https://doi.org/10.3931/e-rara-15217 / Public Domain Mark)]

In Gessner (1565: pp. 166–167) an irregular and a regular petrified echinid, because of their resemblance to sea urchins, *ab Echini marini similitudine,* are referred to as *Echinites.* Several other fossils of the same kind are figured in the *Musaeum Metallicum* (1648: pp. 455–458), under the names of *Variae Echinitis differentiae*, Echinites *magnus*, *Echinitis laevi differentiae*, and *Echinites forma stellari*, practically a sampling of the most common fossil echinoderms. Of particular importance, according to our point of view, is the *Echinites forma stellari*, which is undoubtedly a member of the genus *Scutella* and, as such, a first representation for this type of fossilized flat sea urchins, if not the first ever.

As written earlier for other coherent groups of fossils, even among the echinoderms some have been placed in different categories because of their different aesthetics or similarity to one thing rather than another. In Aldrovandi this is the case of a spatangoid identified with a Scolopendrite (*Musaeum Metallicum*, 1648: p.475), an Eocene Echinocorys from Vicenza described and figured as *Pentaphyllites* because of the similarity of the ambulacra to pentaragued leaves, a *Clypeaster* and a *Spatangus*, probably Pliocene, from the Bologna area, also figured under the designation *Rhodites pentaphyllites*, because their ambulacral impressions would resemble leaves of wild roses (*Musaeum Metallicum*, 1648: pp. 498–499) and three indeterminable specimens in the category of *Astroites* (*Musaeum Metallicum*, 1648: pp. 877–879).

Among the latter is also included the first plate, top left, of the third figure of *De Fossilibus* by Francesco Imperato, which probably belongs to the remains of an echinid (1610: fig. 3).

Three fossilized echinoderms, which were contained in the ninth showcase of the Vatican collection, among other stones of various shapes, are well reproduced in the work of Mercati (1717): the first two come from the Eocene soils of Vicenza, the other, a splendid Clypeaster, from the Pliocene of Siena in the surroundings of San Quirico.

> [...] *Reperitur in Haetruria.* [...] *ex agro Senensi, proximo Sancti Quirici*

The latter, inserted among the *Cryptopetrae with the* name of *Lapis Indicae Cucurbitae similis*, has been subject to a taxonomic revision by Esu and Kotsakis (1981) who identified it as *Clypeaster aegyptiacus* Michelin, 1861 even if they still have some uncertainties about the correspondence or not between the latter and the *C. pliocenicus* (G. Seguenza, 1879).

Other beautiful figures of these fossils are to be found in Agostino Scilla, who illustrates them with a wealth of detail and extreme verisimilitude in twelve full-page plates and two partial ones, of the total twenty eight that accompany his dissertation on *the petrified marine bodies found on land* (1670). Some of the specimens depicted, for the superb aesthetic quality of the engravings, were indicated as a reference in subsequent binomial classifications, especially those reported in the second Latin edition of the work which, precisely because it was written in the learned language used by scholars, was the best known version (*De corporibus marinis lapidescentibus quae defossa reperiuntur*, 1747), although the first draft, precisely because of the remarkable iconographic apparatus, had been circulating for some time among Italian and European scholars. Using the current nomenclature, among the many, we point out *Cidaris cidaris* L., 1758 of plate XXII, Figures 1–3 and plate XXIV, Figures 1–2; *Brissus unicolor* (Leske, 1778) of plate IV, figures 2–3; *Clypeaster altus* Lamarck, 1816 of plate IX, and *Clypeaster marginatus* Lamarck, 1816 shown at the bottom of plate XI. A large part of the Scilla collection, more than two hundred specimens, with the materials used for the engravings of the plates and their preparatory drawings, is still preserved at the *Sedgwick Museum of Earth Sciences in* Cambridge, part of the collection of Sir John Woodward, who had purchased it from the heirs of the naturalist painter in 1717. Among other things, Woodward complained of the imperfect correspondence between the real state of the fossils and their representations, criticizing Scilla because, according to him, he would have embellished the images more than necessary to support his ideas. (We have no evidence on this but we believe it is more likely that the drawings of the plates are correct and some of the fossil specimens have been swapped. This kind of fossils is quite easy to find in the Tropea area of Calabria and it's probable that Scilla had others.) Representations, precisely because of their obvious accuracy and correspondence to reality, have been widely quoted, copied, or plagiarized directly in other naturalistic works, often without giving any credit to the author from Messina (Figs. 20.10, 20.11, 20.12, 20.13, 20.14, 20.15, 20.16, and 20.17).

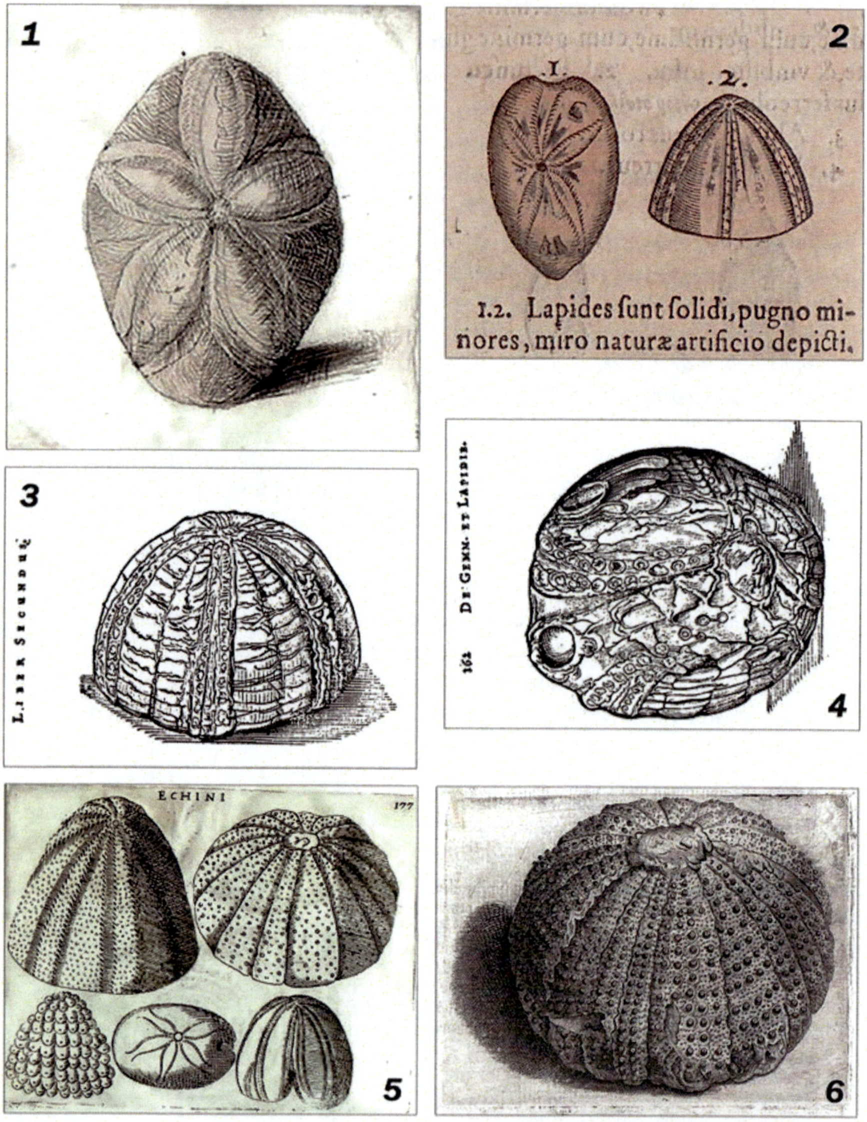

Fig. 20.10 Some representations of fossilized echinids. (1) from an original drawing for the printed illustration of the *Musaeum Septalianum* (Scarabelli 1666). (2) *Echinites*, from Gessner (1565: p. 166). (3–4); echinid similar to that figured by Gessner on p. 166, from the mouth of the Thames, *Inveni autem similem in Anglia ad ostium Thamisis* [...] *qualem et expressit Gesnerus de figura lapid. fol. 166*. (de Laet 1647: pp. 161–162). (5) from Moscardo (1656: p. 177), the one in the lower left corner is covered with bryozoans, probably belonging to the genus *Puellina*. (6) *Lapis* [...] *ab echini marine similitude* (Cerutus and Chioccus 1622: p. 413) [graphic composition Maurizio Forli. (Credit from: (1) https://archive.org/details/museoogaleriaadu00terz_0 No copyright page found (2) https://doi.org/10.3931/e-rara-4176 / Public Domain Mark (3–4) https://doi.org/10.3931/e-rara-35487 / Public Domain Mark (5) https://www.biodiversitylibrary.org/bibliography/157058 (not in copyright) (6) https://doi.org/10.3931/e-rara-50568 / Public Domain Mark)].

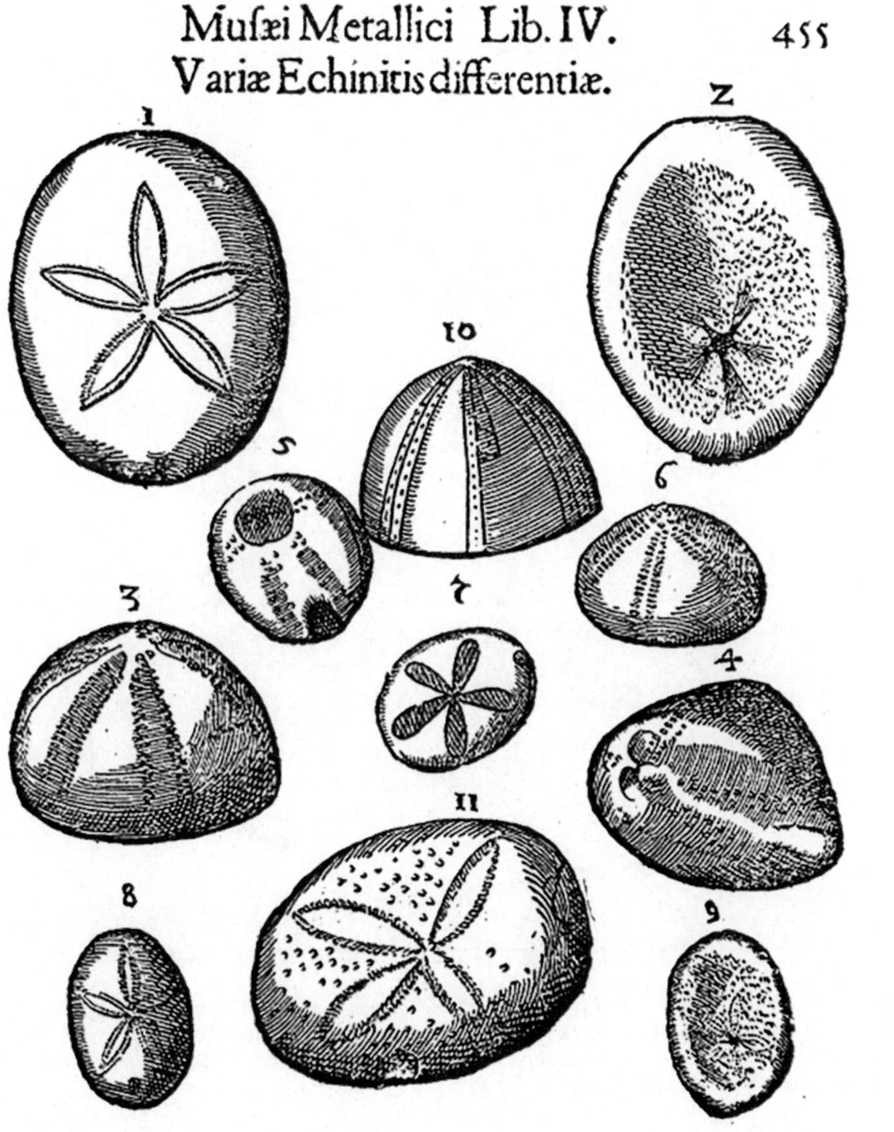

Fig. 20.11 Morphological differences found by Aldrovandi when comparing various species of fossilized sea urchins, and as was then customary, collectively reported under the name of *Echinites* (*Musaeum Metallicum*, 1648: p. 455) (https://doi.org/10.3931/e-rara-13017 / Public Domain Mark)

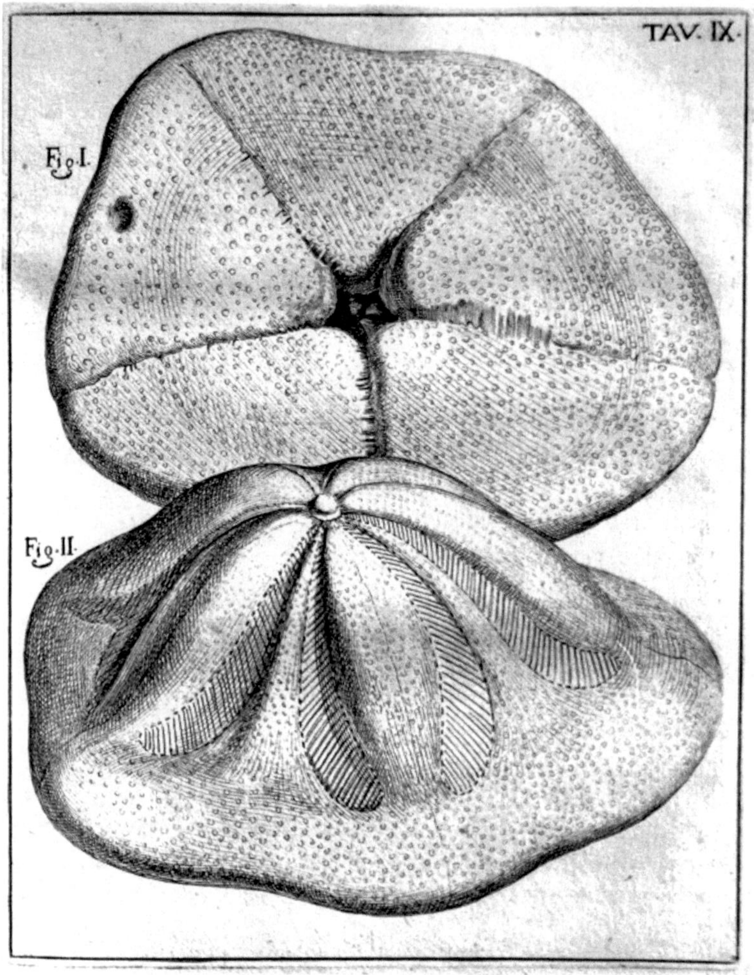

Fig. 20.12 Fig. I. Echino, *called by Aldrovandi, Echinometra, petrified seen below*. Fig. II. *The same seen above. From Malta*. A large Clypeaster taken from Agostino Scilla (*De corporibus marinis lapidescentibus quæ defossa reperiuntur*, 1759: pl. IX) identified as *C. altus* Lamarck, 1816 (personal copy Maurizio Forli)

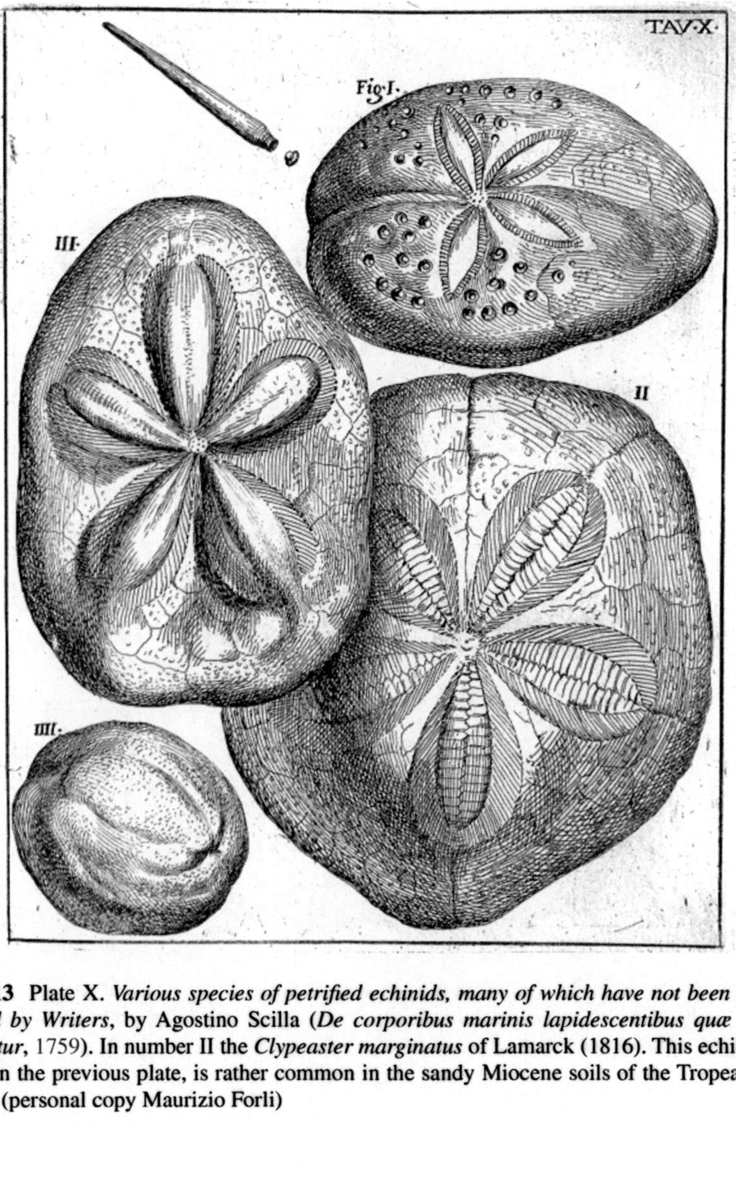

Fig. 20.13 Plate X. *Various species of petrified echinids, many of which have not been hitherto observed by Writers*, by Agostino Scilla (*De corporibus marinis lapidescentibus quæ defossa reperiuntur*, 1759). In number II the *Clypeaster marginatus* of Lamarck (1816). This echinid, and the one in the previous plate, is rather common in the sandy Miocene soils of the Tropea area in Calabria (personal copy Maurizio Forli)

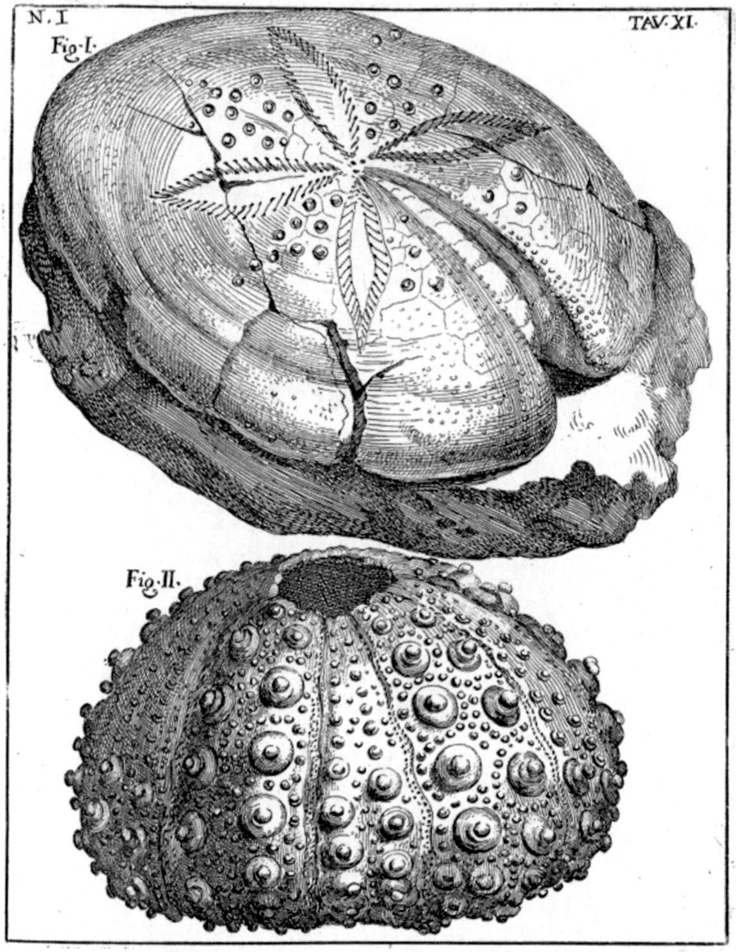

Fig. 20.14 The original description unites the text of this plate and that of the previous one. Agostino Scilla (*De corporibus marinis lapidescentibus quæ defossa reperiuntur*, 1759: pl. XI) (personal copy Maurizio Forli)

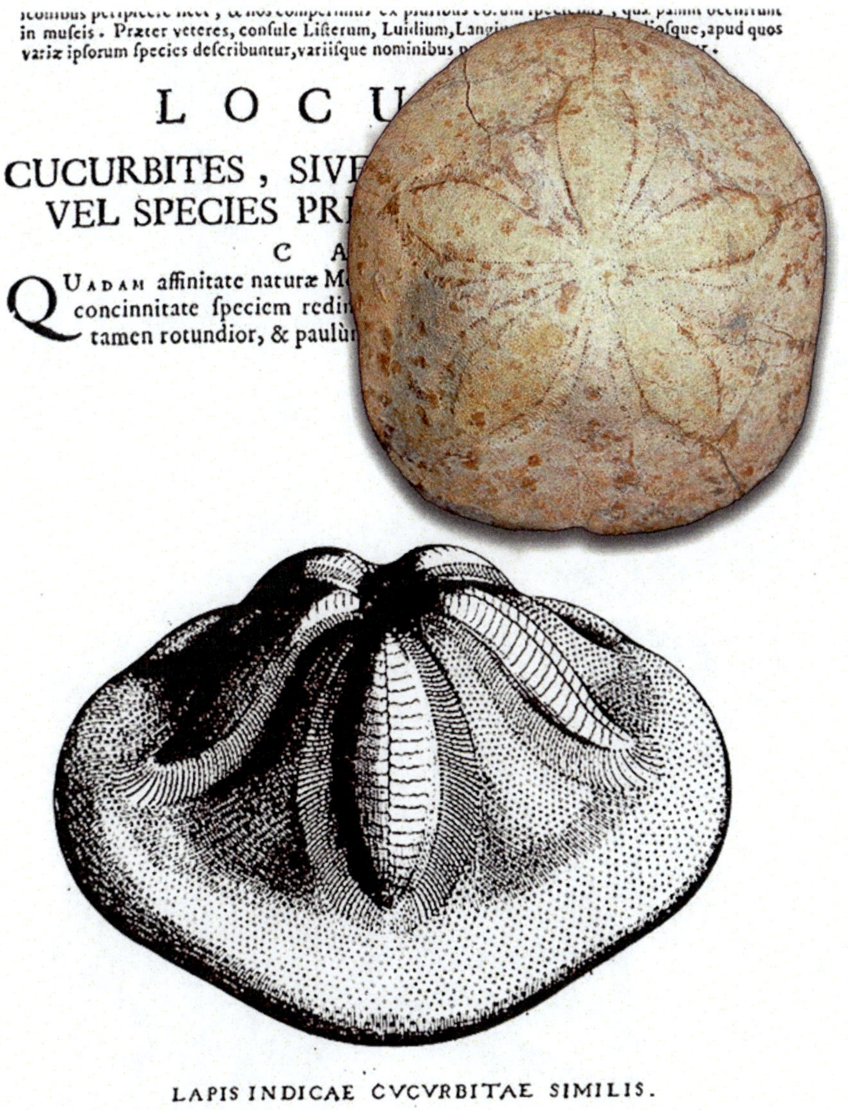

Fig. 20.15 *Cucurbites, sive echites floridus vel species prima cryptopetrae*. Mysterious stone similar in form to a gourd. Michele Mercati (1717: p. 233). This specimen, of Pliocene age, was collected in the vicinity of San Quirico d'Orcia (Siena). On the basis of the picture, Esu and Kotsakis (1981) identified it as *Clypeaster aegyptiacus* Michelin, 1861 even if some uncertainties remained on the possible correspondence, and therefore synonymy, between the latter and *C. pliocenicus* (G. Seguenza, 1879). Above, another specimen from the Tuscan Pliocene, from the surroundings of Montaione (Florence), maximum diameter about 160 mm [photo and graphic composition Maurizio Forli. (Plate credit https://archive.org/details/bub_gb_DC_E6UVWg7EC Public Domain Mark 1.0)]

Fig. 20.16 In the upper left the specimen in the form of a starry echinus, ashy colored with a *foramen in* imitation of that of a *Spatangus*, called by Aldrovandi *Echinites forma stellari* (1648: p. 458). In plate VIII of *De corporibus marinis lapidescentibus, it* is instead illustrated: I. *Petrified echinus of vague Malta spice,* Fig. II. *The same seen below.* Fig. III. *The same broken, showing the cells from within*. Fig. IV. *A small part of the said echinus, observed with the occhialino* (microscope), *showing the mamelets, over which very small thorns played* (text Scilla 1670; pl. Scilla 1759). Bottom right a corresponding fossil specimen of *Scutella* sp., from the French Miocene, diameter~ 85 mm [photo and graphic composition Maurizio Forli. (Plate credit https://doi.org/10.3931/e-rara-13017 / Public Domain Mark)]

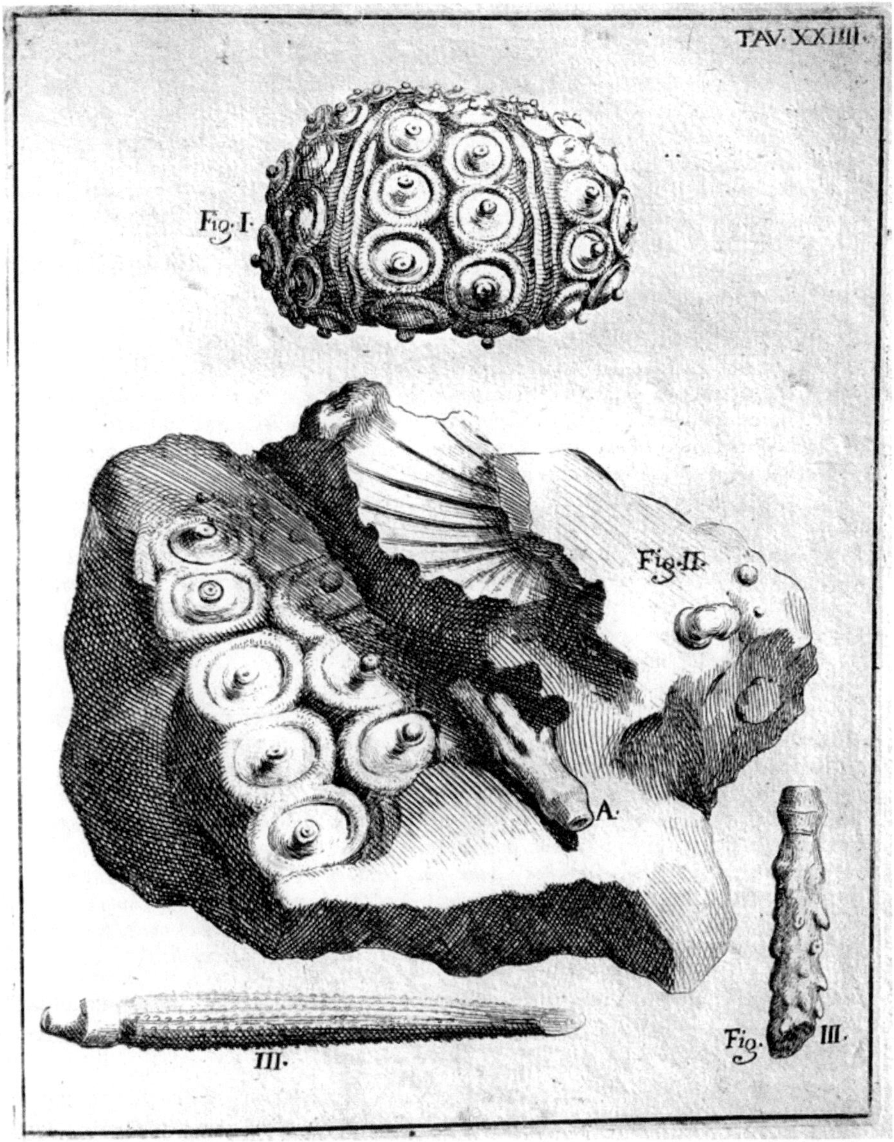

Fig. 20.17 *Stylocidaris melitensis* (Forbes in Wright, 1855), Maltese fossil echinid, illustrated by Scilla (1759: pl. XXIIII) and its particulars, popularly called *Bastoncini di San Paolo* (*St. Paul's rods*), fossil or petrified spines of the sea urchin, and *Mammelle di San Paolo* (*St. Paul's breasts*), the subcircular insertion areas of the tubercle, whose morphology resembles firstly the areola and the human nipple (personal copy Maurizio Forli)

Another curious name given to fossilized sea urchins, especially if small, was Brontie or Ombrie, that is, stones produced by lightning and thunder, precipitated from the sky during storms, similar to belemnites, to the ichthyodontolites of the

Selacians and to some prehistoric stone tools, and for this reason appreciated as protectors from lightning for human beings, animals, and houses.

> *Bronia capitis testudinis similitudivé habet virtus eius est fulminibus resistere* (Camilli Leonardo 1502: XXV)
>
> *La pietra Brontea è simile ai capi delle testuggini. Questa restringe i tuoni, e spenge le cose percosse dai folgori* (Bonardo 1585: p. 21).
>
> *The Brontea stone* (Thunder Stone) *is similar to the heads of tortoises. This restricts thunder, and extinguishes things struck by thunderbolts.*

The images printed in Gessner illustrate what appears to us to be an internal echinoid model by describing it as

> [...] *Ombrorium species à Ioanne Kentmano ad me missa* (Gessner 1565: p. 61).
>
> *Le pietre testugginate, chiamate da alcuni Ombrie, quasi cadute con le pioggie di alto, sono forme ritonde, non di propria vegetazione, ma tali dalla forma precedente impetrata d'incini, & estrici animali marini, che per l'ammarcimento de legamenti spogliate dalle loro spine, e membrane, siano restate in forma di testuggini di edificij, ornate di varij compartimenti, come la proprietà di detti animali apporta. ripiene poi, e trasmutate dall'humor sopravegnente in selce di color bianco, e melato. sono di queste secondo le differenze di detti animali, l'altre dall'una e l'altra parte convesse, e simili à melo, altre à modo di meza sfera, convesse da una sol parte, e dall'altra piane* [...] *resta lo spazio tra le cinque fascie con gli altri proprij ornamenti. Dunque le dette pietre testugginate sono di sustanza selcigne, à quali sono altre simili nella figura, ma di sustanza cementizia. Sono alcuni, che chiamano Ombrie, non queste, ma il Lyncurio da noi detto, indotti dalla forma di saetta quasi dal ciel caduta* (Imperato F. 1599: p. 658).
>
> *Testugginate stones, called Ombrie by some, almost fallen with the rain from above, are round forms, not of their own vegetation, but such as from the previous form impetrated by sea urchins and porcupines, which by the putrefaction of ligaments, stripped of their spines, and membranes, have remained in the form of tortoises of buildings, adorned with various compartments, as the property of said animals brings. They are filled and transmuted by the humor into white and honeycombed flint, some of which, according to the differences between these animals, are convex on both sides and similar to apple, while others are half spherical, convex on one side and flat on the other* [...] *the space between the five bands remains with the other ornaments. So the said testugginate stones are of flint substance, which are similar in shape, but of cement substance. There are some who call these Ombrie, not these, but the Lyncurio we mentioned, induced by the form of lightning almost falling from the sky.*

In this passage Imperato gives an excellent explanation of what these particular turtle shell-shaped stones are and how they are formed, deriving from specimens of echinoids often transformed into siliceous matter.

Even if the ancient knowledge related to the superstitions connected to these objects persisted, from about the second half of the seventeenth century to the first half of the eighteenth century, scholars well realized the impossibility that these materials were created during thunderstorms and recognized their true origin from echinids. Below are two short extracts by different authors, Friedrich Lachmund and Giacinto Gimma: the former still describes them as things in their own right without mentioning their thaumaturgical properties; the latter, writing a few years later, reports the opinions of those who had dealt with them before him but, as for the truthfulness of the properties attributed to them, he is, as is his wont, decidedly incredulous:

> *I. Est ex materia silicea, ex parte candidus, ex parte subfuscus, in apice modulum rosa figura habens, a quo quinquelinea duplicata, qua ex parvu striis transversis constant, descendunt usque in centrum plana superficies. II. Est major semidiaphanus, color fusco, constat ex materia pyrite, in apice modulus cavus exstat, a quo linea ut in priori procedunt. In utroque media spatia eminent* (Lachmund 1669 : p. 22).

> *L'Ombria, dice Plinio, che è pietra, la quale cade colla pioggia, e colle saette, come ancora la ceraunia, e la Bronzia, o Brontea, detta da alcuni Nozia, e che ha il medesimo effetto, che ha la Brontea: e dicono, che ponendola nell'Altare, e ne' Sagrificj degl'idolatri, non ardono. Le numera egli tra le gemme; ma tra le favolose più tosto numerar le possiamo* [...] *Dell'Echinite* [...] *Stimarono alcuni, che dal cielo cadessero, o col fulmine, o colla pioggia; e però alle stesse hanno pur dato il nome di Bronzie, o di Ombrie; da Bronte il tuono, e da Ombres la pioggia derivando.* [...] *ma non è pur dubbio, che si formano queste pietre da una materia sassosa, che essendosi introdotta nella scaglia del Riccio marino, quando n'è uscito il pesce, si è formato dentro quella scaglia, e ne ha presa la figura, e la sembianza, e poi si è indurita, ed impietrita* (Gimma 1730: pp. 203-204).

Pliny says that Ombria is a stone that falls with rain and thunderbolts, as well as Ceraunia *and* Bronzia, *or* Brontea, *called by some* Nozia, which *has the same effect as* Brontea: *and they say that when placed on the altar and in the sacrifices of idolaters they do not burn. He includes them among the gems; but among the fabulous ones we can number them more quickly* [...] *Of the Echinite* [...] *some people estimated that they fell from the sky, or with the lightning, or with the rain; and therefore they have given the name of* Bronzie, *or* Ombrie; *from Bronte the thunder, and from Ombres the rain deriving.* [...] *but there is no doubt that these stones are formed from a stony matter, which having been introduced into the scale of the sea urchin, when the fish came out of it, was formed within that scale, and took the figure, and the appearance, and then hardened and petrified* (Fig. 20.18).

Fig. 20.18 *Ombrorium species à Ioanne Kentmano ad me missa* (from Gessner 1565: p. 61). Ombrie or Bronzie, that is, stones produced respectively by rain and thunder, precipitated from the sky during thunderstorms (https://doi.org/10.3931/e-rara-4176 / Public Domain Mark)

Sea urchin spines may be really annoying in case of stinging, also because quite often, at least for the shallow Mediterranean species, their ends break remaining stuck in the unfortunate victim suffering their insult. But not all radioli are long and pointed, there are some of them with a shape similar to the olive stones, paddle shaped or stick shaped, typical of some genera, mostly belonging to the family Cidaridae. Some of them were already known in the ancient times and utilized in the treatment of the renal diseases, due to their resemblance to small phalluses. They are already cited by Dioscorides and Galen for their medicinal properties in kidney and urinary tract diseases; here follows the translation and adaptation of Dioscorides' text, by Pietro Andrea Mattioli:

> *La Pietra Giudaica è à i tempi nostri familiarissima à tutte le Spetiaerie. E' in uso per rompere le pietre delle reni. Fecene memoria Galeno al IX. delle facoltà de i Semplici, così dicendo: E' una altra pietra nelle virtù sue valorosa, la qual nasce in Palestina di Soria, di colore bianca, di forma di ghianda, con certe linee, come se fussero fatte al torno, Chiamanla dal luogo, ove la nasce Giudaica, & usanla trita in su la pietra d'arrotare, & poscia dandola à bere con tre ciathi di acqua calda, per rompere le pietre della visciga. Ma veramente in coloro, ne i quali noi l'habbiamo isperimentata, mai non ha fatto alcun proffitto per la pietra della visciga: ma in quelle delle reni è veramente efficace* (Mattioli 1548: p. 496).
>
> *The Judaic Stone is in our time very familiar to all drugstores. It is used to break the stones of the kidneys. Galen mentions it in the ninth chapter of the Faculties of the Simple Ones, saying: "There is another stone of great virtue, which is born in Palestine, in Soria, white in colour, in the form of an acorn, with certain lines, as if they were made with the lathe, and is called Judaica from the place where it is born, and is used by crushing it on the stone, and then drinking it with three glasses of hot water, to break the stones of the kidneys. But truly in those in whom we have experienced it, it has never done any good for the stone of the bladder: but in those of the kidneys it is truly effective.*

The images related to the Judaic stones are shown by Gessner who illustrates four varieties (Fig. 20.19). Those relating to the echinid radioli are the first three (1565: p. 129) and similarly to the others transmitted to us by the Swiss author, are reported as such in many other volumes written by authors contemporary or subsequent to him. Currently these particular forms of spines are attributed to echinoderms belonging to the extinct genus *Balanocidaris* Lambert, 1910. In the *Musaeum Metallicum* (p. 711), Aldrovandi depicts eleven different types of remains, some similar to those produced by Gessner, but from observation of the illustrations it is not clear whether all of them can be considered fossil spines of sea urchin (Fig. 20.20). He also reminds us of the Plinian designation of Tecolithi (Pliny XXXV):

> *Some call these stones tecolites, because they cure the bladder: drunk in wine they break the stone* (Domenichi 1844: p. 1336).
>
> *Ritrovasi questa pietra ne' campi della Palestina della Siria, ò Soria, & anche nella Silesia, & chiamasi Giudaica per venir dalla Giudea* [...] *lapis Syriacus, da altri vien chiamata Phenecites, tecolithos, Pyrene* [...] *da Plinio fù detta Tecolythos, quasi volesse inferire Lapidem frangens, per la facoltà che tiene di promover l'orina, & cacciar fuori le pietre, & l'arena, che si trattengono nelle reni* [...] *Preparatione. Prima si preparano sopra il Porfido con acque specifiche, & convenienti, v. g. come d'Ononide, di Sassifraga, di Frache, di Virga Aurea, ò simili. Secondo, si fa il sale. Terzo, il magistero. Quarto l'olio, over liquore: e questo si fa per deliquio* (Passera 1688: pp. 274-275).

This stone is found in the fields of Palestine, Syria, or Soria, & also in Silesia, & is called Judaica because it comes from Judea [...] *lapis Syriacus, by others called Phenecites, tecolithos, Pyrene* [...] *Pliny called it Tecolythos, almost as if he wanted to infer Lapidem frangens, because of its ability to promote urine, & expel the stones & sand which are kept in the kidneys* [...] *Preparation. First of all they are prepared on the porphyry* (fine minced with a pestle) *with specific and suitable waters, such as Ononide, Saxifrage, Frache, Virga Aurea, or similar. Secondly, salt is made. Third, the magisterium. Fourthly oil, or liqueur: and this is made against fainting.*

The following is a medicinal recipe, including the Judaic stones, taken from John Wecker (1660: p. 63):

Take the best Rheubard two drams, Galanga, grains of Paradise*, Anniseed, Fennelseed, Agarick, Mastick*, Cinnamon, of each one dram, Licorish half an Ounce, Jews stone three drams, Mithridate* five drams, Mace* four drams, Cloves half a dram, Aqua vitae one part, Malligoe * two parts. Put all these into a Glass excellent well stopt for fourteen daies; then distill them, let the Patient take a spoonful twice a week upon an empty Stomach.*

(**Galanga*: a root with a pungent taste similar to ginger; **grains of Paradise*: Malagueta pepper, also known as Guinea pepper or Grani del Paradiso; **Mastic*: a vegetable resin obtained from the mastic tree; **Mithridate*: a rather complex pharmaceutical preparation with an antiveneficial function; **Mace*: nutmeg flower; **Malligoe*: white wine of Malaga)

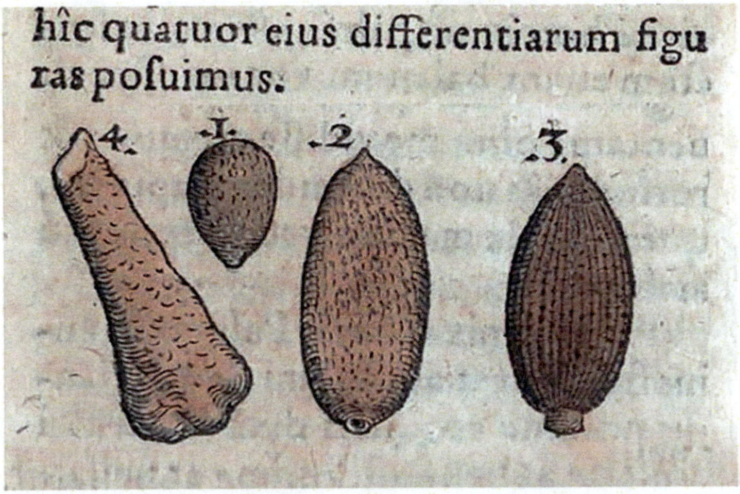

Fig. 20.19 The Judaic Stones illustrated by Gessner (1565: p. 129). The echinoid radioli are numbered from 1 to 3. As with many other illustrations by the Swiss author, these are also found as they are in many other volumes written by authors contemporary with him or later (https://doi.org/10.3931/e-rara-4176 / Public Domain Mark)

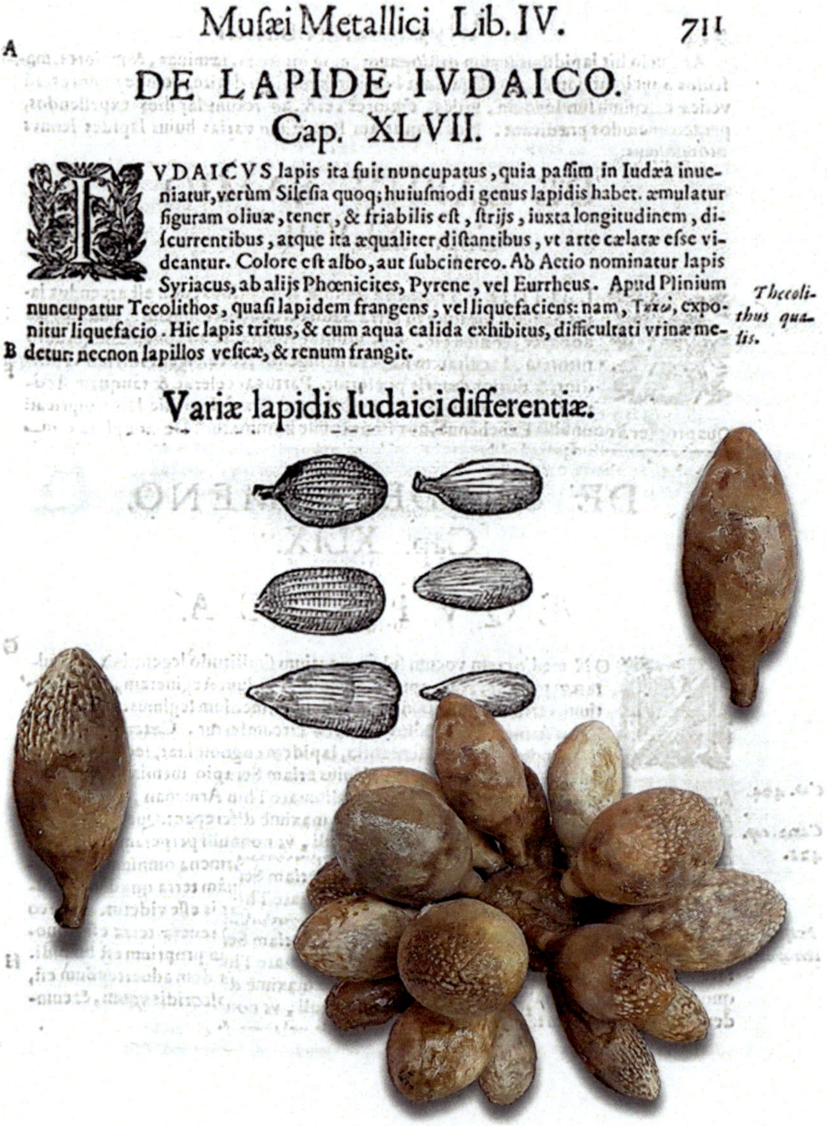
Musæi Metallici Lib. IV. 711

A

DE LAPIDE IVDAICO.
Cap. XLVII.

IVDAICVS lapis ita fuit nuncupatus, quia passim in Iudæa inueniatur, verùm Silesia quoq; huiusmodi genus lapidis habet. æmulatur figuram oliuæ, tener, & friabilis est, strijs, iuxta longitudinem, discurrentibus, atque ita æqualiter distantibus, vt arte cælatæ esse videantur. Colore est albo, aut subcinereo. Ab Actio nominatur lapis Syriacus, ab alijs Phœnicites, Pyrene, vel Eurrheus. Apud Plinium nuncupatur Tecolithos, quasi lapidem frangens, vel liquefaciens: nam, Τήκω, exponitur liquefacio. Hic lapis tritus, & cum aqua calida exhibitus, difficultati vrinæ medetur: necnon lapillos vesicæ, & renum frangit. *Thecolithus qualis.*

B

Variæ lapidis Iudaici differentiæ.

Fig. 20.20 Page from the *Musaeum Metallicum* dealing with the then so-called Judaic Stones (Aldrovandi 1648: p. 711), with the corresponding fossils superimposed: on the right and left two spines and at the bottom a compiled specimen of an echinid of the genus *Balanocidaris*, of which these particular radioles are a characteristic [graphic composition Maurizio Forli. (Credit from: photo Massimo Cresti, plate https://doi.org/10.3931/e-rara-13017 / Public Domain Mark)]

Other fossil radioli, but of elongated shape, coming from the island of Malta, were known under the popular denomination of bastoncini di San Paolo, involved in thaumaturgical rituals connected with the figure of the Saint. The man who sent many fossils to Scylla from Malta, Giovanni Francesco Buonamici, wrote about them, giving his opinion about their origin, diametrically opposed to that of the Messina man

> *... bastoncini di San Paolo, come quì li chiamano, ed altre diverse pietre, che ànno certe figure stravagantissime indeterminate, ed informi, e tali insomma, che non è possibile rassomigliarle a cosa veruna [...] Conchiuderei in somma da tutte le cose suddette, parermi assai più probabile, che tutte queste nostre pietre variamente figurate a somiglianza di animali, e loro parti, altro non siano realmente, che pure pietre di loro specie così formate dalla natura, perché piacque a Dio ...* (Buonamici 1668: p. 165).
>
> *... bastoncini di San Paolo (St. Paul's rods), as they are called here, and other different stones, which have certain extravagant indeterminate and shapeless figures, and such, in short, that it is not possible to resemble anything [...] I would conclude, at last from all the aforesaid things, that it seems to me much more probable that all these stones of ours variously figured in the likeness of animals, and their parts, are not really other than stones of their kind thus formed by nature, because it pleased God ...*
>
> *Egli è un Istrice, fe pur crederà a' suoi occhi; e insiemementе dia un'occhiata all'altro sasso.* B. *ch'è di Malta [...] e se dubitare non si deve, che le colonette, disperse nel tufo [...] siano spine dal vicino guscio calcareo, com'è anche quella, che si vede* A. *della Fig.* II. *Così non si doverebbe determinare altramente de' bastoncini inviatemi, detti volgarmente, e vanamente di S. Paolo .c. essendo manifestissime spine d'Istrice, o maggiore di corpo, o di spezie più terribile de' nostri* (Scilla 1670: p. 148).
>
> *He is a (sea) Porcupine, if you can believe your eyes; and at the same time take a look at the other stone.* B. *which is from Malta [...] and if one should not doubt that the small columns, scattered in the tuff [...] are thorns from the nearby calcareous shell, as is also that which one sees in Fig.* II. In the *same way one should not determine otherwise about the small sticks sent to me, vulgarly and vainly called of St. Paul, since they are very evident thorns of Porcupine, either larger in body or of a more terrible species than ours* (Scilla 1670: p. 148).

In the *Panlessico Italiano* (1839: 669) under the heading St. Paul's sticks we read the following definition: *vulgar name of the fossil or impietrite spines of the sea urchin called by Aristotle echimetra pelagia.* The ones drawn by Scilla belong to the species currently called

> *Stylocidaris melitensis* (Forbes in Wright, 1855).

Another particular of the endoskeleton of the sea urchins, at times described as *St. Paul's mammaries*, is that of the area of the insertion of the tubercle whose morphology is manifestly connected to that of the human breast, simulating the areola and the nipple.

> *L'Istrice Marino d'Imperato, essendo spogliato dalle sue spine, comparisce come un pomo rotondo con molti tubercoli della grossezza di una Lente in tutta la superficie di detto Pomo, che è il corpo di questo Animale Marino: Si pesca vivo ne' Mari della Sicilia, e si trova impietrito nell'Isola di Malta;* [...] *alcuni tuberculi sopra un guscio, che non è altra cosa, che una particula di Istrice Marino d'Imperato, impietrita, che in Malta chiama il Volgo, Tette, ò Mammelle di San Paolo, perché rappresentano due corpi rilevati, come due Mammelle* (Boccone 1697: p. 295).
>
> *The Imperato Sea Porcupine, having been stripped of its spines, appears as a round pommel with many tubercles the size of a lens on the entire surface of said pommel, which is the body of this sea animal: it is fished alive in the seas of Sicily, and is found impetrified in the island of Malta;* [...] *some tuberculi above a shell, which is not other thing, that a particula of Imperato Sea Porcupine, petrified, that the common people in Malta call, Tits, or Mammelle di San Paolo (Saint Paul's breasts) because they represent two bodies detected, as two Mammelle.*

We resort once again to the iconographic apparatus of Agostino Scilla to point out, we believe represented for the first time in ancient printed volumes, a pair of teeth of the buccal apparatus of some species of *Miocene Clypeaster* from Calabria. Another single tooth (*dens Echinitae*), marked with the number 1078, is also in plate VIIII of Lhwyd's *Lithophylacii Britannici ichnographia* (Luidus 1699). A complete Aristotle's lantern, however, referred to a non-fossil specimen, can be seen instead in Aldrovandi, who, completely extracted from the shell, makes it well represented under the echinid from which it comes called *Echinometra maxima pelagica Sardica* (*De Reliquis Animalibus exanguibus libri quator*, 1606: p. 411).

The Cretaceous, the last period of the Mesozoic Era, takes its name from *Craie*, a fine white limestone which is the main component of the marine sedimentary deposits particularly well known and suggestive at Dover and Étretat. The echinids contained in these formations, if not silicified, often retain only their shells and are filled with the "chalky" sediment that includes them. In this particular essence they have been considered a sort of primitive macro lozenges, favoring their use in the medical field as a natural antacid against heartburn, gastroesophageal reflux, and seasickness.

In France and Great Britain, the abundance of specimens has always stimulated popular imagination, and this is how their use as objects endowed with a magical aura is still handed down, often incorporating them in the walls of churches and houses, with superstitious and ornamental functions. As tangible evidence of the use of fossil sea urchins as good luck charms or as protection for certain purposes, from the already mentioned Bellucci collection, we propose one, evidently believed to have medicinal or thaumaturgical powers, which was used to cure lameness in sheep and goats (Fig. 20.21).

Fig. 20.21 Amulet from the Bellucci collection: whitish fossil echinid (Miocene). Campidano di Oristano, Oristano [Bellucci collection, M.A.N.U. (PG) Inv. 720936]. Against the claudication or lameness of goats and sheep, it was passed several times over the sick leg. (Reproduced by kind permission of the Ministry of Cultural Heritage and Activities and Tourism, Regional Directorate of Museums of Umbria—National Archeological Museum of Umbria) (graphic composition Maurizio Forli)

References

Aldrovandi U (1606) De Reliquis Animalibus exanguibus libri quatuor, post mortem eius editi: nemp' De Mollibus, Crustaceis, Testaceis, et Zoophytis. Apud Io: Baptistam Bellagambam, Bononie. https://www.biodiversitylibrary.org/bibliography/119999 (Public Domain)

Bassett MG (1982) Formed Stones, Folklore and fossils. Amgueddfa Genedlaethol Cymru, National Museon of Wales. geological series No. 1. 29 pp.

Aldrovandi U (1648) Ulyssi Aldrovandi patricii Bononiensis Musaeum metallicum in libros IIII : distributum Bartholomaeus Ambrosinus ... labore, et studio composuit cum indice copisosissimo Arcus Antonius Bernia. Bononiae : typis Io. Baptistae Ferronii, 1648. ETH-Bibliothek Zürich, Rar 1027, https://doi.org/10.3931/e-rara-13017 / Public Domain Mark

Boccone PS (1697) Museo di fisica e di esperienze variato, e decorato di osservazioni naturali, note medicinali, e ragionamenti secondo i principij de' moderni. Giacomo Combi, Venetia

Bonardo GM (1585) La minera del mondo dell'ill. sig. Gio. Maria Bonardo Fratteggiano, conte e cavaliero. Nella qual si tratta delle cose più secrete, e più rare de' corpi semplici nel mondo elementare, e de' corpi composti, inanimati, et animati d'anima vegetativa, sensitiva, e ragionevole. Divisa in quattro libri, & per maggior commodità de' lettori ogni capitolo, c'ha potuto riceverlo, ordinato, con ordine d'alfabeto. Mandato in luce da Luigi Grotto, cieco d'Hadria. Fabio, et Agostin Zoppin Fratelli, Venetia

Boodt A Boëtius de (1609) Gemmarum et lapidum Historia. Claudiam Marnium, Hanoviae.. https://hdl.handle.net/2027/gri.ark:/13960/t8w96041q (Public domain)

Boodt A Boëtius de (1636) Gemmarum et lapidum historia. Lugduni Batavorum [Leiden] : ex officina Ioannis Maire, 1636. ETH-Bibliothek Zürich, Rar 1881., https://doi.org/10.3931/e--rara-15217 / Public Domain Mark

Buonamici FG (1668) Lettera missiva del signor Gio: Francesco Buonamici Maltese Dottore di Medicina, Filosofo, e Poeta, diretta ad Agostino Scilla Messinese Pittore ed Accademico della

Fucina detto lo Scolorito. Opuscoli di Autori Siciliani, 1761. 11: 105–195. Stamperia de' Santi Apostoli, Palermo

Camilli L (1502) Speculum Lapidum et D. Petri Arlensis de scudalupis, presbyteri hierosolymitani, Sympathia, septem metallorum ac septem selectorum lapidum ad planetas accredit Magia Astrologica Petri Constantii Albinii villanovensis. Apud Christianum Liebezeit, Hamburgi

Cerutus B, Chioccus A (1622) Musaeum Franc. Calceolari iun Veronensis. Apud Angelum Tamum, Veronae. https://doi.org/10.3931/e-rara-50568 / Public Domain Mark

De Laet J (1647) De Gemmis et Lapidibus, libri Duo. Quibus praemittur Theophrasti, liber de Lapidìbus Graece & latine cum Brevibus Annotationibus. Ioannis Maire, Lugduni Batavorum. https://doi.org/10.3931/e-rara-35487 / Public Domain Mark

Demnard F, Neraudeau D (2001) L'utilisation des oursins fossiles de la Préhistoire à l'époque gallo-romaine. In: Bulletin de la Société préhistorique française. 2001, tome 98, N. 4. pp. 693–715. doi : https://doi.org/10.3406/bspf.2001.12570

Domenichi L (1844) Della Storia Naturale di C. Plinio Secondo, Libri XXXVII. Traduzione di M. Lodovico Domenichi emendata per la prima volta secondo il testo latino con l'aggiunta di un nuovo indice generale. Volume Secondo. Giuseppe Antonelli Ed. Venezia

Esu D, Kotsakis T (1981) Observations on a Pliocene Clypeaster (Echinodermata) figured by M. Mercati. Rendiconti dell'Accademia Nazionale dei Lincei, s.8, 71: 183–190

Gessner C (1565) De omni rerum fossilium genere, gemmis, lapidibus, metallis, et huiusmodi, libri aliquot, plerique nunc primum editi. Tiguri.. https://doi.org/10.3931/e-rara-4176 / Public Domain Mark

Gimma G., 1730. Della Storia Naturale delle gemme, delle Pietre, e di tutti i Minerali, ovvero della Fisica Sotterranea. Tomo II, Felice Mosca, Napoli

Imperato F (1599) Dell'Historia Naturale di Ferrante Imperato napolitano. Libri XXVIII. Nella quale ordinatamente di tratta della diversa condition di miniere, e pietre. Con alcune historie di Piante, et Animali fin'ora non date in luce. Costantino Vitale, Napoli. https://www.biodiversitylibrary.org/item/183177#page/5/mode/1up (not in copyright)

Imperato Fr (1610) De Fossilibus Opusculum. In quo miro ordine continentur Naturalis disciplinae scitu dignissima, eiusque professoribus omnino necessaria; ab alijs minime excogitata. Accesserunt. Dominici Roncalioli, Neapoli. https://archive.org/details/bub_gb_yPMwLx-v1n4kC (Public Domain Mark 1.0)

Lachmund F (1669) Oryktographia Hildesheimensis sive admirandorum fossilium, quae In tractu Hildesheimensi reperiuntur, Descriptio Iconibus Illustrata, Cui addita sunt alia de calculis, de fontibus, &c. Jacobi Mülleri, Hildesheimii. https://wellcomecollection.org/works/rs34c8db (Public Domain Mark)

Luidus E (1699) Eduardi Luidii Lithophylacii Britannici ichnographia : sive lapidum aliorumque fossilium Britannicorum singulari figura insignium quotquot hactenus vel ipse invenit vel ab amicis accepit, distributio classica, [...]. Londini : ex officina M.C ; Lipsiae : sumt. Joh. Ludw. Gleditsch & Weidmann, 1699. Zentralbibliothek Zürich, NG 1883, https://doi.org/10.3931/e-rara-30541 / Public Domain Mark

Karlshausen C, De Putter T (2017) Un oursin pour le dieu. L'oursin de Tjanefer (Turin Suppl. 2761). J Egypt Museum 1:1–12

Mattioli MP (1548) Il Dioscoride dell'eccellente dottor medico m. P. Andrea Matthioli da Siena; co i suoi discorsi, da esso la seconda volta illustrati, et diligentemente ampliati: con l'aggiunta del sesto libro de i rimedi di tutti i veleni da lui nuovamente tradotto, & con dottissimi discorsi per tutto commentato. Vincenzo Valgrisi, at the workshop of Erasmus, Vinegia

Mercati M (1717) Metallotheca Vaticana. Opus posthumum, Autoritate et Munificentia Clementis Undecimi Pontificis Maximi. E tenebris in lucem eductum: Opera autem, & studio Joannis Mariae Lancisii, Archiatri Pontificii illustratum. Ex Officina Jo: Mariae Salvioni Romani, Romae. https://archive.org/details/bub_gb_DC_E6UVWg7EC (Public Domain Mark 1.0)

Moscardo L (1656) Note overo memorie del museo di Lodovico Moscardo nobile veronese. Academico Filarmonico, dal medesimo descritte, Et in Tre Libri distinte. Per Paolo Frambotto, Padoa. https://www.biodiversitylibrary.org/bibliography/157058 (not in copyright)

Passera F (1688) Il Nuovo Tesoro degli Arcani Farmacologici, Galenici. & Chimici, ò Spargirici. Giovanni Parè, Venetia

Scarabelli PF (1666) Museo ò Galeria Adunata dal Sapere, e dallo Studio del Sig. Canonico Manfredo Settala nobile milanese. descritta in Latino dal Sig. Dott. Fis. Coll. Paolo Maria Terzago Et hora in Italiano dal Sig. Pietro Francesco Scarabelli, Fott. Fis. di Voghera. E dal medesimo accresciuta. Per li Figliuoli del qd. Eliseo Viola, Tortona. https://archive.org/details/museoogaleriaadu00terz_0 (no copyright page found)

Scilla A (1670) La Vana Speculazione Disingannata dal Senso. Lettera responsiva Circa i corpi marini, che Petrificati si trovano in varij luoghi terrestri. Andrea Colicchia, Napoli

Scilla A (1759) De corporibus marinis lapidescentibus quæ defossa reperiuntur. Addita Dissertatione Fabii Columnæ de Glossopetris. Joannis Zempel, Romae

Smith WG (1894) Man, the primeval savage. His haunts and relics from the hill-tops of Bedfordshire to Blackwall. Edward Stanford, London. https://wellcomecollection.org/works/sc2hhxdm/items (Public Domain Mark)

Vialli V (1974) Appunti di Paleontologia. Pitagora Editrice, Bologna

Walch JEI (1768-1773) Walch Johann Ernst Immanuel: Die Naturgeschichte der Versteinerungen : zur Erläuterung der Knorrischen Sammlung von Merkwürdigkeiten der Natur. Nürnberg : gedruckt mit Felsseckerischen Schrifften, 1768–1773. ETH-Bibliothek Zürich, Rar 399, https://doi.org/10.3931/e-rara-10246 / Public Domain Mark

Wecker J (1660) Eighteen Books of the Secrets of Art and Nature, being The Summe and Substance of Naturall Philosophy, Methodically Digested. First designed by John Wecker Dr in Physick, and now much Augmented and Inlarged by Dr R. Read. A like work never before in the English tongue. Simon Miller, London

Chapter 21
Crinoidea: Researches among Fairy Coins and Sea Lilies

Abstract Crinoids, commonly known as sea lilies, are marine invertebrates of the phylum Echinodermata. Their bodies are divided into peduncle, theca, and arms. Already present in the Paleozoic, currently about 600–650 recent species are known. Since animals are composed of many segments and plates, in the first printed texts are represented the articles of the peduncles, single or assembled in short strokes, both in starry and round section. The earliest printed figures that we know of are found in Gessner (1565). They received many local designations but were generally called asteriae, Encrinos, Trochites, or Entrochos. Their exact zoological understanding came about around the middle of the eighteenth century. The single articles or fragments of the peduncle were used for medicinal purposes, ingested dissolved in acid drinks or reduced to powder and then dispersed in water, used against poisonings, pestiferous fevers, and hemorrhages. Popular beliefs considered the single five-pointed articles, often assimilated to the Marian cult, able to drive away nightmares and demons, spiders, and other poisonous insects, assuming ambivalent meanings oscillating between Christian and profane beliefs.

Keywords Fossilia · Fossils · History · Crinoidea · Pentacrinus · Asteriae

Crinoids

Crinoids, commonly known as sea lilies or feathered starfish, are marine invertebrates of the phylum Echinodermata, formerly united in the class Crinoidea by Miller (1821) placing all stalked crinoids outside the group of Stelleroidea (sea stars). Distributed in all seas with the exception, apparently, of the Baltic and Black Seas, about 600–650 living species are currently known. The distinctive structures that distinguishes crinoids from other Echinodermata, are: the stalk and the holdfast (a root-like structure that adheres to the substrate), the crown formed by the calyx or theca and the arms. The arms are usually five, thus maintaining the pentaradiate symmetry typical of the members of the phylum.

© The Author(s), under exclusive license to Springer Nature Switzerland AG 2022
M. Forli, A. Guerrini, *The History of Fossils Over Centuries*,
https://doi.org/10.1007/978-3-031-04687-2_21

The stalk (or column or stem) is a flexible structure formed by articulated and superimposed plates, composed of calcite, called columnar articles or columnalia, which are connected by ligamentary tissue. The single articles, with smooth surface or radially striated, which very often are found isolated in the fossil state, may have, depending on the species, different sections, among which the circular and the pentagonal or star shaped are the most common or characteristic. Generally through the "root" of the stalk, the crinoids attach to the substrate. In some species, this happens by means of appendages also formed by mineralized articles, called cirri. In the absence of the stalk, the cirri depart directly from the base of the calyx. Some species of stalked crinoids are permanently anchored to the bottom, while most of the others can move for several meters, still remaining in contact with the seabed. The calyx, or theca, cup shaped, facing upward, is composed by a series of pentagonal plates, rigid or articulated, and by a membrane (*tegmen* or disc), on which the mouth and the anus open. Inside are the other vital organs.

It is from the calyx that the five main arms depart. These are in turn subdivided, giving rise to ten or more appendages, frequently very branched, feather shaped, with hollow pinnules. Arms and pinnules have small brachial plates, arranged in simple or double series. The proximal pinnules or oral pinnules (close to the mouth) are much larger than the others and are equipped with podia, extroversions with sensory papillae; as you move toward the distal portion of the arms, the pinnules become gradually smaller.

All present crinoids are suspensivorous, their food is formed by micro invertebrates and microscopic algae. The arms and the pinnules, placed like a filtering funnel with the mouth turned toward the current, trap the small particles and, through the pedicels forming a fine reticulum, direct them toward the mouth. An analysis of the branching patterns has shown that in the living forms there are various types adapted to take particles of different sizes.

They are benthic animals living on coral reefs and rocky bottoms, from the intertidal zone to the abyssal depths. Some current crinoids, without peduncle, feed in faster currents, attaching themselves to the top of prominent corals or rocks (Figs. 21.1 and 21.2).

Not entering into phylogenetic and taxonomic considerations, we recall that the crinoids appear in the Ordovician, and already in the Middle Ordovician, there are all the Paleozoic subclasses, such as the Disparida, the Cladida, and the Flexibilia. From the Cladida originates in the Triassic the only group of post-Paleozoic crinoids, the Articulata, which include the present forms. The sessile stalked species, that is, those that live anchored to the substrate unable to move, dominant in the Paleozoic, are still well represented until the Oligocene when they begin to decrease in number and to spread more deeply; currently the mobile vagile forms prevail. Fossil crinoids, with more than 6000 described species, constituted the dominant group (already starting from the Upper Ordovician) of echinoderms in the Paleozoic, with a peak in the Lower Carboniferous. The following is the classification referred to the second crinoid that has been illustrated:

Phylum Echinodermata
Subphylum Crinozoa
Class Crinoidea

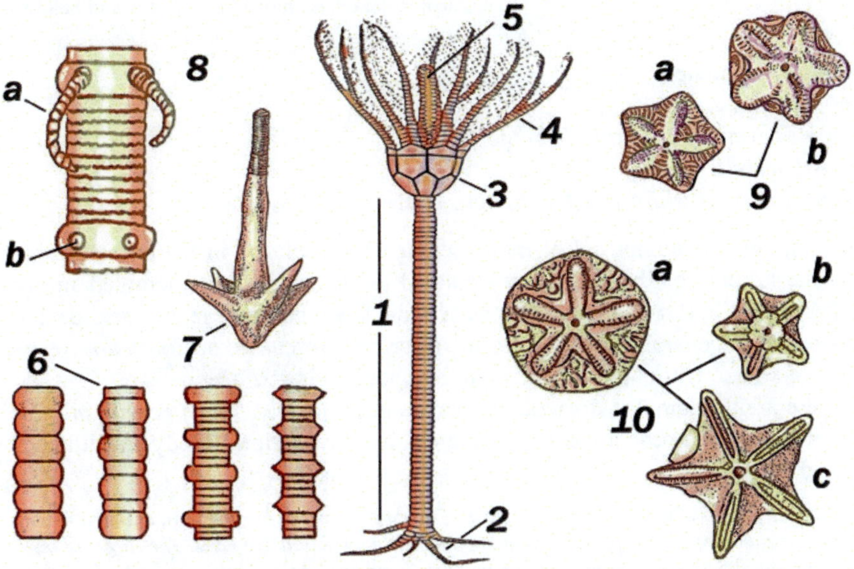

Fig. 21.1 Scheme of crinoid and its parts: (1) Peduncle. (2) Root of peduncle. (3) Calyx or Theca. (4) Arms and Pinnules. (5) Anal pyramid. (6) Different types of peduncles without cirri, in lateral view. (7) Terminal anchor of peduncle of a crinoid of the genus *Ancyrocrinus* (Devonian). (8) Portion of peduncle with cirri, of a member of the genus *Isocrinus* (Triassic). (a) cirrus, (b) cirrus attachment scar. (9) Face of articles: (a) without cirrus. (b) with cirrus scars. (10) Types of articles according to position in the pedicle: (a) columnar article. (b) article of the distal part in *Pentacrinus* sp. (c) article of the proximal part, in the thick species. [Drawing Maurizio Forli modified after Vialli (1974)]

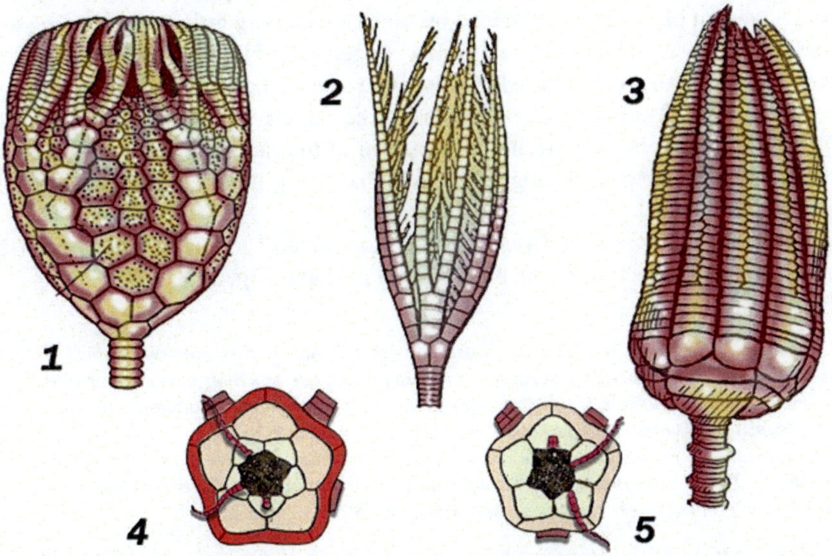

Fig. 21.2 Schematic examples of crinoid calyx types: (1) *Sagenocrinites* sp. (Flexibilia), Devonian. (2) *Dadocrinus gracilis* (Articulata) Triassic. (3) *Encrinus liliiformis* (Articulata) Triassic. (4) Dicyclic base, that is, with two cycles of basal plates below the radial plates. (5) Monocyclic base with only one cycle of basal plates on which another cycle of radial plates is located. The basal plates unite the calyx to the peduncle, the radials support the arms (4–5) bottom view. [Drawing Maurizio Forli modified after Vialli (1974)]

Subclass Articulated
Ordo Encrinida
Family Encrinidae Dujardin & Hupe, 1862
Genus *Encrinus* †
Species *Encrinus liliiformis* † Lamarck, 1801

Being animals composed of many segments and plates, in the earliest printed texts the articles of the stalks are depicted, either single or assembled in short strokes, either in starry or round cross-section. They received many local designations but were generally called *Asteriae, Encrinos, Trochites,* or *Entrochos,* respectively, of course without knowing anything about their true origin.

These are the remains Agricola writes about in his *De Natura Fossilium* (1546) from which we propose a passage in the English translation (Bandy and Bandy JA 1955: p. 98):

> *Just as a species of* entrochos *has the form of a wheel,* encrinos *has the form of a lily. When one angular portion is separated from another one may obtain five lilies. The ridges of one fit into the striae of the other. The five portions have five angles, five sides and a both top and bottom five lilies, hence the Greek name* pentacrinus. *Just as* entrochos *sometimes forms from many* trochites, encrinos *forms from many* pentacrinus. [...] *It is found in the moat of the town of Hildesheim. Stones with palm fronds are found, as Pliny writes, near Munda, Spain* [...].

From this similarity of pentacrinus with lilies, crinoids will later take the name of Stone Lilies and later Sea Lilies.

The first printed figures that we know of are found in Gessner (1565: p. 37) under the heading Asterie (*Asterias separatus, plures còniuncti*) a pentaradiate starry article and a portion of a peduncle formed by nine overlapping articles, belonging to a crinoid of the genus *Pentacrinus*. A few pages later, Entrochos, described as "wheels," are mentioned. It is a radiate disc with a central circular cavity and a fragment of peduncle, formed by three articles of the same type interconnected. About the latter, in the text in support and explanation of the images, the author pointing out that the radial stripes are engraved, says that using it as a seal leaves just the impression of a wheel.

In Rotae figuram formatur Trochitae ab ea ipsa dicti (1565: p. 89) As it says (the name) from itself, Trochite forms the figure of a wheel (Fig. 21.3).

Trochites:

> *C'est une pétrification des articulations de vers de mer rameux, qu'on nomme Tete de-Méduse* [...] *on l'appelle aussi Etolie de mer arbreuse. Les Trochites sont des portions de ses membres* [...] *Peut-etre les divers paquets de rameux forment-ils autant d'animaux, ou de vers unis pour vivre en societé. Une sort de pédicule semble les joindre. Il paroit y avoir autant de tetes & de bouches, que le branches ont d'extremités, rélativement au pédicule qui les unit.* [...] *Rien ne ressemble plus à une plante, & à un polype. Voilà encore un animal - plante, don't l'histoire est assez peu connue* (Bertrand 1763,2: p. 231).

It is a petrification of the joints of branching sea worms, which are called Medusa's Head [...] it is also called Arborescent Sea Star. The articles portions of its limbs [...] It may be that the various branching bundles are formed by many animals, or by

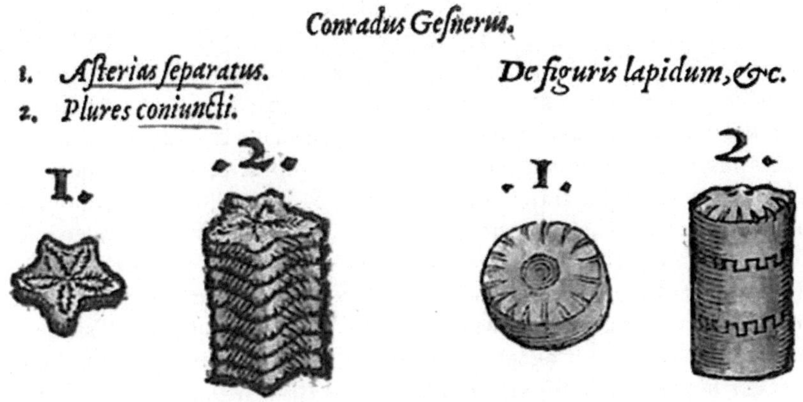

Fig. 21.3 Crinoid articles from Gessner (1565: pp. 37, 89): on the left are *asteriae*, on the right are *entrochi* (https://doi.org/10.3931/e-rara-4176 / Public Domain Mark)

worms united to live in association. A sort of peduncle seems to unite them. They seem to have as many heads and mouths as the branches they have of extremity, relative to the peduncle which unites them. [...] Nothing resembles a plant more than a polyp. Here is another *plant-animal*, whose history is little known.

Helmintholites:

> *Helmintholites o vers pétrifiés* [...] *On comprend sous le nom d'helmintholithes tous les vers de terre & de mer qui sont changés en pierre, ou mineralises, ou agatifiés.* [...] *Nous mettons dans la classe des helmintholithes* [...] *Nous rangeons dans la meme classe les trochites & les entrochites, qui sont des partie ou des articulations des étoiles de mer arbreuses: de meme que les Asteries & les Encrinites* (Bertrand 1763, 1: p. 260).

Helmintholites or petrified worms [...] We include under the name of Helmintholites all worms of land and sea which are changed into stone, or mineralized, or agatized. We place them in the class of Helmintholites [...] we include in the same class the trochites and entrochites, which are parts or joints of arborescent sea stars: as well as the Asterias and Encrinites.

About two centuries passed between Gessner's descriptions and representations and these definitions taken from a dictionary. It would be a compatriot contemporary of Bertrand's who would specify the real structure of the fossils, comparing them to the corresponding living organisms. Understanding the individual fossilized remains of crinoids as part of the body of a single organism was not easy.

Their starry shape, observable on the flat surface of the articles, initially led to their being interpreted and identified with the Asterie, a category in which the radiated calyxes of corals were also included, sharing in this their medical and magical implications. As was the custom, the same Gessnerian images were reproposed many times, for example, we find them in the various editions of Boëtius' treatises (1609: pp. 152, 202; 1636: pp. 300, 411; 1644: pp. 384, 529) or in Johannes De Laet

(1647: p. 139), however, considered to be different from each other. Five-pointed articles continued to be distinguished from circular ones even when the features of their respective stems were compared. So it is also in Aldrovandi where a portion of a circular column is referred to as a *Stelechite*, that is, a twig of petrified wood, lumped together with an antimony mineral for chromatic correspondence, while a stellate is inserted as another, different, in the complex of the Astroites, that is, together with and after the corals and echinoids (1648: pp. 188, 880). Crinoid fossils are more easily found in northern European countries where the Paleozoic and Mesozoic soils containing them emerge more extensively than in Italy. It is therefore in the texts of the writers of these territories that their various structural components are more easily reproduced. A plate in which individual articles and fragments of column are diligently grouped by size is, for example, that found in Frederic Lachmund's *Oryktographia Hildesheimensis* (1669: p. 55). In the text that explains it, the appellation of entrochi and trochites is still widely used, as well as the therapeutic properties attributed to them: *Trochites calculum frangit difficultati urinæ prodest.*

In this function they are assimilated to the radioli of the echinids, the Judaic stones, with which they are sometimes depicted together; reference images for this commonality of theirs may be those found in the second enlarged edition of Ferrante Imperato's *Historia Naturale* (1672: p. 576) and in Mercati's *Metallotheca Vaticana* (1717: p. 227).

Although in a more graphically refined way the usual five-pointed articles, the single one and the corresponding stalk tract, are also included in Plot's *The Natural History of Oxfordshire* (1677: pl. II, Figs. 2–3) in which, however, a chalice is also clearly depicted for the first time (1677: pl. V, Fig. 3).

> *To stones representing the shell-fish of the softer crustaceous kind, such as that tab. 5 fig. 3. in substance and hardness much like a pebble, and of colour yellowish: divided first by five pretty straight lines, adorned on each side with double sets of point, ascending from a protuberant umbilicus in basis of the stone, to another of like form at the top, but foliated round in manner of a rose: and after again subdivided by five other indented lines, terminated before they reach the umbilici; by which means the spaces between these lines are all pentagons; like the outer scales of some sort of tortoise* (1677: pp. 105-106).

The same figure and description are reported the following year in the *Historiae animalium Angliae* of Martin Lister (1678: p. 222, pl. 7, fig. 25) where, however, also this scholar, as the Plot previously, compares it with a figure of Aldrovandian *De Reliquis Animalibus exanguibus* referring, however, to a present echinid cidaroid (1606: p. 407).

> *Juxta Stonor-house agri Oxoniensis inventus est.*

The origin of the crinoid theca immortalized by the two English scholars is Stonor House, ancient and famous residence, also used as a film and television set in some stories, starring among others, James Bond and Inspector Barnaby, another example of lateral thinking involving fossils (Figs. 21.4, 21.5 and 21.6).

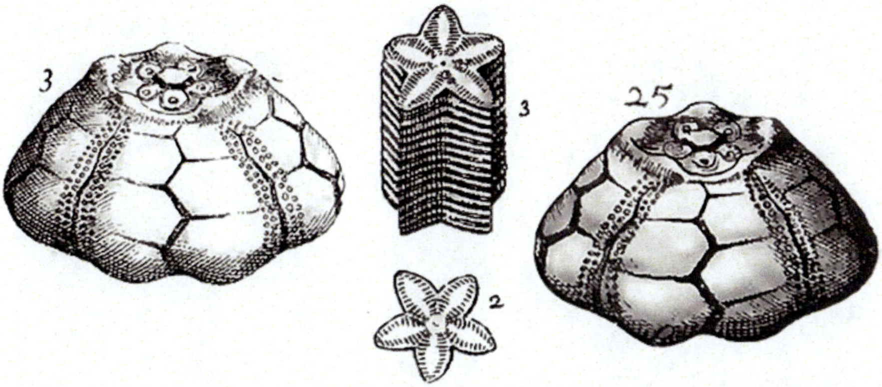

Fig. 21.4 Left, from Plot 1677: pl. V, fig. 3; pl, II, figs 2–3. Right, from Martin Lister 1678: pl. 7, fig. 25. [https://www.biodiversitylibrary.org/bibliography/23488 (not in copyright); https://www.biodiversitylibrary.org/bibliography/65746 (Public domain)]

Fig. 21.5 Peduncle with roots and theca of a crinoid. Lower Jurassic (Ontario, Canada) (photo credit Massimo Cresti)

A few years later, other lithified crinoids from Great Britain are included in the plates accompanying and completing a letter, published in the *Philosophical Transactions* of the Royal Society of London, between Lhwyd and Lister, who a few years earlier (1682) had matched the disconnected and dispersed fragments of crinoids to similar particles of articulated corals.

Peduncles, calyxes, arms, and pinnules are portrayed both in connection and singly, in the same plate in which the first complete trilobite depicted appears. Lhwyd's focus is on this unknown organism, but this does not prevent him from

Fig. 21.6 Fragments of peduncles and single five-pointed articles of *Pentacrinus* sp. (Lower Jurassic). From this similarity of pentacrinus with lilies, crinoids will later take the name of Stone Lilies and later Sea Lilies (photo Maurizio Forli)

commenting briefly on the subjects contained in the other fossil remains depicted, arguing that they are vertebrae and ossicles of organisms that have as close a relationship to starfish as glossopetrae have to shark teeth (Luidus 1698: pp. 279-280 and figs).

This clear conviction of Lhwyd is accepted and supported by other naturalists, such as the Swiss Karl Nicolaus Lang (1708: pp. 64-65), but no one is able to give a certain identity to the heterogeneous remains of these ancient beings. Continuing to consider them similar to our starfish, the German Michael Reinhold Rosinius (1687-1725) published a summa of the knowledge of his time about the Sea Lilies: *Tentaminis de Lithozois ac Lithophytis olim marinis jam vero Subterraneis, Prodromus sive de Stellis Marinis quondam nunc Fossilibus Disquisitio* (1719).

A comparison of the complete fossil specimens with a concretely comparable living marine organism, although the choice would later prove to be incorrect, was proposed by Eberhard Friedrich Hiemer (1682-1727). In describing a slab of oil shale with numerous crinoids of Jurassic age, from near Holzmaden in Germany (1724), he associated them with what seemed to him to be the most similar living thing ever found before, an brittle star with extremely branched arms called Medusa' head, from the Indonesian seas and figured by Rumphius, pseudonym of Georg Eberhard Rumph, a few years earlier (1705: p. 42, pl. 16, a-c). In fact a true Mediterranean crinoid, among other things with a physiognomy similar to the Head of Medusa, illustrated and described many years earlier by Colonna, had gone virtually unnoticed (*Phytobasanos*, 1592: pp. 1-14).

> *Stella marina Altera. Nova et perelegans est huius Stellae forma, ab aliis omnibus differens, nec adhuc descripta: nostro litori frequens* [...].

Different Sea Star. This form of Star is new and very elegant, differs from all others, hitherto undescribed; frequent along our coasts.

This same image is repeated by Ferdinand Jakob Baier (*Monumenta rerum petrificatarum praecipua oryctographiae noricae supplementi*. 1757: p. 11, pl. Vii, figs 1-6) juxtaposed with swimming crinoids of the genus *Saccocoma* from Solnhofen, Germany.

Harenberg Johann Christoph (1696-1774) Germanic theologian and historian, in 1729, published a booklet entitled *Encrinus seu Liliul Lapideum* in which he described in detail the crinoid particles in the ancient manner, Encrinus, Trochites, and Entrochi, reporting the history of publications that had dealt with them until then. In the first half of the eighteenth century there were many researches, but they did not lead to a satisfactory solution that was fully shared by scholars. It is between the second half of this century and the first decades of 1800 that the disputes find their resolution finally establishing the new class Crinoidea.

The capture of some living crinoids was fundamental for the systematic history of this group, one caught by some whalers near Greenland and later described by Ellis (1755) as a "*Clustered Sea-Polype*" and another described in the same year by the Frenchman Jean-Etienne Guettard (1715-1786), preserved together with some corals in the naturalistic collection of Mme de Bois-Jourdain, probably from the Caribbean Sea, called Palmier marin (Palmier marin).

> *Cette découverte a été faite depuis quelque temps, & l'on voit dans le beau & curieux Cabinet de Mme de Bois-Jourdain cet animal, ou une partie de cet animal, assez considérable pour pouvoir nous éclairer entièrement au sujet des pierre étoile & nous fornir beaucoup de lumières sur ce qui regarde l'encrinite, les entroques, les trochites &c.*[...] *il y a plus d'un an ce corps singulier, auquel il donne le nom de Palmier marin;* [...] *il sera facile, au moyen de la description que je vais en donner, de s'assurer que le pierres étoilée ne sont que le types de ce palmier marin [...] l'on ne pourra, à ce que j'espère, s'empêcher de convenir qu'il peut contribuer beaucoup à dèvelopper de plus en plus l'histoire des encrinetes, des entroques &c.* (1755: pp. 224-225)
>
> *This discovery was made some time ago, and this animal, or a part of this animal, large enough to be able to enlighten us completely about the encrinite, the entrochi, the trochitis, etc., can be seen in the beautiful and curious Cabinet of M.* ^{me}*de Bois-Jourdain* [...] *more than a year ago this singular body, to which we give the name of Marine Palm;* [...] *it will be easy, by means of the description that I am going to give, to ensure that the starry stones are but the types of this marine palm* [...] one *will not be able, to what I hope, to do without agreeing that it can contribute a lot to develop more and more the history of encrinites, entrochs etc.*

In 1771, it was Johann Ernst Immanuel Walch, German theologian and naturalist, who published valuable color plates with numerous fossil specimens of *Saccocoma* and stalked crinoids, as an explanation and illustration of the collection of curious natural things of Georg Wolfgang Knorr (Walch 1771: pls XI-XIa-c).

It is Lamarck who, describing the genus *Encrinus* and the only species associated with it, *Encrinus liliiformis*, will recall that "*Les articulations séparées qu'on trouve et qui appartiennent à des tiges d'encrinite, sont connues dans les cabinets d'Hist. Nat. sous les noms de pierre étoilées, de trochites et d'entroques*" (1801: pp. 579-580). Finally, it will be with the treatise of J. S. Miller in 1821, *A natural history of the Crinoidea or lily-shaped animals, with observations on the genera Asteria, Euryale, Comatula, and Marsupites*, that the class Crinoidea will be established, although uncertainties about the systematic organization of these incredible marine animals will remain for several years.

The magical–medicinal virtues of the crinoid particles, singularly described, are summarized by Aldrovandi who, as it almost always happens in the text of the *Musaeum Metallicum*, gives a precise account of them, even if in this case it is rather limited, evidently due to the lack of specific information about them (1648: p. 880).

In particular, the five-pointed Astroites, ingested dissolved in acid drinks or reduced to powder and then dispersed in water, are indicated against poisoning, pestiferous fevers, and hemorrhages; hung in the bedroom they keep away spiders and other poisonous insects, they drive away nightmares and demons. In the popular imagination of the Middle Ages and the Renaissance, these star-shaped ossicles also took on ambivalent meanings, oscillating between Christian and profane beliefs. In France, Spain, and Great Britain, individual stars and their pentagonal columns were thought to be generated in the clouds and sent to earth in the raging of thunder and violent downpours or to be the petrified remains of a shower of falling stars. When present in places of worship, often consecrated to the Madonna, they were associated with her with the most varied reasons among which the one most frequently reported is that they are earthly residues of the stars of her celestial mantle (Fig. 21.7).

The trochites of greater dimensions, for their shape to disc, were instead sometimes joined to the other stones of flattened circular shape, the *numismales*, and

Fig. 21.7 The fragments of articles, especially those with a pentagonal shape, associated with the image of Our Lady, were often considered earthly residues of the stars of her celestial mantle. (Photo and graphic composition Maurizio Forli)

called *Coins of the Fairies* or coins dedicated to some Saint, popular in the various localities of discovery. They were also called Tears of the Giants, Enchanting Stones, and Rosary Beads, the latter definition derived from the use of making rows of them, using the naturally pierced articles, to make devotional crowns (Figs. 21.8, 21.9, 21.10, 21.11, 21.12, 21.13, 21.14, 21.15, and 21.16).

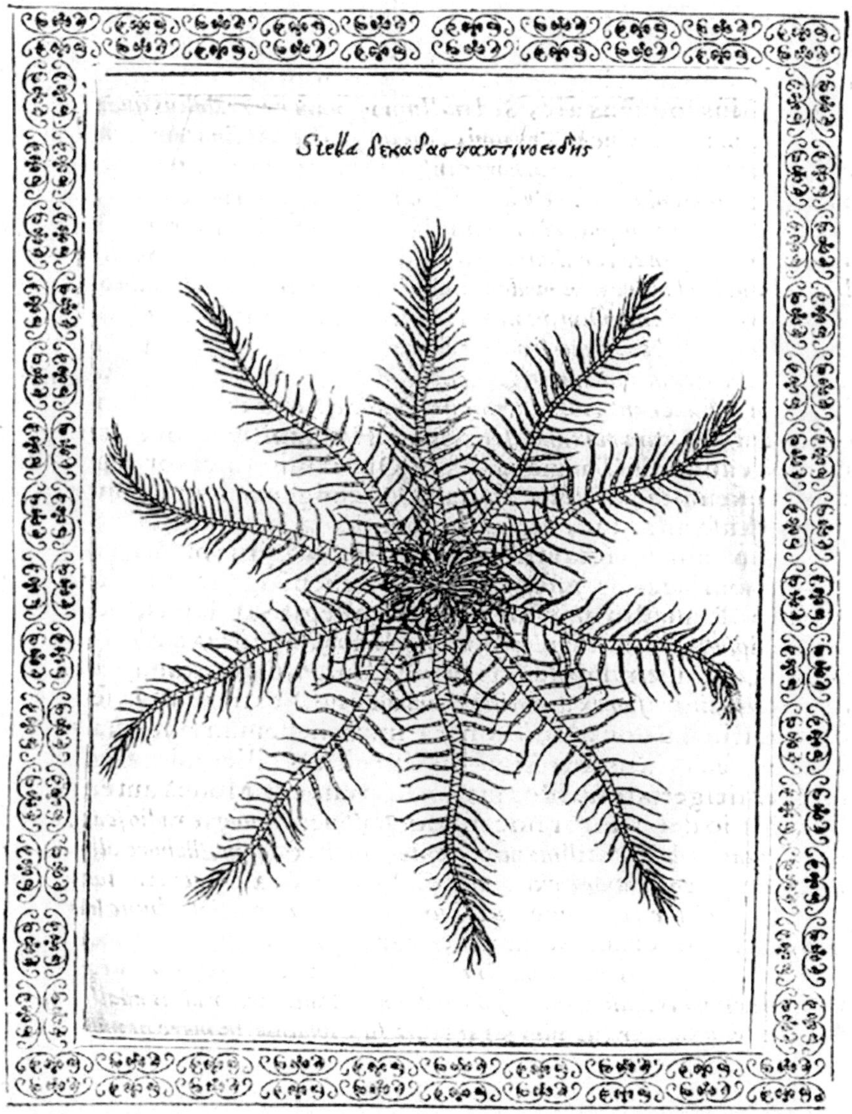

Fig. 21.8 *Stella marina Altera. Nova et perelegans est huius Stellae forma, ab aliis omnibus differens, nec adhuc descripta: nostro litori frequens* [...] *Different Starfish*. This form of Stella is new and very elegant, differs from all others, so far undescribed; frequent along our coasts [Colonna, *Phytobasanos*, 1592: pp. 1–14 (modified)] [https://archive.org/details/gri_33125008412476/page/n155/mode/2up (not in copyright)]

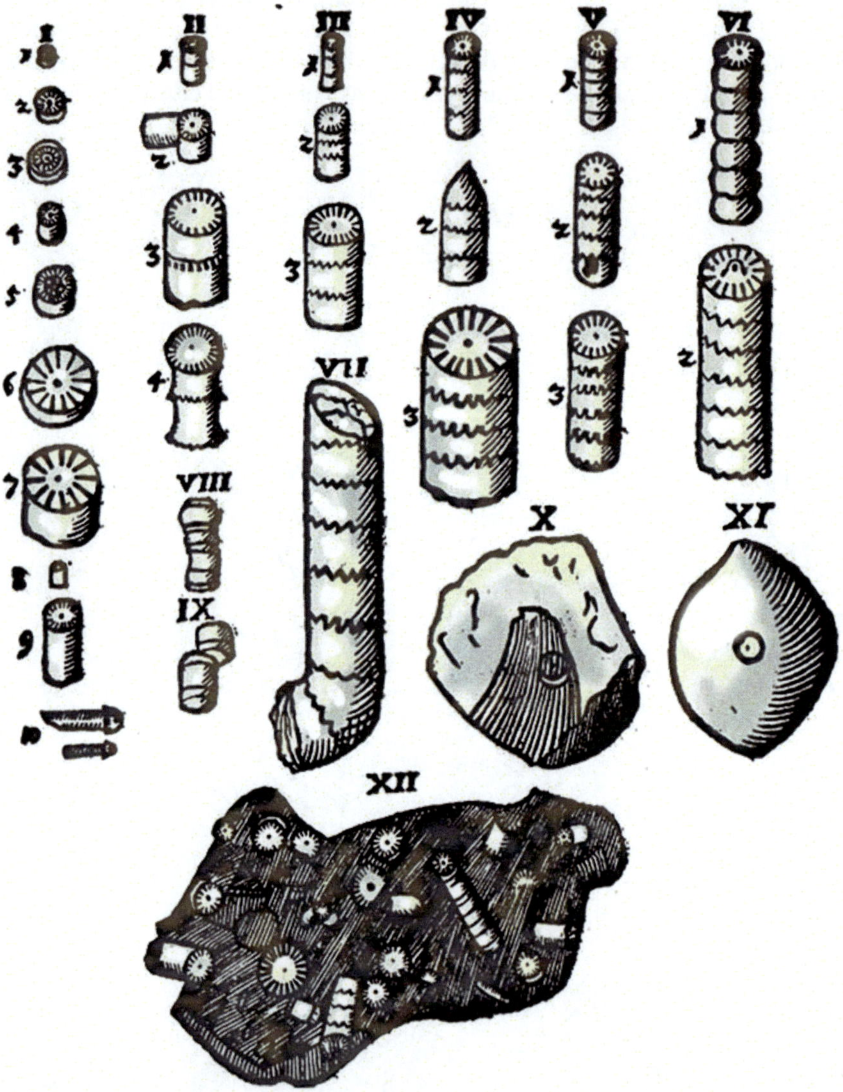

Fig. 21.9 Various forms of *trochites* and *entrochi*, grouped by Frederic Lackmund, according to the shape and length of the discoidal segments (1669: p. 55; ***Oryktographia Hildesheimensis***, Caput XVI, *De Trochite et Entrocho*) (modified) [https://wellcomecollection.org/works/rs34c8db (Public Domain Mark)]

Fig. 21.10 *Entrochi*, segments of radioli grouped among the Judaic stones by Mercati (*Metallotheca Vaticana*, 1717: p. 227) [https://archive.org/details/bub_gb_DC_E6UVWg7EC (Public Domain Mark 1.0)]

Fig. 21.11 Colonnette consisting of a stretch of crinoid peduncle associated with the usual swollen forms of echinoid radioli called Judaic Stones. From Ferrante Imperato (*Historia Naturale*, 1599: p. 660) [https://www.biodiversitylibrary.org/item/183177#page/5/mode/1up (not in copyright)]

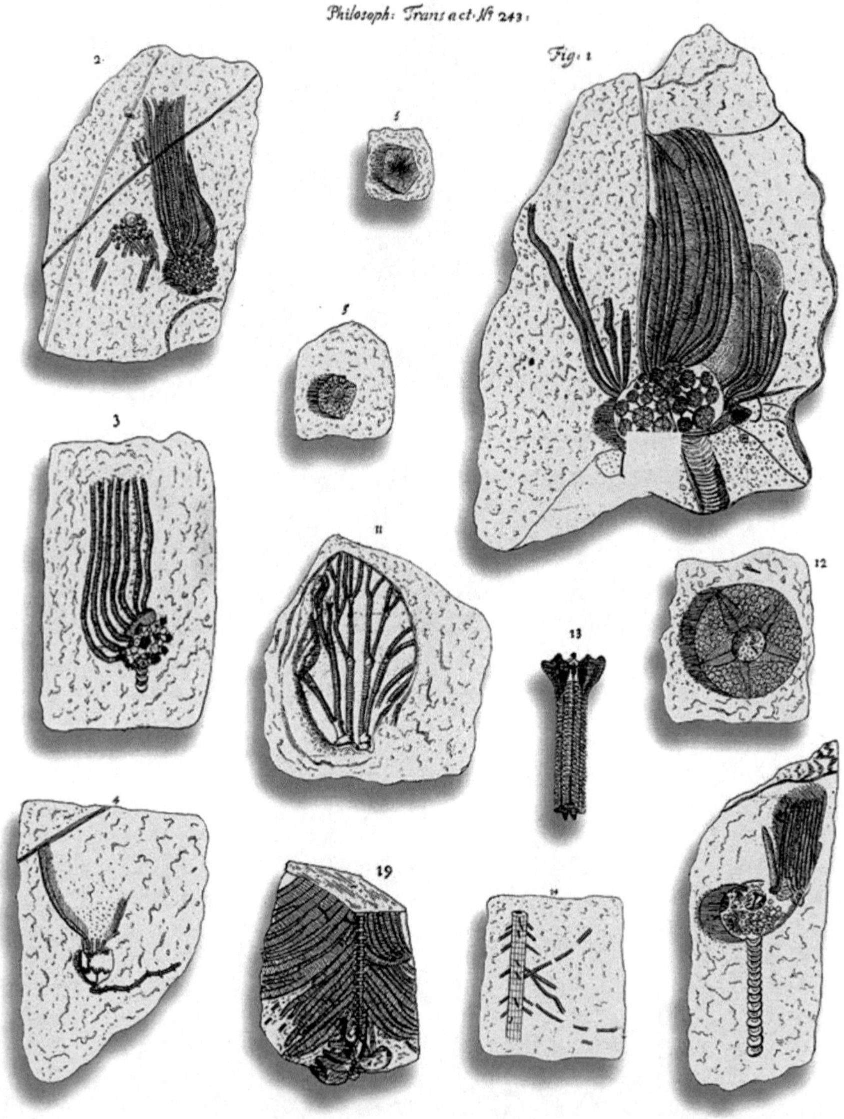

Fig. 21.12 *Figured Stones*, more or less extensive fossil remains of crinoids, then as yet unknown as organisms in their own right, interpreted by Lhwyd (or Luidus) as vertebrae and ossicles of organisms having close relations with sea stars, as well as glossopetrae with shark's teeth (Lhwyd *Phylosophical Transaction*, 1698: pp. 279–280 and figs). (Courtesy https://royalsociety-publishing.org (https://doi.org/10.1098/rstl.1698.0055))

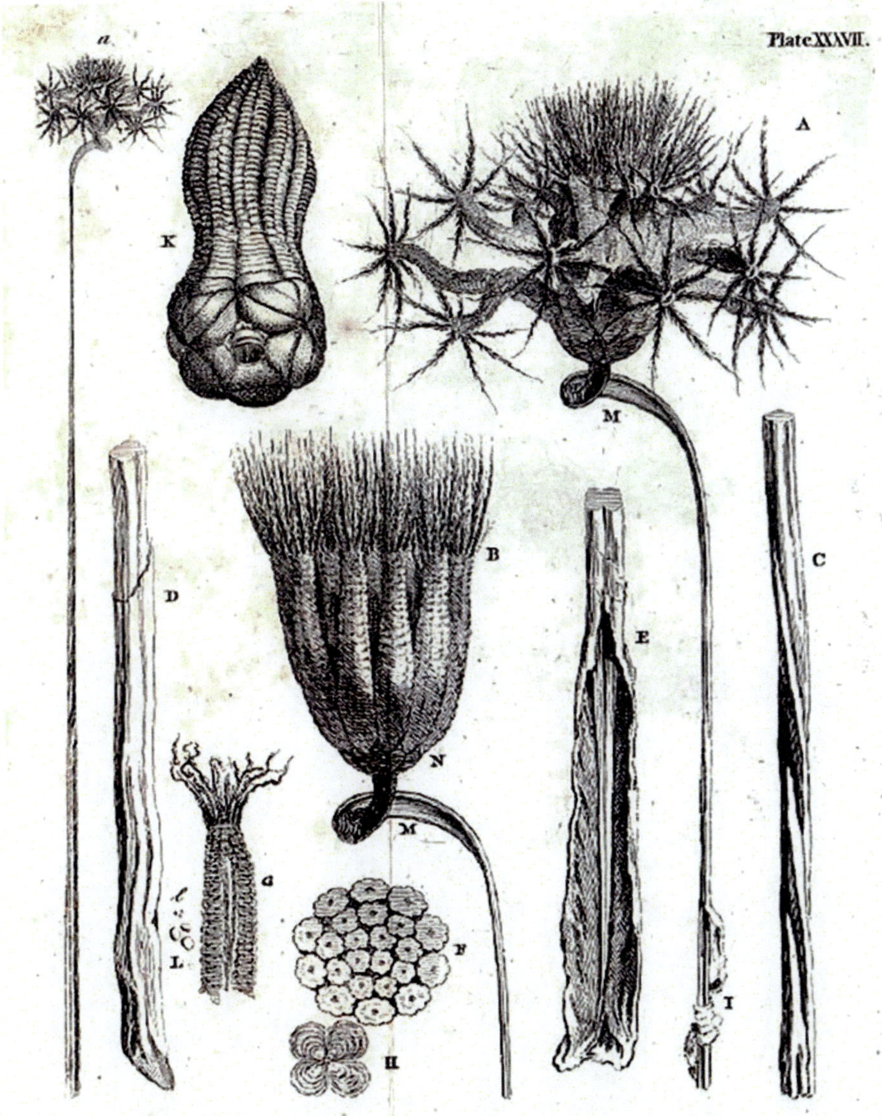

Fig. 21.13 Jean Ellis juxtaposes a fossil goblet with the living crinoid, represented in the same plate, considering them organisms having the same characteristics [...] *in K, it is a figure of* Encrinus *or* Lilium lapideum, *which can only be an exuvia of this animal, a consideration which I submit to the judgment of those who are curious about the study of fossils; for, to the best of my knowledge, they have not yet proposed anything more probable on this subject* (1767: plate XXXVII (modified) (https://doi.org/10.3931/e-rara-23473 / Public Domain Mark)

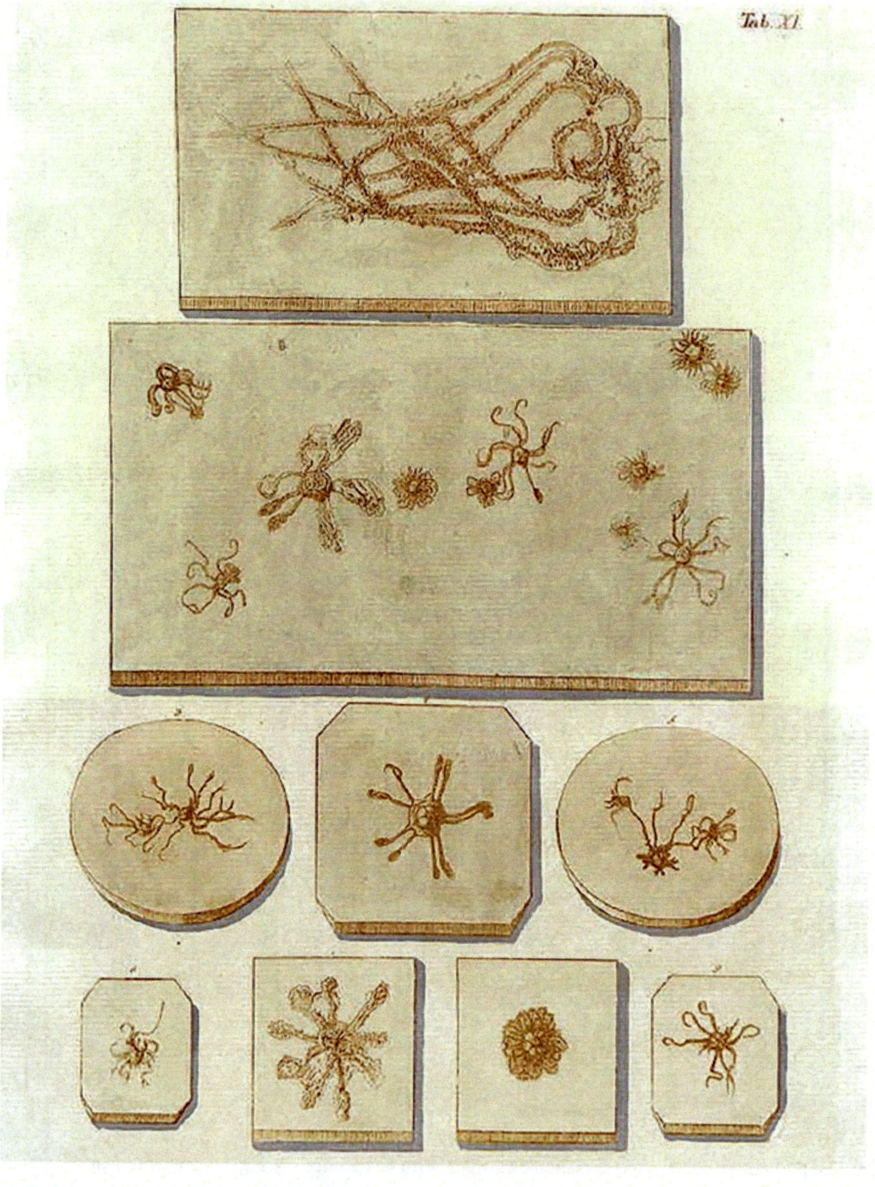

Fig. 21.14 Jurassic of Solnhofen and surroundings (Germany). Larvae of crinoids from Walch (1768-1773: pl. XI). (https://doi.org/10.3931/e-rara-10246 / Public Domain Mark)

Fig. 21.15 Probable juvenile crinoids: *Ascocystites* sp. Paleozoic, Upper Ordovician, Katoua Formation, Morocco (age ~ 450 Ma.). Size of specimen in lower center, without appendages, ca. 60 mm. (Photo credit Maurizio Forli)

Fig. 21.16 *Scyphocrinites elegans* Zenker, 1833; Erfoud, Morocco; Upper Silurian, age ~ 410 Ma. Width of calyx ~ 65 mm. (Photo credit Maurizio Forli)

References

Agricola G (1546) De ortu & causis subterraneorum Lib. V. De natura eorum quae effluent ex terra Lib. IIII. De natura fossilium Lib. X. De veteribus & novis metallis Lib. II. Bermannus, sive de re metallica dialogus. Interpretatio Germanica vocum rei metallicae, addito Indice foecundissimo. Basel: Hieronymus Froben

Aldrovandi U (1606) De Reliquis Animalibus exanguibus libri quatuor, post mortem eius editi: nemp' De Mollibus, Crustaceis, Testaceis, et Zoophytis. Apud Io: Baptistam Bellagambam, Bononie. https://www.biodiversitylibrary.org/bibliography/119999 (Public Domain)

Aldrovandi U (1648) Ulyssi Aldrovandi patricii Bononiensis Musaeum metallicum in libros IIII : distributum Bartholomaeus Ambrosinus ... labore, et studio composuit cum indice copisosissimo Arcus Antonius Bernia. Bononiae : typis Io. Baptistae Ferronii, 1648. ETH-Bibliothek Zürich, Rar 1027., https://doi.org/10.3931/e-rara-13017 / Public Domain Mark

Baier FJ (1757) Ioannis Iacobi Baieri monumenta rerum petrificatarum praecipua oryctographiae noricae supplementi loco iungenda interprete filio Ferdinando Iacobo Bai [...]. Nurembergae [Nürnberg] : in commissione Georgii Lichtenstegeri, 1757. ETH-Bibliothek Zürich, Rar 2333., https://doi.org/10.3931/e-rara-14953 / Public Domain Mark

Bandy MC, Bandy JA (1955) De natura fossilium: textbook of mineralogy. Translated from the first Latin ed. of 1546. Mineralogical Society of America

Bertrand E (1763) Dictionnaire universel des fossiles propres et des fossiles accidentels. A la Haye : Chez Pierre Gosse Junior, et Daniel Pinet, M. DCC. LXIII [1763]. ETH-Bibliothek Zürich, Rar 5720., https://doi.org/10.3931/e-rara-10463 / Public Domain Mark

Boodt A Boëtius de (1609) Gemmarum et lapidum Historia. Claudiam Marnium, Hanoviae. https://hdl.handle.net/2027/gri.ark:/13960/t8w96041q (Public domain)

Colonna F (1592) Phytobasanos, sive Plantarum aliquot Historia, in qua describuntur diversi generis Plantae veriores, ac magis facie, viribùsque respondentes antiquorum Theophrasti, Dioscoridis, Plinij, Galeni, aliorùmque delineationibus, ab aliij hucusque non animadverae. Ex Officina Horatij Salviani, Neapoli. https://archive.org/details/gri_33125008412476/page/n155/mode/2up (not in copyright)

De Laet J (1647) De Gemmis et Lapidibus, libri Duo. Quibus praemittur Theophrasti, liber de Lapidìbus Graece & latine cum Brevibus Annotationibus. Ioannis Maire, Lugduni Batavorum. https://doi.org/10.3931/e-rara-35487 / Public Domain Mark

Ellis J (1755) An Essay towards a natural history of the corallines, and other Marine Productions of the like kind, commonly found on the Coasts of Great Britain and Ireland. Printed for the Author, London. https://www.biodiversitylibrary.org/bibliography/15280 (Public domain)

Ellis J (1767) Versuch einer Natur-Geschichte der Corall-Arten und anderer dergleichen Mer-Cörper, welche gemeiniglich an den Küsten von Gross-Britannien [...] : nebst der Beschreibung eines grossen Büschel-Polypen, welcher in dem Eis-Mere gefangen worden. Nürnberg : bei Gabriel Nikolaus Raspe, 1767. ETH-Bibliothek Zürich, Rar 6161, https://doi.org/10.3931/e--rara-23473 / Public Domain Mark

Gessner C (1565) De omni rerum fossilium genere, gemmis, lapidibus, metallis, et huiusmodi, libri aliquot, plerique nunc primum editi. Tiguri https://doi.org/10.3931/e-rara-4176 / Public Domain Mark

Guettard JE (1755) Mémoire sur les Encrinites et les Pierres étoilées. Dans lequel on traitera aussi des Entroque, &c. Histoire de l'Académie Royale des Sciences, 224–263

Imperato F (1599) Dell'Historia Naturale di Ferrante Imperato napolitano. Libri XXVIII. Nella quale ordinatamente di tratta della diversa condition di miniere, e pietre. Con alcune historie di Piante, et Animali fin'ora non date in luce. Costantino Vitale, Napoli. https://www.biodiversity-library.org/item/183177#page/5/mode/1up (not in copyright)

Imperato F (1672) Historia naturale di Ferrante Imperato napolitano. Nella quale ordinatamente si tratta della diversa condition di minere, pietre pretiose, & altre curiosità. Con varie historie di piante, & animali, ... Venetia https://doi.org/10.3931/e-rara-11175 / Public Domain Mark

Lachmund F (1669) Oryktographia Hildesheimensis sive admirandorum fossilium, quae In tractu Hildesheimensi reperiuntur, Descriptio Iconibus Illustrata, Cui addita sunt alia de calculis, de fontibus, &c. Jacobi Mülleri, Hildesheimii. https://wellcomecollection.org/works/rs34c8db (Public Domain Mark)

Lamarck JBPA de Monet de (1801) Système des Animaux sans vertèbres ou Tableau général des classes, des ordres et des genres de ces animaux; ... L'Auteur, Paris

Lang KN (1708) Historia lapidum figuratorum Helvetiae, eiusque viciniae : in qua non solum enarrantur omnia eorum genera, species et vires aeneisque tabulis repraesentantur, sed insuper adducuntur eorum loca nativa, in quibus [...]. Venetiis : sumptibus authoris, typis Jacobi Tomasini, [1708]. ETH-Bibliothek Zürich, Rar 5680., https://doi.org/10.3931/e-rara-10398 / Public Domain Mark

Lister M (1678) Angliae Historiæ animalium Angliæ, tres tractatus. Unus de araneis. Alter de cochleis tum terrestribus, tum fluviatilibus. Tertius de cochleis marinis. Quibus adjectus est quartus de lapidibus ejusdem insulæ ad cochlearum quandam imaginem figuratis. Joh. Martyn, Londini. https://www.biodiversitylibrary.org/bibliography/65746 (Public domain)

Lhwyd E (1698) IV. Part of a letter from Mr. Edw. Lhwyd to Dr. Martin Lister, Fell. of the Coll. of Phys. and R. S. concerning several regularly figured stones lately found by him. Phylosophical Transaction, pp. 279–280 with figs. https://doi.org/10.1098/rstl.1698.0055

Mercati M (1717) Metallotheca Vaticana. Opus posthumum, Autoritate et Munificentia Clementis Undecimi Pontificis Maximi. E tenebris in lucem eductum: Opera autem, & studio Joannis Mariae Lancisii, Archiatri Pontificii illustratum. Ex Officina Jo: Mariae Salvioni Romani, Romae. https://archive.org/details/bub_gb_DC_E6UVWg7EC (Public Domain Mark 1.0)

Miller JS (1821) A natural history of the Crinoidea or lily-shaped animals, with observations on the genera Asteria, Euryale, Comatula, and Marsupites, illustrated with fifty coloured plates. For the Author by C. Frost, Bristol

Plot R (1677) The Natural History of Oxfordshire, being an Essay toward the Natural History of England. Printed at the Theater, Oxford. https://www.biodiversitylibrary.org/bibliography/23488 (not in copyright)

Rosinius MR (1709) Tentaminis de Lithozois ac Lithophytis olim marinis jam vero Subterraneis, Prodromus sive de Stellis Marinis quondam nunc Fossilibus Disquisitio. Typis Nicolai Sauer, Hamburgi

Rosinius MR (1719) Tentaminis de lithozois ac lithophytis, olim marinis jam vero subterraneis, prodromus : sive, De stellis, marinis quondam, nunc fossilibus. Typis Nicolai Sauer, Hamburgi. https://doi.org/10.5962/bhl.title.14427 (Public domain)

Rumph GE (1705) D'Amboinsche Rariteitkamer, Behelzende eene bescryvinge van allerhande zoo weeke als harde Schaal visschen; te weeten raare Krabben, Kreeften en diergelyke Zeedieren Hoorntjes en Schulpen, die men in d'Amboische Zee vindt: Daar beneven zommige Mineraalen, Gesteenten, ...François Halma, T'Amsterdam. https://www.biodiversitylibrary.org/item/127485#page/79/mode/1up (not in copyright)

Vialli V (1974) Appunti di Paleontologia. Pitagora Editrice, Bologna

Walch JEI (1768-1773) Walch Johann Ernst Immanuel: Die Naturgeschichte der Versteinerungen : zur Erläuterung der Knorrischen Sammlung von Merkwürdigkeiten der Natur. Nürnberg : gedruckt mit Felsseckerischen Schrifften, 1768–1773. ETH-Bibliothek Zürich, Rar 399, https://doi.org/10.3931/e-rara-10246 / Public Domain Mark

Chapter 22
Asteroidea and Ophiuroidea: Starfish and Brittle Stars

Abstract The asteroids and the ophiuroids are exclusively marine benthic animals, living at all depths and at all latitudes, from the mesolittoral zone to the abyssal one. Very active at night, in shallow waters they are disturbed by the sunlight and, during the day, most of them live hiding in the shade. Are known from the Paleozoic to the Present. Because of their physiognomy, mostly five pointed, assimilated to the stars of the celestial vault, they have assumed an infinity of meanings since the most remote times. The very shape of the five-pointed star, in its best-known iconography, circumscribed in a circle, becomes an esoteric symbol which, with the point pointing upward, is an instrument used to invoke benevolent spirits and vice versa, instead, to attract the forces of evil. With doubt, the first, late, figure of a fossilized star is that of an ophiuroids reproduced by Louis Bourguet (*Mémoires pour servir à l'histoire naturelle des petrifications dans les quatre parties du monde : avec figures, et divers indices aussi méthodiques que nécessaires*. A la Haye: chez Jean Neaulme, 1742. ETH-Bibliothek Zürich, Rar 2642, https://doi.org/10.3931/e-rara-16053 / Public Domain Mark, 1742). The influence of their fossil remains in folklore is practically nil.

Keywords Fossils · History · Asteroidea · Ophiuroidea

The stars, the stars in the sky that shine their own light, accompanying the destiny of mankind since time immemorial, have inspired and still inspire all of us, if only for the emotions we feel when we observe them. Asteroids are also those called planets, celestial bodies of various shapes and composition, rocky or metallic, small in size from a few meters to a maximum of about 1000 km in diameter. Both the word *Asteroid* and *Asteroids* have the same meaning: *star shaped*. It is from this extraordinary and varied mixture of feelings that they have taken on an infinite number of meanings since the most remote times. Expressing a desire for light, immortality, ascension to heaven they are positively connoted but, as often happens because of human nature, they can take on other symbolic meanings of opposite sign. Exemplary in this sense, is the star in its most famous iconography, the five-pointed circumscribed in a circle, esoteric symbol that with the tip pointing upward

© The Author(s), under exclusive license to Springer Nature Switzerland AG 2022

M. Forli, A. Guerrini, *The History of Fossils Over Centuries*,
https://doi.org/10.1007/978-3-031-04687-2_22

is an instrument used to invoke the benevolent spirits and placed instead vice versa, to recall the forces of evil. As we have reported just above in the discussion of astroites, those with five rays, in some cases have been believed tangible evidence of stars fallen directly from the vault of heaven or from the Marian mantle. In spite of the complexity of the symbolism linked to the star shape, in the first printed books dealing with naturalia, there are no references to fossil specimens of these marine invertebrates, undoubtedly because of their rarity (Fig. 22.1)

Fig. 22.1 *Sicut superius ita est inferius...* As above so is below... as in the sky so in the sea, this motto is inscribed in the upper cartouche of the illustration, at the base of the title of the first work devoted entirely to the starfish, written by Johann Heinrich Linck, entitled *De Stellis Marinis, Liber Singvlaris* (1733). (Modified image Maurizio Forli after Linck, 1733. https://doi.org/10.3931/e-rara-54925 / Public Domain Mark)

The present species are about 3500; exclusively marine benthic animals, the asteroids, and the ophiuroids live at all depths and at all latitudes, from the mesolittoral zone to the abyssal one. Very active at night, in shallow waters they are disturbed by the sunlight and, during the day, most of them live hiding in the shade. There are some in which the arms are very small, assuming a decidedly pentagonal profile, while others can reach more than 50 distinct arms, assuming a circular perimeter with a maximum diameter of a few decimeters greater than 1 meter. They have no predators but are actively parasitized by many microorganisms, including numerous small gastropod mollusks. They are mainly carnivorous animals and spend most of their time looking for food. They can ingest small prey whole, but they can also evert their stomachs from their mouths, adhere them to even relatively large animals, and digest them externally. Their preferred preys are corals, small crustaceans, and bivalve mollusks, in particular, oysters and mussels, of which they can open the even most resistant shells, utilizing as spreaders the very numerous pedicels, or *podia*, ending with microscopic suckers. These are also useful for moving on surfaces (Fig. 22.2)

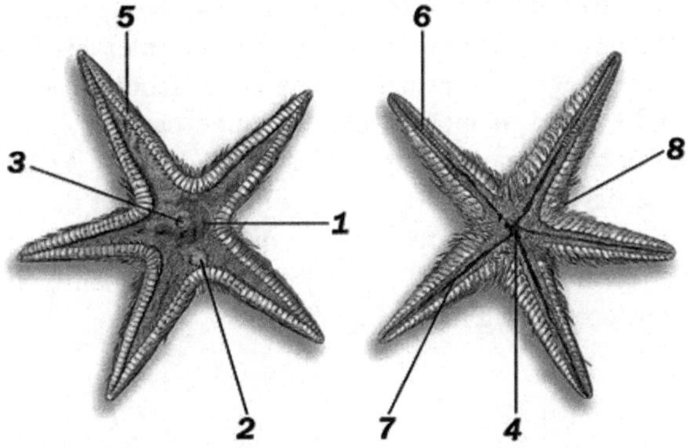

Fig. 22.2 Schematized external morphology of a starfish. Dorsal part: (1) Central disc; (2) Madreporite; (3) Anal opening; (5) Aculei. Ventral part: (4) Oral opening; (6) Ambulacral sulcus; (7) Pedicels; (8) Marginal aculei. (Graphic composition Maurizio Forli, modified image from Linck, 1733, plates 5,6. https://doi.org/10.3931/e-rara-54925 / Public Domain Mark)

Class Asteroidea

It includes species with pentagonal or typically stellate profile, characterized by coalescent arms, fused with each other at the base, which unite gradually forming a central disc; the arms, normally five, in some species can reach even the number of 50. The body, formed by a thick and robust epidermis supported by a complex mobile skeletal apparatus, is more or less flattened and there is a distinction between the lower, or ventral, side with the mouth in the center, and the upper, or dorsal, side, with the pitted madreporic plate connected to the aquiferous system through the petrous canal, and, when present, the anal opening. The general anatomy follows that of other echinoderms; the aquiferous system is constituted by radial canals that extend to the extremities of the arms, supporting symmetrically with respect to them a series of ampullae that continue with a series of terminal pedicels, which, coming out in the ambulacral grooves, have locomotory and sensory functions. The genital organs, digestive system, and coelomic expansions are distributed within the arms.

The respiration occurs through the epidermis of the aboral surface, often provided with protective spines, from which depart the so-called pedicellaria, small prehensile appendages, with protective function.

The peculiar dermaskeleton consists of a system of hundreds of interconnected plates and platelets, called ossicles, which give considerable flexibility to the structure as a whole.

The plates are variously named the ambulacral, symmetrically paired, arranged as an open V downward are the main constituents of the arms; the margino-ventral or inframarginal on the oral face, and the margino-dorsal or supramarginal on the opposite face, serves to consolidate them. There may sometimes exist numerous small plates called intermediate ventrals which fill the intervals between the margino-ventrals and are irregularly arranged. The upper face of the arm, the roof, is occupied by a single series of median plates, called carinal, or by two series of dorso-lateral plates or by a large number of small apical plates. As it can be guessed, the multitude of the ossicles, kept together by the inner and outer organic tissue, easily breaks up after the death of the animal, dispersing in the environment. It is these remains of a few millimeters of size which most frequently are found fossilized in the sediments containing them (Figs. 22.3, 22.4).

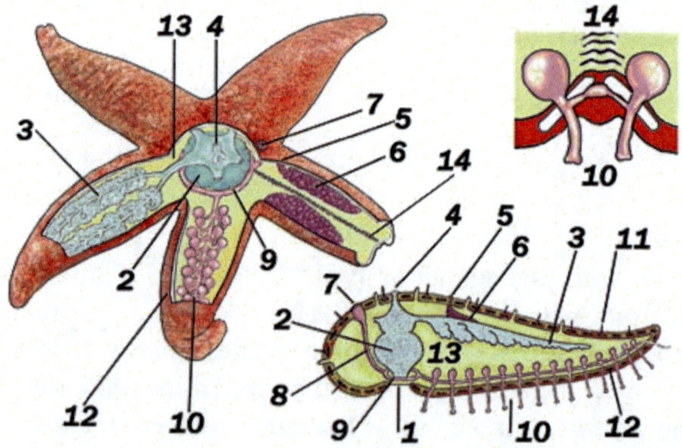

Fig. 22.3 Sketched internal morphology, bottom right, longitudinal section of an arm and top, detail of pedicels: (1) Oral aperture. (2) Stomach. (3) Intestine. (4) Anal aperture. (5) Gonopore. (6) Gonad. (7) Madreporite. (8) Hydrophore canal. (9) Aquiferous ring. (10) Ampullae and Pedicels. (11) Spines and papules. (12) Dermaskeleton. (13) Coelomic cavity. (14) Radial nerve. [Drawing Misa Forli modified after Vialli (1974)]

Fig. 22.4 Central disc and oral aperture of a dry starfish. The pedicels and protective spines are still visible. (Photo Maurizio Forli)

Class Ophiuroidea

Because of theirform they are also called *serpentine starfishes* Their body, covered by a thin semi-transparent epidermal membrane, is provided with a central disc well distinct from the arms which, in number of five, have circular section, are segmented serpentiniform, long and very mobile. The mobility is largely due to the numerical reduction of the marginal plates and other supporting elements compared to the Asteroidea from which they differ also for other characters: for example, the ambulacra are not open furrows but canals closed by special brachio-ventral plates; the coelom is reduced; the ambulacral plates, fused and well articulated with each other, are placed so as to occupy the greatest part of the interior of the arm; the madreporite, instead of being on the dorsal side as in the Asteroidea, is situated, in the adults, on the ventral one; the pedicellaria and not even the anal opening do not exist. The pedicels, coming out from holes placed at the contact with the lateral brachio-lateral plates, do not have ambulatory function but simply sensory, the respiration is provided by evaginations located between the arms and the disc. Both Asteroidea and Ophiuroidea show a surprising capacity of regeneration, giving origin to new specimens starting from portions of the body. In the Asteroidea, for example, from a detached arm, individuals with the so-called *comet shape are* reborn, while among the Ophiuroidea, from a portion of the disc, a complete individual is regenerated.

Furthermore, the ophiuroids have the capacity to resort to the autotomy, that is, to voluntarily detach in case of necessity their arms, partially in segments or entirely. The similarity between the ophiuroid limb and the tail of the lizard is fitting in this case (Fig. 22.5).

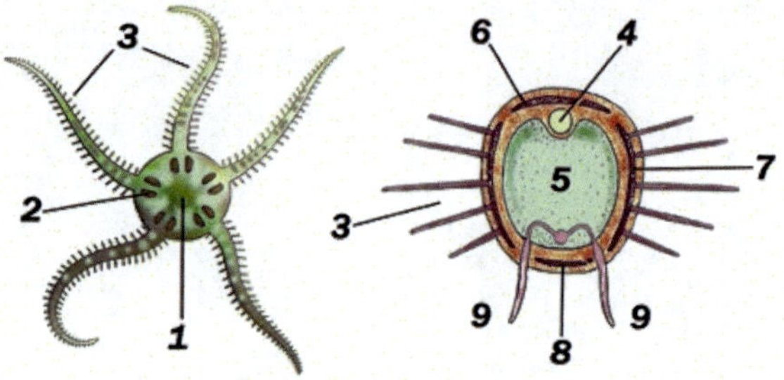

Fig. 22.5 Schematized morphology, external and internal, of an Ophiuroidea: (1) Disc. (2) Radial shields. (3) Arm spines (4) Coelom. (5) Ossiculus. (6) Dorsal arm plate. (7) Lateral arm plate. (8) Ventral arm plate. (9) Pedicels [Drawing Maurizio Forli modified after Vialli (1974)]

The geological role of these organisms is rather scarce, except in particular cases in which their remains pile up on stone surfaces, the discovery of a fossil sea star preserved in toto, generally occurs in exceptional cases. The first asteroids appear in the Cambrian and the first ophiuroids in the Upper Silurian. Both groups show a progressive greater complexity in the skeletal structure from the most ancient species to the most recent and present ones. The ophiuroids are relatively easier to find fossilized integrally because many were, and are, of excavatory habits living on and in the sediments. Among the oldest Italian fossil specimens, we report the personal testimony of Professor Vai "*I found a starfish in the pelagic limestone of the Upper Devonian (Famennian) near the Italian/Austrian border at the edge of Lake Volaia, Carnic Alps in 1963*" and that relating to the discovery of a specimen, with an age of about 240 million years, dating back to the Triassic, the first period of the Mesozoic era, which took place in 2012 in the province of Lecco. As a zoological curiosity we report the classification of what is now considered the largest echinoderm in the world, a starfish with 12 arms that have, from one side to the other, a total length of just over 130 centimeters. It was discovered in the Gulf of Mexico where it lives at depths of 5/600 meters.

Class Asteroidea
Ordo Brisingida
Family Brisingidae
Genus *Midgardia*
Species *Midgardia xandaros* Downey, 1972

Delle stelle marine.

Oltre a queste cose, io veggo gli autori clari per sapienza ammirare la stella che è in mare. Questa è picciolo pescie, et di dentro è carne, et di fuori i ha duro callo. Dicono che è di focosa natura, che abruscia tutte le cose che tocca in mare, et ogni cibo di subito smaltisce. Et per quali esperimenti si sia conosciuto questo non facilmente potrei dire [Brùcioli

Antonio. *Historia Naturale di C. Plinio Secondo nuovamente tradotta di Latino in Volgare toscano*, 1548, X, cap. 60, CCLVI)].

Of the Starfishes.

In addition to these things, I see the wise authors admire the star that is in the sea.This is a small fish, and inside is flesh, and outside has a hard callus. They say that it is of a fiery nature, that burns all the things it touches in the sea, and immediately disposes of all food. And by what experiments this has been known.

Degli animali acquatici et remedii ordinati alle malattie, et prima contro à veleni et velenati animali. Dicano che nò si puo apportare cattivi medicamenti, ò nuocere se la stella marina, impiastrata cò sangue di volpe, è appiccata al cardinale della porta, ò al chiavistello di rame di essa porta [Brùcioli 1548, XXXII, cap. 5, DCCCCVIII)].

They say that it is not possible to bring bad medicine or harm, if the starfish implanted with fox's blood, is stuck to the cardinal of the door, or to the copper latch of the door.

This is how Pliny expressed it, considering starfish "fishes"as inhabitants of the sea, as will also do the sixteenth-century authors who will publish their first printed icons, not renouncing as a habit and necessity, to give indications also on their medicinal and magical powers.

In the volumes of naturalistic character printed during the '500 the illustrations of the living starfishes, meaning with this definition both the Asteroidea and the Ophiuroidea, are rather common, for instance, we find them already in Bellon (*De Aquatilibus*, 1553: pp. 2, 391) described as *Stella solis potius effigiem referens, non edulis* and in Rondelet (*Universae aquatilium Historiae*, 1555), where the representatives of both classes of stars are subdivided in various categories according to the shape. (Fig, 22.6)

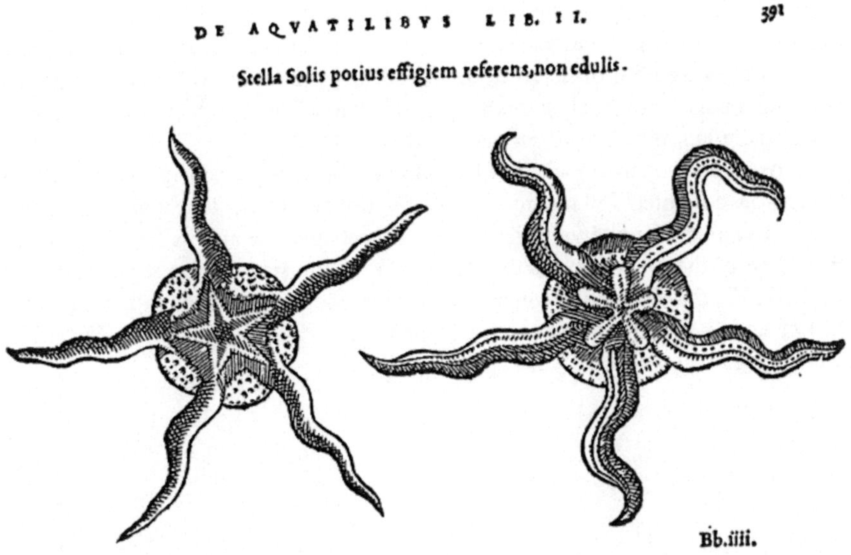

Fig. 22.6 Bellon (*De Aquatilibus*, 1553, 2: p. 391), serpentine stars described as *Stella solis potius effigiem referens, not edulis*. [https://www.biodiversitylibrary.org/bibliography/5765 (Public domain)]

The first extensive treatise devoted exclusively to these fascinating marine invertebrates is that composed by Johann Heinrich Linck (1674–1734), entitled *De Stellis Marinis, Liber Singvlaris*, in which, however, there are no references to fossil specimens, except for those relating to the crinoid ossicles, then believed to be skeleton constituents of Asterias.

The first, late, figure of a fossilized star that we have been able to find is that of an ophiuroids reproduced in plate LIX at number 438 of Louis Bourguet's *Mémoires pour servir à l'histoire naturelle des petrifications dans les quatre parties du monde* (1742).

In consideration of the year of publication of the work, almost 200 years later than the first printed publications, in spite of the efforts made in examining the documentation in our possession in search of representations of these fossil Echinodermata, we still have the doubt that the one we indicate is not really the first illustration of a serpentine petrified starfish.

> *La petite étoile à cinq rayons de la planche LIX mérite l'attention de tous les Curieux. Cette sorte d'animal pétrifié est infiniment plus rare que les écrevisses, les crabes, et les Poissons pétrifié* (Bourguet 1742: pp. 89–90).

A few years later it is Walch who illustrates together five stone specimens with the fossils of two different *Saccocoma* sp., two specimens with Ophiuroidea and one with an Asteroidea with lunate pentagonal profile (*Das Steinreich systematisch entworfen*. 1762: p. 75, pl. II, fig. 1). A specimen morphologically similar to the latter, named *Asteria antiqua*, from Hoburgen in Gotland, is also contemplated in a nineteenth-century publication by von Wilhelm Hisinger:

> *Asteria antiqua* [...] *In Gottlandia ad Hoburg, in saxo arenaceo* [1837: p. 89, pl. XXVI, fig. 6. (*Lethaea Svecica seu Petrificata Sveciae*)].

If ancient texts are lacking on the fossilized remains of starfish, the same cannot be said for their symbology. The starry pentagram consisting of a single intertwined line, considered a symbol of perfection, placed inside a circle, with the five points symbolizing the metaphysical elements of spirit, water, fire, Earth, and air, assumed a mystical meaning of harmony between body and soul.

The subsequent superimposition of a human body on the star thus becomes a representation connoted by meanings closely linked to the physical and divine worlds, with man tending toward the infinity of the universe and its mystical value, Divinity. One of the most famous examples of this interpretation is represented by Leonardo da Vinci's Vitruvian Man, perfectly inscribed both in the circle representing the sky and, in the square, representing the Earth. (Fig. 22.7, Fig. 22.8, Fig. 22.9, Fig. 22.10, Fig. 22.11, Fig. 22.12, Fig. 22.13, Fig. 22.14).

438

Fig. 22.7 From Louis Bourguet: *petite étoile à cinq rayons* [*Traité des pétrifications*, 1742: pl. LIX, fig. 438. (Modified image Maurizio Forli) https://doi.org/10.3931/e-rara-16053 / Public Domain Mark]

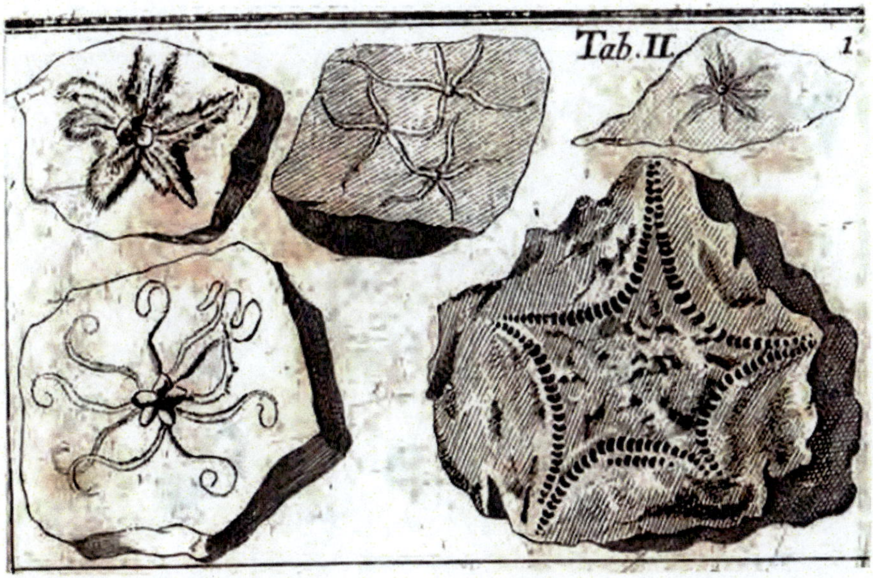

Fig. 22.8 Illustrations by Ernst Emmanuel Walch [1762: p. 75, pl. II, fig. 1. (Modified image Maurizio Forli https://doi.org/10.3931/e-rara-10246 / Public Domain]

Fig. 22.9 Illustration from von Wilhelm Hisinger: *Asteria antiqua* [...] *In Gottlandia ad Hoburg, in saxo arenaceo* [*Lethaea Svecica seu Petrificata Sveciae*, 1837: p. 89, pl. XXVI, fig. 6. (Modified image Maurizio Forli) https://doi.org/10.3931/e-rara-68566 / Public Domain Mark]

Fig. 22.10 Mesozoic Ophiuroidea from Germany (maximum width about 5 cm). (Photo Maurizio Forli, courtesy Carlo Chirli collection)

Fig. 22.11 *Palaeocoma egertoni* (Broderip, 1840) Eype, Bridport, Dorset, Great Britain; Jurassic (Sinemurian) age ca. 190 Ma., size 50 × 35 mm. (Photo credit Geocollection)

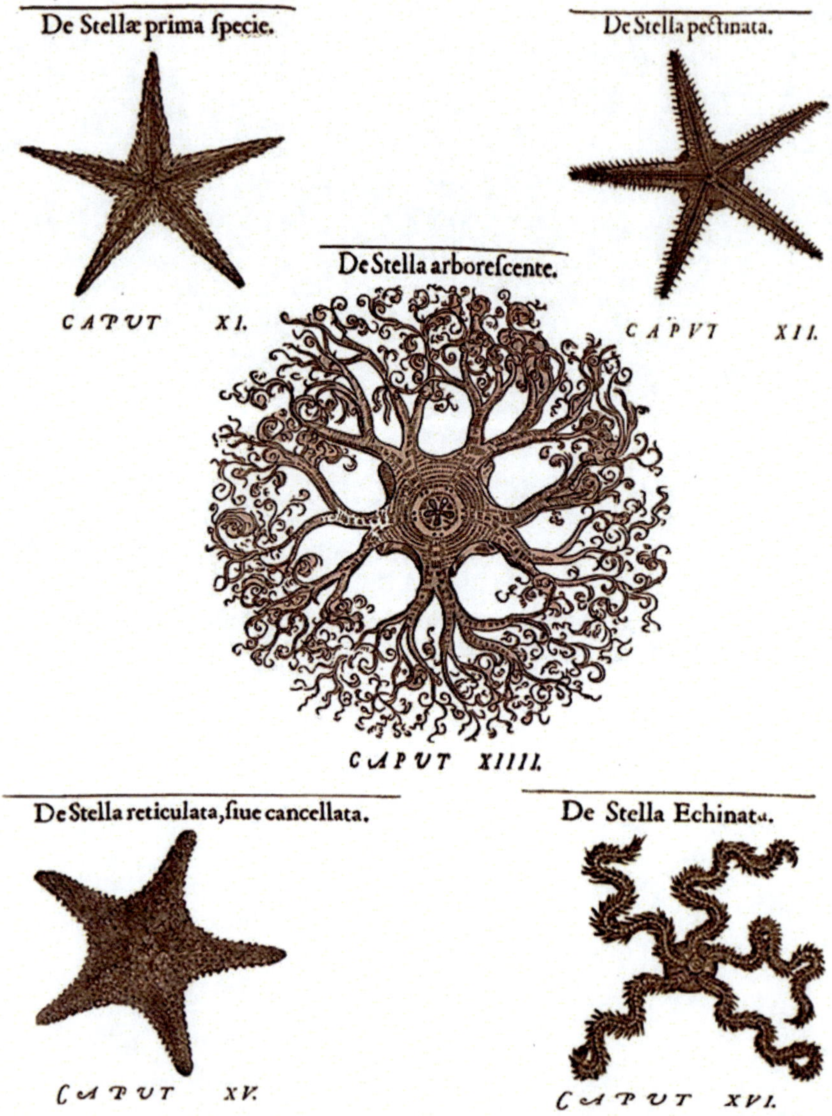

Fig. 22.12 Asteroidea and Ophiuroidea from Rondelet: *Universae aquatilium Historiae pars altera*, (1555): pp. 119–123. [(Modified image Maurizio Forli, https://archive.org/details/ARes071032/page/n13/mode/2up (Public Domain 1.0)])

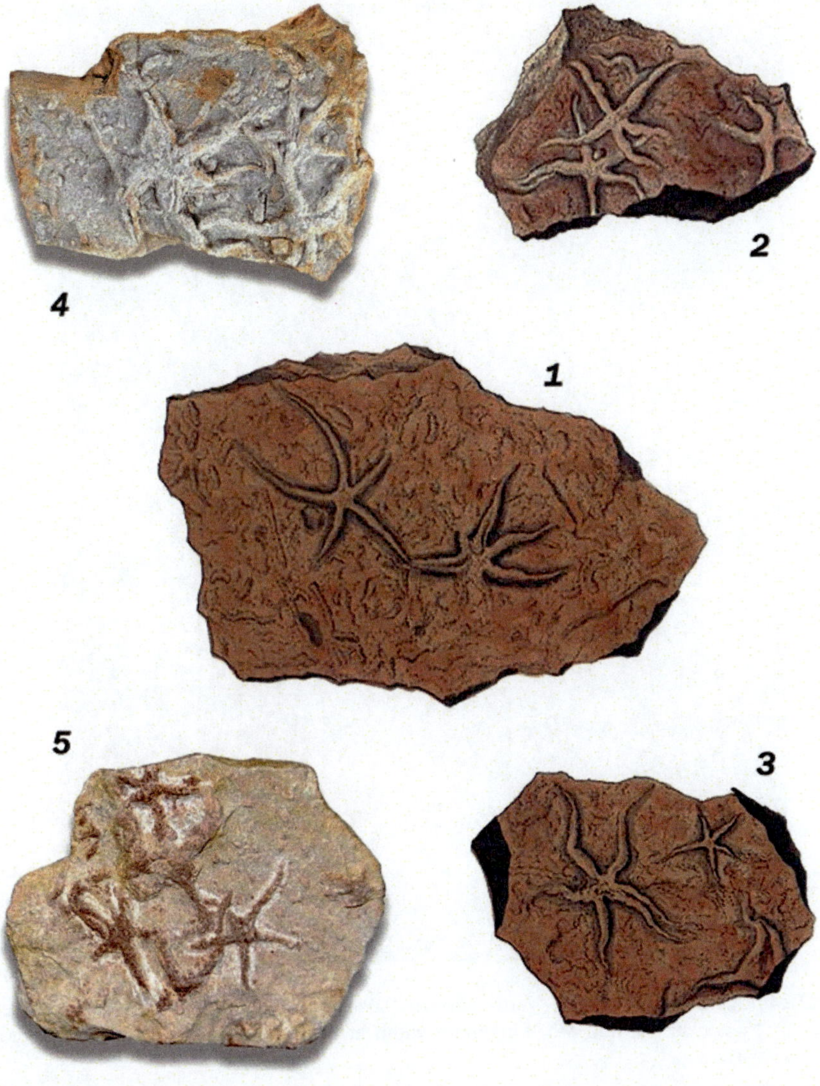

Fig. 22.13 (1–3) Bioturbations of fossil Asterozoa, from Johann Ernst Immanuel Walch 1773: P2, L. Die Naturgeschicte der Versteinerungen [https://doi.org/10.3931/e-rara-10246 / Public Domain]. (4–5) Similar specimens from the French Triassic. (Photo credit Massimo Cresti). (graphic composition Maurizio Forli)

Fig. 22.14 Ophiuroids from the North African Paleozoic. Talkrebt locality, Sahara Desert, Morocco. Ordovician, ca. 450 Ma. Maximum width at extended arms about 130 mm. (Photo Maurizio Forli)

References

Bellon P (1553) De Aquatilibus, Libri duo. C. Stephanum, Parisiis. https://www.biodiversitylibrary.org/bibliography/5765 (Public domain)

Bourguet L (1742) Mémoires pour servir à l'histoire naturelle des petrifications dans les quatre parties du monde : avec figures, et divers indices aussi méthodiques que nécessaires. A la Haye: chez Jean Neaulme, 1742. ETH-Bibliothek Zürich, Rar 2642, https://doi.org/10.3931/e--rara-16053 / Public Domain Mark

Brùcioli A (1548) Historia Naturale di C. Plinio Secondo nuovamente tradotta di Latino in Volgare toscano. Alessandro Brùcioli, Venetia

Hisinger W (1837) Lethaea svecica seu Petrificata Sveciae : iconibus et characteribus illustrata. Holmiae: typis P.A. Norstedt et Filii, MDCCCXXXVII. [1837]. ETH-Bibliothek Zürich, Rar 29431, https://doi.org/10.3931/e-rara-68566 / Public Domain Mark

Linck JH (1733) De Stellis Marinis Liber Singvlaris : Tabvlarvm Aenearvm Figvras Exemplis Nativis Apprime Similes Et Avtoris Observationes / Iohannis Henrici Linckii ... Stellarvm Marinarvm Tabvlae Exemplis [...]. Lipsiae : Apvd Iacobvm Schvstervm, MDCCXXXIII. Zentralbibliothek Zürich, NNN 40 | F, https://doi.org/10.3931/e-rara-54925 / Public Domain Mark

Rondelet G (1555) Universae aquatilium Historiae pars altera, cum veris ipsorum Imaginibus. Matthiam Bonhomme, Lugduni. https://archive.org/details/ARes071032/page/n13/mode/2up (Public Domain 1.0.)

Vialli V (1974) Appunti di Paleontologia. Pitagora Editrice, Bologna

Walch JEI (1768–1773) Walch Johann Ernst Immanuel: Die Naturgeschichte der Versteinerungen : zur Erläuterung der Knorrischen Sammlung von Merkwürdigkeiten der Natur. Nürnberg : gedruckt mit Felsseckerischen Schrifften, 1768–1773. ETH-Bibliothek Zürich, Rar 399, https://doi.org/10.3931/e-rara-10246 / Public Domain

Chapter 23
Scorpions, Ants, and Other Stone Insects. The Understanding of Trilobites Over the Centuries

Abstract Trilobites, characteristic of and exclusive to the Paleozoic Era, were a large group of marine invertebrates, considered among the oldest known arthropods. Because of their external morphological structure that can be divided transversely into three parts or lobes, they have been called trilobites, literally three lobed. The first evidence of contact between man and trilobites comes from a French prehistoric cave, where in 1886, a remnant of a trilobite was discovered perforated on both sides, probably as an ornament. The dating of the layer of finding has been estimated around 15.000 years ago. The trilobites are not described in the ancient Greek lapidaries and consequently also in the following ones there are no indications that with certainty can be traced back to the representation of one of these particular arthropods. As in some regions of Spain, where trilobite fossils are found in abundance, their vestiges are still locally called scorpions or stone ants, it has been proposed that some of the ancient names used, such as in Pliny *Scorpitis, Myrmecitis,* or *Cantharias*, stone in the shape of a scorpion, ant, or beetle, might be referred to the fossils of these arthropods. It is Dr. Edward Lhwyd, who in 1698, publishing part of a correspondence between himself and Dr. Martin Lister, gives us the first illustrations of trilobites. We do not know that in Europe trilobites have been used as a medicinal substance, and even the evidence of their use in magic is extremely limited, leading back only to their function as amulets.

Keywords Fossils · History · Trilobita · Scorpitis · Myrmecitis · Cantharias

The trilobites, characteristic and exclusive fossils of the Paleozoic Era, were a vast group of marine invertebrates and are considered among the oldest known arthropods; except for some related Precambrian forms that are still insufficiently known, they include more than 15,000 described species in the current state of paleontological research.

Their wide diffusion, both spatially and temporally, starts from the Lower Cambrian and ends with the Upper Permian (from ~541 to ~251 million years ago). During their evolutionary history, which lasted a little less than 300 million years,

© The Author(s), under exclusive license to Springer Nature Switzerland AG 2022

M. Forli, A. Guerrini, *The History of Fossils Over Centuries*,
https://doi.org/10.1007/978-3-031-04687-2_23

they underwent a first biological crisis already starting from the Upper Cambrian but, from the following Ordovician, they developed considerably reaching, in this geological period, their most important adaptive radiation. In the following Silurian, Devonian, Carboniferous, and Permian periods, their diversity and diffusion progressively reduced until their total, definitive disappearance at the Permian–Triassic transition, defined by the greatest biological crisis that has ever occurred on our planet.

Trilobites average between 3 and 10 cm in size, with species just over one millimeter in size, as in *Acanthopleurella stipulae* Fortey & Rushton, 1980, to true giant specimens about 70 centimeters long and 40 centimeters wide. The record of maximum size belongs to *Isotelus rex* Rudkin et al., 2003, which to date is the largest known trilobite, with a length of about 72 centimeters.

These animals, which due to their shape and way of life may resemble to some marine crustaceans, sort of small lobsters without claws and antennae, had the inner parts of the body protected on the upper side by a carapace, what in zoology is defined as exoskeleton, that is an outer structure, a more or less rigid and robust shell, with a composition based on calcite, with needle-shaped crystals perpendicular to the surface, with low contents of magnesium, immersed in an organic matrix not well defined, probably chitin or other proteins as in the present arthropods.

It is because of their external morphological structure subdividable, following the longitudinal axis, in three parts or lobes that they have been called Trilobites, literally *three lobed.* As a matter of fact, in the carapace, in a longitudinal sense, there are two axial grooves which subdivide it into two lateral lobes or Pleurae and a central one called also Axial Lobe or *Rachis.* It is, however, odd to note that the same tripartite geometry of the body is evident also in a transversal sense, with a portion called Anterior or *Cephalon,* corresponding to the head and adjacent zone, composed by coalescent segments, that is, attached and fused together, a Median or *Thorax* with mobile segments and a Posterior or *Pygidium*, with fused segments which may end with a central caudal spine which, at times, presents symmetrically on its sides several pairs of marginal spines. In some cases, the pygidium may end with a long axial spine, and in some forms of Olenellidae is composed by a single elongated tergite forming a true telson. While it is true that *pygidium* is generally smaller than *cephalon,* there is variability in this ratio that is also useful for taxonomic purposes. In this regard we can have micropygi trilobites, with *pygidium* much smaller than the *cephalon*, passing through the subisopygi, isopygi, and macropygi, the latter, with *pygidium* larger than the *cephalon.* (fIG. 23.1)

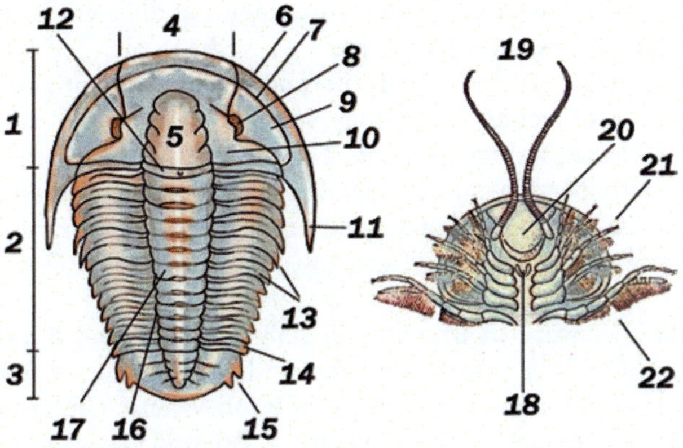

Fig. 23.1 Brief description of the external morphology of a trilobite. Dorsal side: (1) Cephalon. (2) Thorax. (3) Pygidium. (4) Cranidium. (5) Glabella. (6) Anterior flap. (7) Marginal sulcus. (8) Compound eye. (9) Librigena (Mobile cheek). (10) Fixigena (Fixed cheek). (11) Genital tip. (12) Occipital ring. (13) Pleural tip and sulcus. (14) Pleura. (15) Marginal tip. (16) Marginal sulcus. (17) Rachis. Ventral side: (18) Oral aperture. (19) Antennae. (20) Hypostome. (21) Cephalic appendages. (22) Thoracic appendages. [Drawing Maurizio Forli modified after Vialli (1974)]

The mineralized exoskeleton, external and superior, could be smooth or covered by nodes, spines, grooves, ridges, and tubercles of different morphologies and dimensions, while that of the ventral appendages was chitinous and the lower part of the body was covered by a soft membrane, probably similar to that of some present crustaceans. The flattened body with the mouth placed inferiorly and the eyes placed superiorly, the structure and disposition of the ambulacral appendages, indicates that most trilobites burrowed in the marine sediment, some species were even blind and without eyes. Their footprints left in the form of movement tracks, before being associated with their locomotor activity, were interpreted as enigmatic fossils.

Often incomplete or fragmentary fossil carapaces are found, sometimes in considerable number and belonging to the same species, consisting of the remains of the various stages of development, more frequent in the early stages of individual growth, which occurred by changing the coating of the outer and inner surface, with the process that takes the name of *moult*, allowing the body of the animal to grow. In the post-Cambrian trilobites, the ability of many species to roll up on themselves is manifested and diffused. A general morphological description of the upper part of the exoskeleton of an *ideal* trilobite is given in the figure at the beginning of this paragraph.

Trilobites occupied a wide range of environments at different latitudes, inhabiting both shallow waters and tropical reefs as well as polar depths, including a large number of pelagic habitats. Their morphological differentiation is believed to be the expression of a complex ecology, including extremely varied lifestyles, related to very different trophic systems. They fed on mud and microorganisms; they were essentially benthic animals that dug in the sediment, walked, and swam on it. In Europe, the major Paleozoic fossiliferous deposits with trilobite remains are found in Norway and Sweden, Germany, Czech Republic, France, England, and Spain.

As far as the discovery of trilobites in Italy is concerned, the best exposed Paleozoic soils, which have provided interesting fossil faunas, are found in southwestern Sardinia, others in the Carnic Alps, in Friuli Venezia Giulia on the Italian-Austrian border, and finally in Sicily, south of Palermo, between Corleone and Sciacca.

The main orders of trilobites and their chronostratigraphic extension are listed below. According to the proposed taxonomic schemes, some groups of arthropods, which we do not mention here, are considered related to trilobites and included in them:

Agnostida (Lower Cambrian–Upper Ordovician)
Redlichiida (Lower Cambrian–Middle Cambrian)
Corynexochida (Lower Cambrian–Upper Devonian)
Ptycopariida (Lower Cambrian–Upper Ordovician)
Lichida (Middle Cambrian–Middle Devonian)
Asaphida (Middle Cambrian–Upper Silurian)
Odontopleurida (Middle Cambrian–Upper Devonian)
Proetida (Middle Cambrian–Upper Permian)
Harpetida (Upper Cambrian–Upper Devonian)
Phacopida (Lower Ordovician–Upper Devonian)

Finally, the systematic classification of one of the most characteristic and exclusive trilobites of the Lower Cambrian of Sardinia *Dolerolenus zoppii* (Meneghini, 1882), type species of the respective genus, is as follows (Fig. 23.2):

Fig. 23.2 *Dolerolenus zoppii*, specimen from the Paleozoic of Sardinia (Italy). (Photo credit Gian Luigi Pillola)

Phylum Arthropoda
Class Trilobita †
Superfamily Redlichioidea †
Family Dolerolenidae †
Genus *Dolerolenus* †
Species *Dolerolenus zoppii* (Meneghini, 1882) †

The first evidence of contact between man and trilobites comes from a French prehistoric cave in the department of Yonne, in Burgundy, where in 1886, following archaeological research carried out on the site, along with other finds, a remnant of a trilobite was discovered perforated on both sides, probably to make an ornament. Today the cave is called *Trilobite Cave* and the dating of the layer of discovery has been estimated to be around 15,000 years ago, in what is called the Magdalenian or Magdalenian period (from 18,000 to 11,000 years ago) in which this important culture of the Upper Paleolithic, widespread mainly in central Europe, especially in western France and partially in some areas of Spain and Germany, flourished. Since there are no trilobite fossils in the region of the find, it is clear that it was brought there by prehistoric men. After an initial presumed provenience from Germany, recently the most accredited hypotheses indicate its origin in France itself, in the Morvan Mountains, in central France or in the area of the lower part of the Loire Basin, in north-western France (Chlupác 2000; Henry 2001). (Fig. 23.3), (Fig. 23.4), (Fig. 23.5), (Fig. 23.6), (Fig. 23.7).

Fig. 23.3 Locomotion track of a trilobite. Pennsylvanian Inf., Carboniferous, Hale Formation, Cane Hill Member, Washington County. [Photo credit Enrico Bonino collection (Bonino and Kier 2019: p. 117, fig. 88b)]

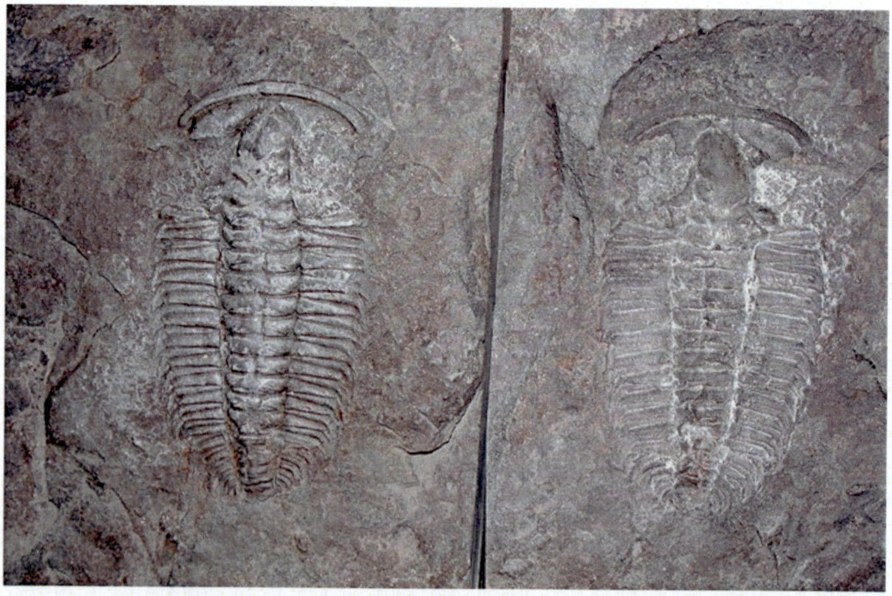

Fig. 23.4 Cast and mold of an indeterminate trilobite from the Chinese Cambrian, probably belonging to the genus *Redlichia*, largest size about 80 mm. (Photo Maurizio Forli)

Fig. 23.5 *Flexicalymene ouzregui* Destombes, 1966, one of the most common trilobites from Morocco, Alnif, Ordovician. Size of the largest specimen, about 80 mm. (Photo Maurizio Forli)

Fig. 23.6 *Phacops rana* (Green, 1832), the "frog-eyed" trilobite, still from the Paleozoic of Morocco, length 140 mm. (Photo Maurizio Forli, courtesy Carlo Chirli collection)

Fig. 23.7 Paleozoic (Cambrian) trilobite from the vicinity of Zaragoza (Spain), probably a species belonging to the genus *Paradoxides*, length about 50 mm. Specimens similar to this one that could have been identified as scorpions or large petrified ants. (Photo Maurizio Forli)

Trilobites are not described in the ancient Greek lapidaries, and consequently there are no indications in the later ones that can be traced to the representation of one of these particular arthropods. Since in some regions of Spain, where trilobite fossils are abundant, their remains are still locally called stone scorpions or stone ants, a few years ago the Spanish authors Liñán and Gozalo (2008) proposed that some of the ancient names used by Pliny, such as *Scorpitis, Myrmecitis,* or *Cantharias*, stone in the shape of a scorpion, ant, or beetle, could be referred to the fossils of these arthropods. The arguments brought in support of this thesis, even if plausible, cannot be verified at the moment as admitted by the writers, even if in the conclusions they take for granted that the ancient Greeks and Romans knew them. Before the description under the present name took definite credence, they were called *Enthomolithus* by some scholars of the eighteenth century, that is, stone insects. Their remains, as well as being relatively rare, are often fragmentary and incomplete as they are almost always parts of the exoskeleton detached during growth (exuviae), by means of a shell moult, so scorpions, ants, and insects are not very recognizable while, for example, fossil dragonfly larvae, often found in the upper Miocene soils, look much more like great ants and are certainly more common and easier to find than trilobites. The same can be said for incomplete remains of crustaceans, such as crabs and shrimps which could also correspond to Plinian descriptions. It is also true that some medieval representations of the scorpion, especially those referring to the corresponding sign of the zodiac, show us segmented and flattened animals in which the bodily appendages are poorly marked and rather indistinct, including the tail with stinger, which in the eyes of our contemporaries may remind us of the trilobites, but it is almost certain that this impression could not be aroused in our predecessors who did not know the trilobites as we know them. (Fig. 23.8)

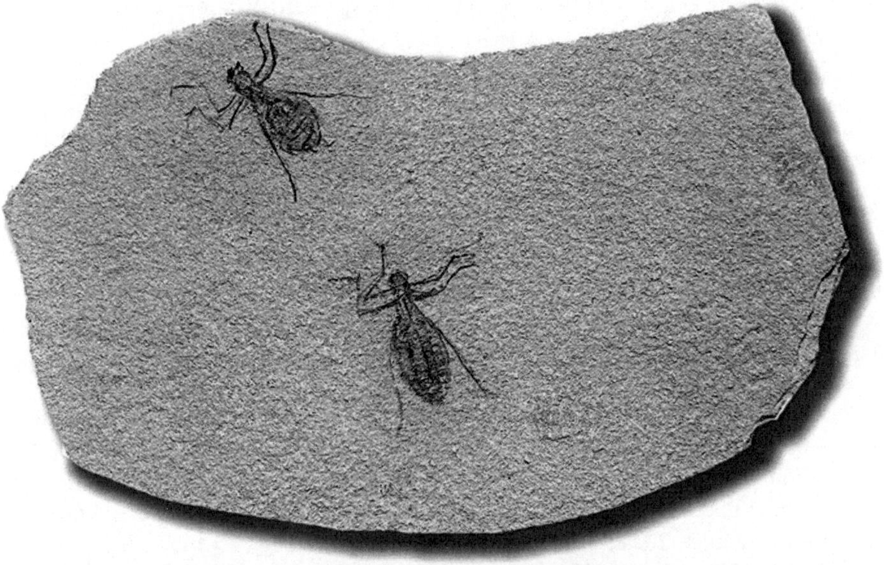

Fig. 23.8 These ant-like fossil insects could also be identified with *Myrmecitis*. Dragonfly larvae, province of Livorno, Tuscany, Upper Miocene. (Photo Maurizio Forli, courtesy Carlo Chirli collection)

That these arthropods were not organisms easy to classify, not remembering much living animals and known to the ancient Greek-Roman world first, to the Latin and Anglo-Saxon world later, is also shown by the definition given in what turns out to be their first certain description and illustration dating back to the late seventeenth century. It is Dr. Edward Lhwyd, also known as Edward Lhuyd or Eduardus Luidius, who, in the proceedings of the Royal Society of London for Improving Natural Knowledge, the Philosophical Transactions, in 1698, publishing part of a correspondence between him and Dr. Martin Lister, gives us the first certain description of trilobites, [from Lhan Deilo in Caermardhishire, the present Llandeilo in South Wales, part of the county of Carmarthenshire, with sediments referable to the Middle Ordovician (~465/460 Ma)], in particular accompanying the largest one with the following indication:

> *... the 15th whereof we found great Plenty, must doubtless be referred to the skeleton of some flat fish;*

The same image and description a little more articulate and thorough is also included in his work of a little later (1699) *Lithophylacii Britannici Ichnographia*, the first illustrated catalog of a public collection published in England, on page 95 and in Figure 2 in the plate on page 92, shows us the same specimen, excellently preserved, presenting it as a skeleton of a fish, practically similar to a sole in the shape of a short and thin ox tongue, whose representation makes us imagine it similar to the *Halibut*, a generic English name which includes large flatfish, some weighing more than 200 kilograms.

> *Iste autem Ichthyomorphos ad planos pisces qui in latus projecti natant, referendus est; quamvis nullae speciei hactenus descriptae per omnia respondeat. At non externa piscis lineamenta quemadmodum vestras lapis Islebiensis; sed ejusdem sceleton, seu* pectinem, *argutè admodum in ambitu striatum exprimit. Figura ad* Buglossum; *seu* Soleam *piscem accedit, nisi quod paulò ovalior sit; ad extremum minùs coarctata; & quod caudà (utpote non piscis perfectus sed sceleton tantùm) omninò careat: unde* Buglossam curtam Strigosam *diximus.*

This fish-shaped thing is traced to a flat fish swimming on the projecting side; however, it does not correspond to any species hitherto described. [...] Figure similar to an *ox tongue*; or it approaches a flattened fish (*sole*), though it is a little oval; [...] hence we call it *scarce, short ox tongue.*

The current scientific name is instead *Ogygiocarella debuchii* (Brongniart, 1822), among other things also illustrated on page 126 of Vai and Cavazza (2003, editors, Four Centuries of the Word Geology); "*Ogygiocarella debuchii*, Middle Ordovician, collected by me in the type area in the early 1980s" (Vai, personal communication, August 2021) (Fig. 23.9), (Fig. 23.10), (Fig. 23.11), (Fig. 23.12), (Fig. 23.13), (Fig. 23.14), (Fig. 23.15), (Fig. 23.16), (Fig. 23.17), (Fig. 23.18), (Fig. 23.19).

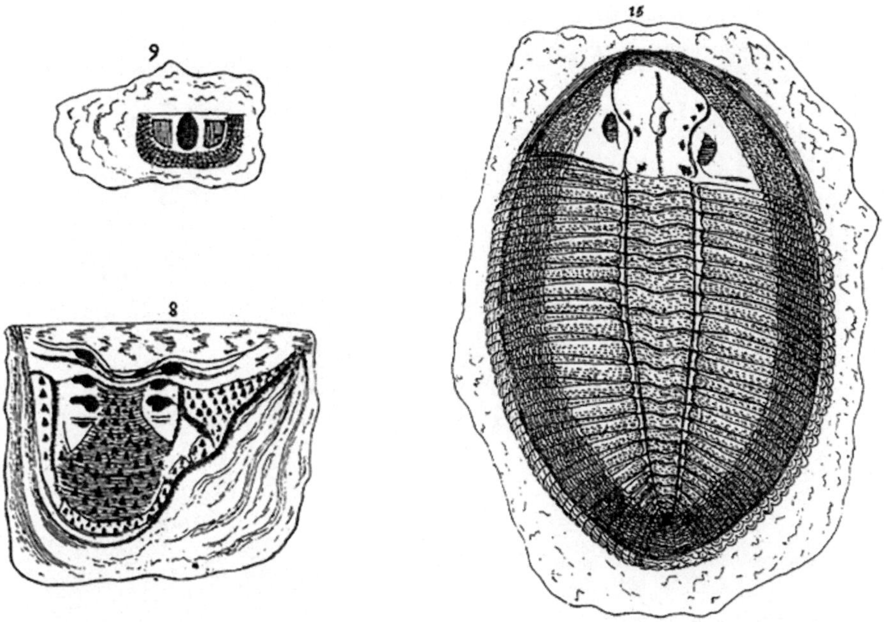

Fig. 23.9 *Philosophical Transactions*, 1698: pp. 279–280. *Part of a Letter from* Mr. Edw. Lhwyd *to* Dr. Martin Lister, *Fell. of the Coll. of Phys. and* R. S. *concerning several regularly Figured Stones lately found by him.* The *Cephalon* of figure 8, to be nitpicky, is the first image of a trilobite published by Lhwyd, while figure 9, the second, is a particular larval form of trilobite, later named *Trinucleum fimbriatum*, also shown in the plate and illustration of page 120 in *Lithophylacii Britannici Ichnographia* (1699) as well as figure 15, the *Ogygiocarella debuchii* (Brongniart, 1822). These two early *Figured Stones* were also unknown to the author who after comparing them with mollusks and crinoid parts (entrochi) concluded by assuming they were exuviae from some marine or terrestrial body. [Courtesy https://royalsocietypublishing.org (https://doi.org/10.1098/rstl.1698.0055)]

Fig. 23.10 Echinites, for the author, remains similar to sea urchins (Echinoderma), which instead, according to the drawing, we attribute to trilobite remains (detail from Hermann 1711: pl. XII, fig. 31). [https://doi.org/10.3931/e-rara-49104 / Public Domain Mark]

Fig. 23.11 Bruckmann, (1730), *Epistola Itineraria XXIII*, pl. I, Fig. III, a folded trilobite named *Polypo Marino petrifacto*. [Modified from: biodiversitylibrary.org/item/37397#page/15/mode/1up (Public domain)]

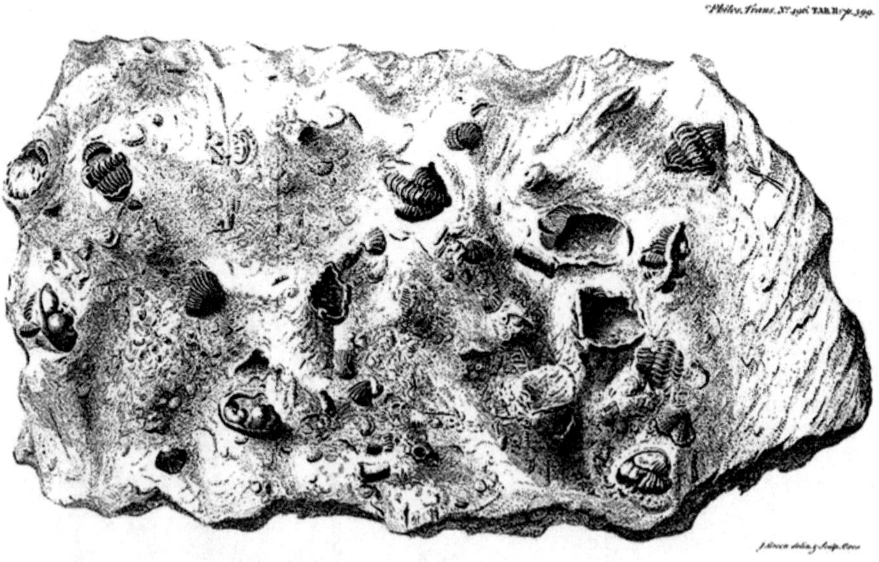

Fig. 23.12 The limestone boulder filled with specimens as described by Charles Lyttelton in his communication published in plate II of the *Philosophical Transactions*, (1750). [Courtesy https://royalsocietypublishing.org (https://doi.org/10.1098/rstl.1749.0112)]

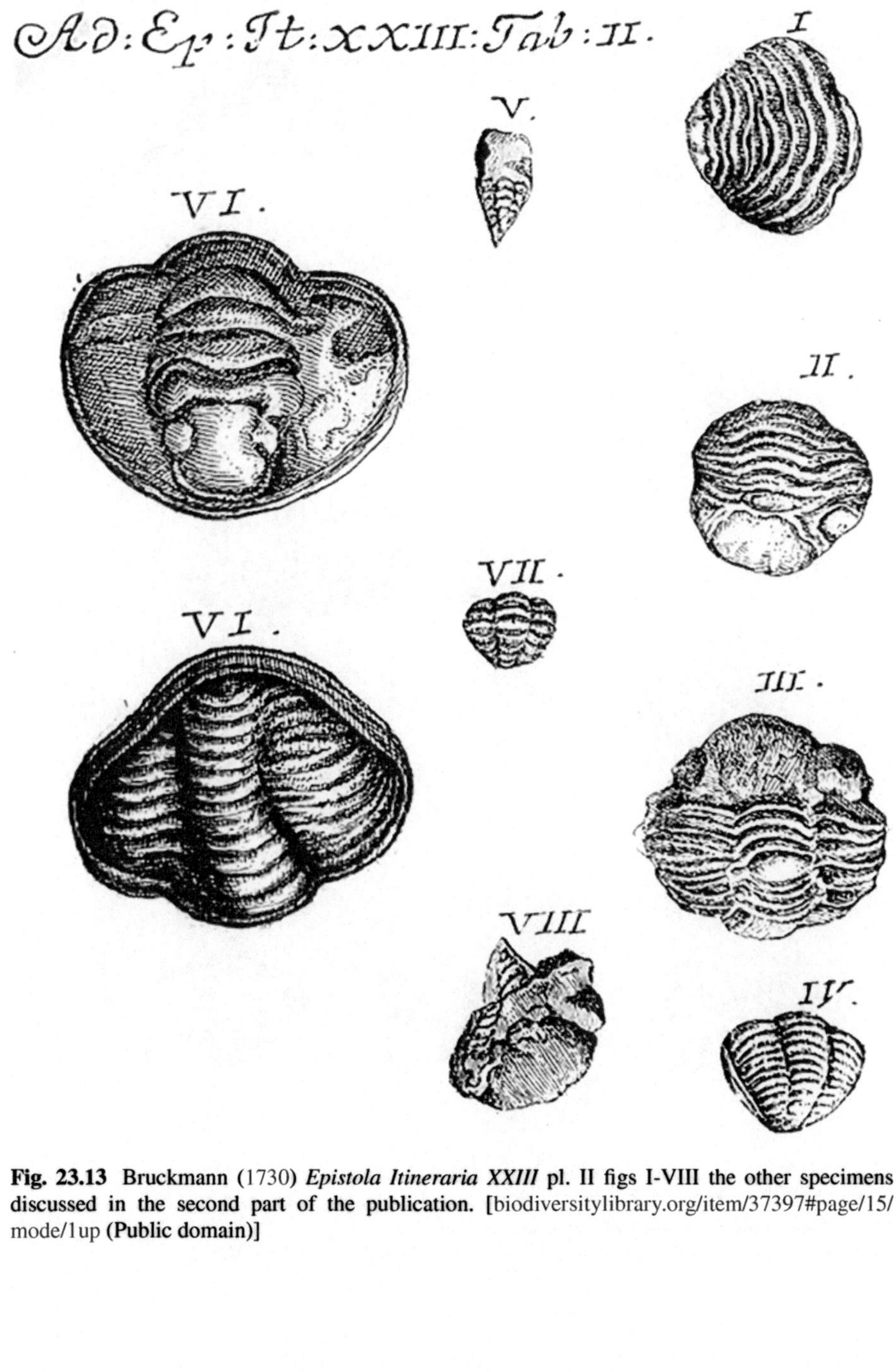

Fig. 23.13 Bruckmann (1730) *Epistola Itineraria XXIII* pl. II figs I-VIII the other specimens discussed in the second part of the publication. [biodiversitylibrary.org/item/37397#page/15/mode/1up **(Public domain)**]

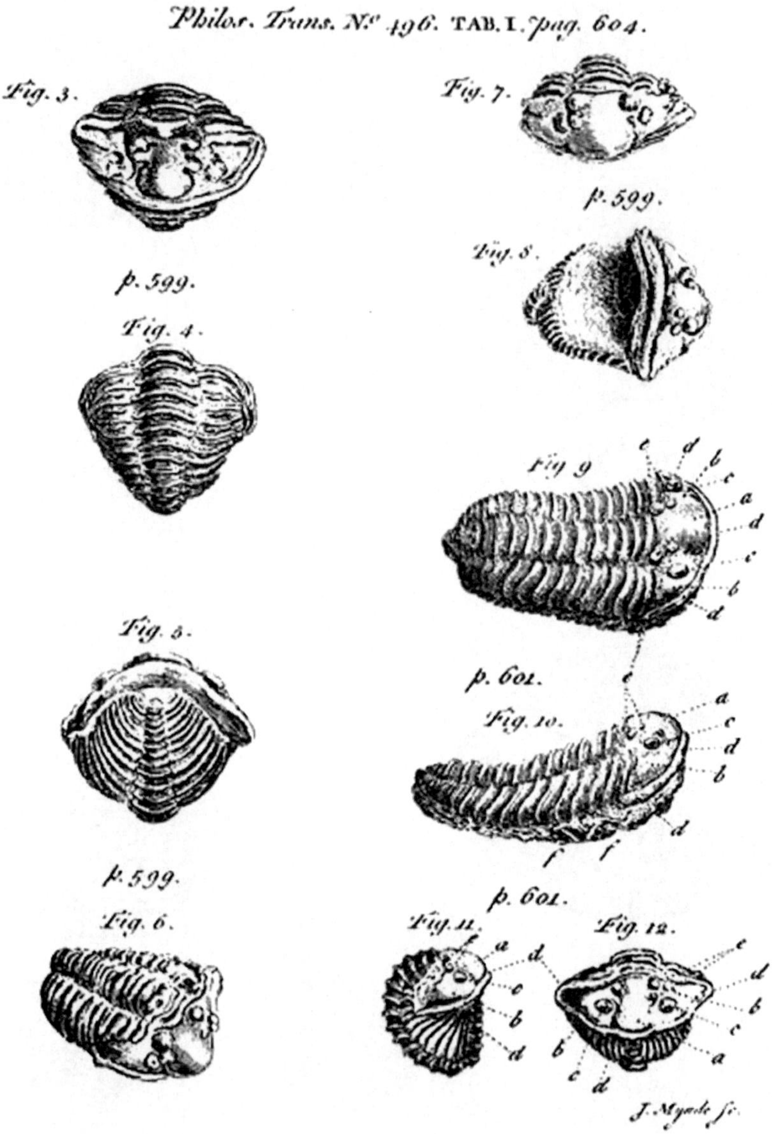

Fig. 23.14 The *Dudley locust*, the unknown petrified insect as interpreted by Lyttelton, whose current scientific name is *Calymene blumenbachii* Brongniart in Desmarest, 1817, found both unwound and coiled, aroused at the time, particular attention and scientific debate. [Courtesy https://royalsocietypublishing.org (https://doi.org/10.1098/rstl.1749.0113)]

Fig. 23.15 More or less in the same years (1740) came out the work *Mineralogia et lithograpica svecana* by Magnus Bromelius (Herrn Magni von Bromell), Swedish physician and paleontologist, in which appear two plates dedicated to trilobite remains, fossilized in blackish schist, identified as vestiges of unknown insects: ***Lapis niger, in quo insectorum imagines & vestigia aureo vel argenteo colore fulgent***. [https://doi.org/10.3931/e-rara-16285 / Public Domain Mark]

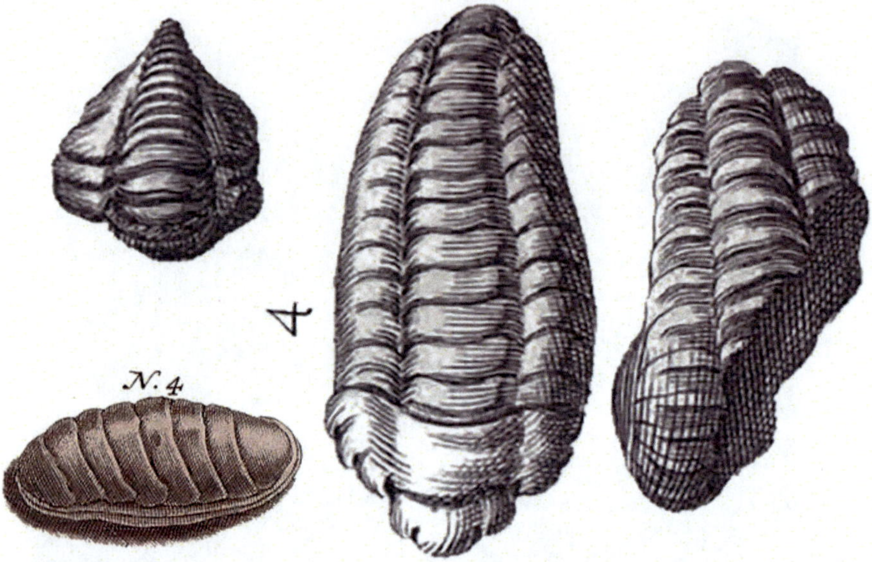

Fig. 23.16 (1) The trilobites from José Torrubia *Aparato para la Historia Natural Española* (1754), juxtaposed with the chiton (2) published in *D'Amboinsche Rariteitkamer* by Georg Eberhard Rumph (Rumphius), a Dutch naturalized German biologist and botanist (1750: folio 18 fig. 4). (Graphic composition Maurizio Forli). [Credit from: (1) https://wellcomecollection.org/works/zn4ugnnt/items?canvas=277 (Public Domain Mark) (2) https://www.biodiversitylibrary.org/item/127485#page/79/mode/1up (not in copyright)]

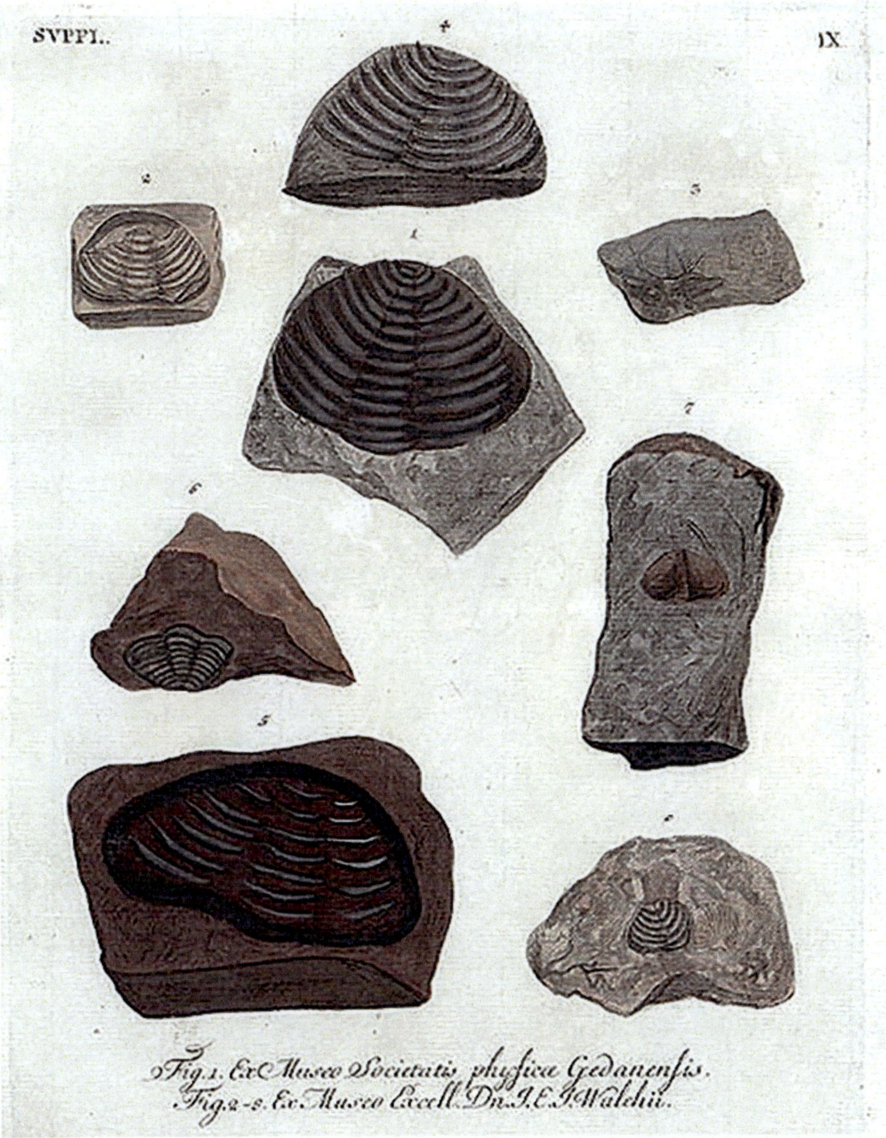

Fig. 23.17 Trilobites from plate IX of Johann Ernst Immanuel Walch in Die Naturgeschichte der Versteinerungen: zur Erläuterung der Knorrischen Sammlung von Merkwürdigkeiten der Natur. Nürnberg. [https://doi.org/10.3931e-rara-10246 **Public Domain]**

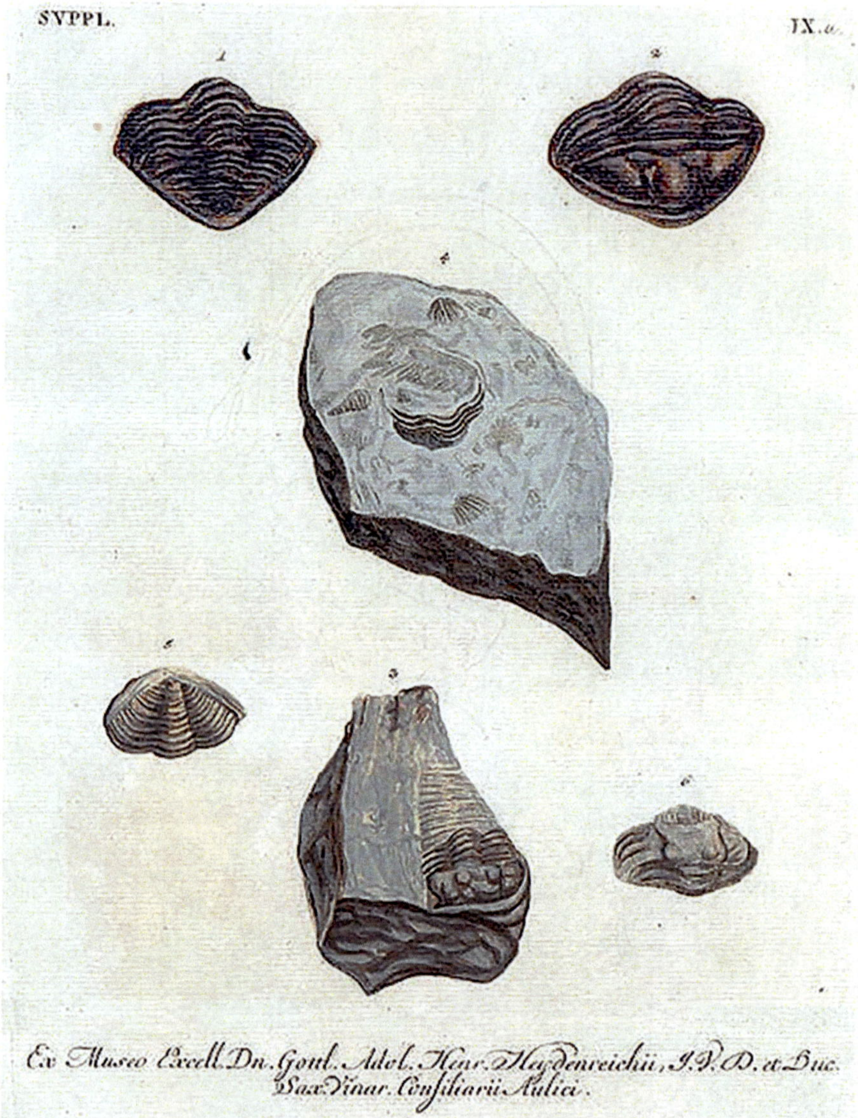

Fig. 23.18 In this plate IXa, the specimens discussed and illustrated by Walch. [https://doi.org/10.3931/e-rara-10246 Public domain]

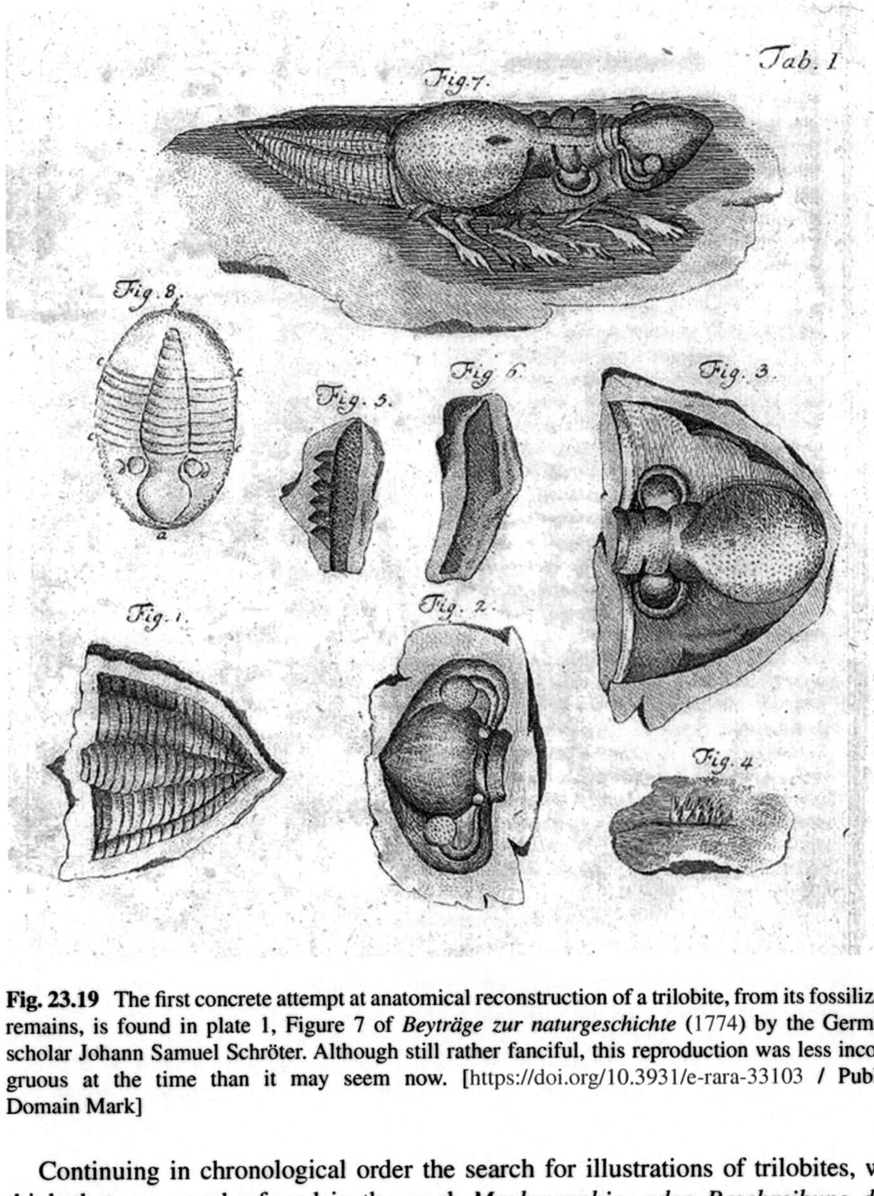

Fig. 23.19 The first concrete attempt at anatomical reconstruction of a trilobite, from its fossilized remains, is found in plate 1, Figure 7 of *Beyträge zur naturgeschichte* (1774) by the German scholar Johann Samuel Schröter. Although still rather fanciful, this reproduction was less incongruous at the time than it may seem now. [https://doi.org/10.3931/e-rara-33103 / Public Domain Mark]

Continuing in chronological order the search for illustrations of trilobites, we think that one can be found in the work *Maslographia, oder, Beschreibung des schlesischen Massel im oelss-bernstädtischen Fürstenthum* by Leohnard Hermann (1711: pl. XII, fig. 31), of the folded remains of thoracic segments and at least a pair of glabellae of *Calymene* (?), identified by the author as *Echinites*.

However, it is another species of trilobite, always coming from England, and more exactly from Dudley, in the West Midlands, which becomes protagonist of the scientific history of these invertebrates. It is described by Charles Lyttelton in 1750, but found in 1749, as a petrified insect, later on called the *Dudley locust*:

> *The curious Fossil I have now the Honour to exhibit to the Society, is as rare as its Figure is elegant; having been mention'd by none of our own Writers who treat on Fossils, and but very*

> *imperfectly describ'd by foreign Lithographists. I discovered a single Specimen of it last Year in the Limestone Pits at* Dadley *in* Worcestershire*; and very lately a large Mass of Limestone full of them in the same place* [...] *Philosophical Transactions*, 1749–1750, pp. 598–600.

This trilobite, whose current scientific name is *Calymene blumenbachi* Brongniart in Desmarest, 1817, found both unrolled and rolled up, arouses particular attention and scientific debate and is from time to time identified as similar to a caterpillar, a marine scolopendra or a "sea piggy," the current *Ligia oceanica* (Linnæus, 1767) also widespread in the Mediterranean. In the annotations to his communication Lyttelton explains the reference to foreign authors who would have improperly dealt with fossils similar to his:

> [...] *Dr.* Bruckmann, *in his* Centuria Epist. Itinerar. Wolffenbuttl (1742, XXIII) *has given several Figures of Petrefactions*, very *much resembling these Dudley Fossils; the first was found at* Steme, *a Village in the Neighbourhood of* Paderborn, *given him by Dr.* Koenig, *which he took for a sort of* Polypus marinus; *he says it is an Animal unknown to him, but he gives those Figures of it, in hopes that some curious Persons, who live near the Sea, may light upon some Animal resembling this. The Body of this stone, he says, has, on each Side three striated Lobes, and three pointed* Appendices *beneath; its inner Substance is white, being* Selenites, *or white Spar; Its Colour on the Outside is every-where brown. His Friend* Linck *had sent him Specimens of these Stones 6 Years before, some modell'd in Wax, others engraved upon Copper.*

In the second edition of the *Lithophylacii Britannici Ichnographia* published in 1760, three figures of trilobites from Dudley are added on page 101, while *O. debuchii*, in this edition, is figured at number 2 of plate 22, the same as the previous illustrations. In the first volume of the *Aparato para la Historia Natural Espagñola* (1754) by the Spanish Franciscan missionary father, José Torrubia (1698–1761), we find their first representations for Spain. At number 4 of the third plate, three specimens of an Ordovician trilobite Calymenidae are depicted, one complete and two fragmentary, compared first in the text (p. 83, n. 96) to petrified crabs and later to the chiton of plate 10, fig. 4 of the work of Rumphius (1705, *D'Amboinsche Rariteitkamer*).

It is Johann Ernst Immanuel Walch who has the merit of introducing the term trilobite in his 1771 work *Die Naturgeschichte der Versteinerungen: zur Erläuterung der Knorrischen Sammlung von Merkwürdigkeiten der Natur*. In the text he also summarizes part of the names used by his contemporaries to describe these strange petrifactions which were still referred to by inappropriate names reflecting the uncertainty and opinions of the different authors who had dealt with them so far:

> *Armata veneris*, *Conchites trilobus*, *Concha triloba rugosa*, *Pectunculites trilobus imbricatus*, *Entomolithus branchiopodis cancriformis marini*, *Eruca*, *Polypus marinus*, *lièvre de mer*.

After discussion and interpretation of the fossil remains at his disposal he also provides valuable illustrations of them in the volume devoted to the plates, VIIId Fig. 17 and in the subsequent series of plates IX, i.e., plate IX, Figs. 1–8, plate IXa, Figs. 1–6, plate IXb, Figs. 1–6, plate IXc, Figs. 1–6, plate IXd, Figs. 1–7, plate IXf, Figs. 1–5. For a more detailed discussion of the chronology of studies concerning trilobites and other aspects of the state of research worldwide, see Mikulic et al. (2007).

The first image of a trilobite in an Italian scientific communication is that of an incomplete specimen from the Lower Silurian of Sardinia, described in an article by Professor Giuseppe Giovanni Antonio Meneghini, physician and naturalist, emeritus professor, then holder of the chair of geology and palaeontology at the University of

Pisa, in a *Supplement to* the *Paléontologie de l'Ile de Sardaigne*, included in the third volume of La Marmora's *Voyage en Sardaigne*, published in 1860. It is the cephalic part of a trilobite determined as belonging to the genus *Dalmanites* Barrande, 1852, from sediments attributed to the upper Silurian floor of Fluminimaggiore, and reproduced at number 5 a-b of the illustrated plate at the end of the volume.

Further important contributions on trilobites, with comprehensive illustrations of fossils from Sardinia, are still the work of Meneghini who, in subsequent years, determines the copious material found. The second contribution with illustrations is given in *Nuovi fossili siluriani di Sardegna* published in the *Memorie della Reale Accademia dei Lincei, Classe di Scienze fisiche, matematiche e naturali*, in 1880. It is a species originally described by Meneghini himself with the name of *Dalmanites lamarmorae* and currently called *Dalmanitina lamarmorai* (Meneghini, 1880). This was followed in 1881 by four works, including *Nuovi Trilobiti di Sardegna* in the *Atti della Società Toscana di Scienze Naturali*, and another three, respectively, one each in the years from 1882 to 1884. However, the most important and complete work, still a reference for scientific research, is *Paleontologia dell'Iglesiente in Sardegna. Cambrian Fauna. Trilobiti*, published in 1888 in the *Memorie del Regio Comitato Geologico d'Italia*. The material object of these first studies, enriched by subsequent collections and recent systematic and stratigraphic revisions (Nicosia and Rasetti 1970, Rasetti 1972, Pill 1991), is preserved in the Paleontological Collection of APAT in Rome, under the name of Meneghini-Rasetti Collection.

The specimens come from Canalgrande in the Iglesiente, a locality included in the Historical and Environmental Geomining Park of Sardinia, where in the Lower Cambrian schists, there is a rich arthropod fauna with over 15 species of trilobites of which the most characteristic are, together with *Enantiapsis enantiopa* (Bornemann, 1891), *Dolerolenus zoppii,* and *Anadoxides armatus* (Meneghini) (Angelelli and Rossi 2014) (Fig. 23.20), (Fig. 23.21).

Fig. 23.20 *Dalmanitina lamarmorai (*Meneghini, 1880), formerly *Dalmanites* sp. in *Supplement alla Paléontologie de l'Ile de Sardaigne*. Left Fig. 5a: *Fragment enclavé dans la roche, vu en dessus*; Fig. 5b: *Le méme vu de flanc*. (Graphic composition Maurizio Forli)

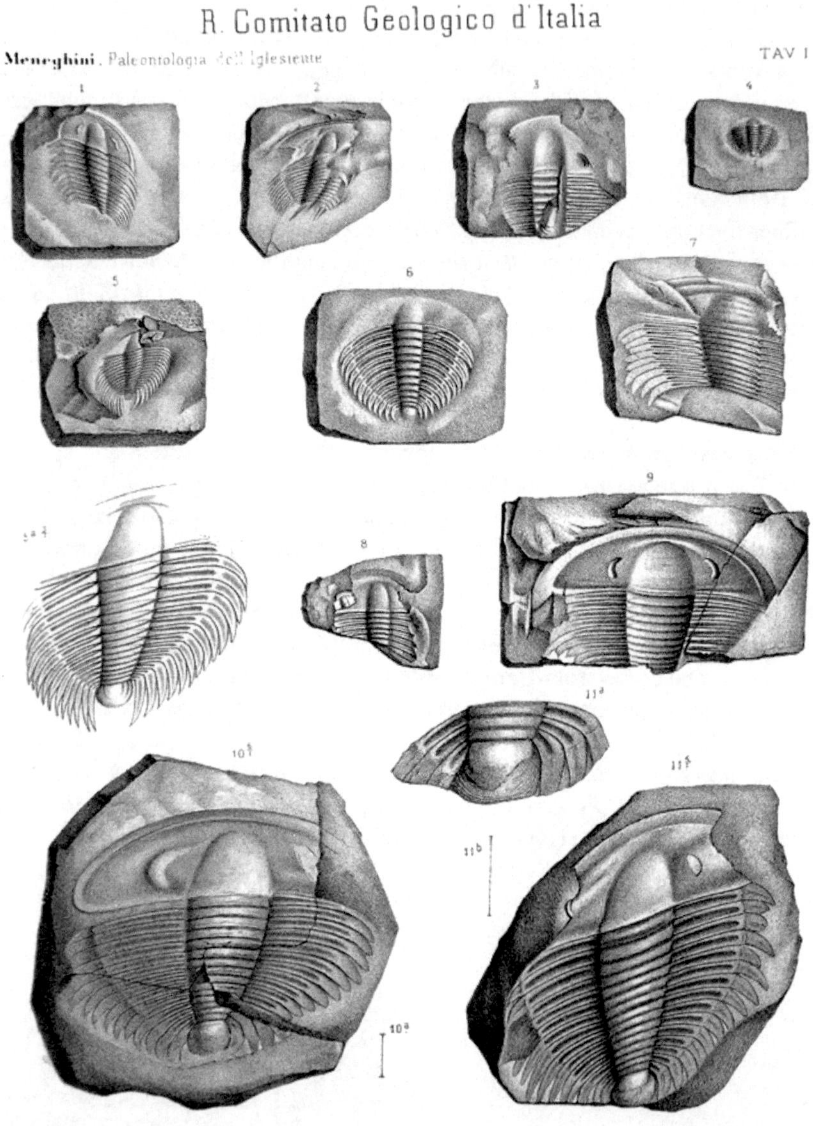

Fig. 23.21 *Paleontology of the Iglesiente in Sardinia. Cambrian Fauna. Trilobites. Memoirs of the Royal Geological Committee of Italy*: the first plate of the study by Giuseppe Meneghini, published in (1888), in which are presented some specimens of *Olenus zoppii*, now called *Dolerolenus zoppii*

The only other Italian localities so far known to have returned trilobite remains are in Sicily and Friuli Venezia-Giulia.

It is in the surroundings of Palazzo Adriano, in the province of Palermo, along the course of the river Sosio and its tributaries, that scattered carbonate blocks are found, the largest of which, the so-called *Solomon's Stone*, is a small relief about 30 meters high with an approximate perimeter of just over half a kilometer. The fossils contained in these particular rocks, including trilobites, which are, however, quite rare compared to other faunas, date back to the upper part of the Middle Permian (Wordian, from ±268 to ±265 Ma) [paleoitalia.org/places/12/palazzo-adriano] (Fig. 23.22).

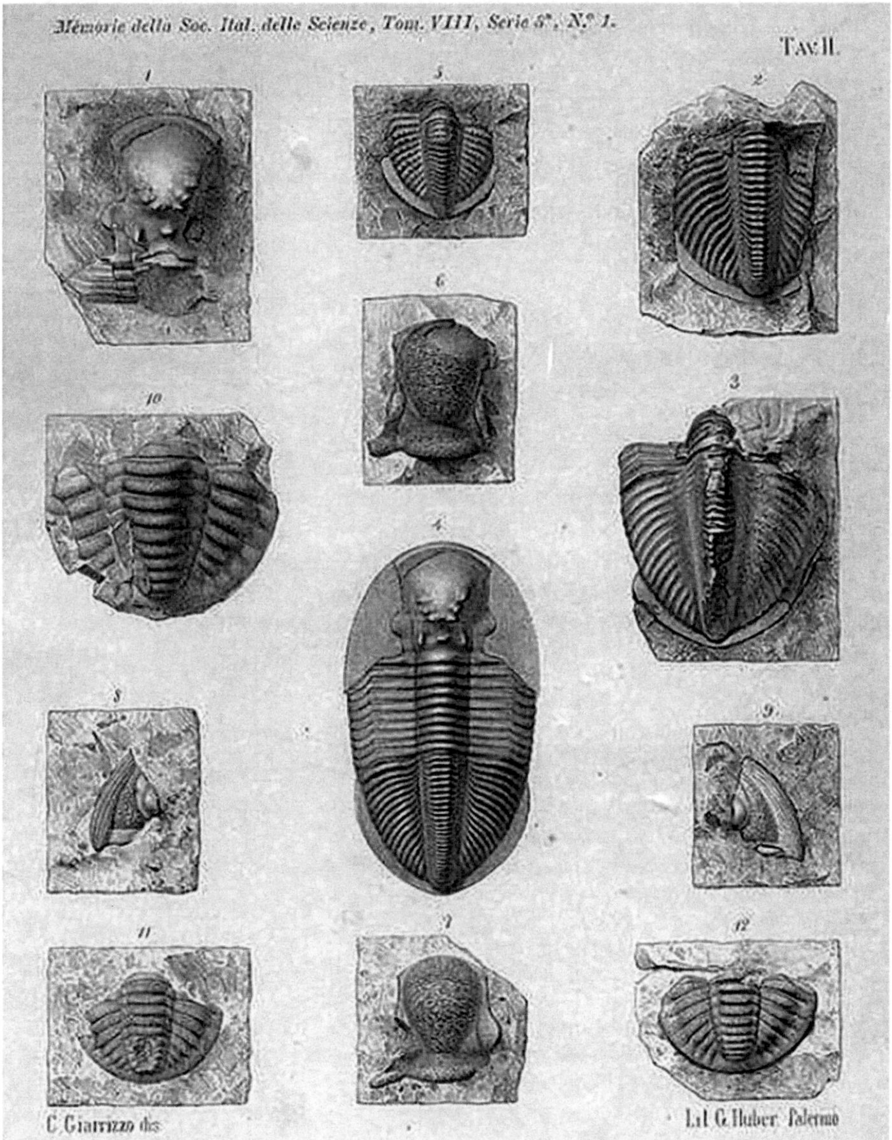

Fig. 23.22 Plate II from *I Crostacei dei Calcari con* Fusulina *della Valle del Fiume Sosio nella Provincia di Palermo in Sicilia*, by Gaetano Giorgio Gemmellaro, published in (1892), in which are illustrated some trilobites characteristic of the Pietra di Salomone, (middle Permian, from ±268 to ±265 Ma): 1–4 *Pseudophillipsia elegans*; 5 *Phillipsia pulchella*; 6–12 *Griffithides verrucosus*

These fossils were discovered by the Sicilian Gaetano Giorgio Gemmellaro, also a physician and naturalist, geologist, and paleontologist, who in 1892 published *I Crostacei dei Calcari con* Fusulina *della Valle del Fiume Sosio nella Provincia di Palermo in Sicily*, in which he treats of trilobites already found by him even previously but until then neither described nor illustrated because of their incompleteness:

> *When, in 1887, I began to illustrate the fauna of these rocks, I already had some specimens of crustaceans, whose state of preservation, however, did not allow the description. Further research has gradually increased the number of these fossil remains, and now in the Geological Museum of the Royal University of Palermo there are so many that it would be unjustifiable not to illustrate them.*

In the Carnic Alps, north of Pordenone and Udine, near the Austrian border, the oldest sedimentary horizons with trilobite remains date back to the Upper Ordovician (460.9 ± 1.6 to 443.7 ± 1.5 Ma). Often these are partial fossils referring to accretionary moults, as in almost all the remains of these arthropods present in the Carnic Palaeozoic rocks, the oldest fossiliferous strata of peninsular Italy, which extend from the Ordovician to the Permian (Venturini, 2006). To the best of our knowledge, to date there are no publications focused exclusively on the trilobite fauna of the area.

We do not know that in Europe trilobites have been used as a medicinal substance and also the evidence of their use in magic is extremely limited, leading back only to their function as an amulet. This use is instead documented for some populations of American Indians and ancient China. At the end of the discussion we report an excerpt of a poem dedicated to a trilobite, which seemed to us rather unusual and curious, written by Ippolito Nievo, the writer and patriot of 1800 (1831–1861), taken from *I Canti del Friuli*.

[full version available at wikisource.org/wiki/Canti_del_Friuli/A_un_trilobite_raccolto_in_Carnia].

> A Trilobite Collected in Carnia*Mentre alla roccia madre, che stretto ti avvolge, l'impronta*
> *del tripartito corpiciattolo strappo,*
> *e nella pietra viva mirabilmente scolpiti*
> *della glabella turgida l'arco miro,*
> *e l'incavata pleura, che un dì l'aura triste bevette*
> *di nessun polline ancor feconda, un grido*
> *misterioso salire par dalla roccia ferita,*
> *dalla roccia pur ora ribaciata dal sole.*
> *Incerto, e dalla mano tremante sfuggonmi e cadono*
> *gli scalpelli, mi arresto. «Qual di secoli, chieggo,*
> *o impietrato vivente, qual mai di secoli corse*
> *catena lunga da quel remoto giorno*
> *in cui la riva fangosa di un tiepido mare segnasti,*
> *vagante artropodo, de l'incerta tua orma?* [...]

References

Angelelli F, Rossi R (2014) Research and studies on the Meneghini – Rasetti historical collection belonging to the ISPRA paleontological collections. Mus Sci Mem 11:49–53

Bonino E, Kier C (2019) Trilobiti, Guida essenziale al riconoscimento e classificazione. In: Bonino E (ed) Back to the Past Museum

von Bromel M (1740) Mineralogia et lithograpica svecana das ist Abhandlung derer in dem Königreich Schweden befindlichen Mineralien und Steinen. Gottfried Riesemetter, Stockholm and Leipzig

Bruckmann FE (1730) Epistola Itineraria XXIII. De Alga Saccharifera, Polypo Marino petrifacto, kakerlacken, frutice koszodrewina et arbore limbowe drewo: ad virum nobilissimum, experientissimum, doctissimum amplissimumque Dominum J.H. Meuschen. Wolffenbuitelae

Chlupác I (2000) Comments on the trilobites from "la Grotte du Trilobite". The Trilobite Papers, 12: 49–50

Henry JL (2001) "La Grotte du Trilobite" encore une fois. The Trilobite Papers, 13, 15

Hermann LD (1711) Maslographia Oder Beschreibung Des Schlesischen Massel Im OelßBernstädtischen Fürstenthum mit seinen Schauwürdigkeiten : Theils Unterschiedlicher so wohl Heydnischer, als Christlicher Antiquit. Brachvogel, Breßlau

Gemmellaro GG (1892) I Crostacei dei Calcari con Fusulina della Valle del Fiume Sosio nella Provincia di Palermo in Sicilia. Memoirs of Mathematics and Physics of the Italian Society of Sciences, 8, series 3: 1–40

Liñán E, Gozalo R (2008) Cryptopalaeontology: magical descriptions of trilobites about two thousand years before scientific references. Cuadernos del Museo Geominero 9:237–241

Luidus E (1699) Eduardi Luidii Lithophylacii Britannici ichnographia : sive lapidum aliorumque fossilium Britannicorum singulari figura insignium quotquot hactenus vel ipse invenit vel ab amicis accepit, distributio classica, [...]. Londini : ex officina M.C ; Lipsiae : sumt. Joh. Ludw. Gleditsch & Weidmann, 1699. Zentralbibliothek Zürich, NG 1883, https://doi.org/10.3931/e--rara-30541 / Public Domain Mark

Lyttelton C (1750) A letter from the rev. Charles Lyttelton LL.D and F.R.S. dean of Exeter to the President, concerning a nondescript petrified Insect. Philos Trans R Soc 46:598–600

Meneghini G (1860) Paléontologie de l'Ile de Sardaigne. Supplément. (31 pp). In: La Marmora, Voyage en Sardaigne, ou Description Statique, physique et politique de cette Ile, avec des recherches sur ses productions naturelles et ses antiquites. Troisieme partie. Frēres Bocca, Turin

Meneghini G (1880) Nuovi fossili siluriani di Sardegna. Memorie della Reale Accademia dei Lincei, Classe di Scienze fisiche, matematiche e naturali, 209–219

Meneghini G (1881) Nuovi Trilobiti di Sardegna. Proceedings of the Tuscan Society of Natural Sciences, Verbatim Proceedings, vols. 1–2, 199–202

Meneghini G (1882) Fauna cambriana dell'Iglesiente inSardegna. Proceedings of the Tuscan Society of Natural Sciences, Minutes, 3: 158–166

Meneghini G (1884) New Cambrian fossils inSardinia. Proceedings of the Tuscan Society of Natural Sciences, Proceedings, 5:7–9

Meneghini G (1888) Paleontology of the Iglesiente in Sardinia. Cambrian fauna, Trilobites Memoirs of the Royal Geological Committee of Italy, 3(2): 1–53

Mikulic D, Landing E, Kluessendorf J (eds) (2007) Fabulous fossils – 300 years of worldwide research on Trilobites. New York State Museum Bulletin 507

Nicosia ML, Rasetti F (1970) Revisione dei trilobiti del Cambriano dell'Iglesiente (Sardegna) descritti da Meneghini. Accademia Nazionale dei Lincei, Memorie, Cl. Sc. Fis. Mat. Nat., ser. VIII,10, II(1): 1–20

Rasetti F (1972) Cambrian trilobite faunas of Sardinia. Accademia Nazionale dei Lincei, Memoirs, Cl. Sc. Phys. Mat. Nat., ser. VIII, 11, II(1): 1–100

Pill GL (1991) Trilobites du Cambrien inferieur du SW de la Sardaigne, Italie. Palaeontogr Ital 78:1–174

Rumph GE (1705) D'Amboinsche Rariteitkamer, Behelzende eene bescryvinge van allerhande zoo weeke als harde Schaal visschen; te weeten raare Krabben, Kreeften en diergelyke Zeedieren Hoorntjes en Schulpen, die men in d'Amboische Zee vindt: Daar beneven zommige Mineraalen, Gesteenten, ...François Halma, T'Amsterdam

Schröter JS (1774) Beyträge zur Naturgeschichte sonderlich des Mineralreichs, aus ungedruckten Briefen gelehrter Naturforscher, und aufmerksamer Freunde der Natur. Altenburg : in der Richterischen Buchhandlung, 1774-1776. Zentralbibliothek Zürich, NG 1867, https://doi.org/10.3931/e-rara-33103 / Public Domain Mark

Torrubia J (1754) Aparato para la Historia Natural Espagñola. Tomo Primero. Contiene muchas disserctaciones physicas, especialmente sobre el Diluvio. Herederos de Dón Agustin, Madrid

Vialli V (1974) Notes of paleontology. Pitagora Editrice, Bologna

Venturini C (2006) Geological evolution of the Carnic Alps, a journey through time Publication no. 48. Municipality of Udine. Editions of the Friulian Museum of Natural History

Walch JEI (1768–1773) Walch Johann Ernst Immanuel: Die Naturgeschichte der Versteinerungen : zur Erläuterung der Knorrischen Sammlung von Merkwürdigkeiten der Natur. Nürnberg : gedruckt mit Felsseckerischen Schrifften, 1768–1773. ETH-Bibliothek Zürich, Rar 399, https://doi.org/10.3931/e-rara-10246 / Public Domain

Chapter 24
Coffee Break–Chelicerata Heymons, 1901 (Maurizio Forli)

Abstract Within the phylum Arthropoda are also included the so-called Chelicerata, including the Eurypterida and Xiphosura. The animals included in the first grouping are better known under the name of marine scorpions, while those characteristics of the second are under the name of Limulus, generally named horseshoe crab. Of both, we give news of their fossil findings interspersed with curious annotations.

Keywords Fossils · Chelicerata · Eurypterida · Xiphosura · Pieter Bruegel the Elder

Within the phylum Arthropoda are also included several other invertebrates apparently very different from each other, the so-called Chelicerata, a group (phylum or subphylum depending on the classification adopted) consisting of animals equipped with an exoskeleton and segmented limbs, which differ from other arthropods for the absence of antennae and the presence of a pair of preoral appendages, called chelicerae. Among the most common animals included in this broad systematic category are spiders and scorpions, known to the vast majority of people. Being a vast set, with wide range of morphological variability, they are usually divided into three classes, Arachnida, Merostomata, and Pycnogonida, and their classification is subject to continuous updates and revisions. A few years ago (2007), Jeffrey W. Shultz, proposed a systematic update on morphological and phylogenetic basis, remodeling the Arachnida, Xiphosura, and Eurypterida, concluding, however, that many aspects still need further and deeper studies. Leaving aside this complex zoological part, we turn our attention to the Eurypterida and the Xiphosura.

The animals included in the first grouping are undoubtedly better known under the name of marine scorpions, while those characteristics of the second are under the name of Limuli. Considering that in the Mediterranean antiquity we have not found news about these fossil animals, we could have not treated this topic, but we like to give some news about them, between the chronicle and the biography. Let us start from the sea scorpions. The first memory I have about them seems to me as old as the collocation of the Paleozoic in the geologic scale in comparison to nowadays. And yet, "only" a little more than half a century has passed. As a child, like many, I watched

© The Author(s), under exclusive license to Springer Nature Switzerland AG 2022
M. Forli, A. Guerrini, *The History of Fossils Over Centuries*,
https://doi.org/10.1007/978-3-031-04687-2_24

or read, depending on the cases, a magazine then in vogue, the Corriere dei Piccoli, otherwise called *Corrierino*. If I remember correctly, it was from those pages that I cut out and pasted into a notebook the cartoons, and their captions, that showed and explained the birth of life on Earth, from the formation of our planet to the most recent periods, with the appearance of man. The ones I still have are a hundred illustrations of various sizes and a full page, with the geological scale and fossil guides, recto verso: one of the ones I liked best is the double one dedicated to a seabed with giant scorpions and other, then, strange beings. I can certainly say that my interest in fossils was born thanks to the reading of this didactic material, which unconsciously, with the passing of time, I transformed into a relic of my early adolescence (Fig. 24.1).

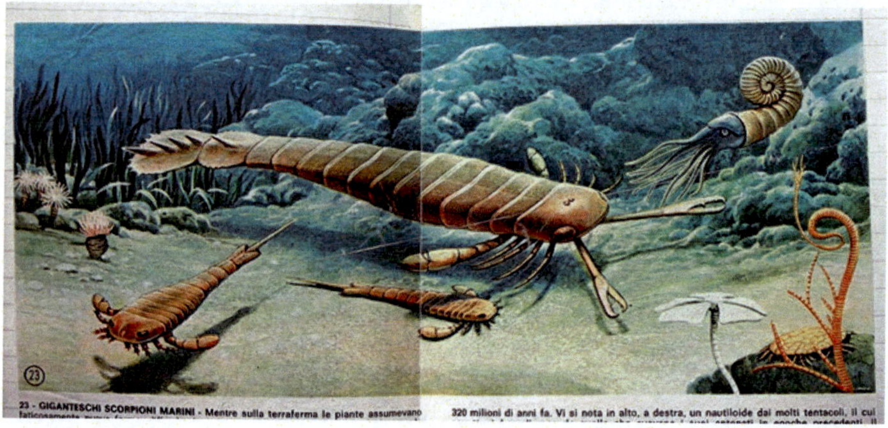

Fig. 24.1 The page dedicated to the "*Giant Sea Scorpions*," one of those that I still consider most fascinating and able to stimulate the imagination about the *Lost Worlds*. (Graphic composition Maurizio Forli)

The certain stratigraphic distribution of the eurypterids is currently between the Ordovician and Permian, with a presence extended for about 200 million years. In Europe their remains are rather rare and localized, scattered in Paleozoic soils from Ukraine to Great Britain. The first specimen discovered, comes from America, more precisely from the surroundings of the town of Westmoreland, in the state of New York, and was described as a torpedo fish by the American Samuel Latham Mitchill (*An account of the impressions of a fish in the rocks of Oneida County, New York*, 1818: p. 291). Below is the introductory period of the communication:

> *The foundation of primitive rock, underlaying the whole superstratum, as far as explored, may be judged of, from the silicious hornblend of the Falls, and of the granular quartz at Utica. Upon this, the secondary layers of limestone, iron ore, and argillaceous shist repose. The latter of these often exhibits, when fractured, the forms of beings that once possessed life. One of the most remarkable of them, is the impression of a fish, resting in a mass of clay slate, from the town of Westmoreland, a few miles north of Hamilton College. It was brought by our worthy colleague, Mr. Clarkson. It is tolerably distinct, except a part of the tail, which is wanting. The length of the figure which remains, is nearly four inches, and the greatest breadth rather more than an inch and a half. The head and shoulders are very stout, but taper away rapidly towards the tail. It evidently belongs to the silure or catfish family.*

A few years later (1825) the naturalist James Ellsworth Dekay redescribed the fossil remains, including them among the arthropods. We use a book by Woodward, to show an image of them (1878) (Fig. 24.2). Marine scorpions are very rare fossils in Italy; the discovery of the first, and at the moment only, eurypterid was made

Fig. 24.2 Three eurypterids attributed by Henry Woodward (1878) to *Pterygotus bilobus* Salter, 1859, in *A Monograph of the British Fossil Crustacea, belonging to the Order Merostomata.* Paleontographical Society in London, 1866–1878. [https://www.biodiversitylibrary.org/bibliography/53733 (Public domain)]

recently. The specimen, posteriorly incomplete and less than four centimeters in length, named *Adelophthalmus piussii* Lamsdell et al. (2013), comes from rocks dated to the Upper Carboniferous, from the surroundings of Pontebba, in the province of Udine (Fig. 24.3).

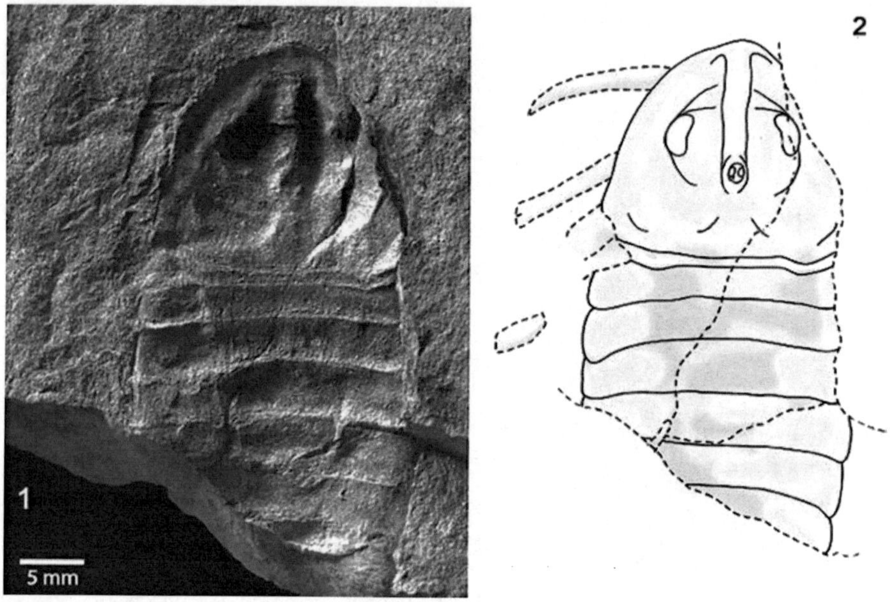

Fig. 24.3 *Adelophthalmus piussii* Lamsdell, Simonetto, Selden, 2013. The first Italian eurypterid from the Upper Carboniferous of the Carnic Alps. [Courtesy Rivista Italiana di Paleontologia e Stratigrafia, 119(2): 147–151 (modified)]

The other protagonist of this "coffee break"is what is traditionally said to be a *living fossil*: the limulus. In seventeenth-century volumes, starting from the concomitant colonial development in East Asia, living specimens are already described and illustrated by naming them *Moluccan Crab*; an example is given by the French-Dutch botanist Carol Clusius, (1526–1609) in his *Exoticarum libri decem* (1605: pp. 127–129). Other representations appear in the *Novus Orbis seu Descriptionis Indiae Occidentalis*, in the *Museum Wormianum* and in *Historiae naturalis de exanguibus aquaticis*, respectively, by De Laet (calling it Ragno Marino, *Araneus marinus*; 1633: p. 56), by Worm (1655: p. 249) and by the Polish Joannes Jonstonus (1665). The first fossilized remains of limulus are made known around the middle of the 1700s, recovered in the Jurassic limestones of Solnhofen. A limestone slab from this famous German locality, still known throughout the world for the beauty and variety of the fossil organisms found there, bearing the imprint of an inverted specimen, with the ventral view exposed, is included at number 2 of plate XIV taken from *Recuil des Monumens des Catastrophes que le globe*

Terrestre à éssuiées, contenant des Pétrifications (1777: pp. 25–26, pl. XIV, fig. 2), a French-language version of Die *Naturgeschichte der Versteinerungen*. These works, begun by Knorr, were completed by Walch (1768–1773).

N. 2. représente un Crabe des Isles Molucques, qui fait la quatriéme des espèces d'Ecrévisses, qu'on retire des carriére mentionnées. J'étois actuellement occupé à dessiner une telle Ecrévisse dans son état naturrel, pour l'insérer dans l'Ouvrage que je publie sous le titre: Deliciae Naturae Selectae ou Auserlesenes Naturalien-cabinet, lorsque j'allais visiter ces Carriéres de marbre dont j'ai parlé ci dessùs. J'étois à peine descendu dans le village, pou me mettre à l'abri de la chaleur, lorsqu'à mon retour l'entendis dire à deux jeunes gens, qu'on venoit de trouver quelque chose de tout à fait curieux. Après m'être informé, qui l'avoit trouvé, je decouvris enfin l'homme duquel j'avois déjà acheté d'autres curiosités. Il me dit que depuis absence il avoit trouvé une tête de mort. Je me moquai de lui, mais en même tems je fus obligé de cacher ma curiosité pour le disposer de me faire voir cette rareté, qu'il n'avoit pas avec lui. Je la lui fis apporter, & au premier coup d'oeil je m'apperçus, que c'étoit un Crabe des Molucques trè facile à reconnaitre, qui outre cela avoit encore à un côté une partie de sa coque naturelle, de couleur brune comme celle de l'Ecrévisse naturelle que je posséde moi même. [...] ces Crabes se trouvent fort rarement [...] Cependant ce n'est pas l'Ecrévisse, qu'on voit sur cette Planche, celle-ci étant prise du magnifique Cabinet de Mr. Beurer.

No. 2. represents a crab from the Moluccan Islands, which is the fourth species of Prawn, which is found from the quarry mentioned. I was momentarily occupied in drawing such a "prawn"in its natural state, in order to insert it in the work which I publish under the title: *Deliciae Naturae Selectae* or *Auserlesenes Naturalien-Cabinet*, when I went to visit these marble quarries of which I have spoken above. I had just gone down into the village, to take shelter from the heat, when on my return I heard two young people say, that they had just found something rather curious. After inquiring, who had found it, I at last found that it was the man from whom I had already purchased other curiosities. He told me that after my absence he had found a skull. I teased him, but at the same time I was obliged to conceal my curiosity to ask him to show me this rarity, which he had not with him. I made him bring it to me, and at the first glance I found that it was a Moluccan crab very easy to recognize, which besides this still had on one side a part of its natural shell, as brown in color as that of the natural crayfish which I myself possess. [...] these crabs are found very rarely [...] However, it is not the "shrimp", which you see on this plate, this is taken from the magnificent cabinet of Mr. Beurer.

Knorr's self-citation refers to the two canonical illustrations, dorsal and ventral, of a present specimen portrayed in plate F from *Deliciae Naturae Selectae* (1766), cited with the name given to it a few years earlier: *Monoculus polyphemus* Linnaeus, 1758.

It was in 1822 that the fossil limulus drawn by Knorr, and published by Walch, was dedicated to the latter under the name of *Limulus walchi*, referring precisely to figure two of plate 14 of the *Monumens des Déluge*. The representation of the find, faithfully copied by Desmarest, is accompanied by another illustration, this time of a caudally incomplete specimen from the collection of the Museum of Natural History in Paris. To confirm the rarity of these findings, the author wants to emphasize that this is the only specimen that he was able to observe from life. And in Italy?

Just as I was writing this literary interlude, in my computer, although old, a scientific article materialized that at first glance I thought it was a joke of some algorithm:

> *An effaced Horseshoe Crab (Arthropoda: Chelicerata: Xiphosura) from the upper Carboniferous of the Carnic Alps (Friuli, NE Italy).* (Selden et al. 2019).

Sometimes coincidences... one summer night many years ago I was going back home with some friends, after having seen one of those cowboys and Indians movies at the open air cinema, with a lot of guns. So, all of a sudden I had the idea to shoot my Colt, made with my thumb and forefinger, at a lamppost... twiu, twiu... to the amazement of all I hit the first shot and turned off the light! Here this article was as likely to be printed at the same time as this one, as the revolvers of the time, to hit the light bulb.

The specimen, from more or less the same locality in the province of Udine reported for the finding of the first Italian Eurypterid, was classified by establishing both a new genus and a new species, named *Stilpnocephalus pontebbanus* Selden et al. 2019.

In the continuation of the article, fossil footprints of other xiphosurs are also depicted, in turn baptized *Kouphichnium* sp. The only similar footprint, which I knew before reading this scientific article, referred to a sample from the Triassic sediments of the Monti Pisani, attributed to a limulus. Not much indeed, but in Tuscany very few people knew it. (Fig. 24.4), (Fig. 24.5), (Fig. 24.6), (Fig. 24.7), (Fig. 24.8), (Fig. 24.9), (Fig. 24.10), (Fig. 24.11).

Fig. 24.4 Above, the rock sample (1) with impressed, upper right side, the semicircular imprint attributed to the Mesozoic limulus. Triassic rocks of Monti Pisani (age ~220 Ma.). [(1) Courtesy Museum of Natural History, University of Pisa (background, photo and graphic composition Maurizio Forli)]

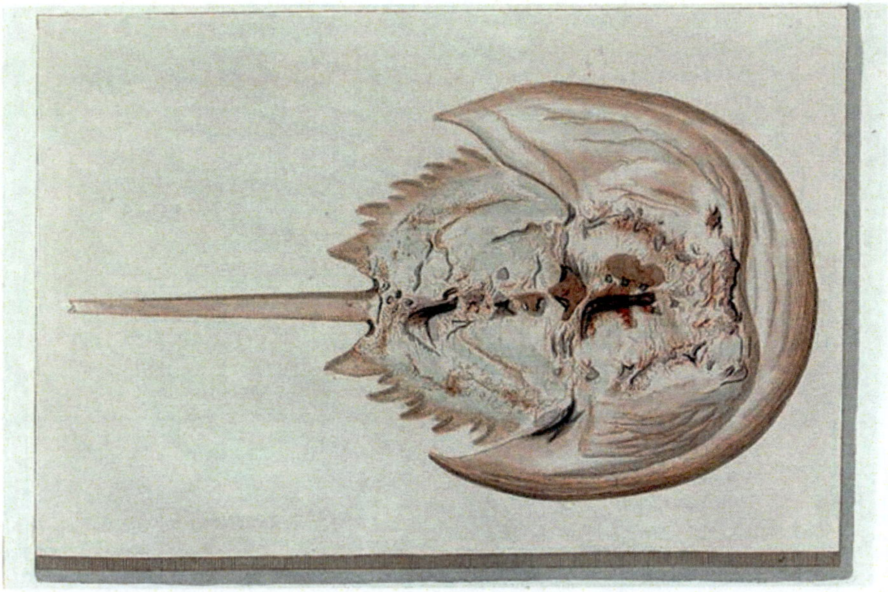

Fig. 24.5 The fossil specimen figured by Walch published also in Knorr (*Recuil des Monumens des Catastrophes que le globe Terrestre à éssuiées, contenant des Pétrifications*, 1777: pp. 25–26, pl. XIV, fig. 2. [Modified from https://doi.org/10.3931/e-rara-10246 / Public domain]

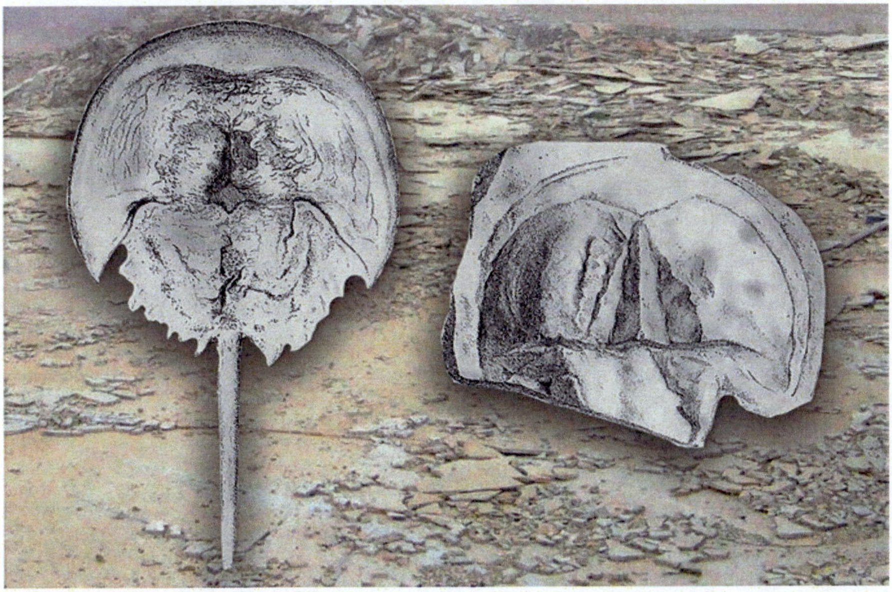

Fig. 24.6 The same as proposed by Desmarest in *Histoire Naturelle des Crustacés Fossiles* [1822: pp. 139–140, pl. XI, Figs 6–7. (Modified from: https://www.biodiversitylibrary.org/bibliography/9102 (Public domain)]. In the background a detail of the soil in a quarry near Eichstaett. (Photo credit Carlo Chirli)

Fig. 24.7 A fossil limulus from the Jurassic locality of Solnhofen (diameter ~100 mm). (Photo credit Massimo Cresti)

Fig. 24.8 A view of the excavations in the limestone quarry near Eichstaett in the Altmuehl valley, not far from the *Bürgermeister Müller Museum* in Solnhofen, where tourists can also go to try and discover some fossils in the so-called lithographic limestone. (Photo credit Carlo Chirli)

Fig. 24.9 Ancient illustrations of living specimens mainly from the Moluccas, the Indonesian archipelago in the Pacific Ocean: (1) *Cancer Moluccanum*, from Clusius 1605: p. 128. (2) *Arachnus marinus*, shown both dorsally and ventrally, from de Laet 1633: p. 56. (3) *Crustatis rotundis, Cancer molucanus*, the same image reproduced by Worm 1655: p. 249. (4) Another similar but more detailed figuration, *Cancer Mollucensis*, prone and supine, by Jonstone 1650: pl. VII, Figs. 1–2. (5) *Moluctische Krebs* or *Monoculos polyphemus* Linnaeus or *Cancer perversus* Rhumphius, from Georg Wolfgang Knorr 1766: pl. F1, figs 1–2. [Graphic composition Maurizio Forli. Credit from (1) https://www.biodiversitylibrary.org/bibliography/7108 (Public domain). (2) https://doi.org/10.3931/e-rara-55254 / Public Domain Mark. (3) https://doi.org/10.5962/bhl.title.63844 (not in copyright). (4) https://doi.org/10.3931/e-rara-52267 / Public Domain Mark]

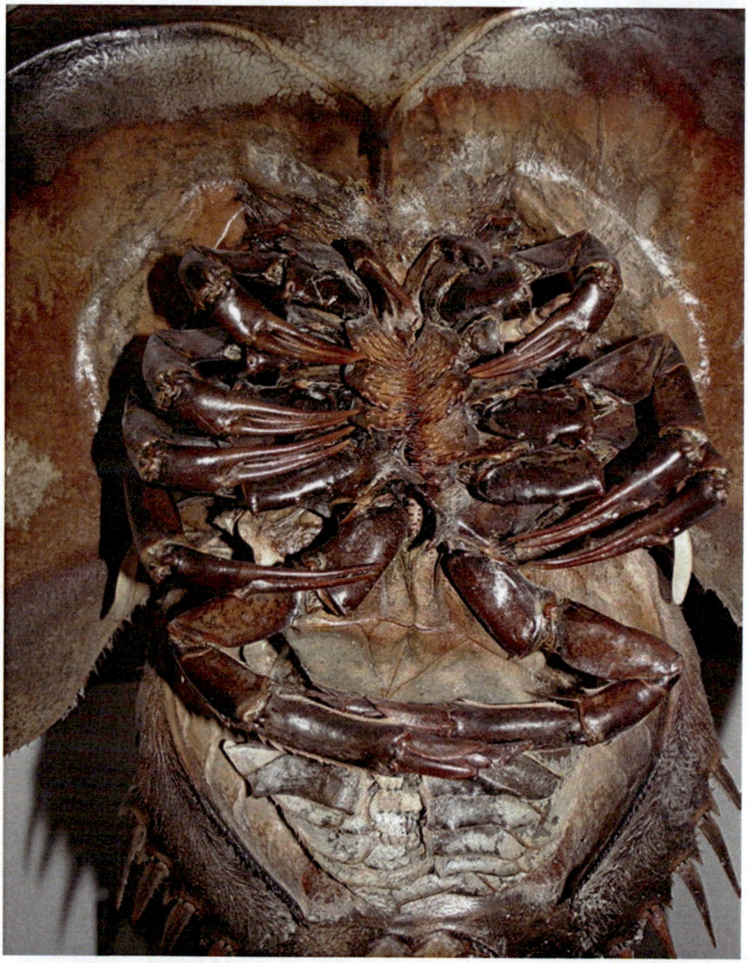

Fig. 24.10 A nightmarish animal or a wonder of Nature? Ventral side of a dried specimen of *Limulus polyphemus* (Linnaeus, 1758) showing the impressive set of appendages, ending with pincers, used both for locomotion and for the capture of preys and their subsequent introduction into the mouth, centrally placed. Note also the sharp spine at the top, also directed toward the oral opening, a further offensive appendage to hold the victims (width of the cephalic cuirass ~400 mm). (Photo Maurizio Forli)

Fig. 24.11 Detail from *The Fall of the Rebel Angels*, a famous oil painting on wood by Pieter Bruegel the Elder. A little to the lower right of the Archangel Michael, who in the center of the painting slays the great demon of the Apocalypse, there is, together with the tangle of other monstrous beings but quite distinct from them, this fantastic figure, halfway between the diabolical and the angelic, whose armor has a shape that closely resembles that of a limulus. Fantasy or reality? Was the artist actually inspired by the characteristic chelicerate arthropod or, once again, does sight (in this case, mine) identify what the mind already knows? (my "concept" of limulus). The work dated 1562 is housed in the Royal Museum of Fine Arts of Belgium, in Brussels. [https://commons.wikimedia.org/wiki/File:Pieter_Bruegel_the_Elder_-_The_Fall_of_the_Rebel_Angels_-_RMFAB_584_(derivative_work).jpg **(Public domain)]**

References

Dekay JE (1825) Observations on a fossil crustaceous animal of the order Branchiopoda. Ann N Y Lyceum Nat Hist 1:375–377

Clusius C (1605) Exoticarum libri decem. Quibus Animalium, Plantarum, Aromatarum, aliorumque peregrinorum Fructuum historiae describuntur: Item Petri Bellonii Observationes, eodem Carolo Clusio interprete. Series totius operis post Praefationem indicabitur. Ex Officina Plantiniana, Raphelengii

De Laet J (1633) Novus Orbis seu Descriptionis Indiae Occidentalis, liber XVIII. Apud Elzevirios, Lugduni Batavorum

Desmarest AG (1822) Histoire naturelle des crustacés fossils. Les Crustacès proprement dits. Imprimerie L.-T. Cellot, Paris

Jonstonus J (1665) Historiae naturalis de exanguibus aquaticis. Libri IV. Johannis Jacobi Schipperi, Amstelodami

Knorr GW (1766) Deliciae Naturae Selectae oder auserlesenes Naturalien-Cabinet welches aus den drey Reichen der Natur zeiget, was von curiösen Liebhabern aufbehalten und gesammlet zu werden verdienet. Nurember

Lamsdell JC, Simonetto L, Selden PA (2013) First Eurypterid from Italy: a new species of Adelophthalmus (Chelicerata: Eurypterida) from the upper Carboniferous of the Carnic Alps (Friuli, Italy). Riv Ital Paleontol Stratigr 119(2):147–151

Mitchill SL (1818) An account of the impressions of a fish in the rocks of Oneida County, New York American Monthly, 3: 291

Selden PA, Simonetto L, Marsiglio G (2019) An effaced Horseshoe Crab (Arthropoda: Chelicerata: Xiphosura) from the upper Carboniferous of the Carnic Alps (Friuli, NE Italy). Riv Ital Paleontol Stratigr 125(2):333–342

Shultz JW (2007) A phylogenetic analysis of the arachnid orders based on morphological characters. Zool J Linnean Soc 150:221–265

Woodward H (1878) A Monograph of the British Fossil Crustacea, belonging to the Order Merostomata. Paleontographical Society in London, 1866–1878

Walch JEI (1768–1773) Walch Johann Ernst Immanuel: Die Naturgeschichte der Versteinerungen : zur Erläuterung der Knorrischen Sammlung von Merkwürdigkeiten der Natur. Nürnberg : gedruckt mit Felsseckerischen Schrifften, 1768-1773. ETH-Bibliothek Zürich, Rar 399, https://doi.org/10.3931/e-rara-10246 / Public Domain

Worm O (1655) Museum Wormianum, seu Historia rerum rariorum, tum Naturalium, quam Artificialium, tam Domesticarum, quam Exoticarum; quae Hasniae, Danorum in aedibus Authoris Servantur. Iohannem Elsevirium, Lugduni Batavorum. https://doi.org/10.5962/bhl.title.63844 (not in copyright)

Chapter 25
Giant Lobsters and *Paguri Lapidei* from the 1500s to the Present, between Folkore and Science

Abstract Following the current taxonomic scheme, the Decapoda are the largest order of the Crustacea Malacostraca represented by crabs, shrimps, and lobsters. The external morphology is common to all of them; in fact, they are provided with a rigid exoskeleton covering the whole body, formed by the cephalothorax and the abdomen. This body structure is subdivided into 19 segments, five cephalic, eight thoracic, and six abdominal, which, all or only some of them, carry each one two paired appendices, more or less deeply modified. Their oldest fossil remains date back to the Paleozoic (Cambrian). The first depiction of a fossil crab was made by Gessner (*De omni rerum fossilium genere, gemmis, lapidibus, metallis, et huius-modi, libri aliquot, plerique nunc primum editi.* Tiguri. https://doi.org/10.3931/e-rara-4176 / Public Domain Mark, 1565); the original specimen is still preserved in the Naturhistorisches Museum in Basel. The dactyl, the mobile pincer of the claws of crabs and larger crustaceans in general, is still appreciated at a popular level for the manufacture of amulets and pendants usually reserved for males. In addition to protecting against the evil eye and jinxes, these amulets are believed to be useful in love, promoting fertility.

Keywords Fossils · Crustacea · History · Gessner · Palaeocarpilius · Amulets

Most of them are exclusively marine, but there are also some freshwater ones, some amphibians, and few others terrestrial. Important morphological differentiations complicate their classification; they tend to have the fusion of some thoracic elements with the cephalic shield. They are subdivided into various subclasses, but we shall treat only the malacostracans and the barnacles. (Fig. 25.1)

© The Author(s), under exclusive license to Springer Nature Switzerland AG 2022

M. Forli, A. Guerrini, *The History of Fossils Over Centuries*,
https://doi.org/10.1007/978-3-031-04687-2_25

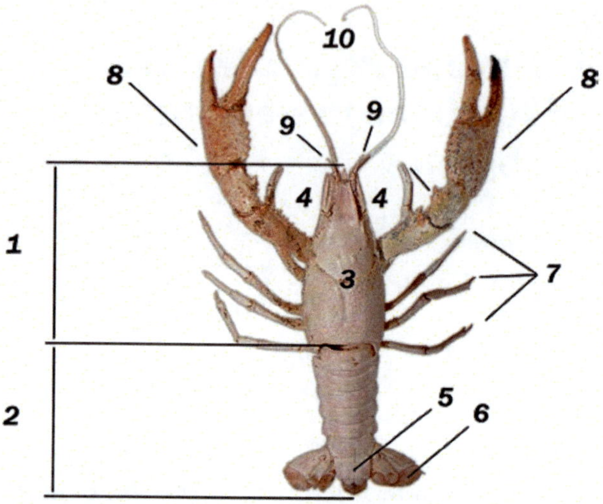

Fig. 25.1 Simplified scheme of the body of an cambarid crayfish, *Procambarus clarkii* Girard, 1852, better known as Louisiana crayfish or Killer crayfish (1988, Lake Massaciuccoli, Pisa, Italy): (1) Cephalothorax. (2) Abdomen. (3) Cervical fold. (4) Eyes. (5) Telson. (6) Uropods. (7) Ambulatory thoracic appendages. (8) Chelipeds. (9) Antennules. (10) Antennae. (Photo and graphic composition Maurizio Forli)

Malacostraca

Crabs: Granciporri and Ranina Crab

Sono le speci de granchi, carabi, astici, maie, paguri, heracleotici, leoni et altre specie ignobili. I Carabi sono differenti dagli altri granchi nella coda. In Phenicia sono chiamati Hippoe, di tanta velocità, che non si possa tenere loro dietro. A granchi è la vita lunga, hanno otto piedi, tutti in obliquo piegati. Il primo pie della femmina è doppio, et del maschio semplice. Oltre à di questo hanno due bracci con forbice dentate. La parte superiore delle quali in questi primi si muove, et la inferiore é in mobile, il destro braccio à tutti è maggiore. I granchi nella paura vanno indietro con pari velocità, che avanti. Combattono fra loro con le corna, come montoni, correndosi in contro. Medicano i morsi dei serpenti. Et passando il sole pel segno del cancro, et il corpo di essi, quando sieno morti in secco, dicano trasmutarsi in scorpioni [Brùcioli A., Historia Naturale di C. Plinio Secondo nuovamente tradotta di Latino in Volgare toscano, (1548: IX, cap. xx, CCLII)].

They are the species of crabs, carabi, lobsters, maias, hermit crabs, heracleotics [the crab *Calappa granulata* Linnaeus, 1758)], *lions and other vile species. The Carabi are different from the other crabs in the tail. In Phenicia they are called Hippoe, of such speed, that it is impossible to keep up with them. Crabs are long lived, they have eight feet, all obliquely bent. The first foot of the female is double, and of the male simple. In addition to this, they have two arms with toothed scissors. The upper part of which moves in these first ones, and the lower one is mobile, the right arm is larger than the others. Crabs in fear go backwards with the same speed as forwards. They fight each other with their horns, like rams, running*

> *against each other. They cure the bites of serpents. And when the sun passes through the sign of Cancer, and their bodies are said to be transmuted into scorpions when they are dead.*

Following the current taxonomic scheme, the Decapoda are the largest order of the Crustacea Malacostraca represented by crabs, shrimps, and lobsters; they include the most specialized and largest forms among crustaceans in general. The external morphology is common to all of them; in fact, they are provided with a rigid exoskeleton covering the whole body, formed by the cephalothorax, derived from the fusion of the cephalic portion with some thoracic segments, and by the abdomen. Another common particularity is to have the body structure subdivided into 19 segments, five cephalic, eight thoracic, and six abdominal, which, all or only some of them, carry each one two paired appendices more or less deeply modified. According to their function, these are ordinarily called antennules, antennae, mandibles, maxillae, thoracic legs, and abdominal appendages. In particular, the abdomen (*pleon*) has six bifid appendages, of which five pairs (*pleopods*) are used to move the water forward toward the gills, enclosed in well-defined chambers at the sides of the cephalothorax, while the last pair of these, taking the name of uropods, constitutes together with the pygidium (*telson*), a caudal fan that acts as a powerful swim fin.

The cephalothorax, protected by a robust carapace, often ends in front with a toothed and sharp rostrum, more or less protruding. In the head, which includes the buccal apparatus, are present the sense organs, in particular, the eyes, compound, spherical, often pedunculated, and mobile, and two pairs of antennae, two main and two secondary, with tactile functions. Laterally to the cephalothorax are placed in symmetrical pairs on both sides, eight pairs of thoracic appendages: the first three form the maxillipeds and the last five are the pereiopods. The maxillipeds are utilized for grasping the food, mincing it and carrying it to the mouth, while the pereiopods, the true and proper legs, with their first pair characteristically ending in a pincer, called cheliped, are mainly dedicated to the locomotion and in some cases also to the swimming. It is from the number and the functionality of the pereiopods, five pairs of thoracic walking limbs, that the name Decapoda comes.

With reference to the crabs, or Brachyura, the abdomen is small and is folded forward, fitting firmly under the cephalothorax. In the male, it has a subtriangular shape, while in the female it is similar to a lozenge, wider and rounded. In spite of this external conformation, the abdominal appendages, but the uropods which have disappeared, are still present even if reduced in size and number. In the females they are utilized for holding the incubating eggs and in the males they have a copulatory function. The most evident external characters of the crabs are the conformation of the carapace and the dimensions, at times exaggerated, of the claws.

Crabs are opportunistic predators, in the sense that they can feed on actively hunted small prey, such as mollusks or worms, or they can feed on carcasses or other organic debris. As there are also some which turn their alimentary preferences to the vegetables, we can say that, in general, they are omnivorous animals. They can walk slowly forward, but usually the lateral movement, at times very fast, is their usual way of locomotion.

This ability of theirs is easily verifiable by direct observation, especially if you try to catch one with your hands in some little puddle of a cliff, only the first attempt is valid! If it succeeds in escaping you will have to desist immediately and if you are so good as to catch it, you will still have to release it without harming it. In Italian "*catch a crab*" in a metaphorical sense an allusion that means that we have made a mistake by misunderstanding one thing for another, is equivalent to taking a blunder, in short, we have not understood anything!

The oldest fossils of arthropods with characteristics similar to the present Crustacea are dated to the Cambrian. Within the Malacostraca, the Decapoda are known starting from the Paleozoic (Devonian), reaching their greatest diffusion starting from the Cenozoic. Among the most famous fossiliferous deposits that preserve them, we remember the lithographic limestone of Solnhofen, of Middle Jurassic age, and the Eocene formations of the Veronese area. Numerous other fossil species with greater biological affinities with the present ones, discovered and studied also recently, come from the Miocenic and Pliocenic soils of Tuscany.

The first famous depiction of a fossil crab is due to Conrad Gessner (*De omni rerum fossilium genere*, 1565: p. 167) who briefly describes it as *Pagurus lapideus, parte supina expressus*.

Certainly coming from the Eocene terrain of the Lessini Mountains in Veneto, together with other finds, it was acquired by the Swiss naturalist doctor during his stay in Venice in 1543.

Although it is sometimes cited as *Palaeocarpilius macrocheilus* (Desmarest, 1817), here is the present classification (Fig. 25.2):

Order Decapoda
Superfamily Carpilioidea
Family Carpiliidae
Genus *Palaeocarpilius* Milne-Edwards 1862
Species *Palaeocarpilius macrochelus* (Desmarest, 1822) †

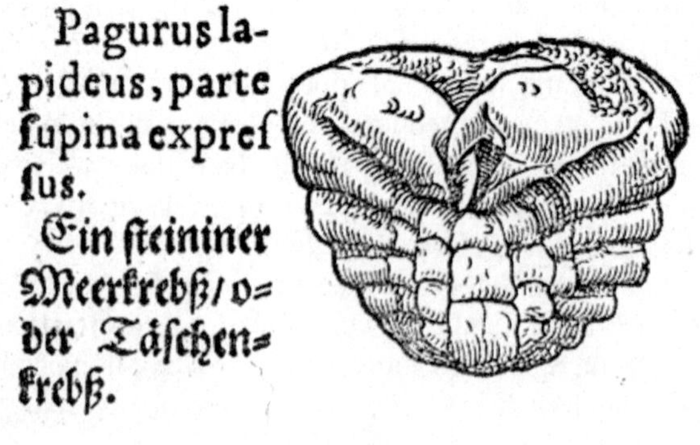

Fig. 25.2 The first representation in a printed book of a fossil crab, *Palaeocarpilius macrochelus* (Desmarest, 1822), from the Eocene of the environs of Verona, illustrated by Gessner in *De omni rerum fossilium genere*, (1565): p. 167. (https://doi.org/10.3931/e-rara-4176 / Public Domain Mark)

Cancer is linked to a zodiac sign which also represents it iconographically. In this type of representation the figure of the crab is sometimes replaced by that of the river shrimp or lobster. The latter, owing to the power of its claws and its larger size than related crustaceans, was sometimes conceived as a sea monster. The biggest lobsters, in extension with the claws forward, may reach measures close to 1 meter of length and a weight of more than 30 kilograms, however, considerable for an invertebrate, but certainly not bigger than a human, as sometimes represented in the past. A couple of these representations, among the best known of the *imago mundi* represented in Renaissance maps, are due to the Swedish humanist and geographer Olav Manson, whose name was Latinized as Olao Magno (1490–1557). In his *Carta marina et Descriptio septemtrionalium terrarum* (1539), a verisimilitude map of Northern Europe, including southern Greenland, among many other terrifying marine creatures more or less fanciful, there are two crustaceans, in the shape of a lobster, of incredible size, one in the act of capturing a castaway with the left claw and another in a fight with a polymorphic sea creature, called by Gessner *marine monstrosus similar to rhinoceroti*. (Fig. 25.3, Fig. 25.4)

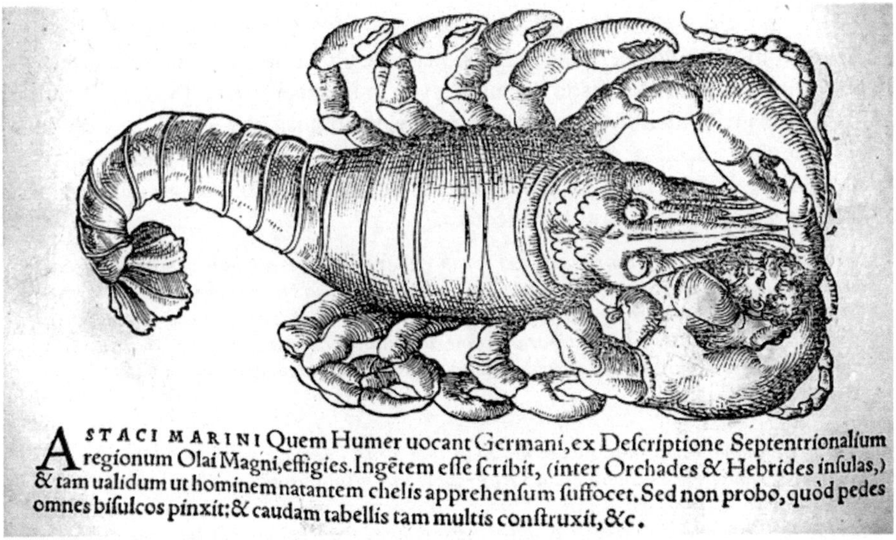

Fig. 25.3 The giant lobster depicted by Olaus Magnus and reproposed by Gessner in his *Nomenclator aquatilium animantium* (1560: p. 212). (https://doi.org/10.3931/e-rara-48090 / Public Domain Mark)

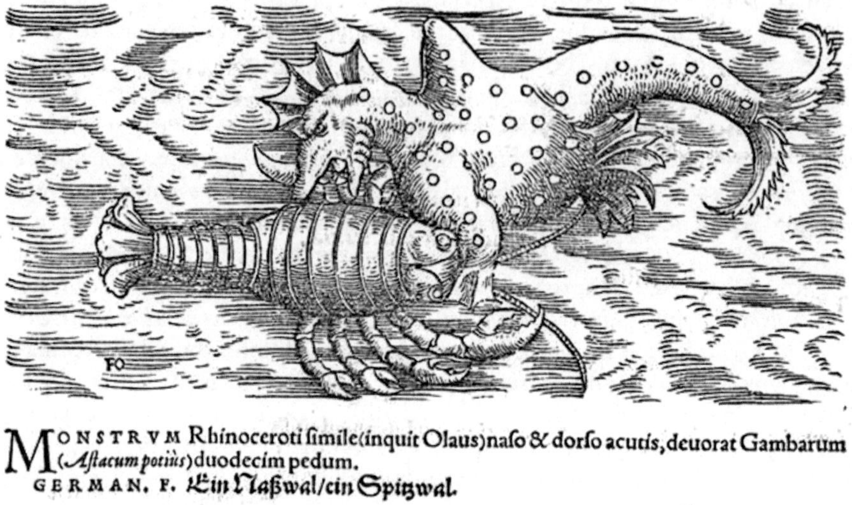

Fig. 25.4 The sea monster defined as similar to a rhinoceros, probably the fantastic elaboration of an orca, while attacking and devouring an equally monstrous shrimp, according to the original writing, equipped with 12 feet, although in the figure there are only eight, taken again from Gessner, *Nomenclator aquatilium animantium* (1560: p. 181). (https://doi.org/10.3931/e-rara-48090 / Public Domain Mark)

Apart from the still uncertain explanation of the phenomenon of petrification and its philosophical and naturalistic meaning, even in Gessner's time, the identification of fossil crabs with their living counterparts posed no difficulty, given their obvious correspondence. Such a singular and consequently "prestigious"find as that of *Pagurus lapideus* could not but arouse interest and a desire to possess it on the part of other contemporary naturalist collectors.

> *Infatti, Gerolamo Fracastoro aveva nella sua collezione un esemplare di granchio, che confluito successivamente nel Museo del naturalista veronese Francesco Calzolari, veniva illustrato nel 1622 da Benedetto Ceruti e Andrea Chiocci come Pagurus venetus Granciporro vocatus. Stando all'illustrazione riportata, poteva trattarsi di un individuo di Palaeocarpilius macrochelus (Desmarest), specie abbastanza comune nei giacimenti eocenici del Veneto* (Beschin et al. 2011).

> *In fact, Gerolamo Fracastoro had in his collection a specimen of crab, which later merged in the Museum of the Veronese naturalist Francesco Calzolari and was illustrated in 1622 by Benedetto Ceruti and Andrea Chiocci as* Pagurus venetus *Granciporro vocatus. According to the illustration, it could be an individual of* Palaeocarpilius macrochelus *(Desmarest), a species quite common in the Eocene deposits of Veneto* (Beschin et al. 2011).

Also in this case, as usual, the image of the crab lying on its ventral side was repeated in other texts more or less contemporary or slightly subsequent to the last Gessnerian literary work. In chronological order of publication, these include *Musaeum Calceolarianum* (1622: p. 430); *Museum Metallicum* (1648: p. 461); *Note Overo memorie del Museo di Lodovico Moscardo* (1656: p. 179); *Musaeum Septalianum* (1666: p. 11); and *Metallotheca Vaticana* (1717: p. 306). (Fig. 25.6)

The progenitor of this genealogy of illustrations, the Gessner specimen, is still preserved at the Naturhistorisches Museum in Basel, along with other surviving finds from the collection served for the illustrations of *De omni rerum fossilium genere* (Etter and Schmidt 2019, 2020). No less important is the Aldrovandian brachiurus, currently on display in Palazzo Poggi, in what, although remodeled, can be considered the oldest naturalistic museum still usable. It is another Eocene specimen from the Lessini Mountains on the border between Verona and Vicenza, belonging to the genus *Harpactoxanthopsis*, *H. quadrilobata* (Desmarest, 1822). Together with the latter there is also a *P. macrochelus* whose dorsal and ventral representation, never included in Aldrovandi's publications, was brought to the attention of the specialized public by Dr. Alessandro Ceregato on the occasion of the printing of the commemorative volume, *Natura Picta. Ulisse Aldrovandi* (2007: p. 605, pl. 57). (Fig. 25.5) to (Fig. 25.11).

Fig. 25.5 Specimen of *Palaeocarpilius* never included by Aldrovandi in his publications [see later the drawing reported in the commemorative volume, *Natura Picta. Ulisse Aldrovandi* (Ceregato 2007). [Photo credit Alessandro Ceregato (permission Sistema Museale di Ateneo Alma Mater Studiorum – Università di Bologna)]

Fig. 25.6 *Harpactoxanthopsis, H. quadrilobata* (Desmarest, 1822): the *Pagurus lapideus*, illustrated by Aldrovandi in the *Musaeum Metallicum* (1648: p. 461), is still preserved and exhibited in Palazzo Poggi, in Bologna, together with many other things that still remain of the original collection of the Bolognese scientist. [Photo and graphic composition Maurizio Forli (permission Sistema Museale di Ateneo Alma Mater Studiorum – Università di Bologna)]

Fig. 25.7 A compiled Eocene crustacean, reconstructed in a naturalistic pose; maximum carapace size ~80 mm. (Photo and graphic composition Maurizio Forli, courtesy Carlo Chirli collection)

Fig. 25.8 *Eriphia cocchii*: the Pliocene ancestor of the present *Eriphia verrucosa* (Forskål, 1775), a crab widespread throughout the Mediterranean, living among rocks at tidal level or slightly deeper [width ~120 mm]. (Photo and graphic composition Maurizio Forli, courtesy Carlo Chirli collection)

Fig. 25.9 (1) *Ilia nucleus* (Linnaeus, 1758), an uncommon crab, characterized by an almost spherical cephalothorax, with average size around 25 mm, with distribution limited to part of the Atlantic Ocean and the Mediterranean Sea. (2) Above, the remains of two specimens from the Pliocene of San Miniato (Pisa). [(1) Photo and graphic composition Maurizio Forli (2) Photo credit Massimo Cresti)]

Fig. 25.10 Other fossil and present crustaceans: (1) top to left: *Goneplax gulderi* Bachmayer, 1953 (Upper Miocene) and (2) *Goneplax rhomboides* (Linnaeus, 1758) Mediterranean; bottom to left: (3) *Maja crispata* Risso, 1827, Mediterranean, (4) two Pliocene congeneric crustaceans). [(2–3) Photo and graphic composition Maurizio Forli (1, 4) Photo credit Massimo Cresti]

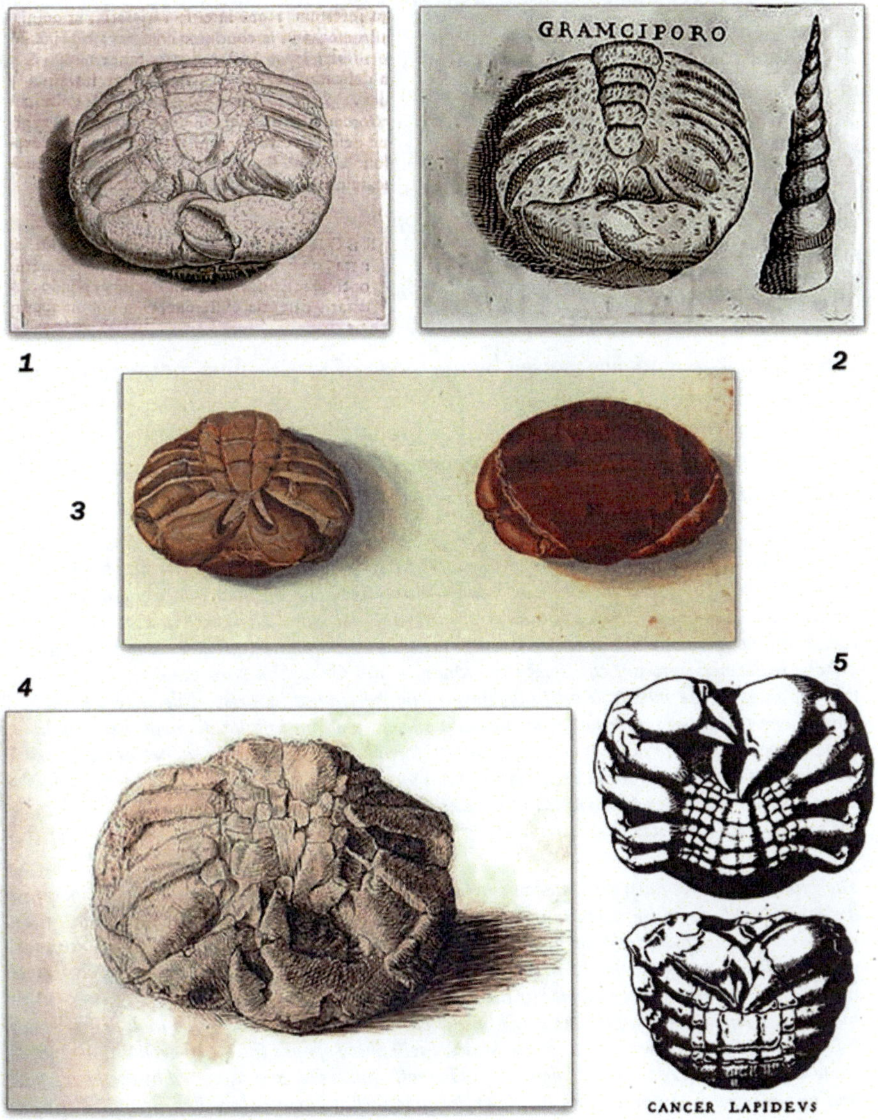

Fig. 25.11 Some of the oldest printed representations of fossil crabs: (1) from Cerutus and Chioccus, *Musaeum Franc. Calceolari iun veronensis* (1622: p. 430), (2) from Moscardo, *Note overo memorie del Museo di Lodovico moscardo nobile veronese* (1656: p. 179), (3) this specimen, included in a plate dedicated to fossilia that has never been included in Aldrovandi's publications, is part of those published in the commemorative volume, *Natura Picta. Ulisse Aldrovandi* (Ceregato 2007: p. 605, pl. 57). (4) Specimen in the museum of Manfredo Settala (1666, 2: p. 11), (5) Mercati, *Cancer Lapideus* (*Metallotheca Vaticana*, 1717: p. 306). [Graphic composition Maurizio Forli. Credit from: (1) https://doi.org/10.3931/e-rara-50568 (Public Domain Mark) (2) https://www.biodiversitylibrary.org/bibliography/157058 (not in copyright) (4) cedoc.mo.it/estense gammah.1.22. https://archive.org/details/museoogaleriaadu00terz_0 (no copyright page found) (5) https://archive.org/details/bub_gb_DC_E6UVWg7EC (Public Domain Mark 1.0)]

Also the two specimens engraved in the work of Mercati are native of the present Natural Park of Lessinia as they were found *in Veronensis agri montibus*.

Two other images of crustaceans are found among the many seventeenth-century volumes we usually consulted in our research: Friedrich Lackmund's *Orictographia Hildesheimensis* and Agostino Scilla's *La vana speculazione disingannata dal senso*. That of the Saxon author, a remnant of a crab seen from the underside, a portion of the abdomen with the attachment and little more than a few legs, is described as a rock *in superficis cerebro similis* (1669: pp. 30–31, fig. II), that of the Messinian is instead a

> *Stone, containing a part of a sea crab, that is a small and a large branch, which strigne a half shell striated. Of Messina* (1670: pl. XIX, fig. 1).

This occasional coincidence caused by the mode of fossilization, not without irony and witticism toward the addressee of the letter responsiva, the form in which the text of *La Vana speculazione* is proposed, is brought as an example of the conviction that the fossils examined were real remains of once living marine organisms, and not random natural productions:

> *Quindi, sì come non dobbiamo determinarli generati nella terra, o ne' laghi; così all'incontro stimar dobbiamo, che questi giunti fossero tanto addentro nelle campagne, e ne' monti della Calabria, insieme con infinite altre cose del mare, per ondeggiamenti terribilissimi* [...] *ed autorità di tanti corpi propriissimi del mare, che giurano di non essere nati, ove li veggiamo. Per l'avvenire l'errore continuo, e populare (qui da noi) nel chiamar bocche le branche grosse, che sembrano tanaglie, del Granchio, sarà condonabile; perciocchè questa, che mostrerò parla da dovero, anche ridotta in sasso. Ella dice, che nel tempo, che si sentiva oppressa, e stretta dalla carica, e da un'infinità di corpi, per rabbia addentò quel, che se le parò avanti, come infatti fortemente l'ha mantenuto, per persuadere lei a mutare opinione. Non ha essa attanagliato una conchiglia striata? si per certo. E per certo anche stimo, che alcuno potrà pretenderla nata nelle rocche delle colline di Messina, senza incorrere nel peccato d'impugnare la verità conosciuta* (Scilla 1670: pp. 137–138).

> *Therefore, just as we must not determine them generated in the land, or in the lakes; so, on the other hand, we must estimate that they came so deep into the countryside, and in the mountains of Calabria, together with countless other things of the sea, because of terrible waves* [...] *and the authority of so many bodies proper to the sea, which swear that they were not born where we see them. For the future, the continuous and popular error (here in our country) in calling mouths the large branches o claws of the crab, which look like tanalia, will be condoned, because this one, which I will show speaks for itself, even reduced to stone. She says, that at the time when she felt herself oppressed, and squeezed by the charge, and by an infinity of bodies, she bit down in anger what came before her, as indeed she strongly maintained it, to persuade her to change her opinion. Did she not grasp a striped shell? Yes, for certain. And I think it is certain, too, that no one will be able to claim that she was born in the rocks of the hills of Messina, without incurring the sin of challenging the known truth*

Below, some illustrations of fossil and present-day "crabs" (Fig. 25.7), (25.8), (Fig. 25.9), (Fig. 25.10).

The dactyl, the mobile pincer of the claws of the crabs and of the bigger crustaceans in general, spider crabs and lobsters, also for its shape comparable to the more common cornets, has been and is still appreciated at popular level for the fabrication of amulets and pendants usually reserved to the males. For this purpose, is frequently used that of the common Mediterranean crab *Eriphia verrucosa* (Fig. 25.12). Some examples of this custom are kept in the Bellucci collection, cited as *shrimp cornets:*

> *cornetto di gambero. – Palena, Chieti. Chèla o pinza di gambero, legata in argento con appiccagnolo. È amuleto contro il malocchio. Cornetto di gambero. Cortona, Arezzo.* (Bellucci G 1898. Amuleti italiani contemporanei).

> dactyl *of shrimp. Palena, Chieti. Chèla or shrimp dactyl, tied in silver with a clasp. It is an amulet against the evil eye. Small horn of shrimp. Cortona, Arezzo.*

Fig. 25.12 On the sides, amulets from the Bellucci collection, in the center, an example of a Picenian bronze pendant/amulet (fourth to fifth century BC), from Ripatransone (AP). (Modified) [Graphic composition Maurizio Forli]

In addition to protecting against the evil eye and jinxes, these amulets are considered useful in love, promoting fertility. Even today, perhaps unconsciously, the dactyls of crabs are used to make pendants with a more ornamental than apotropaic function; our memories of adolescents refer to the summer use of necklaces with similar trophies, shown with a *touch of* vanity to foreign girls who came to spend their holidays at the beach. (Fig. 25.13), (Fig. 25.14).

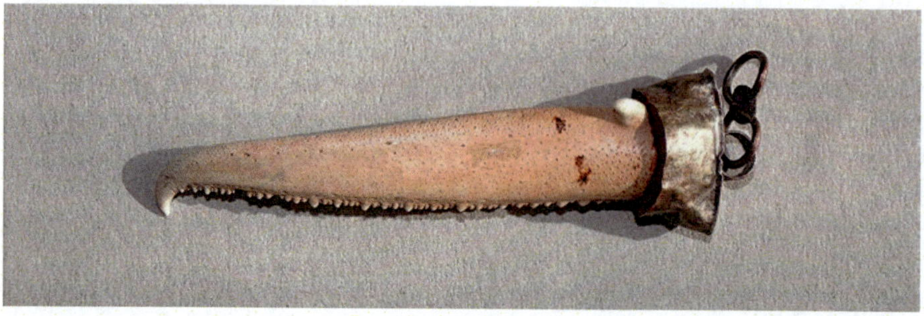

Fig. 25.13 Lobster dactyl mounted with silver alloy hanger, good luck charm of a fisherman from the province of Livorno, 1950s. Length ~75 mm. (Photo Maurizio Forli, courtesy private collection)

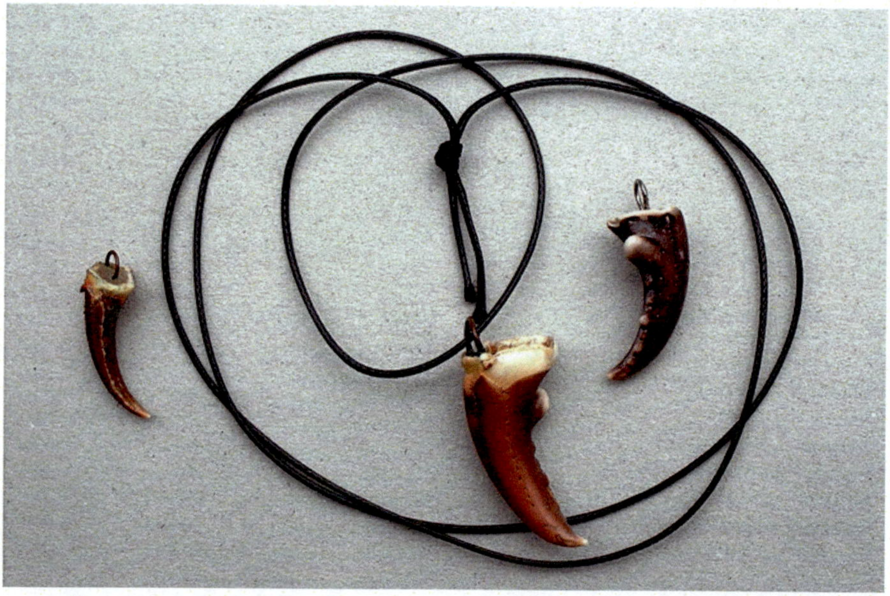

Fig. 25.14 Dactyls of crabs used as pendants in simple summer necklaces (province of Livorno, 1970s). (Photo Maurizio Forli, courtesy private collection)

Fossil decapods do not appear to be used as raw material for medicinal purposes, but their present-day descendants were widely employed in this practice.

De cancri Fluviali

Carcino (Romani cancri) le cenere de questi cancri abbrusciati al peso de doi cucchiari, et uno di gentiana bevuta per tre giorni giova à morsicati da cani rabbiosi. con la decottione del mele mitiga le fenditure de piedi, e del sedere, et il male, che viene a le calcagna ne tempi d'inverno, et i cancri triti, crudi bevuti con il latte d'asino vagliono contra i morsi de serpenti, de falangi, de scorpioni. alessi e mangiati con il loro brodo giovano a la corrottione di tutto il corpo, e che hanno bevuto il lepro marino. triti, e con ocimo ammazzano i scorpioni. i cancri marini sono di minor vertu (Dioscoride fatto di Greco Italiano, Fausto da Longiano 1542: p. 64).

De i Granchi de i Fiumi

La cenere de i cranchi de i fiumi brusciati, data tre di à bere alla quantità di due cucchiari insieme con uno cucchiaro di radice di gentiana, giova efficacemente al morso del cane rabbioso. Impiastrata con mele cotto, mitiga le fissure del sedere, et de i piedi, le bugance, et i cancri triti, et bevuti crudi con latte di Asina, giovano al morso dei serpenti, de ragni che chiamano Phalangi, et alle punture degli Scorpioni. Cotti mangiati con la loro decottione, giovano ò ptisici, et à chi havesse bevuto il Lepre Marino. Messi triti con basilico sopra a gli scorpioni, gli ammazzano. Tutto questo fanno anchora i Marini, ma con assai minore successo [...] *questo tutto disse galeno. Spetie di gambari sono le Locuste, quantunque non abbiano le branche, et similmente i Paguri, i Granchiporri, et le Grancevole. Le pietre che si ritrovano nella testa dei Gambari usano i moderni medici a provocare le pietre delle reni, dandole a bere in polvere, ovunque sia di bisogno. I granchi simili alle Macinette di mare nascono per tutta Thoscana ne i fiumi, et ne i fossati dell'acqua dolce come in lombardia nascono i Gambari* (Il Dioscoride, dell'eccellente Dottor Medico M. P. Andrea Matthioli, 1548: pp. 108–109).

Fluvial cancers

Carcino (Roman crabs): the ashes of these crabs burnt at the weight of two tablespoons, and one of gentian, drunk for three days helps those bitten by rabid dogs. The decoction of honey mitigates the cracks in the feet and bottom, and the pain that comes to the heels in winter, and the crushed, raw crabs drunk with ass's milk help against the bites of snakes, phalanges and scorpions. When boiled and eaten with their broth, they help to corrupt the whole body, and those who have drunk the sea hare. chopped with basil, they kill scorpions. marine crabs are of lesser virtue.

Of the Crabs of the Rivers

The ashes of the crabs of rivers burnt, given for three days to drink at the quantity of two tablespoons together with one tablespoon of gentian root, effectively helps against the bite the rabies of dog. Make a poultice with a cooked honey, mitigates the fissures of the bottom, and of the feet, the chilblains, and the chopped crabs, drunk raw with milk of female donkey, they help to the bite of the snakes, of the spiders that call Phalangi, and to the stings of the Scorpions. Cooked and eaten with their decoction, they are useful for those who have drunk the Sea Hare. Chopped mixed with basil, and put on the scorpions, they kill them. All this is also done by the sea crabs, but with much less success [...] *Galen said all this. Locusts are a kind of lobsters, although they do not have branches, and similarly the Hermit Crabs, the Granciporro (Cancer pagurus Linnaeus, 1758) and the Grancevole (spider Crabs). The stones found in the head of the lobsters are used by the modern physicians to remove the stones of the kidneys, giving them to drink in powder, wherever they are needed. The crabs similar to the macinetta (Carcinus aestuarii Nardo, 1847) are born all over Tuscany in the rivers, and in the ditches of the fresh water, as in Lombardy.*

This is how Fausto da Longiano and Pietro Andrea Matthioli reported from Dioscorides, the posology regarding the administration of crabs and shrimps, the latter enriching it with his own experiences and comments. Obviously also from a culinary point of view, crustaceans were appreciated since ancient times; as a demonstration, the delicious lobsters were commonly included in Roman mosaics among the most valuable subjects of marine fish farms. A number of these considerations, including others naturalistic and of various kinds, can be found in a book entitled *Gammarologia Curiosa* by the German physicist and naturalist Philipp

Jakob Sachs (1627–1672). A sort of compendium of the then knowledge about crabs, shrimps, lobsters, and the like. The treatment of the subject is multifaceted and wide ranging, even going so far as to show a child with feet affected by Ectrodactyly, with the toes fused and separated centrally up to the tarsus, a malformation still referred to today as "lobster claw" (1665: p. 596). (Fig. 25.15)

Fig. 25.15 Illustration after Philipp Jakob Sachs' *Gammarologia Curiosa* (1665: p. 596, pl. X). (Graphic composition Maurizio Forli. Credit: biodiversitylibrary.org/item/209611#page/9/mode/1up Public domain)

Shaped like a cuttlefish bone, seen from both sides, *Sepites pronè, & supinè*, is defined by Aldrovandi as a rare specimen of a fossil crustacean belonging to the family Raninidae, currently classified as *Lophoranina aldrovandii* (Ranzani, 1818). This specimen is also preserved and exhibited at Palazzo Poggi, with other surviving specimens among those represented in the *Musaeum Metallicum* (Aldrovandi 1648: p. 451). In addition to its scientific importance, this is the first illustration of a crustacean belonging to this particular family ofcrabs which in their posture resemble that of a frog. (Fig. 25.16)

Fig. 25.16 *Sepites pronè, & supinè*, the Eocene crustacean currently named *Lophoranina aldrovandii* (Ranzani, 1818), compared with another coeval specimen, also from Tertiary deposits in Veneto (*Lophoranina* sp.). [Modified after: doi.org/10.3931/e-rara-13017 Public Domain Mark (photo credit Massimo Cresti)]

Other engravings of living malacostracans other than the classical crabs are, however, found together with them in the already mentioned encyclopedic works of the mid-sixteenth century: it is not so for the fossils concerning them.

Indeed, three images are considered by Gessner to be lithified shrimp tails, but in this case, instead of abdominal segments of these crustaceans, we are dealing with a partial sequence of chambers of a nautiloid cephalopod (Orthoceratida), with the canal of the siphuncle centrally located, projecting in mamilla fashion, as seen on the left side of the illustrations, rather than a belemnoid phragmocone belemnoid phragmoconus, where this organic structure is placed differently (1565: p. 167). (Fig. 25.17), (Fig. 25.18).

Fig. 25.17 The remains of the orthoceratid nautiloid, likened by Gessner to lithified crayfish tails [1565: p. 167) compared with a similar fossil find from the Silurian-Devonian of Morocco (65 mm long). (Photo and graphic composition Maurizio Forli. Credit from https://doi.org/10.3931/e-rara-4176 / Public Domain Mark)

Fig. 25.18 A fossil decapod from the lithographic limestone of Solnhofen (Jurassic). Length ~120 mm. (Photo Maurizio Forli, courtesy Carlo Chirli collection)

Along with a crab in the usual abdominal pose of the depiction, we find a complete specimen of a shrimp, fossilized on a rocky slab, in the first tome of Johann Jacob Scheuchzer's *Physica Sacra* (1732: p. 210, pl. LVI, figs 63–64).

Other similar findings, in which these particular crustaceans are well recognizable, are part of a couple of plates in *Das Steinreich Systematisch entworsen* by Immanuel Walch (1762: pl. I, fig. 3 *Locusta marina*; pl. IV, figs 1–3); together with many others, among which are noted some specimens of *Eryon* sp., Mesozoic crustacean characteristic of the already mentioned German locality of Solnhofen (Germany), the same reappears in the subsequent *Die Naturgeschichte der Versteinerungen: zur Erläuterung der Knorrischen Sammlung von Merkwürdigkeiten der Natur* (Walch 1771).

To expand the iconography related to malacostracans we also submit these other representations taken from José Torrubia (1754), pertaining to two specimens that we consider "recent," probably only hardened remains of shells incorporated in a consolidated matrix, from Haynan Island (China), therefore not coming from the classical European localities, where the one shown in Figure 1 of plate XIII seems to correspond to a Coconut Crab (*Birgus latro* Linnaeus, 1767). (Fig. 25.19)

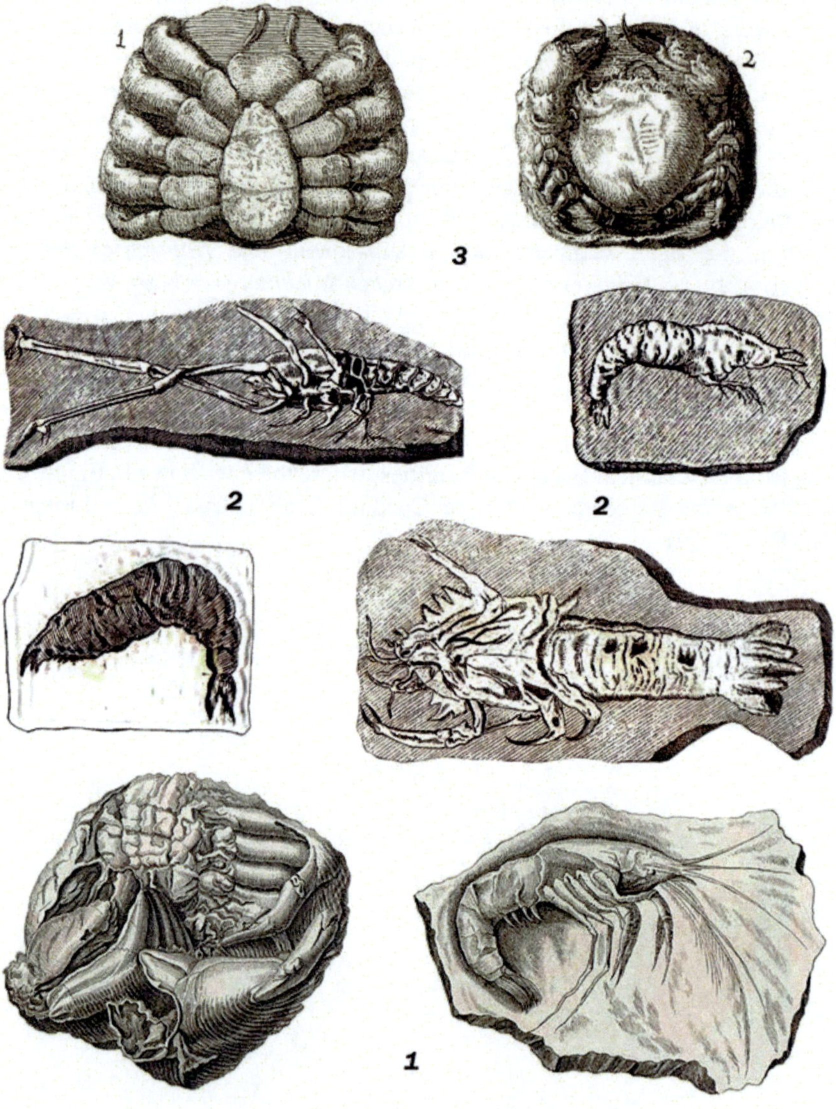

Fig. 25.19 from bottom to top (1) Images from Johann Jacob Scheuchzer's *Physica Sacra* [1732: p. 210, pl. LVI, figs 63–64. (2) *Das Steinreich Systematisch entworsen* by Immanuel Walch (1769: pl. I, fig. 3, pl. IV, figs 1, 3). (3) from José Torrubia, *Aparato para la historia natural española* (1754: pl. XIII, figs 1, 2). [Graphic composition Maurizio Forli modified from: (1) https://doi.org/10.3931/e-rara-1706 / Public Domain Mark (2) https://doi.org/10.3931/e-rara-10246 / Public Domain (3) https://wellcomecollection.org/works/zn4ugnnt (Public Domain Mark)]

We found no “diabolical” references to crustaceans in the books we consulted except for an extravagant and ridiculous image of two dancing crabs, taken from the *Dictionnaire Infernal*, a curious nineteenth-century repertory edited by Jacques Auguste Simon Collin de Plancy: (Fig. 25.20)

> *Crabes. Ces hideux petits habitants de la mer son attachés par quelque lien aux démons des eaux, et, suivant le dire des Écossais riverains, ils dansent au sabbat des sorcières, lorsqu'l se rassemble sur la plage* (1863: p. 188).

Fig. 25.20 Dancing crabs a kind of “*can-can*”. After Jacques Auguste Simon Collin de Plancy, *Dictionnaire Infernal* (1863: p. 188). The inspiration of this pose could derive from the observation of fighting phases between crustaceans, for territorial or reproductive reasons. [Image colour Maurizio Forli. Credit https://archive.org/details/bub_gb_DotbAAAAQAAJ Public Domain Mark 1.0.]

Crabs. These hideous little denizens of the sea are connected by some link with the demons of the waters and dance at the witches’ Sabbath when they gather on the beach, as the inhabitants of the Scottish coasts say.

The following is our pictorial divertissement, a work made by assembling what little we have found, regarding crabs and the like, understood as demonic beings. (Fig. 25.21)

Fig. 25.21 Fake composition: *Crustaceans at the Sabbath*. The two-headed crab in the lower left is the only one included in the original context used as a background that is Cornelius Saftleven's *Le sabbat des sorcières* 1640 [artic.edu/artworks/53495/a-witches-sabbath (modified)]; the small granchiform demon in the lower right is part of a version of The *Temptations of St. Anthony* (1640) by David Teniers [nl.wikipedia.org/wiki/Bestand:David Teniers (II)-The Temptation of St Anthony-WGA22105.jpg (modified)]; the two demons shaped like a lobster at the top are from *The Fall of the Rebel Angels* by Pieter Brughel the Elder. (Modified after: The Fall of the Rebel Angels – Google Art Project.jpg) [Graphic composition Maurizio Forli]

References

Aldrovandi U (1648) Ulyssi Aldrovandi patricii Bononiensis Musaeum metallicum in libros IIII : distributum Bartholomaeus Ambrosinus ... labore, et studio composuit cum indice copisosissimo Arcus Antonius Bernia. Bononiae : typis Io. Baptistae Ferronii, 1648. ETH-Bibliothek Zürich, Rar 1027, https://doi.org/10.3931/e-rara-13017 / Public Domain Mark

Bellucci G (1898) Contemporary Italian Amulets. Descriptive catalogue of the collection sent to the National Exhibition of Turin. Cooperative Typographical Union, Perugia

Beschin C, Angeli D, Zorzin R (2011) The genus Lophoranina Fabiani, 1910 (Crustacea, Decapoda, Brachyura, Raninidae) in the Tertiary of the Lessini Veronesi (northern Italy). Bull Mus Civico Storia Nat Verona 35:33–56

Brùcioli A (1548) Historia Naturale di C. Plinio Secondo nuovamente tradotta di Latino in Volgare toscano. Alessandro Brùcioli, Venetia Brughel

Ceregato A (2007) Natura Picta. Ulisse Aldrovandi. The "fossils" in the closet. Editrice Compositori, Bologna

Ceruti B, Chiocco A (1622) Musaeum Franc. Calceolari iun Veronensis. Apud Angelum Tamum, Veronae

Etter W, Schmidt O (2019) Gessner's fossil crab An icon for the early history of palaeontology. J Hist Collect. https://doi.org/10.1093/jhc/fhaa003

Etter W, Schmidt O (2020) Gessner's fossil crab An icon for the early history of palaeontology. J Hist Collect 32(3):411–430. https://doi.org/10.1093/jhc/fhz039

Gessner C (1560) Nomenclator aquatilium animantium : icones animalium aquatilium in mari et dulcibus aquis degentium, plus quam DCC. cum nomenclaturis singulorum Latinis, Grecis, Italicis, Hispanicis, Gallicis, [...]. Tiguri : excudebat Christoph. Froschoverus ..., 1560. ETH-Bibliothek Zürich, Rar 9734: 1, https://doi.org/10.3931/e-rara-48090 / Public Domain Mark

Gessner C (1565) De omni rerum fossilium genere, gemmis, lapidibus, metallis, et huiusmodi, libri aliquot, plerique nunc primum editi. Tiguri. https://doi.org/10.3931/e-rara-4176 / Public Domain Mark

Lachmund F (1669) Oryktographia Hildesheimensis sive admirandorum fossilium, quae In tractu Hildesheimensi reperiuntur, Descriptio Iconibus Illustrata, Cui addita sunt alia de calculis, de fontibus, &c. Jacobi Mülleri, Hildesheimii

da Longiano F (1542) Dioscoride fatto di Greco Italiano. To the end of which are affixed his tables ordered, with certain warnings, and treatises necessary, for the same subject. Curtio Troiano di Navò, Venetia. https://archive.org/details/hin-wel-all-00000399-001

Manson O (1539) Carta marina et Descriptio septemtrionalium terrarum, ac mirabilium rerum in eis contentarum diligentissime elaborata, Veneciis

Mattioli MP (1548) Il Dioscoride dell'eccellente dottor medico m. P. Andrea Matthioli da Siena; co i suoi discorsi, da esso la seconda volta illustrati, et diligentemente ampliati: con l'aggiunta del sesto libro de i rimedi di tutti i veleni da lui nuovamente tradotto, & con dottissimi discorsi per tutto commentato. Vincenzo Valgrisi, at the workshop of Erasmus, Vinegia

Mercati M (1717) Metallotheca Vaticana. Opus posthumum, Autoritate et Munificentia Clementis Undecimi Pontificis Maximi. E tenebris in lucem eductum: Opera autem, & studio Joannis Mariae Lancisii, Archiatri Pontificii illustratum. Ex Officina Jo: Mariae Salvioni Romani, Romae, 395 pp

Moscardo L (1656) Note overo memorie del museo di Lodovico Moscardo nobile veronese. Academico Filarmonico, described by the same, and in three distinct books. For Paolo Frambotto, Padoa

de Plancy JASC (1863) Dictionnaire infernal: répertoire universel des êtres, des personnages, des livres... qui tiennent aux esprits, aux demons, aux sorciers, au commerce de l'enfer, aux divinations, aux maléfices, a la cabale et aux autres sciences occultes, aux prodigies, aux impostures, aux superstitions diverses et aux prognostics, aux fait actuels du spiritisme, et généralement a toutes les fausses croyances merveilleuses, surprenantes, mysterieuses et surnaturelles. (6e éd.) par J. Collin de Plancy

Sachs PJ (1665) Gammarologia sive Gammarorum vulgo Cancrorum consideratio Physico, Philologico, Historico, Medico, Chimica. Francofurti & Leipzig

Scilla A (1670) La vana speculazione disingannata dal senso. Lettera responsiva Circa i corpi marini, che Petrificati si trovano in varij luoghi terrestri. Andrea Colicchia, Napoli

Scilla A (1759) De corporibus marinis lapidescentibus quæ defossa reperiuntur. Addita Dissertatione Fabii Columnæ de Glossopetris. Joannis Zempel, Romae

Scheuchzer JJ (1732) Physica Sacra. Tomus I. Augustae Vindelicorum & Ulme

Scarabelli PF (1666) Museo ò Galeria Adunata dal Sapere, e dallo Studio del Sig. Canonico Manfredo Settala nobile milanese. descritta in Latino dal Sig. Dott. Fis. Coll. Paolo Maria Terzago and now in Italian by Mr. Pietro Francesco Scarabelli, Ph.D. Physicist of Voghera. And increased by the same. For the sons of qd. Eliseo Viola, Tortona. https://archive.org/details/museoogaleriaadu00terz_0

Torrubia J (1754) Aparato para la Historia Natural Espagñola. Tomo Primero. Contiene muchas disserctaciones physicas, especialmente sobre el Diluvio. Herederos de Dón Agustin, Madrid

Walch JEI (1762) Das Steinreich systematisch entworfen. J. J. Gebauer, Halle

Walch JEI (1768–1773) Walch Johann Ernst Immanuel: Die Naturgeschichte der Versteinerungen : zur Erläuterung der Knorrischen Sammlung von Merkwürdigkeiten der Natur. Nürnberg : gedruckt mit Felsseckerischen Schrifften, 1768–1773. ETH-Bibliothek Zürich, Rar 399, https://doi.org/10.3931/e-rara-10246 / Public Domain

Chapter 26
Cirripedia. A Strange Story: from Ducks to Barnacles

Abstract The name of this subclass of Phylum Arthropoda, from the Latin *Cirrus*, comes from the particular conformation of the multiarticulated bifid appendages, present in number of six pairs on the thorax of the individuals, which, when projected externally to the carapace, resemble to a lock of curls. These marine invertebrates have a deeply modified morphology if compared to the typical scheme of the crustaceans and of the arthropods in general, for instance, the characteristic segmentation of the body does not appear. Many of them have an exoskeleton formed by calcareous plates, easily fossilizable. The paleontological classification is based on internal and external structures. The ancestral barnacles have appeared from the lower Paleozoic. The balanomorph barnacles were called with the popular name of Acorns of the sea, following strange translations in which our invertebrates were associated with buds or fruits similar to palm dates or acorns of various plants and colors. The first printed illustration of living specimens is due to Rondelet (1555), while what is generally considered the oldest figure of a fossil barnacle with a complete crown is that expressed by Baier (1708). Within the Cirripedia, in the Family Lepadidae, the species have the body formed by plates supported by a peduncle through which they attach themselves to the substrate, usually floating objects, for example, tree trunks or keels of ships. During the Middle Ages, due to this characteristic, they gave credence to the theory that the Lepadidae underwent a singular metamorphosis, starting from their initial status of arboreal fruit up to the status of flying aerial vertebrates, originating sea ducks or geese.

Keywords Fossils · History · Balanidae · Lepadidae · *Lepas anatifera*

The name of this subclass of Phylum Arthropoda,, from the Latin *Cirrus*, comes from the particular conformation of the multiarticulated bifid appendages, present in number of six pairs on the thorax of the individuals, which, when projected externally to the carapace, resemble to a lock of curls (Fig. 26.1). These marine invertebrates have a deeply modified morphology if compared to the typical scheme of the crustaceans and of the arthropods in general, for instance, the characteristic segmentation of the body does not appear. They live attached to a substrate by means

© The Author(s), under exclusive license to Springer Nature Switzerland AG 2022

M. Forli, A. Guerrini, *The History of Fossils Over Centuries*,
https://doi.org/10.1007/978-3-031-04687-2_26

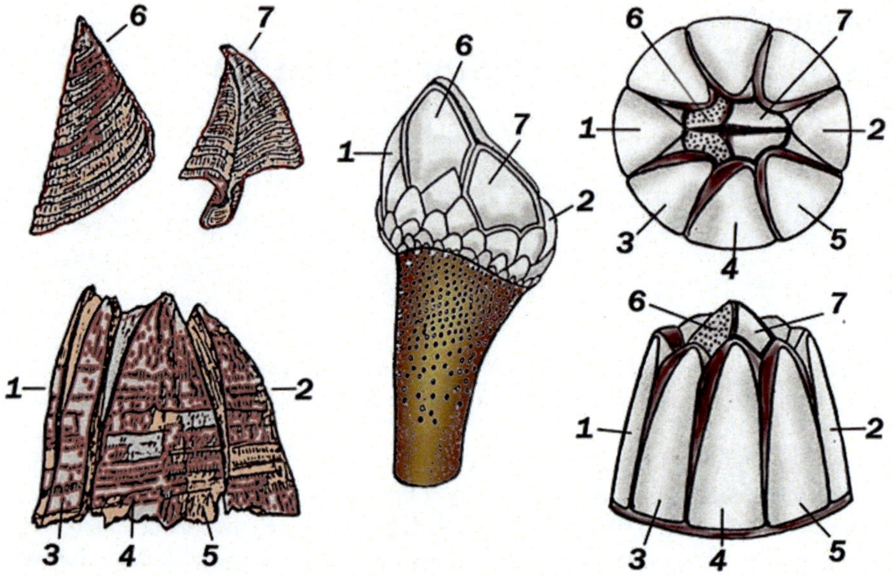

Fig. 26.1 From left to right, simplified arrangement of skeletal plates in barnacles: (1) Carena. (2) Rostrum. (3) Careno-lateral plate. (4) Lateral plate. (5) Rostro-lateral plate. (6) Tergum. (7) Scutum. [Drawing Maurizio Forli modified after Vialli (1974)]

of a peduncle or adhere by means of their subcircular chitinous and calcareous base. Many of them are equipped with an exoskeleton made of calcareous plates, easily fossilizable, on whose internal and external structure the paleontological classification is based (Fig. 26.1).

Most barnacles belong to the coastal environment, mainly occupying the mesolittoral zone between the low and high tide levels. For this reason, they are of great palaeoecological importance, in particular, for the reconstruction of ancient coastal lines. On the contrary, their contribution in stratigraphy is not very significant, although ancestral barnacles have appeared since the Lower Paleozoic. Representatives of the genus *Balanus* are mentioned from the Middle Jurassic to the Upper Jurassic.

Family Balanidae Leach, 1817
Subfamily Balaninae
Genus *Balanus* Da Costa, 1778

The group of these barnacles has an exoskeleton, of truncated cone shape, formed by a series of plates, which surround and protect the body of the animal, forming a circular wall, welded together and at the base of attachment to the substratum, which in the physiognomy remind small craters. This conformation, called *calcareous wall*, is formed by six plates which fit together thanks to their lateral extensions. The wall is a rather rigid complex whose main plates are divided into scuta, terga, rostrum, and keel, while the internal cavity is occluded at the top by an opercular structure formed by four mobile plates, united by a chitinous membrane. The walls of the plates consist

Fig. 26.2 A fossil barnacle retaining *scutum* and *tergum* internally. Pliocene, from the vicinity of Certaldo (Florence), base diameter 40 mm. (Photo and graphic composition Maurizio Forli)

of two calcified laminae, one external and the other internal, in the midst of which are the parietal canals separated from other by septa. The inner or outer faces of the walls may be traversed longitudinally by grooves and ribs. The opercular plates, considered paleontologically very important for the classification, are the *Scutum* and the *Tergum*, which are not always easy to find, because they tend to detach and disperse during the fossilization phase (Fig. 26.2).

> ΒΆΛΑΝΟΣ, *en Latin Glans, ainsi nommé de la figure d'un Glan de chesne, mis par Aristote entre les poissons couverts de test dur. Athenèe en fait de deux sortes, l'une de ceux qu'on appellee en France é en Bretagne Poulcepieds, que I'ai fait pourtraire attachés contre le roc, semblables aux Glans de Chesne: car comme le premier bout des Glans est lis, é venant en pointe, le dernier bout rude, ainsi les Poulcepieds du dernier bout du quel ils tienent contre le roc, sont aspres é rudes.* (*La Secondè Partie de L'Histoire Entiere des Poissons*, 1558: p. 21. Laurent Joubert: Rondelet, translated from Latin into French)
>
> *Balanos*, in Latin *Glans*, so called from the figure of an oak acorn, put by Aristotle among the fish covered with hard remains. Athenaeus makes them of two kinds, large and small. I too find two kinds, the one of these which is called in France and Brittany Pollicipes (leaf barnacle), I have had portrayed attached against the rock, resembling oak acorns: for as the beginning of acorns is smooth, and ends in a point, the back rough, so the Pollicipes from the back which they hold against the rock, are rough and rugged.

Sea acorns, this was the popular name with which the balanomorph barnacles were called, following strange translations where our invertebrates were associated to buds and fruits similar to palm dates, to acorns of various plants and colors, to Egyptian balms and perfumes, with which they had in common only the etymologic derivation. We think more correct, morphologically speaking, their definition as *Glans*, resembling to the summit part of the male sexual organ, seen also that often the plates have externally a red/pink color.

> *Balanitae duo genera habent, subvirides, et Corinthii aeris similitudine. Illa Copto, haec ex Troglodytica veniens, medias secante flammea vena.* (Pliny, XXXVII: LV)

> *Of the Balanite there are two kinds, somewhat green, and it has similarity of copper corinthio. That comes from Coptic, this from the Troglodytic region, segantela pel mezzo una vena infiammata* (Brùcioli A. *Historia Naturale di C. Plinio Secondo nuovamente tradotta di Latino in Volgare toscano*, (1548: libro Trigesimo Settimo, cap. 9, MLX)]

> *Del phenico balano Et calamo odorato In Egytto è una palma, che si chiama Adipos di odore di melo cotogno, senza alcuno legno dentro, et questa ha uso simile, et prossimo al myrabolano, negli unguenti. Et cosi poco avanti che cominci à maturarsi, quello che rimane si chiama Phenico balano, et ha del nero, et inebria quegli che lo mangiano* [...] (Ibidem, libro Duodecimo, cap. 22, CCCLIX)

> *In Egytto there is a palm tree, which is called Adipos with the odour of quince, without any wood inside, and this has similar use, and close to myrabolano, in ointments. And so just before it begins to ripen, what remains is called Phenico balano, and is black, and inebriates those who eat it* [...]

The representations of Rondelet are the first for living specimens, reported also by Gessner and Aldrovandi, among others. As he writes in his text there are two types placed among the testacei, the first one described as above by us, Sea Acorn, in chapter XXIX, while that of the second type, after a couple of pages, shows an aggregate of seven individuals, more similar to olives than to acorns, from the top of which come out tripartite feathered tufts (Rondelet, *Universae aquatilium Historiae pars altera*; 1555: pp. 28, 30). In Aldrovandi (1606: pp. 522–525), in addition to the iconography of the French author, new different icons are introduced, representing six different colonies of balani, some still attached to parts of bivalves, differentiated by shape, size, and color. Among other things, the appellation of *Polycipedes* is reintroduced, recalling their presumed resemblance to the big toe, to be attributed instead, without doubt, to representatives of the other Order of Cirripedia, *Scalpelliformes* Buckeridge & Newman, 2006, in which is currently placed the Family *Pollicipedidae* with the species *Pollicipes pollicipes* (Gmelin 1790), widespread along the Atlantic coasts of Europe, up to Brittany, and of North Africa, also consumed as a gastronomic delicacy (Figs. 26.3 and 26.4).

Scarce news about balani are those offered by other seventeenth-century naturalists, limiting themselves to a few words of description:

> *I Balani, overo Ghiande Marine sono così chiamate per la similitudine, che hanno con la Ghianda di Quercia: nascono sopra de' sassi, sopra de mituli, e sopra de Petuncoli, come si vede dalla sopra posta figura. Pullulano in quantità, ma sempre uniti insieme: sono di color bianco, che tira al violaceo con alcune linee, overo canaletti, et hanno un solo forame per chiascheduno nella sommità.* (Moscardo 1656: pp. 209–210)

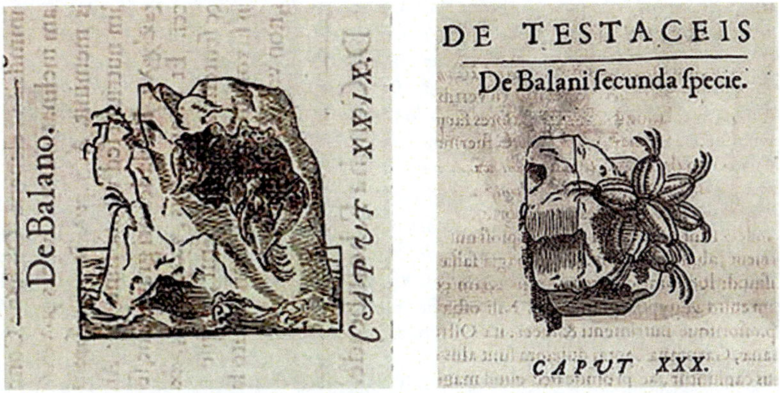

Fig. 26.3 Illustrations from Guillaume Rondelet, *Universae aquatilium Historiae pars altera* 1555: pp. 28, 30 (https://archive.org/details/ARes071032/page/n13/mode/2up. Public Domain 1.0)

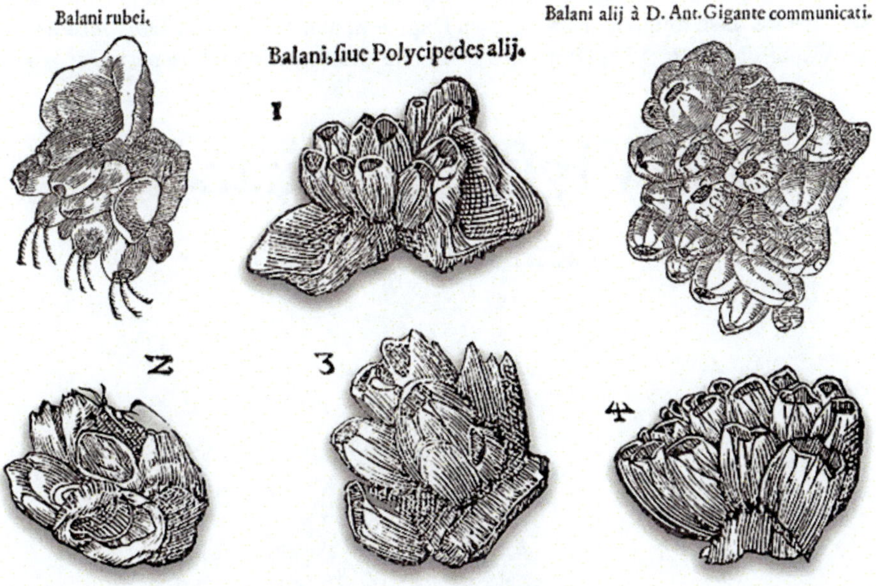

Fig. 26.4 Barnacles from Aldrovandi, *De Reliquis Animalibus exanguibus*, 1606: pp. 522–525. (Graphic composition Maurizio Forli. Credit https://www.biodiversitylibrary.org/bibliography/119999. Public Domain)

The Balani, or Sea Acorns, are so called because of the similarity they have with the Oak Acorn: they grow on rocks, on mussels, and on Petunculus, as can be seen from the above figure. They pullulate in quantity, but always united together: they are white in colour, tending to violet with some lines, or small canals, and they have only one hole for each one in the summit.

> *Gusci chiamati col nome generico Ballani, che significa tuttociò, che da Noi si spiega col nome di Ghianda: Nascono molti insieme, come si vede nella figura, sopra i sassi, legni, e gusci de' Testacei d'ogni sorte. Sono per ordinario di colore cinereo, ò avvinato. Ballani diversi da' primi, simili alla ghianda nella forma. Hanno il forame più degli altri. Sono lavorati a bugne, come i meloni, e di color rosso, e smorto. Nascono sopra gli scogli.* (Buonanni 1681: p. 152, pl. 4, figs 14–15)

> *Shells called by the generic name of Ballani* (Acorn Barnacles), *which means everything that is explained by us by the name of Acorn: they are born in many places, as can be seen in the figure, on the stones, wood and shells of Testacei of all kinds. They are usually of an ashy or* red wine color. *Ballani different from the first, similar to the acorn in form. They have the foramen more than the others. They are worked with ashlars, like melons, and of a red and dull colour. They are born on the rocks.*

Despite the numerous works dedicated to fossils from the '500 onward, strangely enough there are no published images of fossilized barnacles until the early '700; for example, in the *Musaeum Metallicum* and in the *Metallotheca Vaticana* we would expect to find them because there are often represented images of Pliocene shells on which the *Acorn Barnacles* could have settled but it is not so. The only reference we have found is from a crown fragment, named *Pinnites Molybdoides* by Aldrovandi (1648: p. 170) (Fig. 26.5). Among the finds from this period it is not

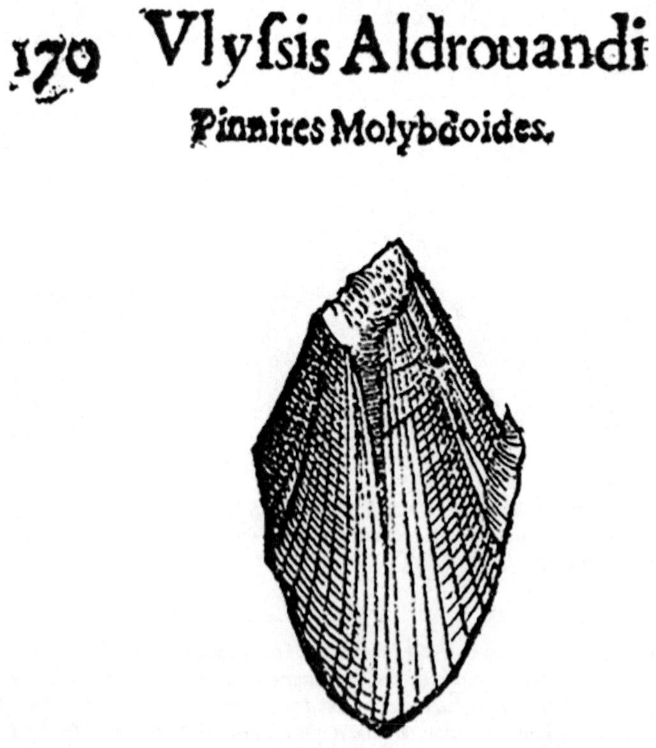

Fig. 26.5 Brocchi (1814: p. 597) reports under the name *Lepas tintinnabulum* L. a partial remnant of a fossil barnacle taken from the *Musaeum Metallicum* of Ulisse Aldrovandi (1648: p. 170) originally described as the name *Pinnites Molybdoides*. A fin-shaped stone with the color of lead

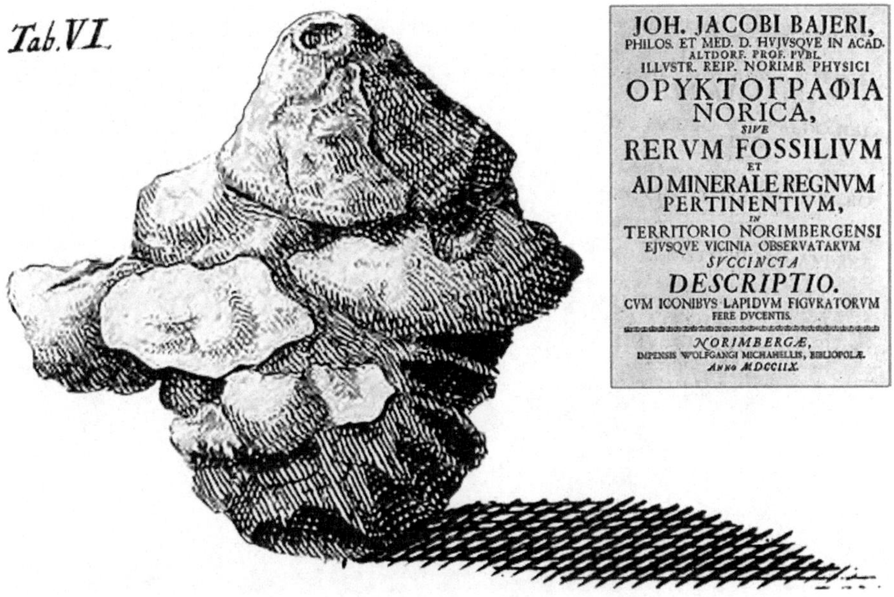

Fig. 26.6 Baier, *Oryctographia Norica* (1708: pl. VI, fig. 13). (Graphic composition Maurizio Forli. Credit https://doi.org/10.3931/e-rara-18618. Public Domain Mark)

difficult to find their solitary aggregates on the ground or adhering to pebbles marking marine levels close to ancient coastlines. It is therefore quite surprising that these petrified organisms have not been illustrated, similarly to the living ones, instead known, as we have seen, since Rondelet (1555). What is generally regarded as the earliest figure of a fossil Acorn Barnacles with a complete crown is that expressed by Baier in his *Oryctographia Norica* (1708: p. 72, pl. VI, fig. 13) (Figs. 26.6, 26.7, and 26.8):

> *Balanum quidem unicum fossilem, hactenus in Norico nostro mihi visum, exibeo in Tab. VI. Fig. 13*

The prints relatively nearest to us, devoted to single individuals of fossilized Acorn Barnacles, which to our taste appear the best up to the time of their popularization, are those in color published by James Sowerby in his monumental work *The Mineral Conchology of Great Britain* (1812: pp. 192–193; 1834: pl. LXXXIV, figs 1–2) (Figs. 26.9, 26.10, and 26.11).

Figures 13 and 14 of the previous figures illustrate a particular species, *Coronula diadema* (Linnaeus, 1767), which, although belonging to the Balanomorpha, does not fall zoologically into the same superfamily as the more common Balanidae. This one, instead of living on hard substrata in the marine zone included between the levels of the low and high tide, as the common barnacles do, has adapted to live on a relatively soft substratum and in continuous movement, the skin of some big cetaceans nesting of preferences on the head and on the fins of the whales, in

Fig. 26.7 A barnacle's colony fixed on an oyster valve (measuring ~90 mm), Pliocene, age ~3.5 Ma. In addition to the arrangement of plates to form the crown, a partial section of a plate is visible in the upper left corner. (Photo Maurizio Forli)

particular, the humpback whales, parasitizing them. Their structure suggested an apt title for a scientific article concerning the discovery of a specimen from the Tuscan Pleistocene: *Large kings with small crowns,* with an obvious and witty allusion to the large cetaceans and the royal attribute recalled by the specific name (Dominici et al. 2011) (Fig. 26.12).

Typical representatives have been reported since the Miocene. They are barnacles whose body has a peduncle by means of which they attach to a substratum which can be either fixed or mobile, meaning with this, floating objects both natural and artificial, for instance, tree trunks or keels of ships. The peduncle supports the crown, or *capitulum*, formed by five plates which protect internally the body. These plates are composed by one *carina*, two *scuta,* and two *terga*, held together by chitinous bands. Other genera of lepadids have a greater number of plates (Figs. 26.13 and 26.14).

Family Lepadidae Darwin, 1852
Genus *Lepas* Linnaeus, 1758
Species *Lepas anatifera* Linnaeus, 1758

During the Middle Ages it was believed that the pelagic gooseneck barnacle, nowadays classified among the sessile benthic marine invertebrates, underwent a singular metamorphosis, starting from their initial state of arboreal fruit to the state of flying aerial vertebrates, originating sea ducks or geese.

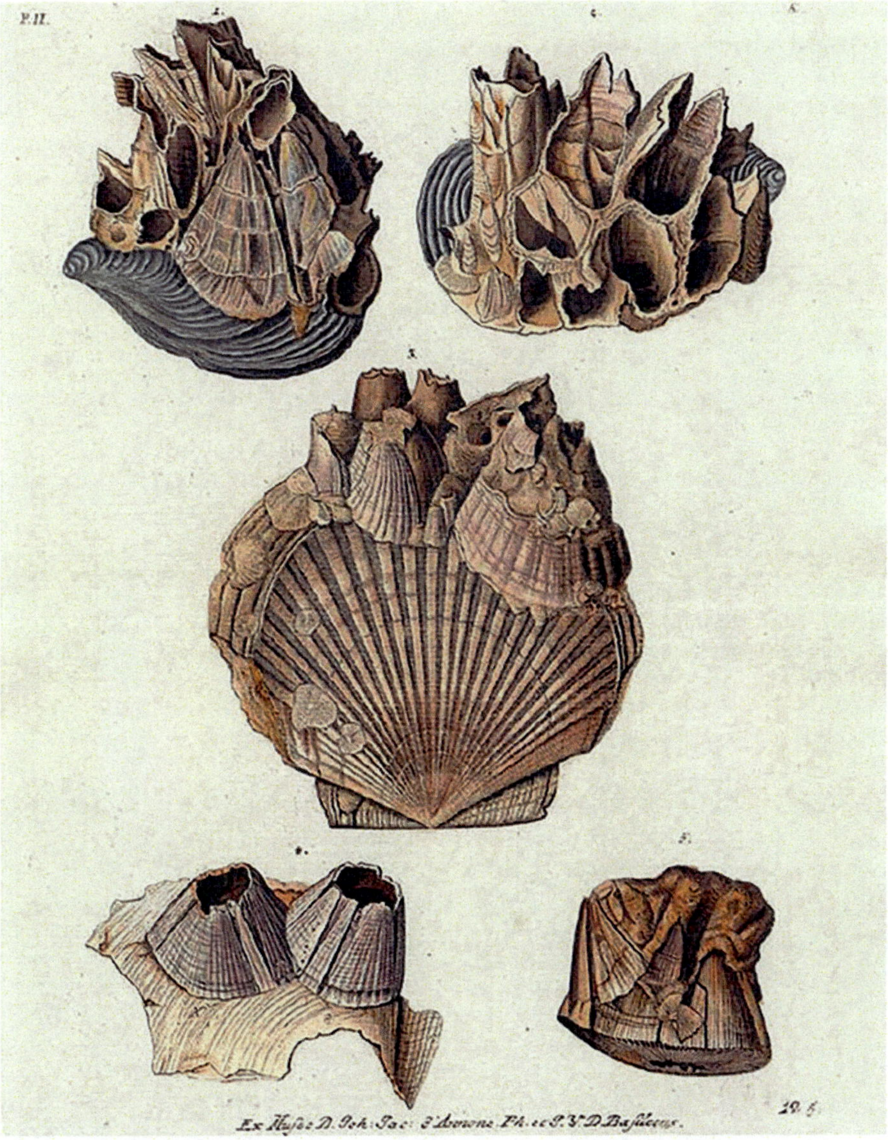

Fig. 26.8 Barnacles encrusted fossil shells, plate II from Walch (1768–1773), former Museum Johann Jakob d'Annone, Basel. (ETH-Bibliothek Zürich, Rar 399, https://doi.org/10.3931/e-rara-10246. Public Domain)

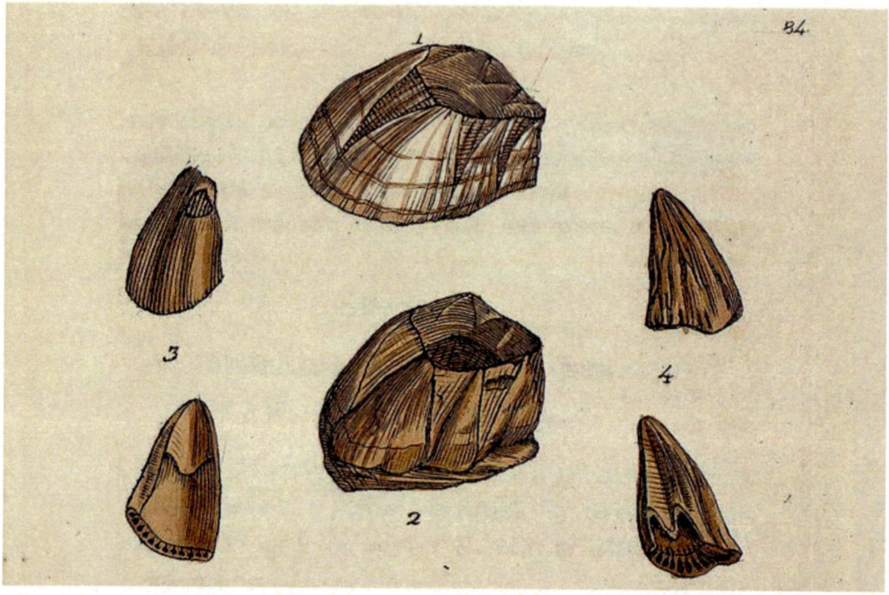

Fig. 26.9 From *The Mineral Conchology of Great Britain* (1812: pp. 192–193; 1834: pl. LXXXIV, figs 1–2) by James Sowerby [https://www.biodiversitylibrary.org/bibliography/14408 (not in copyright)]

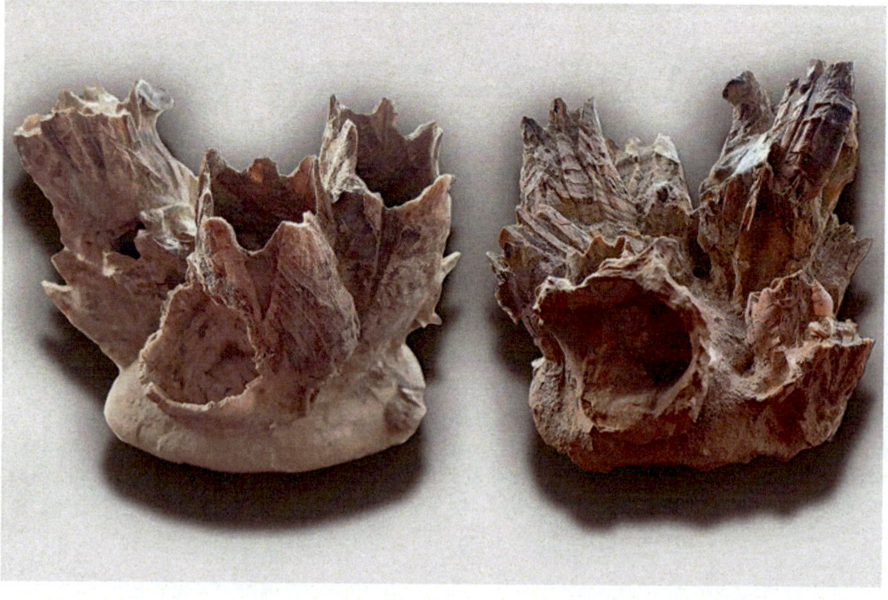

Fig. 26.10 Lower Pleistocene Tuscan barnacles attached to the shell of an echinid of the genus *Spatangus*, evidently covered post mortem. Anterior and posterior views, base size 85 mm. (Photo and graphic composition Maurizio Forli)

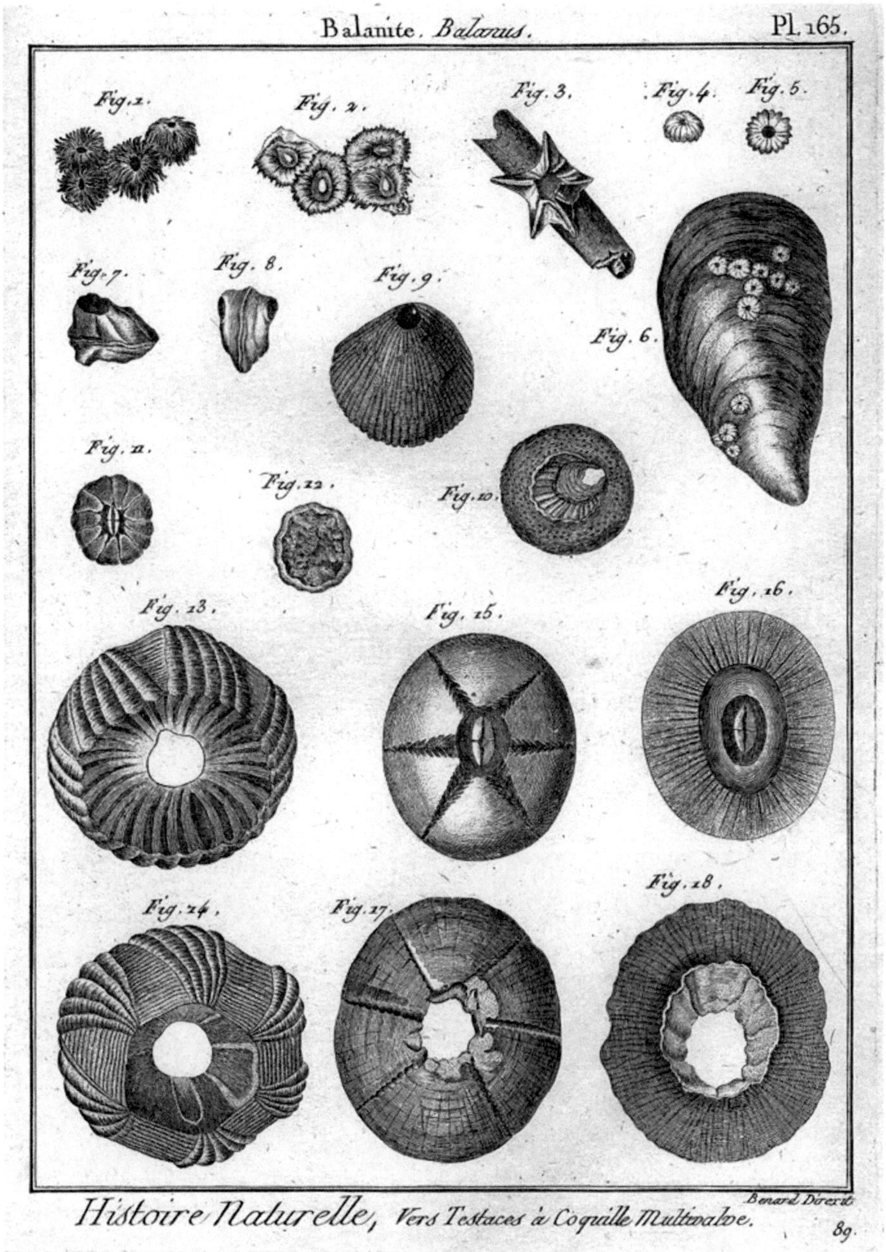

Fig. 26.11 *Multivalve-shell testaceous worms*: *Balanite, Balanus*. One of the plates relating to these characteristic crustaceans, still considered zoologically akin to mollusks, taken from *Tableau encyclopédique et méthodique des trois règnes de la nature contenant l'helminthologie, ou les vers infusoires, les vers intestins, les vers mollusques, & c.* (Bruguiére 1792, personal copy Maurizio Forli)

Fig. 26.12 Top and right, the specimen of *Coronula diadema* (Linnaeus, 1767) described and discussed in Dominici et al. 2011; maximum size about 70 mm (modified); left, a replica of a similar specimen, exhibited in the cetacean gallery of the Museum of Natural History of the University of Pisa. In the background part of the relevant explanatory panel. (Graphic composition Maurizio Forli. Credit from: courtesy Museo di Storia Naturale dell'Università di Pisa, photo Maria Teresa Tommasi and Stefano Dominici)

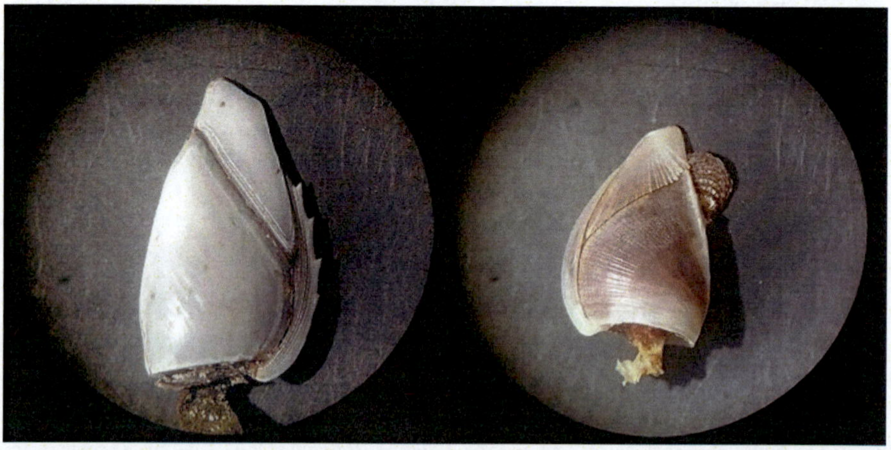

Fig. 26.13 Two dried specimens with the characteristic arrangement of the plates protecting the body; in the specimen on the right (*Lepas pectinata* Spengler, 1851) some partially exposed cirri can still be seen. Excluding the portion of peduncle, the height of the largest specimen (*Lepas anatifera* Linnaeus, 1767) is ~15 mm. (Photo credit Daniele Pagli)

Fig. 26.14 Fragment of a tree, about half a meter long, where hundreds of Lepas had settled, washed up near Populonia (Tuscany). (Photo Maurizio Forli)

> *Conca Anatifera. Nome dato alle Anatife e particolarmente all'A. laevis, dai primi conchiologi e che viene da due nomi latini, Anas, Anitra, e ferre, portare, cioè conchiglia che porta un'Anitra, denominazione che prese origine da un'opinione volgare degli abitanti della Scozia, i quali credevano che le Oche e le Anitre, nascessero da tali conchiglie.* [...] *Dicevasi che l'Anatifa era un frutto che cresceva sulla spiaggia del mare, il quale pervenuto a maturità, cadeva nell'acqua ed aprivasi poscia per lasciar uscire dalla sua guscia, secondo gli uni, la specie d'Oca chiamata Bernacla o Bernicla in Iscozia, Anas bernicla, o, secondo gli altri, la Folaga, Anas nigra. Alberto il Grande confutò nel decimoterzo secolo simile assurdità, al pari di altri dotti ne' secoli seguenti; e tuttavia si è trovato, dice Cuvier, nel decimosettimo chi fu tanto ardito per sostenerla.* (AA.VV.; Dizionario Classico di Storia Naturale. Prima Edizione Italiana. 1831: p. 372)

> *Conca Anatifera. Name given to the pelagic gooseneck barnacle and particularly to the A. laevis, by the first conchiologists and that comes from two Latin names, Anas, Anitra, and ferre, carry, that is shell carrying an Anitra, a name that originated from a vulgar opinion of the inhabitants of Scotland, who believed that the Geese and the Ducks, were born from such shells.* [...] *It was said that the Anatifa was a fruit that grew on the beach of the sea, which once ripe, fell into the water and then opened to let out of its shell, according to some, the species of goose called Bernacla or Bernicla in Iscozia, Anas bernicla, or, according to others, the Coot, Anas nigra. Albert the Great refuted this absurdity in the thirteenth century, as did other scholars in the following centuries; and yet we have found, says Cuvier, in the seventeenth who was so bold as to support it.*

The fact that these invertebrates are almost always found adhering to wooden fragments beached by the wave motion supported the opinion that they were "animate fruits." For further clarification of this fabulous belief, we report some illustrations which clarify the presumed modalities of occurrence of this singular process. The classification of the members of the family Lepadidae has always been rather contrasted and diversified. Leaving aside the fantastic interpretations, they were included in the multivalved marine mollusks, akin to the chitons and lithophagous mollusks, such as the *Pholas*. It is in Bruguière that we find their first distinction as organisms totally different from mollusks and partly from balanids.

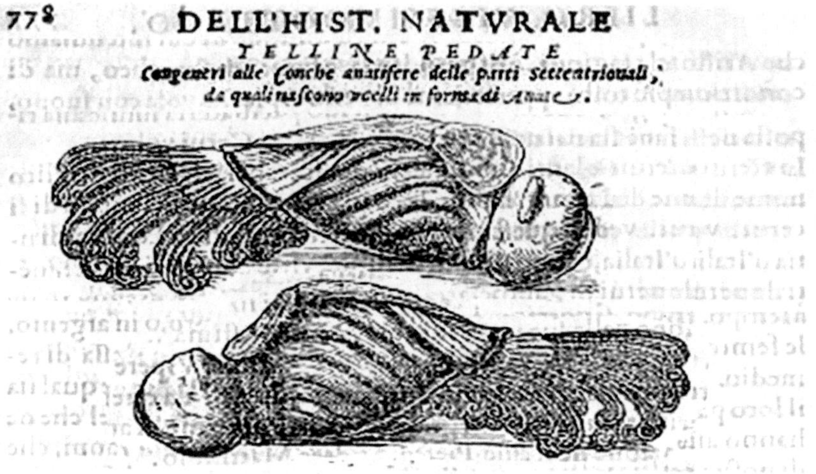

Fig. 26.15 The Pedate Telline, that is, "with feet," that is the shells of the common *Lepas*, as illustrated by Ferrante Imperato in his *Dell'Historia Naturale* (1599: p. 778) [https://www.biodiversitylibrary.org/item/183177#page/5/mode/1up (not in copyright)]

ANATIFES, vulgairement conques anatifères ou pouce-pieds, Anatifa; NOB.

GENRE DE COQUILLES MULTIVALVES, qui a pour caractère une coquille fixée, formée de plusieurs valves articulées, réunies par des membranes et soutenues sur un pédicule tendineux, cylindrique et flexible; une overture longitudinale. (1792: p. 60)

These systematic indications are still used to demonstrate their correct designation and definition (Fig. 26.15).

Here is the description and the news given by Ferrante Imperato about the *Telline Pedate,* that is, *Pedunculated Tellin* and the common Tellins bivalves (*Dell'Historia Naturale*, 1599: 772).

Sono le Telline communemente conosciute in uso di cibo, separate dal luogo ove si trovano, e bivalve: a quali nella contenenza del corpo sono simili le affisse: quantunque nel resto delle condizioni del corpo siano dissimili. percioche in vece di due gusci n'hanno cinque, di pezzo commessi, con sustanza membranosa trameza. dunque le due bande che rispondono alli due gusci delle communi Telline ciascuna è di due pezzi, e la minore delle due è verso l'appuntato. il quinto pezzo simile alla tagliatura dell'unghia humana, stà nella commissura anteriore tra l'una e l'altra banda. sono tutte le Telline di forma amendolare, ma con maggior somiglianza le affisse, o pedate, che nello stremo più ritondo con lo piede menbranoso aderiscono a legni, & spesso con lo piè attaccato ad altro piede, quasi rami de sterpi. dunque le nostrali hanno semplicemente vita nel modo di piante animali. le forastiere del paese Settentrionale contengono concetto di animali volatili, che cascando dopo la maturazione nell'acqua crescono in forma di anate, non altrimenti, che a noi molti insetti volatili provengono da madri diverse prodotte da piante viventi, quasi lor proprij parti [...]

Tellins are commonly known as food, separated from the place where they are found, and bivalves: to which in the containment of the body the pedunculated are similar, although they are dissimilar in the rest of their bodies, because instead of two shells they have five pieces of jointed shells, with a membranous texture. Therefore the two bands (sides) which correspond to the two valves of the common Tellins are each made of two pieces, and the smaller of the two is towards the pedunculus. the fifth piece, similar to the cut human nail, is in the anterior commissure between the one and the other band.

all the Telline are of an almond shape, but with greater similarity the pedunculate, which in the most rounded end with the membranous foot adhere to woods, and often with the foot attached to another foot, almost like branches of brushwood. therefore the our simply have a life in the manner of animal plants. The foreigners of the northern country contain the concept of volatile animals, which, falling after ripening in the water, grow in the form of ducks, not otherwise than in our case many volatile insects come from different mothers produced by living plants, almost like their own parts [...]

As can be easily understood, the careful Neapolitan naturalist had not overlooked the morphological differences between the shells of bivalves and those of lepadids, including anatomical differences and their different *modus vivendi*. Writing instead of these organisms as plants–animals shows that the knowledge of the mode of reproduction and development of these and other invertebrates was still rudimentary. Unlike his friend Fabio Colonna, who had decisively refuted this theory (1592), he does not deny the possibility that the specimens coming from northern countries could give rise to birds (Figs. 26.16, 26.17, 26.18, and 26.19).

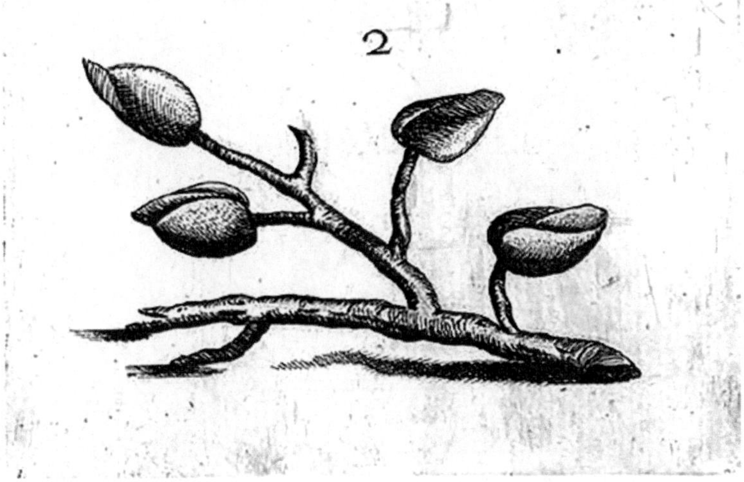

Fig. 26.16 Illustration taken from Filippo Buonanni (1681) in which the lepas appear to arise and develop directly from the twigs, like fruits and leaves, supported by the petioles. (Modified from *Ricreatione dell'occhio e della mente nell'osservation'delle chiocciole*. https://doi.org/10.3931/e--rara-39893. Public Domain Mark)

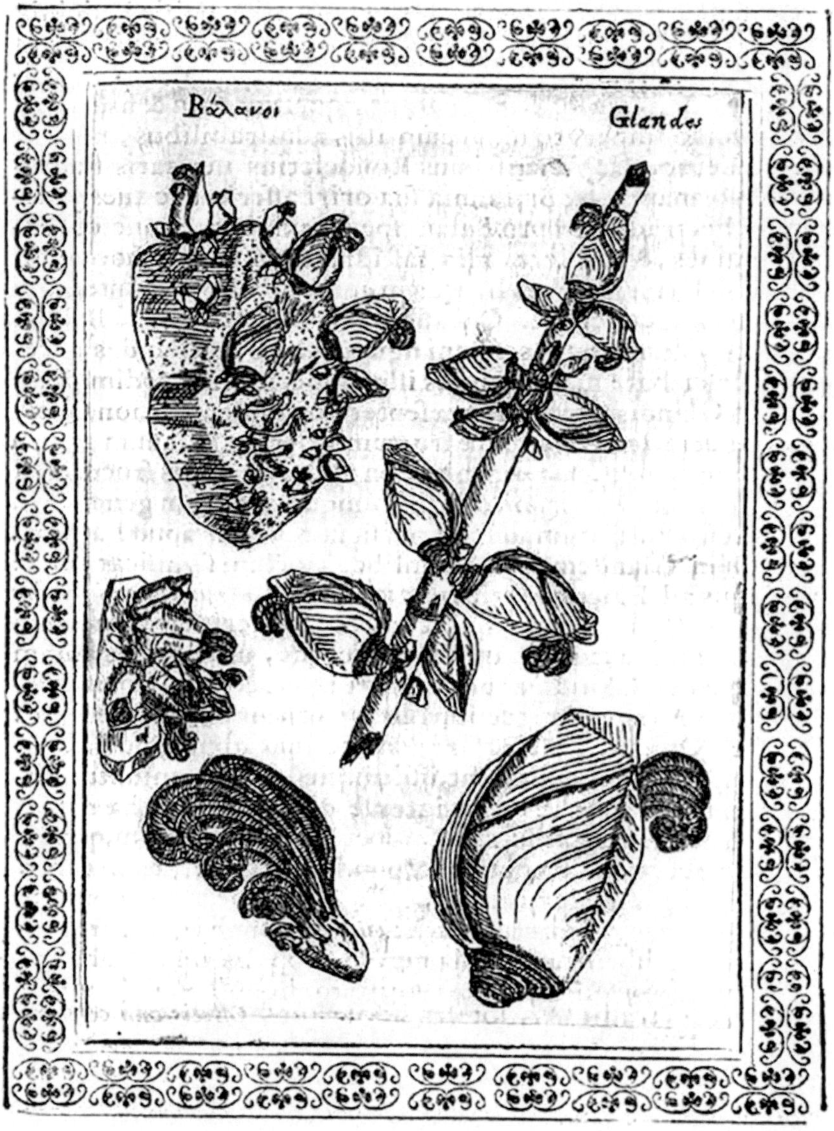

Fig. 26.17 *Conchas vulgò Anatiferas, non esse fructus terrestres, neque ex ijs Anates oriri: sed Balani marinae speciem.* The shells commonly called Anatiferous, are not a terrestrial fruit, nor from them ducks are born, but marine species of balani. Fabio Colonna, *Phytobasanos* (1592: pp. 14–17) [https://archive.org/details/gri_33125008412476/page/n155/mode/2up (not in copyright)]

Fig. 26.18 *Claxis generatio secundum aliquos*. The alleged transformation from fruits into ducks, according to some, recounted and illustrated by Ulysses Aldrovandi (1603: p. 174, *Ornithologiae tomus tertius*; 1606: p. 543, *De reliquis animalibus exanguibus* (modified). Permission Sistema Museale di Ateneo Alma Mater Studiorum – Università di Bologna)

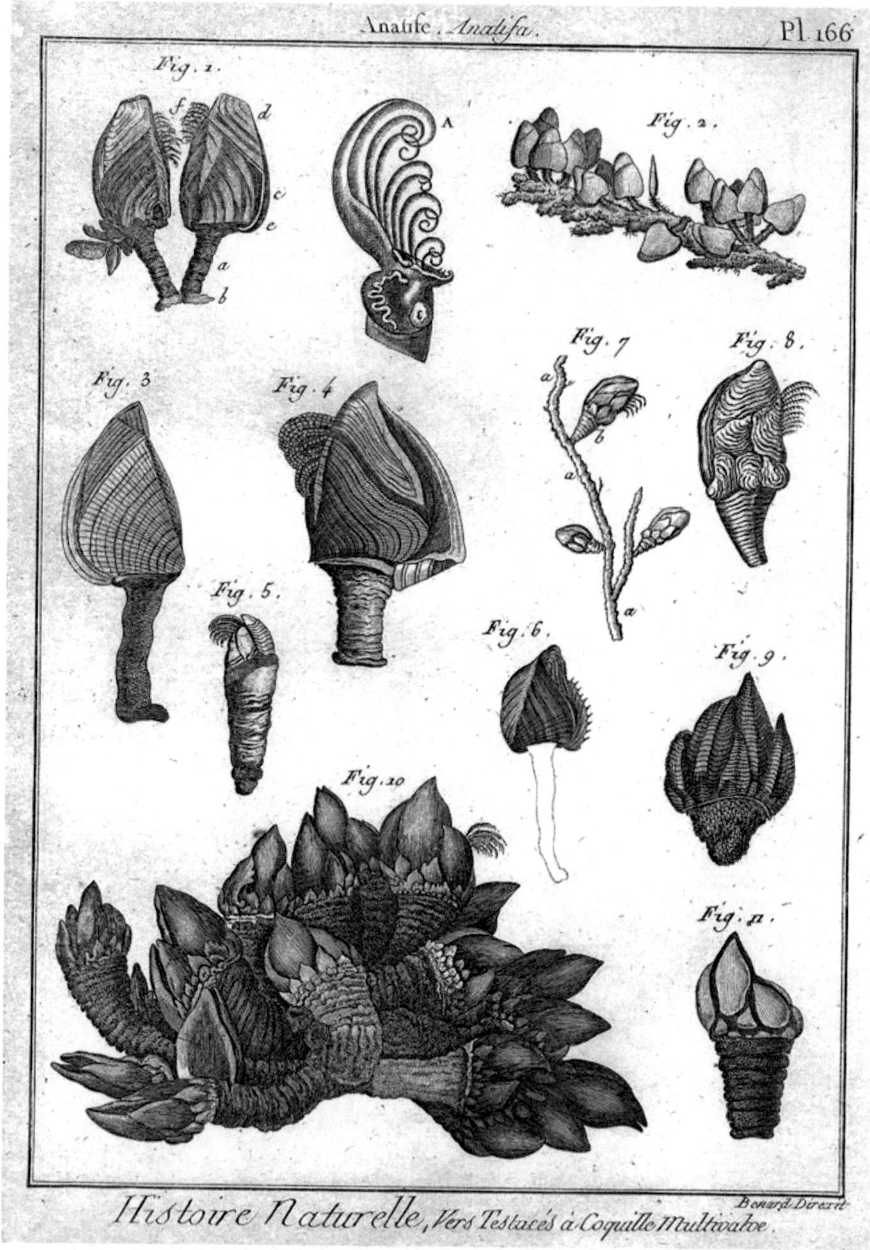

Fig. 26.19 *Multivalve shell testaceous worms: Anatife, Anatifa.* Another plate taken from *Tableau encyclopédique et méthodique des trois règnes de la nature contenant l'helminthologie, ou les vers infusoires, les vers intestins, les vers mollusques, & c.* (Bruguière 1792), in which are depicted representatives of *Lepas* and *Pollicipes*, presently placed in different families, still placed in the same Class. (Personal copy Maurizio Forli)

References

AAVV (1831) Dizionario Classico di Storia Naturale. Prima Edizione Italiana

Aldrovandi U (1603) Ornithologiae tomus tertius, ac postremus. Ad eminentissimum principem Alexandrum Perettum ... Cum indice copiosissimo variarum linguarum. Bonon.sumptibus M. Antonij Berniae bibliopol, Bonon. https://amshistorica.unibo.it/34#

Aldrovandi U (1606) De Reliquis Animalibus exanguibus libri quatuor, post mortem eius editi: nemp' De Mollibus, Crustaceis, Testaceis, et Zoophytis. Apud Io: Baptistam Bellagambam, Bononie. https://www.biodiversitylibrary.org/bibliography/119999. Public Domain

Baier JJ (1708) Oryktographia Norica sive rerum fossilium et ad minerale regnum pertinentium, in territorio Norimbergensi eiusque vicinia observatarum [...]. Norimbergae: impensis Wolfgangi Michahellis, 1708. ETH-Bibliothek Zürich, Rar 3707. https://doi.org/10.3931/e-rara-18618/. Public Domain Mark

Brocchi GB (1814) Conchiologia fossile subapennina con osservazioni geologiche sugli Apennini e sul suolo adiacente. Milano. Vol. 1: i-lxxx, 1-56, 1-240; vol. 2: 241-712, 16 pls. https://www.biodiversitylibrary.org/item/43889#page/5/mode/1up (not_in_copyright)

Brùcioli A (1548) Historia Naturale di C. Plinio Secondo nuovamente tradotta di Latino in Volgare toscano. Alessandro Brùcioli, Venetia

Bruguière M (1791) or (1792). Tableau encyclopédique et méthodique des trois règnes de la nature contenant l'helminthologie, ou les vers infusoires, les vers intestins, les vers mollusques, &c. Septième Livraison. Panckoucke, Paris

Buonanni F (1681) Ricreatione dell'occhio e della mente nell'osservation' delle chiocciole: proposta a' curiosi delle opere della natura. In Roma: per il Varese: a spese di Felice Cesaretti ..., Zentralbibliothek Zürich, NNN 713 I G, https://doi.org/10.3931/e-rara-39893. Public Domain Mark

Colonna F (1592) Phytobasanos, sive Plantarum aliquot Historia, in qua describuntur diversi generis Plantae veriores, ac magis facie, viribùsque respondentes antiquorum Theophrasti, Dioscoridis, Plinij, Galeni, aliorùmque delineationibus, ab aliij hucusque non animadverae. Ex Officina Horatij Salviani, Neapoli. https://archive.org/details/gri_33125008412476/page/n155/mode/2up (not in copyright)

Dominici S, Benvenuti M, Bartalini M, Balestra B (2011) Large kings with small crowns: a Mediterranean Pleistocene whale barnacle. Boll. Soc. Paleontol. Ital. 50(2):95–101. https://doi.org/10.4435/BSPI.2011.10

Imperato F (1599) Dell'Historia Naturale di Ferrante Imperato napolitano. Libri XXVIII. Nella quale ordinatamente di tratta della diversa condition di miniere, e pietre. Con alcune historie di Piante, et Animali fin'ora non date in luce. Costantino Vitale, Napoli. https://www.biodiversitylibrary.org/item/183177#page/5/mode/1up (not in copyright)

Joubert L (1558) La Secondè Partie de Histoire Entiere des Poissons, composèe premierement en Latin par maistre Guilaume Rondelet Docteur en Medicine en l'université de Monpelier. Maintenant traduite en Francois sans avoir rien omis estant necessaire à l'intelligence d'icelle. By Mace Bonhome, Lyon

Moscardo L (1656) Note overo memorie del museo di Lodovico Moscardo nobile veronese. Academico Filarmonico, dal medesimo descritte, Et in Tre Libri distinte. Per Paolo Frambotto, Padoa. https://www.biodiversitylibrary.org/bibliography/157058 (not in copyright)

Rondelet G (1555) Universae aquatilium Historiae pars altera, cum veris ipsorum Imaginibus. Matthiam Bonhomme, Lugduni. https://archive.org/details/ARes071032/page/n13/mode/2up. Public Domain 1.0

Sowerby J (1812) The Mineral Conchology of Great Britain; or coloured figures and descriptions of those remains of testaceous animals or shells, which have been preserved at various times and depths in the earth. Vol. I. Benjamin Meredith, London. https://www.biodiversitylibrary.org/bibliography/14408 (not in copyright)

Sowerby J (1834) Systematical, stratigraphical, and alphabetical indexes to the first six volumes of the Mineral Conchology. Printed by Richard Taylor, Red Lion Court, Fleet Street, London. https://www.biodiversitylibrary.org/bibliography/14408 (not in copyright)

Vialli V (1974) Appunti di Paleontologia. Pitagora Editrice, Bologna

Walch JEI (1768–1773) Walch Johann Ernst Immanuel: Die Naturgeschichte der Versteinerungen: zur Erläuterung der Knorrischen Sammlung von Merkwürdigkeiten der Natur. gedruckt mit Felsseckerischen Schrifften, Nürnberg. ETH-Bibliothek Zürich, Rar 399, https://doi.org/10.3931/e-rara-10246. Public Domain

Postface (Andrea Guerrini)

As a dear friend of ours, a botanist with a passion for antiquities, used to say [Acciai Bruno, Montemurlo (Prato, Italy)], we always go back to Pliny…

> Gaius Plinius Caecilius Secundus, known as Pliny the Younger, writing about the habits of his uncle, Pliny the Elder, relates that in the summer, during their free time, while they were reading some book, he, besides making observations and drawing passages from them to quote, used to say that no text was *so bad that it was not useful in some aspect*. (*Epistularum libri*, III, 5)

If you have read these lines, there are probably some aspects of our work that have stimulated your curiosity, leading you to explore a world, at least this was our intention, which for a modern geologist or naturalist may make you smile, but which in Europe, in those centuries steeped and permeated by Christian religion and Aristotelian or Neoplatonic philosophy, represented the reality of things, plausible explanations of the phenomena observed in nature, also expressed by "objects" generally endowed with magical and sometimes therapeutic properties.

The impetus to write this text also came from the countless phone calls between Maurizio and me, who had already been trying his hand at this research for some time, where every now and then the discussion fell on the fact that in reality even today, as mentioned at the beginning, even in a hyper-technological world with a high level of scientific development, vestiges of this magical, unscientific vision remain.

Just think of the power attributed to some stones and many other phenomena that are on the borderline between religion and superstition, leaning much more towards the latter. I always remember, to make a personal quotation too, my grandmother who, although in good faith, in case of some malaise subjected me to an atavistic rite that began by reciting verses in Latin with the rosary, with which she chased away the evil eye whose presence was previously verified, pouring oil into the water contained in a dish where the "eyes" manifested themselves, proportional to the entity of how much they wanted you bad. In today's phenomenology of these aspects, thus perpetuating also the part reserved for charlatans, there are dozens of

© The Author(s), under exclusive license to Springer Nature Switzerland AG 2022

M. Forli, A. Guerrini, *The History of Fossils Over Centuries*,
https://doi.org/10.1007/978-3-031-04687-2

magicians, fortune tellers, and others who, certainly in bad faith, pontificate in some private television and not only.

Here, the difference is that the Renaissance authors mentioned above (and also my grandmother) were serious people, not to be considered superficially but children of that time, where reality was related to the supernatural, a sort of augmented reality, larger but considered possible.

Index

© The Editor(s) (if applicable) and The Author(s), under exclusive license to Springer Nature Switzerland AG 2022
M. Forli, A. Guerrini, *The History of Fossils Over Centuries*,
https://doi.org/10.1007/978-3-031-04687-2

L

M

N